MINDING THE BRAIN

Minding the Brain

Models of the Mind, Information, and Empirical Science

Edited by
Angus J. Menuge, Brian R. Krouse,
and Robert J. Marks

Seattle Discovery Institute Press 2023

Description

Is your mind the same thing as your brain, or are there aspects of mind beyond the brain's biology? This is the mind-body problem, and it has captivated curious minds since the dawn of human contemplation. Today many insist that the mind is completely reducible to the brain. But is that claim justified? In this stimulating anthology, twenty-five philosophers and scientists offer fresh insights into the mind-brain debate, drawing on psychology, neurology, philosophy, computer science, and neurosurgery. Their provocative conclusion? The mind is indeed more than the brain.

Copyright Notice

Library Cataloging Data

Minding the Brain: Models of the Mind, Information, and Empirical Science, edited by Angus J. Menuge, Brian R. Krouse, and Robert J. Marks

Cover design by Brian Gage.

488 pages, 7.5 x 9.25 inches

Library of Congress Control Number: 2023941347

ISBN: 978-1-63712-029-3 (paperback), 978-1-63712-032-3 (hardcover), 978-1-63712-030-9 (Kindle), 978-1-63712-031-6 (EPUB)

BISAC: PHI015000 PHILOSOPHY/Mind & Body

BISAC: SCI089000 SCIENCE/Life Sciences/Neuroscience

BISAC: MED057000 MEDICAL/Neuroscience

BISAC: PSY051000 PSYCHOLOGY/Cognitive Neuroscience & Cognitive Neuropsychology

Publisher Information

Discovery Institute Press, 208 Columbia Street, Seattle, WA 98104

Internet: discoveryinstitutepress.com and mindingthebrain.org

Published in the United States of America on acid-free paper.

First edition, September 2023.

Advance Praise

The mind-body problem lives! A stimulating collection of contemporary perspectives on a perennial conundrum.

—**Gregory Chaitin**, algorithmic information theory pioneer; author, *Building the World from Information & Computation*

Materialism about the mind is a deeply entrenched assumption, so much so that alternative viewpoints are shrugged aside as inconsequential. *Minding the Brain* challenges that mind-set, but not by giving a single, knock-down refutation of materialism or a single, obviously superior alternative. Instead, it presents a kaleidoscopic array involving multiple objections and multiple alternatives, authored by highly competent thinkers from neuroscience, consciousness studies, computer science, information theory, and philosophy. Both materialists and anti-materialists who want to understand the mind should not miss this book.

—**William Hasker**, Emeritus Professor of Philosophy, Huntington University

Minding the Brain is an imposing assemblage of cutting-edge criticisms of materialist views of the mind while advancing compelling alternative accounts of consciousness. The chapters on information, computation, and quantum theory are groundbreaking, advancing serious unacknowledged problems for materialism that must be contended with.

—**Brandon Rickabaugh**, Assistant Professor of Philosophy, Palm Beach Atlantic University; Franz Brentano Fellow in the Metaphysics of Mind, The Martin Institute

Written by renowned experts in different fields of science and philosophy, *Minding the Brain* provides a thorough, multifaceted, and insightful analysis of the age-old mind-body problem. It is well known that even an apparently simple inanimate entity like a sandpile may present a complex, non-linear, and chaotic dynamic which cannot be predicted by the individual properties of its constituting elements. With a unique common thread, the essays in this anthology elegantly expose reductionism for what it truly is, a simplistic endeavor grounded on the scientific materialism creed which, on the topic of the mind-body problem, tries to explain all the complexity of higher-order cognitive phenomena exclusively through reference to the most basic physico-chemical interactions within its underlying biological strata. Such a myopic and simplistic naturalistic approach is not only intellectually disappointing but also inherently flawed, ultimately falling short of the awe-inspiring grandeur of the life of the mind as we all know and experience it. Try explaining the totality of the delightful experience of reading this academic masterpiece through a mathematical equation!

—**Tobias A. Mattei**, MD, FACS, Assistant Professor of Neurosurgery, St. Louis University School of Medicine

Minding the Brain is an important book on substance dualism that comes with breadth, depth, and insight. It incorporates a number of fields of study and academic disciplines; it is up-to-date and rigorous in its presentation and argument; and it is fresh, thoughtful, and thought-provoking. I am pleased to see this robust defense of substance dualism that pushes back against the dominant view of naturalism in the academy as well as alternative views that likewise attempt to avoid the explanatory power of substance dualism and its important implications.

—**Paul Copan**, Pledger Family Chair of Philosophy and Ethics, Palm Beach Atlantic University; coeditor, *The Naturalness of Belief: New Essays on Theism's Rationality*

Minding the Brain is a very up-to-date anthology on the body-mind problem. The editors have assembled a team of excellent scholars from philosophy, neuroscience, psychology, computer science, quantum physics, and mathematics. Together they provide a very strong, cross-disciplinary, and cumulative argument for the need of non-material explanations of human characteristics such as consciousness, will, feelings, and creativity. A recurrent theme of several chapters is the importance of information as a mediator between the non-material and material. The book is a must-read for anyone who wants to understand why purely physical accounts of the mind have failed, and that alternative dualistic or idealistic theories are more credible than ever. I'm sure *Minding the Brain* will simulate many interesting discussions and much further research.

—**Ola Hössjer**, Professor of Mathematical Statistics, Stockholm University

Minding the Brain is an intriguing and comprehensive anthology. This thought-provoking collection delves into the realms of philosophy of mind, neuroscience, psychology, and the intersections of information, computation, and quantum theory. The book presents a diverse range of perspectives and arguments, providing readers with a rich exploration of the mind-body problem and the nature of consciousness.

The book begins with an introductory chapter by the editors, setting the stage for the subsequent discussions. Angus J. L. Menuge's chapter on declining physicalism and resurgent alternatives offers a compelling examination of philosophical viewpoints surrounding the mind. J. P. Moreland's contribution on neuroscience and the metaphysics of consciousness and the soul raises intriguing questions about the nature of consciousness and its relationship to the brain.

One of the highlights of this book is the section dedicated to the philosophy of mind, where different perspectives such as substance dualism, idealism, and physicalism are thoroughly explored. Stewart Goetz and Charles Taliaferro present a robust defense of substance dualism, while Douglas Axe offers a commonsensical defense of idealism. These chapters provide readers with a deep understanding of the philosophical underpinnings of different theories of mind.

The exploration of neuroscience and psychology in the anthology is equally engaging. Michael Egnor's chapter on neuroscience and dualism challenges the prevailing materialistic view, while Cristi L. S. Cooper's discussion on free will and the limitations of Libet experiments offers a fresh perspective on agency and determinism. Joseph Green's chapter on the limitations of cutting-edge neuroscience prompts readers to critically examine the current state of the field.

The book also studies the fascinating relationship between information, computation, and quantum theory. Bruce L. Gordon's chapter on consciousness and quantum information offers intriguing insights into the potential role of quantum processes in understanding consciousness. Additionally, Winston Ewert's discussion on the human mind's sophisticated algorithm presents a compelling argument about the nature of human creativity and its computational basis.

Overall, *Minding the Brain* is an excellent compilation of diverse perspectives on the mind-body problem. The book covers a wide range of topics and offers deep insights into the crossroads of philosophy, neuroscience, psychology, and quantum theory. Readers with an interest in the nature of consciousness, the mind-brain relationship, and the limits of empirical science will find this book to be a valuable resource. The contributors present rigorous arguments and engage in thought-provoking discussions, making this book a must-read for those seeking a deeper understanding of the complexities of the mind and human-level intelligence.

—**Lipo Wang**, Associate Professor of the School of Electrical and Electronic Engineering, Nanyang Technological University, Singapore

Minding the Brain is a fascinating look at the relationship between conscious experience and the three-pound mass of neurons resting in one's skull. Scholars from different fields address the challenge of understanding the immaterial mind using a materialist framework, and they make the case that a multidisciplinary approach is required to unravel this enigma. What follows is a tour de force of philosophy, neuroscience, and computer science that presents non-materialist solutions to the mind-brain problem. Anyone who has wondered if people are more than a pile of atoms should read this book.

—**Andrew Knox**, MD, MS, Assistant Professor of Neurology,
University of Wisconsin School of Medicine and Public Health

CONTENTS

1. Introduction

Angus J. L. Menuge, Brian R. Krouse, and Robert J. Marks

Is your mind the same thing as your brain? Or are there aspects of mind that are external to the biology of the brain? This question, referred to as the mind-body problem or the mind-brain problem, has been debated for centuries and has captivated curious minds since the dawn of human contemplation. What is the relationship between our mental life and physical body? Intuition suggests our subjective experience of the world is tightly bound up with our physical bodies. Exactly what kind of beings are we, with both a personal mental life and a corporeal nature, somehow all wrapped up in one?

Fresh insights into the mind-brain debate are the subject of this anthology. Analysis is presented from a spectrum of expertise including psychology, neurology, philosophy, computer science, and neurosurgery. Although there are differences in details, all agree there is evidence that the mind is, indeed, more than the brain.

In our modern age, full of science and technology, physical existence often appears to be the most substantial and "real" aspect of the world. After all, the technology that permeates our lives has been made possible by humanity's progress in understanding and manipulating the material world, including our own bodies (and brains). In light of these technological wonders, it may seem plausible to assume that physical existence constitutes the most fundamental layer of reality, and everything else, including our mental lives, is built upon that foundation.

Yet we often take our mental lives for granted. Upon reflection, however, we recognize they possess unique characteristics that do not align well with a materialist framework. These include the inherent subjectivity of our sensory experiences (i.e., pain cannot be ownerless—it must belong to someone), our ability to employ abstract logic and mathematics to explain the workings of the natural world, our capacity to envision a future state and then actualize it in reality, and—perhaps the most distinctive feature—the sense of being a consistent entity, an "I" that persists over time, at the center of our mental activities. It is challenging to comprehend how an arrangement of impersonal matter could give rise to an agent with these distinctively mental attributes.

Delving deeper, we realize that these facets of our mental life are actually more immediate and tangible to us than the external world. Our perception of the world is entirely mediated through our senses. Moreover, the practice of science itself is reliant on our mental capability to employ abstract logic and mathematics. Similarly, the engineering of technology hinges on our mental ability to design solutions to problems and manifest those solutions in the physical world.

Considering that our mental lives possess attributes that are not evidently rooted in physicality, and in fact hold a more immediate place in our lives than the physical world, could it be that our initial assumption—that the physical world is the most fundamental or the most "real" aspect of existence—is incorrect?

Nevertheless, our bodies certainly seem to belong to the physical world. This raises the question: how can we be both mental agents and possess a physical body that occupies time and space? How can two such vastly distinct aspects coexist within a single entity?

Overview

This anthology brings together twenty-three scholars and twenty-five chapters (three of the chapters online only) to explore the mind-body (or mind-brain) problem from diverse perspectives. The target audience of this anthology is academic but multi-disciplinary. Both philosophy and various scientific fields have a lot to contribute to this topic. Unfortunately, technical jargon often creates a barrier to understanding for those outside a specific field. For instance, many neuroscientists may struggle to comprehend a contemporary philosophy of mind journal, and the reverse is also true. In order to reach a broad academic audience, the authors featured in this anthology strive to present cutting-edge philosophical and scientific ideas in an accessible manner. Ideally, by minimizing jargon and providing clear definitions for key terms, the chapters can be understood by non-specialists in the respective fields, enabling valuable interdisciplinary dialogue on this fascinating topic.

The organization of this volume loosely groups chapters written by philosophers in the first two units and those penned by scientists in the last two. Generally speaking, the editors have observed that philosophers tend to tread lightly when discussing scientific matters, while scientists often show the same caution regarding philosophical matters. This is perhaps not surprising, given the extensive scholarship involved in each field and the necessity for academics to specialize in order to develop depth and rigor. However, the editors firmly believe that robust engagement with the topic at hand requires a fusion of astute philosophical thinking and meticulous empirical analysis. In line with this belief, the editors have strongly encouraged the contributing philosophers to consider how their work might be constrained by recent scientific findings, or motivate certain scientific practices or hypotheses. Similarly, they have urged the scientists to reflect on the philosophical presuppositions and implications of their research.

The diversity of viewpoints in the philosophy of mind is vast, necessitating a focused approach in a volume like this—especially when considering the interplay of scientific findings with philosophical perspectives. This volume specifically develops a certain subset of non-materialist philosophical frameworks for reasons that are succinctly summarized below and elaborated in detail in Unit 1. In short, our philosophical exploration (in Unit 2) prioritizes several forms of dualism and idealism.

Units 3 and 4 invite authors to discuss scientific findings from various disciplines relevant to the mind-body debate. Unit 3 includes chapters on neuroscience, psychology, social psychology, and near-death experiences. Unit 4 groups chapters on information theory, quantum theory, computer science, and mathematics.

The intent of this volume is not to advocate for a particular approach to the mind-body problem across all of these academic disciplines. Indeed, the featured authors do not all concur on all points, and neither the editors nor the publisher endorse every aspect of every chapter. Rather, the objective is to aggregate and showcase a broad spectrum of non-materialist perspectives and insights on the mind-body issue, drawing from a range of disciplines, both philosophical and scientific. The editors hope that these diverse contributions will inspire future interdisciplinary scholarship in a similar spirit.

Unit 1: Background

Explorations of the mind-body problem are inevitably situated within a metaphysical framework, and these typically posit the fundamental substances or types of "stuff" that exist in reality. As one might imagine, analyzing the mind-body problem within a specific metaphysical framework requires careful philosophical examination, often involving technical concepts and arguments. This type of work is best suited for academic philosophers, who are equipped with the necessary expertise and training to engage with these intricate issues. Accordingly, the first two units of this anthology feature chapters written by these philosophers.

As mentioned above, this anthology specifically emphasizes a diverse array of non-materialist approaches to the mind-body problem. In Unit 1, three chapters are dedicated to justifying the volume's focus on these non-materialist frameworks.

In Chapter 2, entitled "Declining Physicalism and Resurgent Alternatives," Angus Menuge traces the historical trajectory of physicalism in the philosophy of mind. Over the last century, the prevailing approaches to the mind-body problem have been grounded in materialist metaphysics, which asserts that matter (or more accurately matter-energy) is the sole fundamental type of substance in existence. Menuge recounts the numerous theories which have endeavored to explain the emergence of mental phenomena from impersonal matter, with each successive theory aiming to rectify the limitations or pitfalls of its predecessors.

Despite the persistent dominance of materialist metaphysics in academia, some scholars contend that attempts to account for mental phenomena in purely physical terms over the past century have largely been unsuccessful. Moreover, the recurring pattern of shortcomings in these approaches suggests a potential fundamental problem with materialist metaphysics itself. This circumstance has sparked renewed interest in investigating non-materialist solutions to the mind-body problem, such as dualism or idealism, many of which trace their roots to ancient philosophies. These frameworks have been adapted and refined to address modern critiques, resulting in a vibrant exploration of non-materialist viewpoints on the mind-body problem.

In Chapter 3, "Neuroscience and the Metaphysics of Consciousness and the Soul," James (more often "J. P.") Moreland sheds additional light on the historical developments observed by Menuge by arguing that materialist metaphysics cannot be an adequate foundation for the mind in principle, whereas some form of substance dualism is not only up to the task but can interact productively with the empirical sciences by generating testable research programs—even though Moreland argues that neuroscience will always underdetermine the metaphysics.

In Chapter 4, "Methodological Naturalism and the Mind," Robert Larmer addresses a prevalent concern about the constraints of scientific inquiry. Larmer scrutinizes the common notion that science must strictly adhere to methodological naturalism, an approach committed to explaining all events solely through physical causes. He contends that this presupposition is unwarranted and can hinder the fair evaluation of effective theories. By restricting science to physical explanations, methodological naturalism may obstruct the investigation of non-physical causes (such as mental or rational factors) that could be operative in the world. Consequently, this approach could risk constraining science's ability to fully comprehend the authentic nature of various phenomena.

Unit 2: Philosophy of Mind: Dualism, Idealism, and Physicalism

The first four chapters in Unit 2 offer an introduction to several important non-materialist philosophical models that address the mind-body problem. These chapters are designed to familiarize readers with various schools of thought, assuming minimal prior knowledge. For further exploration, each author has written comprehensive book-length treatments of their respective subjects, which are referenced in the endnotes of each chapter.

In Chapter 5, "Substance Dualism," Stewart Goetz and Charles Taliaferro present a concise survey of the history of thought on the human soul as a distinct substance, spanning Plato, Augustine, and Descartes. The authors then contrast substance dualism with property dualism, of which Aristotle was arguably one of the earliest proponents, which denies a substantial soul in favor of a dualism of mental and physical properties within a fundamentally material human being.

While the problem of causal interaction between body and soul can motivate some philosophers to prefer the latter, the authors argue that property dualism does not in fact escape the challenge of considering how our mental and physical aspects (whether properties or substances) are causally related. Finally, the authors consider the Libet experiments (which are also the subject of Chapter 14), and how these empirical results relate to the consideration of substance and property dualism.

In Chapter 6, "Mere Hylomorphism and Neuroscience," James Madden unpacks the essential concepts in the holistic system that is hylomorphism. Like substance dualism, the Aristotelian doctrine of hylomorphism also has its roots in antiquity. Madden explains that hylomorphism is not primarily formulated to address the mind-brain problem but is rather a distinct philosophy of nature which can be fruitfully applied to the mind-brain problem. On hylomorphism, humans (as well as plants and animals) have "souls"—but this term is not used to refer to a spiritual substance, as in substance dualism. Rather, using terminology that can be easily misunderstood by modern ears, hylomorphism views substantial entities as a composite of "soul" and "matter."

By "soul" is meant the "form"—a term that refers, rather abstractly, to that which, when combined with matter, makes a substantial entity (e.g., a living organism) the thing that it is, rather than something else. The soul is the "principle of actuality" that, when combined with "matter" makes the substantial entity what it is. "Matter," here, is not meant to refer to material "stuff," but instead is the "principle of potentiality"—that which has the potential to be a particular substantial entity, when combined with the right form.

While all living things have a soul in this sense, the souls of plants, animals, and humans have different capabilities. Madden explains how, on the view of hylomorphism, the distinctly human capability to think about and grasp universal truths implies that the human soul is "uniquely separable from matter"—putting hylomorphism in a category that is clearly distinct from materialism. Madden then proceeds to explore how a non-materialist, non-dualist conception of the mind and the brain is situated within this framework. Madden concludes that hylomorphism and neuroscience can work together in partnership towards the holistic objective of "the full actualization of *human-being*."

Douglas Axe introduces us to the concept of idealism in Chapter 7, "Of Thinkers, Thoughts, and Things: A Commonsensical Defense of Idealism." Axe approaches idealism, which is typically associated with the philosopher George Berkeley, both from a commonsense perspective and with modern scientific content that directly challenges physicalism. Axe structures his presentation around the ideas of "Thinkers, Thoughts, and Things," examining the nature of each through introspection and compelling, yet highly accessible argumentation. Ultimately, Axe concludes that Berkeleyan idealism provides a metaphysical framework that not only addresses the mind-body problem and aligns with the Christian Scriptures, but also solves deep challenges posed by modern physics that neither physicalism nor dualism can resolve.

In Chapter 8, "Mind over Matter: Idealism Ascendant," Bruce Gordon also defends what he terms neo-Berkeleyan ontological idealism. Gordon's analysis of the subject takes a detailed and philosophically sophisticated approach. After introducing his proposed framework, Gordon summarizes and critiques the dominant physicalist and dualist alternatives. He examines eliminativist, reductivist, and non-reductivist varieties of physicalism, finding problems with each.

He then considers property dualism, substance dualism, and hylomorphism, and although he finds flaws in each of these positions, substance dualism fares the best in his view. In the end, Gordon, like Axe, concludes that neo-Berkeleyan idealism is preferable because it avoids the philosophical challenges that other frameworks face and has superior explanatory power in accounting for some of the more puzzling aspects of modern physics.

The final four chapters in Unit 2 are chosen to enhance the reader's immersion in the field of philosophy of mind. Delving into distinct topics, these chapters uncover additional challenges for a physicalist account of the mind and simultaneously bolster the case for one or more of the non-materialist options.

In Chapter 9, "The Simple Theory of Personal Identity and the Life Scientific," Jonathan Loose explains how the very act of engaging in the scientific process has implications for the mind-brain problem. To conduct scientific observation and reasoning, a person must not only have a unified consciousness at any given moment in time but also endure over time with a persistent personal identity. Loose examines why the fact of our unified field of consciousness, which is immediately manifest to all of us upon introspection, is difficult to accommodate on the view of physicalism, and instead points to substance dualism. Further, he demonstrates that the kind of observation and the type of reasoning that scientists routinely employ relies upon the unified conscious field that humans experience.

Addressing the subject of personal identity over time, Loose compares the robustness of the simple view, which aligns with substance dualism, to the challenges faced by the complex view. The complex view, grounded in materialism, relies on psychological and physical continuity, and ultimately falls short in accounting for personal identity. As he showed regarding the unified conscious field, Loose shows that the observational and rational processes employed by a scientist rely on the persistence of personal identity over time. Loose concludes that, given an esteem for science, one should prefer substance dualism over materialism.

Mihretu Guta explores the concept of mirror neurons in Chapter 10, "Mirror Neurons, Consciousness, and the Bearer Question." Guta clarifies that mirror neurons, a category of brain cells, are considered by some neuroscientists to play a crucial role in reflecting another person's mental state. These neurons activate when an individual observes someone else engaging in goal-oriented actions. Guta notes that while mirror neuron theory has received considerable positive attention, it also has its critics. However, Guta observes that this scientific criticism has missed a prior and more basic metaphysical issue: whether the functional properties attributed to mirror neurons require acknowledgment of an irreducible consciousness and its bearer (i.e., the self or person). Guta conducts a detailed analysis of this metaphysical issue, and in so doing exhibits how important it is to apply careful metaphysical reasoning in order to draw valid neuroscientific conclusions.

Guta pursues a more purely philosophical kind of project in Chapter 11, "In What Sense is Consciousness a Property?" Guta considers how the metaphysics of properties, which has been an important and controversial issue in philosophy since antiquity, might be applied to the topic of mental properties. In particular, Guta asks how consciousness might be considered a property. After surveying the major schools of thought on properties, Guta notes that consciousness, with its irreducibly subjective nature, is difficult to locate within any of the traditional theories. Motivated by this challenge, Guta proposes and discusses what he calls the bearer-dependent model of consciousness. Guta concludes that given his bearer-dependent model, the most promising framework to make sense of consciousness as a property is one that takes a realist conception of properties, and makes an ontological distinction between physical and non-physical substance.

In Chapter 12, titled "Subject Unity and Subject Consciousness," Joshua Farris delves into the nature of the unity observed in our conscious experience and its implications for the mind-body problem. Farris initially criticizes various physicalist and non-reductive physicalist theories of the mind, demonstrating their inadequacy in explaining the special kind of unity present in our phenomenal consciousness. He then examines several specific substance dualist models, or "obscure dualisms," recently proposed by contemporary philosophers, and reveals their inability to account for the self's transparency. Ultimately, Farris argues that a neo-Cartesian form of substance dualism best explains the phenomenal unity of consciousness.

Unit 3: Neuroscience and Psychology

WHILE PHILOSOPHY is foundational when addressing the mind-body problem, many fields of modern science are also implicated in this discussion. Neuroscience has made it abundantly clear that there is a close connection between our brain's physical functioning and our mental capabilities. Psychology, which focuses on mental well-being, is also closely linked to the mind-body problem. Unit 3 features contributions from neuroscience and psychology scholars who consider the relationship between the scientific findings of their field and the nature of the mind and brain. Additionally, a noted theologian and philosopher compiles and examines the evidential case for near-death experiences in the unit's final chapter.

In Chapter 13, "Neuroscience and Dualism," Michael Egnor begins by examining the materialist philosophical preferences prevalent in twentieth-century neuroscience, and provides an overview of idealist and dualist views of the mind. Egnor then proposes an approach for empirically testing these metaphysical frameworks. Within this context, Egnor evaluates several prominent neuroscience experiments, such as Roger Sperry's research on split-brain patients, Wilder Penfield's cortical stimulation experiments, and Benjamin Libet's study of brain activity before decision-making. Ultimately, Egnor concludes that materialism struggles to explain crucial empirical findings, while both idealism and dualism are more consistent with the scientific evidence. In particular, Egnor favors Thomistic dualism (i.e., hylomorphism) as the most suitable metaphysical framework for neuroscience.

In Chapter 14, titled "Free Will, Free Won't, and What the Libet Experiments Don't Tell Us," Cristi Cooper delves into the Benjamin Libet experiments (briefly discussed earlier in the volume by Goetz and Taliaferro, and in the previous chapter by Egnor). Published in 1983, the Libet experiments continue to captivate those interested in neuroscience and free will. These experiments studied a spike of neural activity, known as the readiness potential, that occurred just before human subjects decided to press a button. Many people, particularly in popular science coverage, interpreted the results as evidence against free will, arguing that the readiness potential indicates our brains "decide" before we consciously do. However, it is less well known that Libet himself interpreted his results differently, and that many scientists have further investigated the readiness potential since 1983.

After summarizing several key studies published on this topic in the intervening years, Cooper contends that the popular interpretation of the readiness potential as a clear refutation of free will is actually weakened by subsequent research. Cooper concludes with a cautionary message for scientists (and their popular interpreters) not to overextend the implications of the research. Finally, Cooper encourages future researchers to investigate the neuroscience of free will, as it remains an open question.

Joseph Green offers a broad perspective on neuroscience and the mind-body problem in Chapter 15, "On the Limitations of Cutting-Edge Neuroscience." He observes that rapid advancements in neuroscience, amplified by popular media coverage, have fostered heightened expectations about our current understanding of the brain and our capacity to manipulate it using engineering techniques. Green evaluates the state of the field, celebrating areas of remarkable technological progress while also highlighting current limitations within neuroscience, particularly regarding our limited understanding of neural circuit dynamics. He then explores how the philosophy of mind could guide neuroscientists, and in turn, how neuroscience might help inform philosophers. In conclusion, Green advocates for a more philosophically cautious approach for neuroscientists, promoting humility and an overall agnosticism since current neuroscience itself warrants no specific metaphysical stance.

In Chapter 16, "Revising Our Pictures of Emotions," Natalia Dashan and David Gelernter explore the nature of human emotions through the lens of affective psychology. They investigate the significance of feelings and emotions in human cognition, aspects that have often been overlooked by artificial intelligence researchers and computationalists since Alan Turing's time. By analyzing several fictional case studies, Dashan and Gelernter demonstrate the central role our conception of emotions plays, affecting our emotional experiences and responses and thereby influencing our sense of reality and fundamental behavioral patterns. They examine various mental metaphors that people live by, some more accurate than others. Ultimately, Dashan and Gelernter leave readers with an appreciation for how these frames for understanding emotions serve as lenses through which we perceive and experience reality.

In Chapter 17, "A Case for the Relational Person," Eric Jones examines two opposing perspectives on the concept of personhood in social psychology, the atomistic/egoistic view and the relational view. The former views the person as determined, atomistic, and explicable in terms of the propagation of genetic material. The relational view highlights the fundamental dependency of individuals on their relationship with others, in the course of human development and in the context of what makes for a fulfilling life.

Upon reviewing relevant social psychology research, Jones concludes that the atomistic view does not adequately explain the data, while the relational view is broadly supported by the evidence. Due to the shortcomings of the atomistic model, Jones explores the role of metaphysics, suggesting that a materialistic metaphysical framework may limit social psychologists to the less effective atomistic/egoistic model. In contrast, a non-materialist metaphysical framework can provide the necessary ontological resources to support the more successful relational model. Jones ultimately posits that the findings of social psychology might warrant a preference for a non-materialist metaphysics of personhood.

In Chapter 18, titled "Evidential Near-Death Experiences," Gary Habermas explores the evidential support for and potential implications of such experiences. He begins by distinguishing between near-death experiences (NDEs) that provide captivating narratives but lack verifiable elements, and those that involve "corroborated veridical recollections." Habermas assembles and categorizes a substantial number of NDEs that feature corroborated observations made by individuals during their NDEs, which would have been impossible to perceive from their physical location using their ordinary senses. He then criticizes various explanations for this data from both naturalist and non-naturalist perspectives, such as those involving extra-sensory perception.

In conclusion, Habermas argues that the considerable number of high-quality evidential NDE cases gathered offers a persuasive case for interpreting these reports as genuine, veridical experiences of the individuals involved. As such, these NDEs appear to have metaphysical implications, indicating that people's souls or minds might be separable from their physical bodies, while amazingly retaining some ability to perceive "sensory" information (e.g., take in visual perceptions as if they were using their eyes, which they are not) and even move about in space.

Unit 4: Information, Computation, and Quantum Theory

Since Claude Shannon pioneered the field of information theory and Alan Turing developed a comprehensive theory of computation, numerous scholars have utilized these theories to decipher the nature of the mind and brain. Among physicalists, this approach led to views such as functionalism and the computational perspective of the mind. However, as Angus Menuge shows in his previously mentioned Chapter 2, these views fall short of providing a convincing description of the mind.

Nevertheless, these disciplines do offer potent tools to enhance our understanding of the human mind, especially when combined with a metaphysical framework that isn't restricted to reducing the mind to merely organic computation, as physicalism must. Unit 4 compiles several chapters that leverage insights from information theory, computer science, quantum theory, and mathematics to interrogate the mind-body problem.

In Chapter 19, "Information and the Mind-Body Problem," Angus Menuge investigates the idea that information, given its dual existence in both abstract and concrete forms, could serve as an effective means of clarifying the dualist interaction between mind and body. Menuge acknowledges that while physicalism struggles to account for core aspects of the mind, such as subjective consciousness and intentionality, dualism also encounters a significant hurdle in explaining mind-body interaction.

He uses these criticisms as a springboard for his project to provide a non-physicalist explanation of mind-body interaction. Drawing upon the inherent properties of information, Menuge introduces the Command Model of Action (CMA), a model designed to explain how non-physical mental intentions can cause physical effects, and the Signal Model of Sensation (SMS), a model that describes how nerve signals can lead to subjective experiences. Menuge concludes by exploring how the CMA and SMS models could interact with various non-materialist philosophies of mind, and how these models might inform scientific research.

Bruce Gordon delves deep into the intersection of quantum theory and idealism in Chapter 20, "Consciousness and Quantum Information." He starts by unpacking the fascinating findings of quantum physics, showing how these discoveries have painted a picture of a natural world that is surprisingly devoid of material substances. Gordon contends that this understanding propels us towards theistic quantum idealism. This perspective suggests that our perception of an external physical reality is, in essence, our subjective experience of God's thoughts. Moreover, he presents evidence from quantum cosmology supporting the broad view that reality is a single, timeless mental act through which God conceived and brought forth the universe. He outlines a framework for a quantum-informational neuroscience and discusses its role in studying neural systems and their connection to conscious experiences.

To answer the question of how conscious minds interact with the world, Gordon turns to the metaphysical basis of theistic conscious realism. In closing, he explores a variety of additional topics, including near-death experiences, theistic beliefs about life after death, and a broad approach to scientific practice in light of this conception of reality.

Chapter 21, by Eric Holloway and Robert J. Marks II, is titled "Human Creativity Based on Naturalism Does Not Compute." In this chapter, the authors pose the intriguing question of whether the physical, human brain is capable of creating the large volumes of creative prose that humans regularly produce. Holloway and Marks analyze this problem by conducting an informational analysis of a simpler problem, namely the probability that *any* meaningful phrase can be generated by chance. Explaining essential concepts such as active information and the conservation of information along the way, the authors reach the remarkable finding that the entire universe's informational capacity is far exceeded by the demands of a single book. Thus, Holloway and Marks conclude that human literary creativity cannot be explained by merely naturalistic computational brain activity.

Chapter 22, by Winston Ewert, is entitled "The Human Mind's Sophisticated Algorithm and Its Implications." In this chapter, Ewert explores the extent to which the human mind can be compared to a computer. Setting aside the question of phenomenal consciousness for the moment, Ewert concentrates on the cognitive problem-solving abilities of the human mind. To facilitate his analysis, Ewert introduces the concept of the halting problem, a unique computational task that involves determining whether a given program or procedure will cease or loop indefinitely.

Ewert demonstrates that the halting problem is logically equivalent to a variety of essential human tasks, including mathematical reasoning, pattern detection, prediction making, and item searching, implying that findings about the halting problem could have wide-ranging implications. Importantly, it has been proven that no program can possibly exist that solves the halting problem for every possible problem. Ewert likens human cognition to the ability to solve the halting problem for a limited range of programs.

After addressing some criticisms of his stance, Ewert extends his analysis, concluding that humans will never create an artificial intelligence (AI) that can match human intelligence, nor will an AI system ever self-generate another AI system superior to itself. Ultimately, Ewert suggests that his conclusions point to an origin of the human mind in a form of intelligence that is non-computational, transcending the constraints of the halting problem.

In Chapter 23, "Mathematical Objects are Non-Physical, So We Are Too," Selmer Bringsjord and Naveen Sundar Govindarajulu consider our ability to understand and interact with logico-mathematical objects. The authors provide several examples of these objects, including the Quicksort algorithm—a well-known algorithm in computer science which sorts a list of ordered objects (e.g., numbers)—and *modus tollens*, an inference schema in propositional logic, i.e., "If P, then Q. Not Q. Therefore, Not P." As the first step in their two-part argument, Bringsjord and Govindarajulu argue that these objects—distinct from their concrete embodiments, for example, in a programming language—are indeed non-physical. In the second step, the authors argue that our *understanding* of these objects indicates that we are more than merely material, biological computing machines; we must possess at least some immaterial aspect.

In Chapter 24, titled "Can Integrated Information Theory or the Theory of Cognitive Consciousness Explain Consciousness?," Naveen Sundar Govindarajulu and Selmer Bringsjord examine two proposed theories aimed at scientifically explaining consciousness: Integrated Information Theory (IIT), conceptualized by neuroscientist and psychiatrist Giulio Tononi and extensively promoted today by neurophysiologist and computational neuroscientist Christof Koch, and the Theory of Cognitive Consciousness (TCC), which is currently developed by Govindarajulu and Bringsjord. IIT aims to provide a scientific explanation of phenomenal consciousness—our subjective experience of the world—whereas TCC focuses on cognitive consciousness, which has to do with the contents and structure of our cognition.

The authors first elucidate what is typically meant by a "scientific explanation," applying this understanding to the task of scientifically explaining consciousness. Armed with this framework, Govindarajulu and Bringsjord embark on their analysis of the two theories. They conclude that IIT is arguably based on some debatable axioms and postulates, and that the success of this theory remains to be definitively determined. Lastly, the authors assess their own theory, TCC, arguing that it is an extraordinarily productive model for guiding future engineering work aimed at constructing high levels of cognitive consciousness in artificial systems.

In Chapter 25, "How Information Realism Dissolves the Mind-Body Problem," William Dembski concludes the anthology by integrating ideas from information theory together with metaphysical concepts, thereby introducing the notion of informational realism. Dembski describes informational realism as the belief "that the defining characteristic of reality is the ability to exchange information." Informational realism adopts a minimalist ontology; it neither prescribes nor excludes specific metaphysical substances such as matter or spirit. Instead, it posits that the fundamental entities of reality are information sources that generate information, which is then received by other sources; these sources gain their reality through this reciprocal exchange of information.

Dembski contrasts major historical perspectives with informational realism, including Aristotelian hylomorphism, Plato's theory of forms, and Berkeleyan idealism. Dembski underscores that informational realism's primary virtue lies in not setting rigid metaphysical claims a priori, instead leaving room for philosophical and scientific exploration. He criticizes materialism specifically for its a priori insistence on certain types of information sources, which could potentially limit our understanding of reality's true nature. By avoiding a priori presumptions, Dembski suggests, we are free to examine the nature of the informational exchanges and thereby draw inferences about the metaphysical nature of the information sources.

Conclusion

From the foregoing summary of this volume, we can see that there is ample evidence across a spectrum of specialties that the mind is more than the brain.

In his book *A Brief History of Time*, physicist Stephen Hawking claims nothing in physics is ever proven. We simply accumulate evidence. Drop a pencil and watch it fall. This is additional evidence for something called "gravity." Similarly, this anthology has not proven that aspects of the mind are disjoint from the corporeal brain, but has presented strong evidence of such separation. This central theme has herein been articulated through the lens of diverse philosophical, medical, mathematical, psychological, and scientific perspectives. Over the past century, materialism has predominantly guided our approach to these topics, yet has repeatedly failed to adequately account for the core attributes of the mental. Evidence for a non-materialist account is accumulating.

As articulated above, several compelling non-materialistic models are available as alternatives to physicalism. As exhibited by the chapters in this anthology, these alternatives exhibit remarkable potential in elucidating the nature of the mind and brain. The shared conviction of this anthology's editors and contributing authors is that these frameworks warrant further exploration in a multidisciplinary manner, fostering promising lines of inquiry for a new generation of philosophical and scientific discovery.

Acknowledgments

The editors would like to extend their deepest gratitude to everyone who contributed to the assembly of this anthology. We owe our thanks to Stephen Meyer for conceiving of and encouraging this project. The Bradley Center commissioned this anthology, and for that we are exceedingly grateful. In the early stages of brainstorming and team building, both Brian Miller and Cristi Cooper played crucial roles. Daniel Reeves was immensely helpful in planning and executing a conference for the authors to collaboratively discuss and refine their ideas. Katherine West was of great help in many administrative matters. Our copy editor, who wishes to remain anonymous, admirably rose to the challenge of the Herculean task of copy-editing all of the chapters and preparing the proofs for publishing. Lastly, and certainly not least, our profound thanks go to John West, whose skill and expertise ensured that every plan was effectively put into action.

Unit 1: Background

2. Declining Physicalism and Resurgent Alternatives

Angus J. L. Menuge

1. Introduction

The late nineteenth and early twentieth centuries saw the rise of *analytic philosophy*, an approach to philosophy that focused on the careful analysis of the meaning of language. The movement included *Logical Positivism*, an austere school of thought that disparaged traditional metaphysics and claimed that statements are literally meaningful only if they are true by definition or verifiable by observation.

Philosophers eventually rejected Logical Positivism,[1] but analytic philosophy remains the dominant Anglo-American paradigm. In this tradition, responsible philosophy must be constrained by a scientifically informed view of the world. Wilfrid Sellars spoke for many when he asserted, "Science is the measure of all things."[2] By "science," they meant such paradigmatic natural sciences as physics and chemistry.

These philosophers typically concluded that substance dualism, the assumption that there are mental as well as physical substances, cannot be taken seriously. This was not only because of the mind-body problem raised for René Descartes's version of substance dualism (how could substances of such fundamentally different kinds interact if they had no shared medium?), but also because natural science has no room for souls, or for the goals of rational agents, since it does not recognize immaterial beings or final causes.

Human beings, it was thought, must be *naturalized*—made part of the same nature that includes rocks, rivers, and ravens. Moreover, since the world revealed by natural science is a physical world, it must be possible to understand human beings in physical terms. With the rejection of Cartesian egos (immaterial mental subjects), philosophers turned to the more scientifically tractable tasks of explaining behavior and understanding the relation between mental and physical states and events. Enter *physicalism*, the sustained attempt to explain the human mind in ways that are compatible with scientific materialism.

The history of physicalism is one of extraordinary diversity: a wide variety of theories, with multiple versions, have jockeyed for dominance. Yet it is also a tale of persistent failure. One physicalist theory after another has either ignored or falsified the central characteristics of consciousness, intentionality, and rationality that define our mental life. We will begin by tracing the history of physicalism from the early varieties of behaviorism to the present day, making the case that physicalism is now entering a period of paradigm crisis (section 2). The crisis centers on a basic dilemma for physicalism: the stricter accounts that remain faithful to the core doctrines of physicalism seem obviously inadequate, but more promising, relaxed accounts are no longer obviously physicalist at all.

This crisis has converged with two other currents of change with the surprising result that alternatives to physicalism are now taken seriously (section 3). First, while physicalists understood the mental in terms of states of an organism, it could not be denied that these states belonged to a particular mental life, so some explanation is needed for the existence and character of mental subjects. But it turns out that physicalist accounts of subjectivity are woefully inadequate. Second, around the same time there was an explosion of research centered on historical and new accounts of the soul, which revealed the large number of options available for dualists.

As a result of these three factors—physicalism in paradigm crisis, a renewed focus on the self, and new research on the soul—alternatives to physicalism were put back on the table for serious discussion. These alternatives included several varieties of substance dualism, hylomorphism (which understands a human being as the combination of matter and form), and idealism, which denies the existence of mind-independent matter and reduces reality to minds and their ideas. Serious attempts were made, both to address well-known philosophical objections to these alternatives and to show how they inform current work in science, including neuroscience and a variety of practical therapies. We will end by briefly considering some of the reasons scientists should welcome these developments (section 4).

2. Physicalism: From Dominance to Paradigm Crisis

The foundation of physicalism is a materialistic view of substances. What is a substance? Within ontology, the branch of philosophy which studies kinds of being, substances are understood as persistent things that can exist on their own, as opposed to modes (ways those things can be) which depend on substances for their existence. For example, a cat is a substance that can persist over time, sometimes in a hungry or a purring mode, sometimes in other modes. Yet while the substance can persist with or without these modes (it might be hungry or not), the modes depend on the substance for their existence (e.g., a cat's hunger could not exist all by itself without the cat).

Materialism is a *monistic* view: there is only *one* kind of substance, material substance, though that thing may exist in many modes (e.g. living or non-living, conscious or unconscious). What this means is that all of the things in reality are of the kind recognized by physical science, and at least in principle, they can be understood in completely impersonal terms.

Materialism focuses on the material cause (what something is made of) and the efficient cause (what brought something into being), and insists in both cases that the cause must always be something physical. But it is in tension with final causes (which give the purpose for which an effect exists), since matter seems void of goals. While some materialists would allow that final causes somehow emerge from matter, expressed in the goals and purposes of living organisms and rational agents, no natural science discerns these causes operating at the level of basic physics and chemistry.

Due to its materialist ontology, physicalism rejects all kinds of non-material substances, including human souls and God. As a result, physicalist philosophy of mind rejects any appeal to some occult ego or self and tries to explain cognition in terms of ultimately physical causation and processes. Minds cannot be observed in the sense required by physical science. In this sense, something qualifies as observable only if it is publicly accessible (many people could observe it), so private introspection of one's own mind does not count, and it is impossible for one person to introspect the mind of another. So if cognition can be understood in physicalist terms, it must be explained by the causal relations between observable phenomena.

It is not surprising then that the early versions of physicalism (which I call "*reductive physicalism*") attempt, in various ways, to reduce our talk about mental states to talk about physical states. We will consider several of the leading versions of reductive physicalism and show that they all suffer from several common flaws. Recognition of this fact motivated the rise of more sophisticated views, which, while insisting that all substances are material, allow the existence of irreducibly non-physical properties ("non-reductive physicalism"). While these theories are more plausible than the varieties of reductive physicalism, I will argue that they still fail to capture a number of the most important aspects of human cognition.

2.1 Reductive Physicalism

Since physicalism hopes to explain cognition in terms of scientifically observable phenomena, it is not surprising that one of the earliest physicalist proposals was *behaviorism*. There are different versions of behaviorism, but the common denominator is that we can understand attributions of mental states to an individual in terms of that individual's behavior.

On one version ("logical behaviorism"), to say that someone is in pain *means* that they are disposed to produce appropriate pain behavior, such as withdrawing, wincing, crying out, etc. Other behaviorists admitted this does not seem right because the meaning of the word "pain" does seem to include the idea that pain feels a certain way, which cannot be captured by how pain may cause us to behave. Nonetheless, these behaviorists agreed that, regardless of the meaning of the word "pain," what identifies a state as a pain state is that it disposes us to behave in certain ways. The apparent advantage of behaviorism is that it provided an account of how we could learn to speak about mental states like pain without having direct access to other people's minds. But behaviorism of both kinds soon fell to a number of decisive objections.

First, a false assumption of behaviorism is that mental states can be understood atomistically, so that one can identify the dispositions to behave of each state independently of other states. But this is not true. We cannot simply say that the belief that it is raining disposes an individual to stay indoors or go out with an umbrella or coat, because this assumes that the individual wants to stay dry: if she wants to get wet, she will behave differently. Likewise, a pain may not dispose us to cry out because we are afraid of being hurt or because we want to seem strong. Thus, we cannot speak unconditionally of what a particular mental state disposes us to do: we must consider the whole network of mental states that interact with that state. Behavior becomes intelligible only when we consider an individual's mental states holistically, as an integrated whole, and not atomistically, as if they were independent and unrelated parts.

Secondly, and in part because of the holism of the mental, there is no behavior that is always necessary or sufficient for the possession of a mental state. As Hilary Putnam pointed out, Spartans or super-Spartans may train themselves to manifest no pain behavior at all (not even the willingness to say, calmly, "I am in pain") despite experiencing excruciating agonies, because they do not want to show any sign of weakness. So pain behavior is not necessary for being in pain. And method actors hoping for an Oscar and deceitful athletes hoping for a penalty can produce behaviors (grimacing, crying, etc.) indistinguishable from those of someone in pain though they do not feel a thing. So pain behavior is not sufficient for being in pain either.[3]

Thirdly, what is most obviously left out of any behavioral account, even when individuals do produce the normal expected behavior, is any explanation of what it is like to be in pain. The experience of pain—a *quale*—has a particular subjective character, and while two individuals may produce indistinguishable pain behavior, it could be that one experiences pain and the other does not, and even if both experience a qualitatively identical pain, still it is not possible for the individuals to share the same, numerically identical, pain. That is, there are two distinct pains experienced by the two individuals, and two things cannot be one thing.

But it is arguable that the experience of pain (what it is like to feel pain) is *intrinsic* to pain—it is what pain *is*—whereas pain behavior is extrinsic to pain—it only concerns what pain normally *does*. And in general, we cannot define the nature of a state by its effects. For example, what ice *is* cannot adequately be defined by its effect on roads (making them slippery) or on drinks (making them cooler), etc.

Recognition of this fact led physicalists to the conclusion that mental states must be understood, not in terms of behavioral effects, but rather as states with their own identity. These states may or may not produce typical behaviors, but their identity does not depend on whether or not they do. Since physicalism requires these states to be understood in scientifically observable terms, it is not surprising that the next version of physicalism, *identity theory*, proposed that mental states just are physical states of an individual's brain.

Early on, J. J. C. Smart proposed *type identity theory*, according to which each type of mental state is identical to some type of physical state.[4] It was suggested that being in a pain type of state was identical to being in the physical state of C-fibers firing, where "C-fibers firing" is a placeholder for whatever neuroscientists discover that all pain has in common neurologically. But the type identity of pain with a neurological state requires that there is one type of neurological state that is always correlated with pain in all creatures that experience it. This, it was pointed out, is empirically implausible. Many animals that feel pain have very different nervous systems from humans.[5] And even within the human species, there may be marked differences between brains (developmental differences, deficits, etc.) making it unlikely that all humans that are in pain are in exactly the same type of state.

A second iteration was *token identity theory*, according to which each individual pain state is identical with some neurophysiological state. This allows that pain states may be "realized" differently in different species and even in different individuals of the same species. But the problem with token identity theory is that if pain is multiply realized—realized by different physical states—then we still need an illuminating account of what all these states have in common that makes them pain states.

The answer cannot be that they *are* all pain states, since what we want to know is what makes a state a pain state rather than some other sort of state. So it seems that the token identity theorist would have to appeal to some shared physical property of all of the realizing states, and this faces two objections. First, if there is such a property, then token identity theory collapses into type identity theory, since this property will characterize a type of physical state all pain states have in common. But we just saw it is implausible that there is such a type. Second, nothing about any physical property entails the existence of a conscious state of pain. And this leads to the most fundamental objection to identity theory.

This objection is that identity theory violates the plausible principle of the indiscernibility of identicals. According to this principle, to say that *x* is identical to *y* means that, for any property P, *x* has P if, and only if *y* has P. Thus, to say that water is identical to H_2O means that water has a property if, and only if, H_2O has that property. In other words, it could not be that H_2O has a property that water lacks or that water has a property that H_2O lacks. But water is not identical to H_2SO_4 (sulfuric acid) because they have different properties. (No one is advised to try to determine those properties at home!) The problem for identity theory is that mental and physical states seem to have categorically different properties. Thus, pain has a subjective character (how pain feels, or what it is like to feel pain), but as described by physical science, no state of the brain has, or entails, a subjective character.

Further, a pain state can have content; e.g., a runner feels a sharp pain in her toe due to an ingrown toenail. The pain thus has intentionality—it is directed toward an object. Likewise, one can believe that the Danube runs through Budapest or desire a vacation in Budapest. So beliefs can be about other things (the Danube, Budapest) and can have a propositional content (that the Danube runs through Budapest). But as described by physical science, no physical state is about anything beyond itself or has any propositional content. Yet subjectivity and intentionality seem to be intrinsic to mental states; they would not be the mental states that they are without these properties. So if subjectivity and intentionality are intrinsic to mental states but are not intrinsic to any physical states, mental states cannot be identical to physical states.

About the same time that identity theory fell into disrepute, the new paradigm of cognitive science emerged. In the 1970s and 1980s, impressed with the powers of digital computers, many assumed some version of the computational theory of mind, according to which the mind is a sophisticated kind of computer. The great appeal of this view is that computational states seem to be like mental states in that they can be understood at an abstract level, allowing a great deal of variation in how the states are physically realized.

If one cannot identify mental states with obviously physical states, perhaps the reason is that the physical states are not being described at a sufficiently high level of abstraction. This suggested a new theory, *functionalism*, which claimed it could overcome the deficiencies of both behaviorism and identity theory. According to functionalism, we should understand a mental state not simply in terms of its behavioral output, but also, like a computational state, in terms of its input and interaction with other states. Thus, functionally considered, pain is a state that plays the complex role of mediating certain kinds of input (e.g., impending or actual tissue damage), interactions with other states (e.g., having or not having a stoic disposition) and certain kinds of output.

In this way, functionalism, unlike behaviorism, takes account of the holism of the mental, since the output of a pain state depends on its interaction with other states, and unlike identity theory, it does not require pain to be realized in any particular physical way, so long as the state plays the appropriate causal role.

However, critics soon realized that functionalism is only a sophisticated, partly internalized version of behaviorism. Instead of focusing exclusively on an individual's outer bodily behavior, functionalism also considers the behavior of internal states of the brain, including their interaction with one another. But this still means that a mental state is being defined extrinsically—by what is done to it, or by what it does—and not intrinsically, by what that state is. As a result, functionalism does no better than identity theory in accounting for the intrinsic subjectivity and intentionality of mental states.

A robot can surely be programmed so that when the robot is damaged, the interaction of its internal states produces behavior appropriate for someone in pain. Yet this provides no reason for us to conclude that the robot is in pain, or knows what it is like to feel pain. Likewise, John Searle pointed out that a computer system might be designed that would satisfy a functionalist theory of understanding Chinese, but without understanding anything.[6] All the system would need is a means of mapping input symbols representing questions to output symbols representing answers, and this could be done via a stored database that matched questions and answers by their form, or syntax, but without any understanding of the meaning, or semantics, of those questions and answers. This is because there is no reason to think that the computer's states have intentionality (those states are not intrinsically about anything and do not have a meaning for the computer).

More generally, Ned Block pointed out that if all that matters to a mind is having the right functional roles, then if a crowd of billions of people emulated all the interconnections and signals between the neurons of a brain, we would have to conclude that the crowd has its own consciousness and mental states over and above those of each individual member of the crowd.[7] But we judge this to be obviously false. No amount of extrinsic complexity captures the internal character of what it is like for a subject to be in a mental state.

Thus, mental states cannot be reduced to behavioral dispositions, brain states, or even abstract functional roles. What is missing from all of these accounts is a recognition of the peculiarly *mental* character of mental states. If this is not reducible to anything physical, even when the physical is abstractly described, there seem to be only two remaining options.

One can take the extreme position of *eliminative materialism*, championed by Paul and Patricia Churchland, which claims that we should conclude from the irreducibility of commonsense beliefs and desires to anything physical that they do not really exist![8] In their view, all of cognition can be understood in terms of transitions between neural activation patterns in the brain, reducing human beings to organic computers. But this is highly implausible, and seems to be incoherent. It is implausible because no physical description of those neural activation patterns entails that we have subjective conscious states with intentional content. This would seem to imply that consciousness is an illusion, yet this is an incoherent idea, because only a conscious subject could be subject to an illusion, and illusions are intentional states.

Furthermore, in a world of passive transitions from one neural activation pattern to the next, there are no goals, and we cannot identify anything that counts as a reason for an action. Without desires and beliefs, such as the desire to test a theory and the belief that the theory predicts a certain observable event, science itself is no longer intelligible as a rational activity.[9] As a result of these and many other difficulties, most philosophers of mind reject eliminative materialism.

The other and much more popular option is to embrace a weaker form of physicalism that allows irreducible mental properties in its ontology. Maybe it is just a fact that, under the right circumstances, purely physical systems generate special mental properties. This option is known as non-reductive physicalism.

2.2 Non-reductive Physicalism

If strong (reductive) versions of physicalism fail to do justice to the mental, some physicalists thought it would be worth considering whether one could maintain the basic core commitments required to be a physicalist while allowing that mental states do not reduce to the physical. Many concluded that this was feasible because, they argued, it is enough for physicalists to insist on two related doctrines.

First, physicalists must maintain the *causal closure of the physical world*, i.e., that any physical effect has a sufficient physical cause.[10] Without this assumption, it could be that some effects are the result of immaterial entities like the soul or God.

Second, physicalists must hold that mental properties are completely determined by, and dependent on, physical properties. Donald Davidson suggested that this idea could be captured through the notion of *supervenience*.[11] While different forms of supervenience have been proposed, the basic idea is that the physical determines the mental in the sense that there can be no mental difference without a physical difference.

Thus, if two individuals X and Y are physical duplicates, X and Y must have all and only the same mental properties. It could not be, for example, that X was thinking of Budapest, but Y was thinking of Gary, Indiana. Supervenience is an asymmetric dependence relation, which means that though physical duplicates could not have different mental properties, it could be that the same mental property supervenes on different physical properties in different individuals. In that sense, mental states are "multiply realizable" by the physical states of brains.

However, bare supervenience does not seem to provide an adequate theory of the mind. First, if supervenience is a fact, it is not a fact that we can claim to know *a priori*. As David Chalmers pointed out, it is conceivable that there are physical duplicates of us in another possible world that are "zombies" who have no mental life at all.[12] And it is also conceivable that there are "invert" worlds, in which our physical duplicates have different mental states; for example, whenever I feel pain, my physical duplicate feels pleasure, and vice versa. If the intuitions driving these thought experiments are correct, what they show is that the relation between physical and mental properties is a contingent one, and if so, the claim that supervenience captures that relation requires both evidence and an explanation.

But there seems no way to establish supervenience as a fact by scientific evidence, since given the immense complexity of the brain and the marked genetic and developmental differences between individuals (not to mention the implications of quantum mechanics and chaos theory) there is no practical possibility of producing two physical duplicates to see if their mental states coincide. And if supervenience is a contingent fact, it is a remarkably odd one, which surely cries out for explanation. There is nothing evidently in common between being in a particular physical brain state and believing that the Green Bay Packers are playing badly or desiring to eat cheese curds. Why brain states are correlated with any mental states, or with the particular mental states that they are, seems utterly mysterious.

Further, as Jaegwon Kim has repeatedly argued, supervenience seems incompatible with the commonsense idea of mental causation—that being in a mental state, such as pain, can have effects, including other mental states (such as the belief that I am in pain) and behavior. This is because the physical base-states on which mental states supervene seem to exclude the possibility that the mental *qua* mental has any distinctive causal role (the exclusion argument).

To see this, consider any case of mental causation. Suppose mental state M causes a further mental state M*. For example, suppose having a headache (M) causes my belief (M*) that I need to take an aspirin. By hypothesis, M is completely determined by some physical base-state P, and M* is completely determined by some physical base-state P*. Given the assumed priority of the physical over the mental, M* cannot exist without its base P* (or some alternative base, which we may assume is not present), so M must cause M* by causing P*.

However, physicalism is also committed to the causal closure of the physical, which implies that every event has a purely physical cause. So, given the dependence of M on P it is natural to say that P causes P*, and hence that P causes M*. For without P, M would not be there, and hence P* and M* would not be there, so it appears that P causes P*, and hence M*. But assuming we do not allow systematic overdetermination (with many effects having *both* a sufficient mental cause and a sufficient physical cause), if P causes M*, and P has ontological priority over M, then M cannot also be the cause of M*: M is excluded. These relationships are summarized in Figure 2.1. So, if we grant physicalism, there can be no *distinctively* mental causation.

While some contest Kim's exclusion argument, it has proved remarkably difficult to evade, and if correct, it implies that non-reductive physicalism supports epiphenomenalism. Epiphenomenalism is the implausible claim that our mental states do not cause our behavior. According to a proponent of this view, Daniel Wegner, mental states are causally impotent previews of what our brain will make our body do, implying that free will is an illusion.[13]

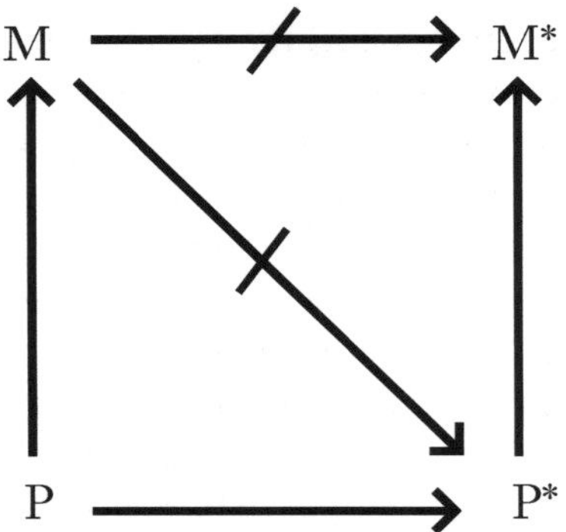

Figure 2.1 The Exclusion Problem for Supervenience. M is excluded by P, which causes M* by causing P*.

Key:
1) M and M* are mental properties.
2) P and P* are physical properties.
3) Vertical arrows signify supervenience.
4) Horizontal arrows signify causation.
5) Crossed arrows are excluded causal pathways.

But epiphenomenalism seems an absurd doctrine for a number of reasons. If it is true, no one could communicate the idea, since that would require a mental idea to cause physical speech or writing, that is, mental causation. And, as Richard Swinburne has pointed out, there seems to be no way of providing scientific evidence for epiphenomenalism.[14] For to show that an individual's mental state does not cause behavior, we first need to know that the individual is in that mental state, and we can only know that if the individual's statement that he is in that mental state is reliably caused by that mental state. Thus, any experiment designed to test whether epiphenomenalism is true seems tacitly to assume that epiphenomenalism is false!

In an attempt to avoid these problems with supervenience, some physicalists, like John Searle, have embraced a richer notion of *emergence*. Defenders of emergence emphasize the fact that when systems are sufficiently complex, new and unexpected properties emerge. For example, while a single H_2O molecule does not exhibit liquidity, that property emerges when we have enough H_2O molecules in the right temperature range.

Some emergentists hold that the emergent properties are entirely determined by lower-level properties, and so they accept supervenience. Others claim that emergent properties are radically new and deny supervenience. While there are different models of how emergence works, the basic idea is that the relation between physical base-states and mental states is a causal one. The causal relation is offered as an *explanation* of the dependence of the mental on the physical.

In Searle's view, consciousness "is not ontologically reducible to physical microstructures," due to its first-person ontology.[15] However, he claims that consciousness is still *causally* reducible to the brain, in the sense that it "has no causal powers beyond the powers of the neuronal (and other neurobiological) structures."[16] Yet Searle has also attempted to argue that his idea of emergence is compatible with a robust understanding of mental causation and even libertarian free will (where an agent's decision is not caused by prior states of the brain), suggesting that the mind has new powers of its own. So he seems torn between a form of emergence that affirms the supervenience of the powers of consciousness and one that grants consciousness independent powers.[17]

However, it seems that emergentism along these lines faces a dilemma. Either the causal powers of mental properties are wholly determined by the physical base-states that cause them (supervenience) or they are not. If they are, then emergent physicalism runs into the same exclusion problem as supervenience that we saw earlier. Jaegwon Kim has explicitly defended that claim.[18] The only difference we need make in Figure 2.1 is to substitute "emergent determination" for supervenience. If this is right, the causal dependence of mental states on physical states claimed by emergence will still lead to epiphenomenalism—the denial that mental states have any effects.

On the other hand, if an emergentist insists that mental properties do have novel effects, then this violates the core physicalist commitment to the causal closure of the physical, because there are physical effects that do not arise from physical causes. And now there are all kinds of puzzles. It is plausible that we recognize that two properties are distinct because they have different causal powers; thus, gravitation is not the same thing as magnetism (as Kepler may have thought). So if we insist that mental properties are distinct from and irreducible to any physical properties (property dualism), that appears to imply that mental properties do have causal powers different from any physical properties, in which case the causal closure of the physical is false. And it is by no means clear that these new causal powers can be understood as powers of anything physical.

As we will see, property dualism seems to be an unstable position, because once we pay close attention to the nature of mental properties, it does not seem that they can be properties of a physical object. As one obvious example, conscious mental states all have the property of subjectivity. But how can subjectivity be the property of a purely impersonal entity like the brain as it is described by various physical sciences? It seems, rather, that if mental states have subjectivity it is because they are states of a mental subject, which makes sense if they belong to a mental rather than a physical substance.

The upshot is that like their reductive predecessors, liberalized, non-reductive forms of physicalism are also highly problematic. If they are faithful to the core physicalist doctrines of the causal closure of the physical and the complete dependence of the mental on the physical, then it is hard to avoid the unpalatable doctrine of epiphenomenalism. On the other hand, if it is asserted that mental states have autonomous mental powers, then it becomes increasingly difficult to prevent a slide from physicalism to substance dualism (or something so close as to make physicalism an uninteresting claim).

This state of affairs represents a severe paradigm crisis in physicalism. Those theories that are clearly faithful to physicalist commitments appear to be obviously false. But those theories which are more promising are not, in the final analysis, obviously physicalist. And, with some exceptions, the overall trajectory of physicalist thought seems to be moving in a dualist direction. This trend has been intensified by renewed interest in the nature of mental subjects and in personal identity at and over time.

3. Convergent Currents: The Return of the Subject

While the physicalist paradigm fixated on an understanding of mental states and events, ignoring mental subjects as something too much like Cartesian egos, the widespread recognition of the irreducibility of mental properties has caused many philosophers of mind to reconsider. Surely, it cannot seriously be denied that pains, beliefs, and desires belong to subjects. While a neuron can remain intrinsically the same if it is removed from a brain or even if it is inserted in someone else's brain, it does not make sense to suppose that a pain or a thought could exist outside of a mind, or be transferred as the very same mental state to another person's mind. On any plausible view, mental states and events are part of a particular person's mental life, and so an adequate philosophy of mind must give some account of the nature and possibility of that life.

Similarly, it seems obvious that acts of reasoning require a mental subject that persists over time. If I am to reason to a conclusion, I must be the same subject who entertains all of the premises at time t that draws the conclusion at some later time $t+k$. For otherwise, no individual can be credited with an act of reasoning. First, if we suppose that at time t, each premise belongs to a different subject, we cannot explain why it is reasonable for any of these subjects to draw a conclusion. Thus, if at time t, subject S1 believes A = B (but not B = C) and subject S2 believes B = C (but not A = B), neither S1 nor S2 has a sufficient reason to draw the conclusion that A = C. So it must be one subject that unifies both beliefs at the same time. Further, this subject must persist over time. For suppose subjects are only instantaneous and that the subject S1 that entertains all of the premises at t is not the same as the subject S2 that thinks of the conclusion at $t+k$. Then S1 does not survive long enough to draw the conclusion, and S2 cannot be credited with drawing the conclusion either, since S2 never knew the premises.

Without the assumption of unified, persistent subjects, it appears that human rationality, which includes scientific investigation, does not make sense. So it seems that everyone, physicalists included, must provide some account of the nature and possibility of mental subjects. Recognition of this fact has led to an explosion of physicalist theories of the self. Physicalists hope to account for the emergence of a *first-person perspective*, a perspective allowing one to think of oneself as oneself at and over time, for example, when I look forward to a walking holiday in Bonnie Scotland next summer. This looking forward involves my thinking of what I myself will do at a later time, and we normally suppose that I may live to see whether this thought is true, requiring me to be the same mental subject next summer that is thinking of the holiday now.

Thus, physicalists have recently given considerable attention to explaining the possibility of a first-person perspective, including the unity of consciousness at a time[19] and the persistence of selves over time.[20] Two leading theories are animalism and constitutionalism. *Animalists*, like Peter van Inwagen[21] and Eric Olson,[22] maintain that persons are human animals, so that the unity and persistence of mental subjects are grounded in the unity and persistence of particular living human organisms.

Constitutionalists, like Lynne Rudder Baker[23] and Kevin Corcoran,[24] disagree with animalists. They deny that human persons are human animals, because, they claim, persons and animals have different persistence conditions. In order to be a person one must be a human animal but also have intentional states and a first-person perspective, which some human beings lack. Thus, a human may persist when a person does not. Constitutionalists therefore claim that persons are not identical to, but are constituted by, human animals. As an analogy, a statue of David may be constituted by a lump of marble (that is what it is made of), but it is not identical to the lump of marble, because if the marble is pulverized, marble persists, but the statue of David does not. Similarly, a person is constituted by a human animal, but if the human animal loses a first-person perspective, the animal remains, but the person does not.

However, both animalism and constitutionalism face serious problems.

Animalism does not give a plausible account of the origin of intentionality and a first-person perspective. Neither of these properties is recognized by biology or any other physical science. So no account of us as living organisms implies that we have the mental life that we do.

Second, it is not obvious that animalism is physicalist. If we are to ground mental subjects in human organisms, we first need to show that the unity and persistence of human organisms can be given a physicalist account. But there are good reasons to think that it cannot. As Dean Zimmerman points out, when we look at living systems, including a living human brain, we see that at the level of atoms, they resemble clouds: they are constantly in flux, with particles of matter being added or lost over time.[25] As a result, there are many clouds of atoms that can coexist with a living human organism both at and over time. At time *t*, my body may consist of a cloud C of atoms, but there are many smaller clouds $C_1, C_2, C_3, C_4, C_5, \ldots C_n$ containing a subset of the atoms of C, that could also compose a living body. We know that many clouds can do this, because over time, we can gain or lose some particles, forming different clouds, and each of those clouds composes a living body.

Furthermore, we also know that over time there can be huge changes in the composition of these clouds, so there is no single, constant subset of particles that is always required for the living organism to exist. But that also means that even at a single time *t*, there are many clouds in the same body (many subsets of its particles) that could compose a human organism. For example, at time *t*, the clouds containing one, two, or three fewer atoms than the total cloud may be sufficient to compose a living organism. So if the identity of a living organism is determined by the physical, it seems that at time *t*, we should expect that I am *many* living organisms, not *one*. This is because each of these clouds can compose a living organism, and it is also not the case that there is one, fixed subset of particles that must always exist within each of these clouds. So one cannot simply strip away "unnecessary" particles and say that those that remain are the ones that matter, since the particles in the residual subset could also be replaced and therefore are not necessary either. Likewise, due to the constant flux in the atoms of my body over time, the cloud of atoms in my body will change, from C at time *t* to C* at some later time *t+k*. If the identity of a living organism were determined by the atoms composing the body, we would predict that C and C* are different organisms, particularly as it may be that, over sufficient time, there is no core subset of particles shared by C and C*. Thus, construed as a physicalist theory, animalism fails to explain why I am just one organism at and over time, and without the unity and persistence of a single organism, animalism cannot explain the unity and persistence of mental subjects.[26] Thus, to the extent animalism seems plausible, it appears tacitly to assume a non-physicalist account of organismal identity.[27]

Constitutionalism also faces numerous serious problems.[28] Perhaps the most fundamental is its claim that a first-person perspective is merely a mental *property*. This is because the property, having-a-first-person-perspective, is impure. A pure property does not presuppose the existence of any particular object. Thus, the property of being-red does not presuppose the existence of any red objects (it can be defined as a certain wavelength of light even if no object reflects that wavelength). By contrast, an impure property is a property that presupposes the existence of a particular object. Thus, the property being-ten-feet-from-the-Statue-of-Liberty is impure because it makes reference to the Statue of Liberty. But likewise, the property having-a-first-person-perspective references a person. It is not hard to see that only persons can have first-person perspectives. But if that is right, having a first-person perspective presupposes that a person exists, in which case it cannot be the case that having a first-person perspective *makes* someone a person.[29]

There are several other physicalist accounts of the self, but the common weakness is that they do not account for the irreducibility of subjectivity. The single, unified, persistent consciousness of an individual at and over time is hard for physicalists to account for, because it is not predicted by the changing aggregate of parts that makes up the brain. Given that brain, one would predict that thoughts, like neurons (or regions of the brain), are separable parts of the mind, but they are not. Though one may have diverse thoughts at and over time, those thoughts belong to just one mind: they cannot exist in another mind, and their thinker does not change despite extensive changes in the brain over time.[30]

Another current that has encouraged greater attention to the self is the emergence of able new defenders of Cartesian dualism, as well as renewed interest in a variety of alternatives. Cartesian dualism holds that mind and matter are substances of fundamentally different kinds. A material substance is by nature extended and located in space, and is divisible into parts, at least by God. But a mental substance is by nature neither extended nor located in space and is simple in the sense that, although it has many faculties and modes, it cannot be decomposed into separable parts. For Descartes, a person is the union of a material body and an immaterial soul, and, though we do not know how, these two substances causally interact, volitions in my mind causing limb movements, nerve signals from my body causing pain in my mind.

On behalf of Cartesian dualism, Richard Swinburne has offered a sophisticated improvement on Descartes's argument for the possibility of disembodied existence, and also shows how a Cartesian dualist may respond to the "pairing problem."[31] The pairing problem asks how a particular soul is paired with a particular body, such that when my soul wants to raise my arm, my arm goes up, not someone else's, and when I bruise my shin, my soul feels pain, not someone else's. Swinburne argues that no law can explain why a particular soul is caused to exist by a particular body, but souls have a primitive "thisness" that individuates them and explains their causal powers. Swinburne also develops ingenious thought-experiments to show that physicalism cannot account for our intuitions about personal identity.

In one, Swinburne imagines that diseased parts of a brain are gradually replaced by parts from other brains (suppose one-tenth is replaced every year over many years). He argues that because the process is gradual, even if every part of the original brain is eventually replaced (after ten years), it is conceivable that the same person remains, in which case personal identity cannot be determined by the physical characteristics of the brain.[32]

In another thought-experiment, Swinburne considers the implications of the fact that a person can survive with only one cerebral hemisphere, and that it may become possible to transplant a hemisphere from one patient to another. He imagines what would happen if the left hemisphere of one individual X were transplanted into the skull of another individual Y, both of whose hemispheres had been removed, and connected to the lower part of Y's brain, while the right hemisphere of X was transplanted into the skull of another individual Z, both of whose hemispheres had been removed, and was connected to the lower part of Z's brain.[33] If that happened, and Y and Z remain alive, since Y and Z both share large portions of X's brain, we would certainly expect both Y and Z to have some of X's memories and character traits, and that both might naturally claim to be X.

However, two things cannot be one thing, so it cannot be the case that *both* Y and Z are X, and it cannot be the case that either Y or Z individually is two people. What seems conceivable is that X survives as Y (so Y is the same person as X and Z is not), or that X survives as Z (so Z is the same person as X and Y is not), or that X does not survive as either Y or Z. But if that is the case, then what it is to be the same person as X "is not analyzable solely in terms of having some or all of the earlier [X's] brain or similar memories and character to those of the earlier [X]."[34] For if it were, both Y and Z could be X, and we know that is impossible. This argues that continuity is not sufficient for personal identity: we need something else, a "thisness," and that must derive from something non-physical, the soul.

Moreover, non-Cartesian dualists have other resources for handling objections to substance dualism.[35] For example, many dualists would deny Descartes's claim that souls cannot be located in space. For Augustine and Kant, the soul is present in space, but in a different way from anything physical. While a physical object excludes any other physical object from the same space, and is only partly present in any of its parts, the soul is present throughout the body (wherever there is sensation) and is wholly present in each part of it. Since souls and bodies share the common medium of space, it is less mysterious how they can causally interact (though, as I argue in a later chapter, more needs to be said about how they do so).

Yet other dualists have followed Aristotle and Aquinas, who deny that the soul is a substance in the same sense (a thinking thing) advocated by Augustine and Descartes. For Aristotle and Aquinas, the soul is a substantial form, and the substance proper is the person, a union of this form with matter. The form is what makes a person a rational animal, and, at least in Aquinas's formulation, though this form is incomplete without matter (one does not have the full person), still it may subsist in a disembodied state. The advantage of this view is that the mind-body problem does not arise (at least in its classical form), because it is the whole person (body and soul) that acts as one organism. In addition to defenders of classical Thomism, like Ed Feser,[36] J. P. Moreland has defended a modified version of the Thomistic view.[37] Along similar lines, E. J. Lowe also advocated a holistic form of dualism, which recognized persons as psychological substances distinct from their bodies, yet also possessing physical properties.[38]

Still others defend the idea of emergent substance dualism. Unlike the emergent physicalist, who claims only that mental properties may emerge from a physical brain, the emergent substance dualist argues that a mental substance can itself emerge. Defenders of this position, like William Hasker, contend that this version of dualism makes most sense of the close dependence of mental function on the brain, while also explaining the unity of consciousness.[39]

Finally, various forms of idealism are making a comeback (see the chapters by Doug Axe and Bruce Gordon in this volume).[40] Idealism is the view that all that exists are minds and their contents. For example, Berkeleyan idealism maintains that the only kinds of being are spirits (either an infinite spirit, God, or finite spirits, like human persons) and ideas. In this view, there is no such thing as mind-independent matter. This does not mean there is no such thing as physics: a physical law is simply a regular train of thoughts that God thinks. And "things," like tables and chairs, can really exist outside of human minds. But they do not exist outside of all minds, but rather exist in God's mind.

Some philosophers have argued that idealism is superior to both physicalism and dualism in its ability to solve the mind-body problem.[41] The reason is that there is a natural affinity between minds and ideas (they are both immaterial), so their interaction is not between fundamentally different kinds of being. On this view, bodies are bundles of ideas under the governance of a spirit, and involuntary sensations reflect the independence from our will of ideas thought by God.

4. Why Should Scientists Welcome These Developments?

THIS RENAISSANCE of non-materialist views of persons that affirm the human soul is exciting for philosophers, but is it something scientists should take seriously? I think there are three main reasons for an affirmative answer. First, the practice of science assumes the existence of persistent mental subjects, which is best explained by substance dualism (or idealism). Second, the well-known scientific objection against recognizing the existence of the soul, that it violates the conservation of energy, can be answered. Third, there are several empirical arguments, and though they are insufficient to demonstrate a metaphysical conclusion, they do provide significant evidence most naturally accounted for by the existence of the soul.

4.1 The Practice of Science

Any attempt to use science to discredit the existence of mental subjects is fatally flawed because the bedrock data for all science comes from observation, which presupposes the existence of conscious subjects. The idea that the findings of physical science are unproblematic but mental subjects are questionable ignores the fact that our only access to physical phenomena is via the minds of scientists. Thus, as Charles Taliaferro points out, one "cannot presume to have any clearer understanding of nonmental physical phenomena than [one] does of [the] concepts, reasons and reasoning, grasping entailment relations, reliance on experience and observations that go into the practice of the sciences."[42] Since concepts, reasons, grasping entailments, and experience all seem to be mental phenomena, and all are required by the practice of science that investigates physical phenomena, we are not within our rights to assume that we have more reliable access to physical phenomena than we do to the mind.

Furthermore, scientific inquiry assumes that it is one and the same conscious subject that has a research question, and persists over the time necessary to answer that question. How can a scientist claim to discover the answer to his question, or to verify or falsify a prediction that he made, if he is not the very same person that asked the question or made the prediction? For example, consider François Englert and Peter Higgs, who predicted the existence of the Boson nearly fifty years before its existence was confirmed. When these scientists became Nobel Prize winners in 2013, everyone assumed that the very same persons receiving the prize made the prediction decades before. Yet due to the constant flux of matter in our physical bodies and brains over time, physicalist approaches to personal identity find it very difficult to justify this assumption (for more on this, see Jonathan Loose's chapter in this volume).

What is more, as Daniel Robinson argues, neuroscience in particular has implicitly dualist commitments, because the correlation of brain states with mental states would be a waste of time if we did not have *independent* evidence that these mental states existed.[43] It would make no sense, for example, to investigate the neural correlates of pain if we did not have independent evidence of the existence of pain from the subjective experience of what it is like to be in pain. This evidence, though, is not *scientific* evidence: it depends on introspection (the self becomes aware of its own thoughts and experiences), which again assumes the existence of mental subjects. Further, as we saw above, Richard Swinburne has argued that scientific attempts to show that mental states are epiphenomenal are self-refuting, since they require that mental states reliably cause our reports of being in those states. The idea, therefore, that science has somehow shown the irrelevance of the mind to explaining behavior is seriously confused.

4.2 Answering Scientific Objections to the Soul

One of the most popular scientific arguments against the soul is that, were souls to interact with the physical world, they would violate the conservation of energy law. While this argument is routinely cited as a decisive reason to reject dualism, it faces a number of serious objections. One often hears that energy conservation is universal in its application to physics and that it is among the best-confirmed scientific laws. However, even if that were the case, it does not follow that mind-body interaction cannot occur, for, as E. J. Lowe points out, "the empirical evidence in support of that principle… is all drawn from… *purely physical* systems…, whereas… mind-body interactions are not of that kind."[44] If substance dualism is true, we cannot assume that mind-body causal interaction operates in the same way as purely physical causal interaction.

Secondly, even within physics, energy conservation does not account for all causal interactions. As Robin Collins points out, within general relativity energy conservation does not apply to the causal powers of gravitational fields. This is because "no local concept of stress-energy (and hence energy or momentum) can be defined for the gravitational field in general relativity."[45] Yet "Gravitational fields… clearly have real effects on matter, even though… energy cannot be said to be conserved."[46] As a result, "A dualist could argue that… the notion of energy simply cannot be defined for the mind, and hence one cannot even apply the principle of energy conservation to the mind/body interaction."[47] If the conservation principle does not apply to all purely physical interactions, why should it apply to mind/body interaction?

Further, the assumption that causation must involve a transfer of energy can be questioned, not only because mind-body interaction might not require an energy transfer, but also because there are examples of apparent causation without such transfer even in physics. In quantum physics, John Bell showed that there are apparent causal correlations between distant particles that cannot be explained by an energy transfer because the signal would have to travel faster than the speed of light.[48] If these correlations are causal, it follows that causal interactions do not require energy transfer; if they are not causal, then because there are reliable, systematic mind-body correlations, it is most reasonable to conclude that some non-causal principle explains this correspondence (Malebranche's parallelism and Leibniz's pre-established harmony would be examples), which is still compatible with substance dualism.

Finally, the energy conservation principle only holds in a closed system, but is silent about what happens in an open system. So if the soul *does* produce movement by generating energy, then if the principle is true, we should conclude that the body is an open system. If it is objected that this would still involve new energy being fed into the physical world as a whole, the dualist can reply that this further shows that the physical world is an open system, something no theist that believes in miracles would deny.[49]

In sum, the principle of energy conservation cannot be used to exclude the causal influence of the soul without further assuming that: (1) the principle applies to mind-body interaction; (2) causal influence must be via an energy transfer; and (3) the body is a closed system. But there are good reasons to doubt all three of these assumptions, and so the energy conservation objection to dualism does not have the force its proponents claim.

4.3 The Empirical Evidence for the Soul

As many authors in this volume argue, it is generally impossible to prove a metaphysical position by empirical arguments: typically, the data underdetermine the metaphysical positions so that the data does not decisively favor one position over another. However, one can make the case that the empirical data does not rule out dualism (or idealism) and that, in some cases, the data are more naturally explained by some version of dualism (or idealism) than they are by materialism.

"Split-brain" cases, where the cerebral hemispheres are disconnected, have often been used as arguments for materialism, based on the idea that dividing the brain divides the conscious subject. But on closer inspection of the data, these cases actually seem to favor dualism (or idealism). In his pioneering work, Roger Sperry severed the corpus callosum that connects the cerebral hemispheres, and there is a significant impact on how a subject responds to stimuli. The operation has the consequence that if stimuli are presented to the left visual field, the response only comes from organs controlled by the right hemisphere, and if stimuli are presented to the right visual field, the response only comes from organs controlled by the left cerebral hemisphere, which usually includes speech. So one might see something in both visual fields, but be able to name the object in the right but not the left visual field. If the corpus callosum had not been severed, information would have been shared by the hemispheres, and the subject would be able to name the objects in both visual fields.

At first sight, this might suggest that in dividing the brain, the conscious subject itself has been divided, that we now have two subjects and not one, and hence the materialist assumption that the character of consciousness is fully explicable by the brain would be vindicated. However, as Michael Egnor points out in this volume, this conclusion is too hasty: while split-brain patients do exhibit some disunity at the level of perceptual and motor tasks, they continue to exhibit a unified self-awareness and capacity to reason.

Most notably, as Tim Bayne points out, split-brain patients seem quite able to carry on their everyday lives: "They can cook, cycle, swim, and play the piano, and naïve observers are rarely aware that they suffer from cognitive impairments."[50] Thus, models that claim that the subject itself has been divided seem impossible to square with the actual lives of split-brain patients. Two different conscious subjects controlling the same body would not be able to ride a bicycle or play the piano! Yet this recalcitrant unity of the subject seems very unlikely on materialist premises, since the brain is an aggregate of separable parts, and so if all of our psychological life were explained by the character of the brain, such a major division of the brain would naturally predict a fundamental division of the mind that is not observed. However, the persistent unity of the mind makes sense if there is a simple (indivisible) mental subject, which is the claim of dualism (and idealism).

Another pioneer in neuroscience, Wilder Penfield, discovered unexpected evidence for dualism. By using local anesthetic and electrodes, Penfield was able to document hundreds of thousands of cases in which he stimulated the brain to induce an automatic response. He was able to induce movements, sensations, emotional interpretations, and memories. But he also discovered two extraordinarily robust limitations to his experiments.

First, he discovered that when he induced a response in the patient, the patient knew that the response was caused by someone else and was not his own action. The patient made a clear distinction between his own choices and actions, and the passive results of an external agency.[51]

Second, despite numerous attempts, Penfield could not cause a patient to make a decision or to form an abstract thought, suggesting the existence of independent free will and intellect.[52] To be sure, a materialist might reply that Penfield simply failed to find the right parts of the brain to stimulate, but the available data does suggest a fundamental dichotomy between automated capacities that are passively stimulated, and active powers of the mind which are within an agent's control, and this is more naturally explained by dualism (or idealism) than by materialism. See Egnor's chapter in this volume for a fuller exploration of this argument.

A third source of empirical evidence for dualism derives from the systematic scientific study of near-death experiences (NDEs). These cases involve reports by reviving or resuscitated patients whose content apparently derives from a time when there was no measurable brain activity. While there is a considerable body of literature here, and not all of the reports are of equal scientific value, it is possible to isolate a significant number of cases that are especially hard for materialism to explain.

As Gary Habermas argues in this volume, there are many cases of "evidential NDEs," where patients report independently verifiable facts which they could not have acquired on the basis of a functioning brain from their location (e.g., on an operating table). Thus, there are patients who upon revival can accurately describe the serial number on top of medical machines above their head, or know facts about the placement of items which were not physically accessible to them even if their brains and senses had been working, which they were not. Since the content of these experiences is veridical, it is not plausible that these are hallucinations or confabulations of the waking brain. In response to one challenge after another, Habermas shows that there are NDEs that meet the most stringent demands of skeptics. Together, these constitute important evidence for the dualist (or idealist) claim that at least some powers of the mind are independent of brain function.

In addition, four of the most impressive recent empirical evidences for the existence of the soul derive from cognitive behavioral therapies, downward suppression studies, the placebo effect, and psychoneuroimmunology.[53]

Jeff Schwartz studied Obsessive Compulsive Disorder (OCD), in which patients experience a repeated urge to do something, such as lock a door or turn off the oven, which they know does not need to be done.[54] His cognitive behavioral therapy for these patients required them to use conscious attention to refocus their minds on an alternative behavior whenever they felt the urge to do the futile action. What he discovered was that conscious attention was able to exploit neuroplasticity to overcome the "brain lock" of their condition, so that the urge diminished. The significance of these and related experiments targeting various phobias and behavioral problems is that they rely on the active, top-down causal power of conscious attention to change neural pathways, something hard to square with the physicalist idea that our thoughts are passively determined by physical events.

Mario Beauregard reviewed many similar experiments, including ones in which patients could control negative emotions (such as fear and sadness) by a conscious choice to downwardly suppress them.[55] He also noted the significance of the placebo effect. A physically ineffective drug or procedure may still be used to benefit the patient's health, because the patient trusts it.[56] This means that the conscious act of trust has a physical effect on the body's ability to recuperate. And there have been many scientific studies in the area of psychoneuroimmunology, which show that helping patients to be calm affects their immune system, and hence their ability to fight diseases and recover from medical procedures.[57] The fact that stress, a psychological state, causes negative changes to the cardiovascular system, blood chemistry, and immune system, while a calm state of mind causes corresponding positive changes, makes sense if the mind has a top-down causal influence on the body, but is hard to square with the materialist assumption that our mental life is simply the result of an organic computer.

All of these cases provide evidence of the mind's active power to affect the state of the human body. This makes sense if there are mental substances, as dualism and idealism claim, but seems inexplicable if the mind is entirely determined by physical causal processes.

5. Conclusion

PHYSICALISM IS in a period of paradigm crisis, since neither reductive nor non-reductive versions are adequate to explain important mental phenomena, and even more sophisticated accounts do not make sense of conscious subjects. Happily, there has been a resurgence of interest in the soul, and important work has countered standard objections and shown ways in which the soul illuminates recent advances in psychology and medicine. As you will further learn from many other chapters in this volume, this is a great time for scientists to explore the implications of non-physicalist theories for their work.

Notes

1. Two of the most famous reasons for this rejection are that Logical Positivism is self-refuting and that it does not adequately account for the character of scientific theories. It is self-refuting, because the statement of logical positivism is neither true by definition nor verifiable by observation, making it literally meaningless; it fails to account for the character of scientific theories, because scientific theories appeal to theoretical entities (e.g., forces) which are neither required by logic nor directly observable.
2. Wilfrid Sellars, *Empiricism and the Philosophy of Mind* (Cambridge, MA: Harvard University Press, 1997), 83.
3. Hilary Putnam, "Brains and Behavior," in *Analytical Philosophy*, ed. R. J. Butler (Malden, MA: Wiley-Blackwell, 1968), 1–19.
4. See J. J. C. Smart, "Sensations and Brain Processes," *Philosophical Review* 68 (1959): 141–156.
5. See David Lewis, "Mad Pain and Martian Pain," in *Readings in the Philosophy of Psychology*, ed. Ned Block, vol. 1 (Cambridge, MA: Harvard University Press, 1980), 216–222.
6. John Searle, "Minds, Brains and Programs," *Behavioral and Brain Sciences* 3 (1980): 417–457.
7. Ned Block, "Troubles with Functionalism," in *Minnesota Studies in the Philosophy of Science* IX, ed. by C. Wade Savage, 261–325 (Minneapolis, MN: University of Minnesota Press, 1978).
8. See, for example, Paul and Patricia Churchland's *On the Contrary: Critical Essays, 1987–1997* (Cambridge, MA: MIT Press, 1998).
9. For a more thorough exposition of the problems facing eliminative materialism, see Angus J. L. Menuge, *Agents Under Fire: Materialism and the Rationality of Science* (Lanham, MD: Rowman and Littlefield, 2004), ch. 2.
10. For a more precise statement of the principle, see Jaegwon Kim, *Philosophy of Mind*, 3rd ed. (Boulder, CO: Westview Press, 2011), 214.
11. Donald Davidson, "Mental Events," in his *Essays on Actions and Events* (New York: Oxford University Press, 1980).
12. David Chalmers, *The Conscious Mind: In Search of a Fundamental Theory* (New York: Oxford University Press, 1996).
13. Daniel Wegner, *The Illusion of Conscious Will* (Cambridge, MA: MIT Press, 2002).
14. Richard Swinburne, *Mind, Brain, and Free Will* (New York: Oxford University Press, 2013), 117–123.
15. John R. Searle, *Freedom and Neurobiology* (New York: Columbia University Press, 2007), 50.
16. Searle, *Freedom and Neurobiology*, 50.
17. For a critique of Searle's naturalistic theory of libertarian free will, see Angus J. L. Menuge, "Neuroscience, Rationality and Free Will: A Critique of John Searle's Libertarian Naturalism," *Philosophia Christi* 15, no. 1 (2013): 45–60.
18. Jaegwon Kim, "Emergence: Core Ideas and Issues," *Synthese* 151 (2006): 547–559.
19. See, for example, Michael Tye, *Consciousness and Persons: Unity and Identity* (Cambridge, MA: MIT Press, 2003), and Tim Bayne, *The Unity of Consciousness* (Oxford: Oxford University Press, 2012).
20. For example, see *Personal Identity: Complex or Simple?*, eds. Georg Gasser and Matthias Stefan (Cambridge: Cambridge University Press, 2012).
21. Peter van Inwagen, "A Materialist Ontology of the Human Person," in *Persons: Human and Divine*, eds. Peter van Inwagen and Dean Zimmerman, 199–215 (New York: Oxford University Press, 2007).
22. Eric T. Olson, *What Are We? A Study in Personal Ontology* (Oxford: Oxford University Press, 2007).
23. Lynne Rudder Baker, *Naturalism and the First-Person Perspective* (New York: Oxford University Press, 2013).
24. Kevin Corcoran, *Rethinking Human Nature: A Christian Materialist Alternative to the Soul* (Grand Rapids, MI: Baker Academic, 2007).
25. Dean Zimmerman, "From Experience to Experiencer," in *The Soul Hypothesis: Investigations into the Existence of the Soul*, eds. Mark C. Baker and Stewart Goetz, 168–196 (New York: Continuum, 2011).
26. It is here interesting to note that this is one reason why Aristotle and Aquinas believed one needed a sensitive soul to account for the identity of living organisms. This soul provided a unifying and persistent substantial form that accounted for the oneness of an organism at and over time despite the variability of its matter. Since these forms are not determined by matter, but rather, they determine it, it is not plausible to claim that substantial forms are themselves physical.
27. For further critique of animalism, see Stewart Goetz, "Against Animalism," in *The Blackwell Companion to Substance Dualism*, eds. Jonathan J. Loose, Angus J. L. Menuge, and J. P. Moreland, 307–315 (Oxford: Wiley-Blackwell, 2018).

28. See Ross Inman, "Against Constitutionalism," in *The Blackwell Companion to Substance Dualism*, eds. Jonathan J. Loose, Angus J. L. Menuge, and J. P. Moreland, 351–367 (Oxford: Wiley-Blackwell, 2018).
29. For a fuller development of this argument, see Angus Menuge, "The First-Person Perspective is Not a Mere Mental Property," *Philosophia Christi* 20, no. 1 (2018): 67–72.
30. For a thorough defense of the dualist implications of the unity of consciousness, see J. P. Moreland, "Substance Dualism and the Unity of Consciousness," in *The Blackwell Companion to Substance Dualism*, eds. Jonathan J. Loose, Angus J. L. Menuge, and J. P. Moreland, 184–207 (Oxford: Wiley-Blackwell, 2018).
31. See Richard Swinburne's *Mind, Brain, and Free Will*, and also his "Cartesian Substance Dualism," in *The Blackwell Companion to Substance Dualism*, eds. Jonathan J. Loose, Angus J. L. Menuge, and J. P. Moreland, 133–151 (Oxford: Wiley-Blackwell, 2018).
32. Richard Swinburne, "Cartesian Substance Dualism," 149.
33. Richard Swinburne, "The Argument to the Soul from Partial Brain Transplants," *Philosophia Christi* 20, no. 1 (2018): 13–19.
34. Swinburne, "Argument to the Soul from Partial Brain Transplants," 15.
35. See Stewart Goetz and Charles Taliaferro, *A Brief History of the Soul* (Malden, MA: Wiley-Blackwell, 2011).
36. Edward Feser, "Aquinas on the Human Soul," in *The Blackwell Companion to Substance Dualism*, eds. Jonathan J. Loose, Angus J. L. Menuge, and J. P. Moreland, 88–101 (Oxford: Wiley-Blackwell, 2018).
37. See J. P. Moreland, *The Soul* (Chicago, IL: Moody Press, 2014), and also Moreland's "In Defense of a Thomistic-like Dualism," in *The Blackwell Companion to Substance Dualism*, eds. Jonathan J. Loose, Angus J. L. Menuge, and J. P. Moreland, 102–122 (Oxford: Wiley-Blackwell, 2018).
38. E. J. Lowe, *Personal Agency: The Metaphysics of Mind and Action* (Oxford: Oxford University Press, 2008).
39. See William Hasker, *The Emergent Self* (Ithaca, NY: Cornell University Press, 1999), and "Souls, Beastly and Human," in *The Soul Hypothesis: Investigations into the Existence of the Soul*, eds. Mark C. Baker and Stewart Goetz, 202–217 (New York: Continuum, 2011).
40. For a nice survey of this idealistic comeback, see James S. Spiegel, "Berkeleyan Idealism and Christian Philosophy," *Philosophy Compass* 12, no. 2 (February, 2017), available online at: https://onlinelibrary.wiley.com/doi/abs/10.1111/phc3.12400.
41. For example, see Charles Taliaferro, "Idealism and the Mind-Body Problem," in *Idealism and Christian Philosophy*, eds. Steven. B. Cowan and James S. Spiegel (New York: Bloomsbury, 2016).
42. Charles Taliaferro, "Substance Dualism: A Defense," in *The Blackwell Companion to Substance Dualism*, eds. Jonathan J. Loose, Angus J. L. Menuge, and J. P. Moreland, 43–60 (Oxford: Wiley-Blackwell, 2018), 46.
43. See Daniel N. Robinson, "Minds, Brains, and Brains in Vats," in *The Soul Hypothesis: Investigations into the Existence of the Soul*, eds. Mark C. Baker and Stewart Goetz, 46–67 (New York: Continuum, 2011).
44. Lowe, *Personal Agency*, 60.
45. Robin C. Collins, "The Energy of the Soul," in *The Soul Hypothesis: Investigations into the Existence of the Soul*, eds. Mark C. Baker and Stewart Goetz, 123–133 (New York: Continuum, 2011), 127.
46. Collins, "Energy of the Soul," 129.
47. Collins, "Energy of the Soul," 130.
48. Collins, "Energy of the Soul," 132.
49. See Alvin Plantinga, *Where the Conflict Really Lies* (New York: Oxford University Press, 2011), 79.
50. Tim Bayne, *The Unity of Consciousness* (Oxford: Oxford University Press, 2010), 199.
51. Wilder Penfield, *The Mystery of the Mind* [1975] (Princeton, NJ: Princeton Legacy Library, 2015), 76.
52. Penfield, *Mystery of the Mind*, 77–78.
53. For a fuller exposition of these cases, see Angus Menuge, "Is Downward Causation Possible? How the Mind Can Make a Physical Difference," *Philosophia Christi* 11, no. 1 (Summer 2009): 93–110.
54. Jeffrey M. Schwartz and Sharon Begley, *The Mind and the Brain: Neuroplasticity and the Power of Mental Force* (San Francisco, CA: Harper, 2002).
55. Mario Beauregard, "Mind Does Really Matter: Evidence from Neuroimaging Studies of Emotional Self-Regulation, Psychotherapy and Placebo Effect," *Progress in Neurobiology* 81, no. 4 (March 2007): 218–236.
56. Arthur K. Shapiro and Elaine Shapiro, *The Powerful Placebo: From Ancient Priest to Modern Physician* (Baltimore, MD: Johns Hopkins University, 1997).
57. See, for example, Mary Jane Ott, Rebecca L. Norris, and Susan M. Bauer-Wu, "Mindfulness Meditation for Oncology Patients: A Discussion and Critical Review," *Integrative Cancer Therapies* 5, no. 2 (2006): 98.

3. Neuroscience and the Metaphysics of Consciousness and the Soul

J. P. Moreland

1. Introduction

The great Presbyterian scholar J. Gresham Machen once observed, "I think we ought to hold not only that man has a soul, but that it is important that he should know that he has a soul."[1] From a Christian perspective, this is a trustworthy saying. Though not unique in this regard, Christianity is a dualist, interactionist religion in this sense: God, angels and demons, and the souls of men and beasts are immaterial substances that can causally interact with the world.

Specifically, human persons are (or have) souls that are spiritual substances that ground personal identity in a disembodied intermediate state between death and final resurrection.[2] Clearly, this was the Pharisees' view in intertestamental Judaism, and Jesus (Matt. 22:23–33; cf. Matt. 10:28) and Paul (Acts 23:6–10; cf. 2 Cor. 12:1–4) side with the Pharisees on this issue over against the Sadducees.[3]

Besides biblical teaching, property and substance dualism are the commonsense views held by the overwhelming number of human beings now and throughout history. As Charles Taliaferro points out, this is widely acknowledged by physicalists, including Michael Levin, Daniel Dennett, David Lewis, Thomas Nagel, J. J. C. Smart, Richard Rorty, Donald Davidson, and Colin McGinn.[4] Throughout history, most people have been substance and property dualists. Thus, regarding the mind/body problem, physicalist Jaegwon Kim's concession seems right: "We commonly think that we, as persons, have mental and bodily dimension.... Something like this dualism of personhood, I believe, is common lore shared across most cultures and religious traditions."[5]

People do not have to be taught to be dualists, as they must if they are to be physicalists. Indeed, little children are naturally dualists. Summing up research in developmental psychology, Henry Wellman states that "young children are dualists: knowledgeable of mental states and entities as ontologically different from physical objects and real [nonimaginary] events."[6]

Nevertheless, most scientists and philosophers of mind are physicalists of one form or another regarding the ontological status of consciousness and its possessor/unifier. The main reason for this is the view that neuroscientific discoveries have overwhelmingly supported physicalism. According to Nancey Murphy, "science has provided a massive amount of evidence suggesting that we need not postulate the existence of an entity such as a soul or mind in order to explain life and consciousness."[7] This evidence consists of the fact that "biology, neuroscience, and cognitive science have provided accounts of the dependence on physical processes of *specific* faculties once attributed to the soul."[8] Elsewhere she claims: "My argument in brief is this: all of the human capacities once attributed to the mind or soul are now being fruitfully studied as brain processes—or, more accurately, I should say, processes involving the brain, the rest of the nervous system and other bodily systems, all interacting with the socio-cultural world."[9]

I demur regarding these physicalist claims. Accordingly, in what follows, I first lay out a philosophical lexicon of concepts which are of crucial relevance to the mind/body problem, and which need to be clearly understood to appreciate the arguments for substance dualism. Then I state the reasons for taking consciousness to be non-physical, state the reasons for taking the possessor/unifier of consciousness to be a substantial soul, and close with a section showing that when it comes to the ontological status of consciousness and the soul, neuroscience is simply irrelevant. Space forbids me from analyzing physicalist critiques of and alternatives to dualist views of consciousness and the soul. But where appropriate, I offer sources where these can be found. It is enough to get before the reader the topics to follow. Let us begin in earnest, then, and see what we can discover.

2. A Philosophical Lexicon of Concepts Crucially Relevant to the Ontological Status of Consciousness and Its Possessor/Owner

WHENEVER A group of scholars come together from different disciplines in order to cooperate in integrative projects, the coterie immediately faces a problem: different fields often use terms, concepts, and distinctions in different ways. As a result, they often talk past each other without knowing it. In light of this problem, I shall set out a philosophical lexicon of important concepts and distinctions relevant to philosophy of mind. It will help the reader to have this lexicon for the arguments to follow, and I encourage readers to refer back to it as needed.

- *Part/whole relations* are important for treatments of substances, and there are two kinds of parts relevant to our discussion: *separable* and *inseparable*.
  1. *p* (e.g., a table leg) is a separable part of some whole *W* (e.g., a table). We can formally define any separable part *p* as follows: *p* is a particular and *p* can exist if it is not a part of *W*. Examples of separable parts include table legs (can exist on their own, apart from tables) and electrons (can exist on their own, apart from atoms). Separable parts usually make up *mereological aggregates*—see below.
  2. *p* is an inseparable part of some whole *W*. We can formally define any inseparable part *p* as follows: *p* is a particular and *p* cannot exist if it is not a part of *W*. The existence and nature of an inseparable part is dependent on and determined by the whole of which it is a part. The whole is ontologically prior to inseparable parts.
- *Simplicity*. Setting aside properties, there are two main ways something can be simple in the sense relevant to philosophy of mind: by being uncomposed (i.e., not composed) of separable parts or by being metaphysically indivisible. I use "metaphysically indivisible" to mean what

many philosophers say by "indivisible in thought." If something is metaphysically indivisible, then it is spatially unextended. Something could be metaphysically divisible (e.g., it has spatial extension with spatial regions within, say, a left and right half) but not *physically divisible* (if, say, such division annihilated the whole), but not conversely. Moreover, all particulars that are metaphysically indivisible are uncomposed of separable parts, but not conversely (an extended whole with no separable parts could still be divided). According to our usage, a substance with inseparable parts and a number of properties/attributes is simple. Such a substance is not a derivative whole built up out of ontologically prior separable (e.g., atomistic) parts.

This discussion of simplicity is crucial for what follows. The advocate of the soul holds that it is simple in virtue of not being an aggregate of separable parts. But it is complex in virtue of having several different attributes and inseparable parts (e.g., faculties such as the mind and will).

- A *substance* can be formally defined as an essentially characterized particular that (1) has (i.e., exemplifies) properties but is not had by something more basic than it; (2) is an enduring continuant that remains literally the same substance through change in accidental properties; and (3) may have inseparable parts (modes) but is not composed of separable parts.

Substances are fundamental wholes that have two important features. First, their (inseparable) parts are grounded in the metaphysically prior wholes in that the wholes give existence to and ground the nature of its parts. The parts are what they are in virtue of the wholes of which they are parts. An important implication for biological research is this: Do the various cells, organs, and systems of living organisms get their identity, nature, and function from the whole organisms to which they belong, or is the converse the case? Or are organisms *mereological aggregates* (see below) such that their various parts are ontologically prior to their wholes, and those wholes are derived from and dependent on the separable parts that compose them?

In order to understand what dualists mean when they claim that the soul is a substance, it is important to keep the notions of a *substance* and *stuff* completely distinct. This distinction can be illustrated by two ways of understanding this proposition: "Mary had a little lamb." If "lamb" is used to refer to a substance, then we can ask, "Where is the little fella?" Here, "lamb" refers to an individual enduring unit, a substance. But if "lamb is used to refer to stuff, then we can ask, "How much did she eat? Nine ounces?" Here, "lamb" refers to an amount, or quantity of massy material. Virtually all substance dualists hold that the soul is a substance and is not constituted by stuff.

- A *material substance* can be formally defined as a substance that is (1) spatially located, extended, capable of movement, and in its entirety incapable of being located in more than one place at once; (2) metaphysically divisible; (3) essentially characterized by the actual and potential properties of an ideal chemistry and physics.

- A *mereological aggregate*, or ordered collection of separable parts (from *meros*, meaning "part," usually "separable part"), can be formally defined as a *particular whole* that is constituted by at least substantial separable parts and external relation-instances between and among those substantial separable parts. (An external relation, like "being on top" or "to the left of," is such that things can cease standing in that relation to each other and still exist. For example, a lamp can be moved from the top of a desk to below the desk, and the lamp and desk still exist.) The individual structure of an object is nothing but an ordered mereological aggregate itself, composed of all the individual relations instanced between and among the object's separable parts.

  A mereological aggregate such as a chair, a molecule or atom, or an organism's body is on physicalist accounts subject to *mereological essentialism*: an aggregate's parts are essential to it such that it could not have had different parts than the ones it actually has and, *a fortiori*, cannot gain or lose a part and remain literally the same object. If a chair's parts are replaced with different ones, it is not the same chair in the strict, philosophical sense.

  The distinction between a substance and a mereological aggregate is important for one of the arguments for the soul below. Basically, if I am a substance that remains literally the same "I" through certain changes, including part replacement of my body or brain, then, since my body or brain is an aggregate of atoms, molecules, and cells that does not remain the same through part replacement, I cannot be my body or brain.

  The so-called scientific image of the world, based on the atomic theory of matter, depicts everything above atomic simples (ultimate constituents of matter not composed of further separable parts—if such there are) as mereological aggregates, e.g., atoms, molecules, ordinary-sized objects like rocks and tables, and, for some, cells and living organisms.

  For substances, the whole is ontologically prior to its inseparable parts. For mereological aggregates, the separable parts are metaphysically prior to the wholes they compose. This distinction between substances and mereological aggregates will play a critical role later in an important argument for the soul's reality.

- A *spiritual substance* can be formally defined as a substance that is (1) metaphysically indivisible in being (though it may be fractured in functioning; example: metaphysically speaking, a multiple-personality person is still one, single substantial substance, but may function is an abnormal, divided way); (2) not spatially extended (though most characterizations hold that it is spatially located); and (3) essentially characterized by the actual and potential properties of consciousness and, for some, life.

  It would be instructive to compare the definitions of a spiritual substance, a material substance, and a mereological aggregate.

  A *soul* is an example of a spiritual substance. The above definition of spiritual substance allows for the possibility—which I take to be actual—of animal souls.

- *Phenomenal Consciousness* (hereafter, simply *consciousness*). The nature of consciousness is pretty commonsensical. Suppose you are in the recovery room immediately after surgery. You are still deeply under anesthesia. Suddenly and somewhat faintly, you begin to hear sounds. It is not long until you can distinguish two different voices. You begin to feel a dull throb in your ankle. The smell of rubbing alcohol wafts past your nose. You remember a childhood accident with the same smell. You feel an aversion towards it. You feel thirsty and desire a drink. As you open your eyes to see a white ceiling above, you begin to think about getting out of the hospital. What is going on? The answer is simple: you are regaining consciousness.

Note two things about this example. First, whereas any physical object (state, process, property, relation) can and only can be completely described from within a third-person perspective, descriptions of states of consciousness require an approach from within the first-person point of view. Second, states of consciousness are best defined *ostensively*: by citing or pointing to specific examples. In fact, a fairly good characterization of consciousness is this: consciousness is what you are aware of when you engage in first-person introspection. Both of these observations are exactly what the dualist approach to consciousness would predict.

In addition to ostensive definitions, three further traits seem to uniquely characterize conscious states. For all conscious states C:

1. There is a *what-it-is-like* to C, an experienceable texture, tone, or quality to C. Thus, a state of pain, tasting cherry pie, hearing a musical chord, being angry, desiring a drink of water, or thinking that snow is white all have a distinctive what-it-is-like that (i) is essential to each state and makes it the kind of state it is; (ii) is knowable by direct introspective awareness or focus.
2. They exhibit *private access*; in contrast to all physical entities, including brain states, which are publicly accessible in that everyone has equal access to them, the subject who possesses C has a way of knowing it not available to anyone else: by private, introspective awareness.
3. They have *intentionality* (an exception may be states like itches and pains which have causes but seem to lack intentionality): ofness, aboutness, directedness towards an object. One has a sensation of the lamp, a thought about London, a belief or desire about/for such and such.

At least five kinds of conscious state exist:

1. A *sensation* is a state of awareness or sentience, e.g., a conscious awareness of sound or pain. Some sensations are experiences of things outside me, like a tree or the color red. Others are awarenesses of states within me, like pains. Emotions are a subclass of sensations and, as such, they are forms of awareness of things. I can be aware of something in an angry way.
2. A *thought* is a mental content that can be expressed in an entire sentence. Some thoughts logically imply other thoughts. For example "All dogs are mammals" entails "This dog is a mammal." If the former is true, the latter must be true. Some

thoughts don't entail, but merely provide evidence for other thoughts. For example, certain thoughts about evidence in a court case provide grounds for the thought that a person is guilty. Thoughts are the sorts of things that can be true or false, reasonable or unreasonable, and can stand in logical or evidential relations to other thoughts or beliefs.

3. A *belief* is a person's view, accepted to varying degrees of strength, of how things really are.
4. A *desire* is a certain felt inclination to do, have, or experience certain things or to avoid such.
5. An *act of will* is a choice, an exercise of power, an endeavoring to act, usually for the sake of some purpose.

- *The Soul:* In the history of philosophy, there have been, broadly speaking, two different ways to define a soul:[10] one follows Descartes, and is called the Cartesian view of the soul, and the other follows Aristotle and/or Aquinas and is called the Aristotelian or Thomistic view (I will use the latter term) of the soul. There are more views than these, but they are the main ones on the table.
- *The Cartesian Soul*: An immaterial, spiritual substance that has (exemplifies) and unifies actual/potential conscious properties/states. The body is a strictly physical object (most likely, a mereological aggregate).
- *The Thomistic Soul:* An immaterial, spiritual substance that has (exemplifies) and unifies actual/potential conscious properties/states *and animates/diffuses/enlivens the body.* The Cartesian soul is the possessor/unifier of conscious states, and that's all. The Thomistic soul does this as well, but also contains vital powers for making the body alive. For the Cartesian, the human body is entirely physical; for the Thomist, the human body is an en-souled biological structure. The Thomistic view is best explained with an example. Suppose I have salt and a glass of water. If I put the salt in the water (and it dissolved), I would have salted water. If I could take the salt out of the water, there would be plain water left. In the example, the salt = the soul, the salted water = the body, and the plain water = a human corpse. Note that the body is not merely a physical object. A body without a soul is, indeed, a mere physical object and, as such, it is a mere corpse. It was a body, but no longer is. The soul can exist without the body just as salt can exist without water. But a body cannot exist without the soul, just as salted water cannot exist without salt. Remember: *For the Thomist, the body includes its physical, biological aspects—organs, cells, and so forth—but also includes the soul, fully present at every point where the body exists in order for it to be a body.*
- *The Mind.* The notion of the mind is a very loose concept, and to understand its meaning, one must be very careful to examine the context in which it is being used.
  1. For the Cartesian, the mind is the same thing as the soul (the terms can be used interchangeably) and is equal to the person. The mind (person) is what contains all five kinds of conscious states, and not just thoughts. If a Cartesian wants to limit the use of "mind" to include only thoughts and beliefs, then he or she needs to say so.

2. The Thomistic use of "mind" needs a good bit of clarification. We have already seen that both the Cartesian and Thomistic soul contain the actual conscious states occurring in that soul. But, in addition to its states, at any given time, both kinds of soul have a number of capacities (aka, dispositions, potentialities, untriggered causal powers) that are not currently being actualized or utilized. To understand this, consider an acorn. The acorn has certain actual characteristics or states; e.g., a specific size or color. It also has a number of capacities or potentialities that could become actual if certain things happen. For example, the acorn has the capacity to grow a root system or change into the shape of a tree. Likewise, the soul has capacities. I have the ability to see color, think about math, or desire ice cream even when I am asleep and not in the actual states just mentioned.

   Capacities come in hierarchies. There are first-order capacities, second-order capacities to have these first-order capacities, and so on, until ultimate capacities (those that constitute my essential "humanness," those I have simply in virtue of being human) are reached. For example, if I can speak English but not Russian, then I have the first-order capacity for English as well as the second-order capacity to have this first-order capacity (which I have already developed). I also have the second-order capacity to have the capacity to speak Russian, but I lack the first-order capacity to do so. Higher-order capacities are realized by the development of lower-order capacities under them. An acorn has the ultimate capacity to draw nourishment from the soil, but this can be actualized and unfolded only by developing the lower capacity to have a root system, then developing the still lower capacities *of* the root system.

   The adult human soul has literally thousands of capacities within its structure. But the soul is not just a collection of isolated, discrete, randomly related internal capacities. Rather, the various capacities within the soul fall into natural groupings called *faculties* of the soul. In order to get hold of this, think for a moment about this list of capacities: the ability to see red, see orange, hear a dog bark, hear a tune, think about math, think about God, desire lunch, desire a family. The ability to see red is more closely related to the ability to see orange than it is to the ability to think about math. We express this insight by saying that the abilities to see red or orange are parts of the same faculty, the faculty of sight. The ability to think about math is a capacity within the thinking faculty. In general, *a faculty is a compartment (inseparable part; mode) of the soul that contains a natural family of related capacities.*

   So far, Cartesian and Thomistic dualists agree. However, as noted above, the Cartesian limits the soul's faculties to those of consciousness and, accordingly, either identifies the mind with the soul (where "mind" refers to a conscious substance) or with that faculty of the soul with the powers of having thoughts, beliefs, and engaging in reasoning.

For many Thomists, the soul has a number of faculties/modes/ inseparable parts within it. But these extend beyond those relevant to consciousness and include those powers for developing, unifying, and grounding the living functions of the body. An interesting research program would be a careful study of the relationship between the essence of a Thomistic soul and the contemporary notion of biological information.

In any case, the soul is mereologically simple with respect to not having separable parts (it is a substance, not a mereological aggregate), but metaphysically complex, containing a number of different faculties and dispositional or actualized properties. On this view, the mind is a faculty and inseparable part of the person or soul, alongside the faculties of will, emotion, sensation, and so on.

3. *The Physicalist Use of "Mind."* When physicalists use the term "mind" they are not talking about a substance or a faculty. Rather "mind" is a loose way of referring to the various conscious states possessed by an organism's brain. Or it may be limited so as to refer to a subset of an organism's conscious states, viz., its thinking or belief states. For most physicalists, these conscious states are physical. Thus, "mind" is just a term to refer to various brain states that "count" as being mental.

- *The Law of Identity* (also known as *Leibniz's Law of the Indiscernibility of Identicals*). This law states that it is necessarily the case that, for any entities *x* (Mark Twain) and *y* (Samuel Clemens), if *x* and *y* are identical (they are really the same thing, there is only one thing you are talking about, not two), then any truth that applies to *x* will apply to *y* as well and conversely. This suggests a test for identity: if you could find one thing true of *x* not true of *y*, or vice versa, then *x* cannot be identical to (be the same thing as) *y*. Further, if you could find one thing that could *possibly be* true of *x* but not of *y* (or vice versa), even if it isn't actually true, then *x* cannot be identical to *y*.

  An important implication of the law of identity says that it is necessarily the case that if *x* is identical to *y*, then it is not possible for x not to be identical to y. There is no possible world where *x* exists and is not identical to *y*. Mark Twain may happen to be six feet tall, but the very man we call "Mark Twain" is necessarily identical to the man we happen to call "Samuel Clemens." A thing's self-identity is a necessary feature of it. Note that just because *x* (e.g. a feeling of empathy) is functionally or causally dependent on *y* (e.g., the firing of mirror neurons), it does not follow that *x* is identical to *y*. These points about identity and the necessity of identity are at the very core of the case that consciousness and the self are not physical.

3. The Nature of Consciousness

THE NATURE of consciousness is pretty commonsensical. As we saw above from our example of waking up from surgery, it can be defined ostensively. Some philosophers go further and define conscious states in one of three ways. A state C is a conscious state when (1) there is a what-it-is-like to C, or (2) C's subject has private access to C, or (3) C has intentionality (ofness or aboutness).

You will recall from the philosophical lexicon above that there are at least five kinds of conscious states (also called mental states): sensations, thoughts, beliefs, desires, and acts of free will.

Property dualists (those who believe that mental properties or states are non-physical and irreducibly mental) argue that mental states are in no sense physical since they possess *five* features that do not characterize physical states:

a. there is a raw qualitative feel or a "what-it-is-like" to having a mental state such as feeling a pain or being me; but there is no what-it-is-like for being a negative charge, a methane molecule, or a computer, or for any entirely physical property/state, substance or mereological aggregate;
b. at least many mental states have intentionality—*ofness* or *aboutness*—directed towards an object (e.g., a sensation is *of* the lamp, a thought is *about* Kansas City); physical states cause other physical states, but no physical state is *about* another physical state;
c. mental states are inner (they occur inside of me; I have or exemplify them), private (I have private access to my mental states) and immediate to the subject having them (I am directly aware of them, and not by means of some other state; but I am indirectly aware of all physical states by means of my mental states—I see the brain state, etc.);
d. mental states require a subjective ontology—namely, mental states are necessarily owned by the sentient subjects who have them; mental states do not float around without belonging to someone; but brain states may obtain with no subject having them;
e. mental states fail to have crucial features (e.g., spatial extension, being a structural property composed of separable parts like the state of being a neuron) that characterize physical states and, in general, cannot be described using physical language. And mental states have crucial features (e.g., a thought can have the property of being true or false, of being normative—I *should* have this thought, given my other thoughts about the evidence) that brain states lack.

Space forbids me from presenting arguments against and physicalist alternatives to the property-dualist view of conscious states. I have done that elsewhere and must leave it at that.[11] Here, I want to turn to the issue of the self, soul, or "I."

4. The Nature of the Soul

When dealing with questions about the nature and reality of the soul, there are at least four issues that we must keep before us. What is the nature of:

1. the possessor/unifier of consciousness?
2. the entity that employs the first-person indexical term "I" to refer to that same entity, and thus is both a subject and an object of the reference?
3. the thing that has a first-person point of view, a perspectival vantage point?
4. the thing that exercises responsible agency and is an appropriate object of praise and blame?

As we noted in the philosophical lexicon above, there are different ways to define the soul, but there is a generic way to define it, and to define substance dualism: The soul is an immaterial or spiritual substance that is characterized by and unifies mental properties or states and that animates or enlivens its body; generic substance dualism is the view that the "I" or self is an immaterial or spiritual substance with the attributes of consciousness and which is not identical to anything physical, including one's body or any of its parts (e.g., its brain).

5. Arguments for the Existence of the Soul

HERE ARE five (or six) arguments for the existence of the soul:

Argument #1: Our Basic Awareness of the Self

Stewart Goetz has advanced the following type of argument for the nonphysical nature of the self, which I have modified:[12]

1. I am essentially an indivisible, simple, spiritual substance.
2. Any physical body is essentially a divisible or complex entity (any physical body has spatial extension or separable parts).
3. The law of identity (if *x* is identical to *y*, then whatever is true of *x* is true of *y* and vice versa).
4. Therefore, I am not identical with my (or any) physical body.
5. If I am not identical with a physical body, then I am a soul.
6. Therefore, I am a soul.

Premise (2) is pretty obvious, and (5) is commonsensical. The body and brain are complex material objects made of billions of parts—atoms and molecules. Premise (3) is the law of identity I introduced above. Regarding premise (1), we know it is true by introspection. When we enter most deeply into ourselves, we become aware of a very basic fact presented to us: We are aware of our own self (ego, "I," center of consciousness) as being distinct from our bodies and from any particular mental experience we have, and as being an uncomposed, spatially unextended, simple center of consciousness. In short, we are just aware of ourselves as simple, conscious things. This fundamental awareness is what grounds my properly basic belief (a belief it is rational to have that is not based on other beliefs) that I am a simple center of consciousness. On the basis of this awareness, and premises (2) and (3), I know that I am not identical to my body or my conscious states; rather, I am the immaterial self that *has* a body and a conscious mental life.

An experiment may help convince you of this. Right now I am looking at a chair in my office. As I walk toward the chair, I experience a series of what are called phenomenological objects or chair representations. That is, I have several different chair experiences that replace one another in rapid succession. As I approach the chair, my chair sensations vary. If I pay attention, I am also aware of two more things. First, I do not simply experience a series of sense-images of a chair. Rather, through self-awareness, I also experience the fact that it is I, myself, who has each chair experience. Each chair sensation produced at each angle of perspective has a perceiver who is I. An "I" accompanies each sense-experience to produce a series of awarenesses—"I am experiencing a chair sense-image now."

I am also aware of the basic fact that the same self that is currently having a fairly large chair experience (as my eyes come to within twelve inches of the chair) is the very same self as the one who had all of the other chair experiences preceding this current one. Through self-awareness, I am aware of the fact that I am an enduring I who was and is (and will be) present as the owner of all the experiences in the series.

These two facts—I am the owner of my experiences, and I am an enduring self—show that I am not identical to my experiences. I am the conscious thing that has them. I am also aware of myself as a simple, uncomposed and spatially unextended center of consciousness. In short, I am a mental substance. Moreover, I am "fully present" throughout my body; if my arm is cut off, I do not become four-fifths of a self. My *body* and *brain* are divisible and can be present in percentages (there could be 80 percent of a brain present after an operation), but *I* am an all-or-nothing kind of thing. I am not divisible; I cannot be present in percentages.

Argument #2a: The Indexical "I" and the First-Person Perspective

Consider the following argument:

1. If I were a physical object (e.g., a brain or body), then a third-person physical description would capture all the facts that are true of me.
2. But a third-person physical description does not capture all the facts that are true of me.
3. Therefore, I am not a physical object.
4. I am either a physical object or a soul.
5. Therefore, I am a soul.

A complete physical description of the world would be one in which everything would be exhaustively described from a third-person point of view in terms of objects, properties, processes, and their spatiotemporal locations. For example, a description of an apple in a room would go something like this: "There exists an object three feet from the south wall and two feet from the east wall, and that object has the property of being red, round, sweet..."

The first-person point of view is the vantage point that I use to describe the world from my own perspective. Expressions of a first-person point of view utilize what are called indexicals—words like "I," "here," "now," "there," and "then." "Here" and "now" are "where" and "when" I am; "there" and "then" are "where" and "when" I am not. Indexicals implicitly or explicitly refer to me, myself. "I" is the most basic indexical, and it refers to my self that I know by acquaintance with my own self in acts of self-awareness. I am immediately aware of my own self and I know to whom "I" refers when I use it: It refers to me as the self-conscious, self-reflexive owner of my body and mental states.

According to a widely accepted form of physicalism, there are no irreducible, privileged first-person perspectives. Everything can be exhaustively described in an object language from a third-person perspective. A physicalist description of me would say, "There exists a body at a certain location that is five feet eight inches tall, weighs 160 pounds," and so forth. The property dualist would add a description of the properties possessed by that body, such as the body is feeling pain or thinking about lunch.

But no amount of third-person description captures my own subjective, first-person acquaintance of my own self in acts of self-awareness. In fact, for any third-person description of me, it would always be an open question as to whether the person described in third-person terms was the same person as I am. I do not know my self *because* I know some third-person description of a set of mental and physical properties and also know that a certain person satisfies that description (namely, me). I know myself as a self immediately through being acquainted with my own self in an act of self-awareness. I can express that self-awareness by using the term "I."

"I" refers to my own substantial soul. It does not refer to any mental property or bundle of mental properties I am having, nor does it refer to any body described from a third-person perspective. "I" is a term that refers to something that exists, and does not refer to any object or set of properties described from a third-person point of view. Rather, "I" refers to my own self, with which I am directly acquainted and which, through acts of self-awareness, I know to be the substantial, uncomposed possessor of my mental states and my body.

Argument #2b: Unity of Consciousness and the Soul

A related argument has been offered by William Hasker:

1. If I am a physical object (e.g., a brain or a body), I do not have a unified visual field.
2. I do have a unified visual field.
3. Therefore, I am not a physical object.
4. I am either a physical object or a soul.
5. Therefore, I am a soul.

To grasp the argument, consider one's awareness of a complex fact, say, one's own visual field, consisting of awareness of several objects at once, including a number of different surface areas of each object. Now, one may claim that such a unified awareness of one's visual field consists in the fact that there are a number of different physical parts, each of which is aware only of part of and not the whole of the complex fact. Indeed, this is exactly what physicalists say. We now know that when one looks at an object, different regions of the brain process different electrical signals that are associated with different aspects of the object (e.g., its color, shape, size, texture, location).[13] However, this claim will not work, because it cannot account for the fact that there is a single, unitary awareness of the entire visual field.[14] There is no region in the brain that "puts the object back together into a unified whole." Only a single, simple mental substance can account for the unity of one's visual field or, indeed, the unity of consciousness in general.

Argument #3: The Modal Argument

The core of the modal argument for the soul is fairly simple: I am possibly disembodied (I could survive without my brain or body), my brain or body are not possibly disembodied (they could not survive without being physical), so I am not my brain or body; therefore (since I am either a soul or a brain or body), I am a soul. Let's elaborate on the argument.

Thought experiments have rightly been central to debates about personal identity. For example, we are often invited to consider a situation in which two persons switch bodies, brains, or personality traits or in which a person exists disembodied. In these thought-experiments, someone argues in the following way: Because a certain state of affairs *S* (e.g., Smith existing disembodied) is conceivable, this provides justification for thinking that *S* is metaphysically possible. Now if *S* is possible, then certain implications follow about what is or is not essential to personal identity (e.g., Smith is not essentially a body).

We all use conceiving as a test for possibility or impossibility throughout our lives. I know that life on other planets is possible (even if I think it is highly unlikely or downright false) because I can conceive it to be so. I am aware of what it is to be living and to be on earth and I conceive no necessary connections between these two properties. I know square circles are impossible because it is inconceivable, given my knowledge of being square and being circular. To be sure, judgments that a state of affairs is possible or impossible, grounded in conceivability, are not infallible. They can be wrong. Still, they provide strong evidence for genuine possibility or impossibility. In light of this, I offer the following criterion:

> For any entities *x* and *y*, if I have good grounds for believing I can conceive of

x existing without *y* (e.g., a dog existing without being colored brown) or vice versa, then I have good grounds for believing *y* (being brown) is not essential or identical to *x* (being a dog) or vice versa.

Let us apply these insights about conceivability and possibility to the modal argument for substance dualism. The argument comes in many forms, but it may be fairly stated as follows:[15]

1. The law of identity: If *x* is identical to *y*, then whatever is true of *x* is true of *y* and vice versa.
2. I can strongly conceive of myself as existing disembodied. (For example, I have no difficulty believing that near-death experiences are possible; that is, they *could* be real.)
3. If I can strongly conceive of some state of affairs *S* (e.g., my disembodied existence) that *S* possibly obtains, then I have good grounds for believing of *S* that *S* is possible.
4. Therefore, I have good grounds for believing of myself that it is possible for me to exist and be disembodied.
5. If some entity *x* (for example, "I") is such that it is possible for *x* to exist without *y* (for example, my brain or body), then (i) *x* (for example "I") is not identical to *y* (my brain or body), and (ii) *y* (my brain or body) is not essential to *x* (me).
6. My body (or brain) is not such that it is possible to exist disembodied, i.e., my body (or brain) is essentially physical.
7. Therefore, I have good grounds for believing of myself that I am not identical to my body (or brain) and that my physical body is not essential to me.

A parallel argument can be advanced in which the notions of a body and disembodiment are replaced with the notions of physical objects. So understood, the argument would imply the conclusion that I have good grounds for thinking that I am not identical to a physical object, nor is any physical object essential to me. A parallel argument can also be developed to show that possessing the ultimate capacities of sensation, thought, belief, desire, and volition are essential to me—that is, I am a substantial soul or mind, and I could not exist without the ultimate capacities of consciousness.

I cannot undertake a full defense of the argument here, but it would be useful to a say a bit more regarding (2). There are a number of things about ourselves and our bodies of which we are aware that ground the conceivability expressed in (2). I am aware that I am unextended (I am "fully present" at each location in my body, as Augustine claimed; I occupy my body as God occupies space, by being fully present throughout it). I am also aware that I am not a complex aggregate made of substantial parts or the sort of thing that can be composed of physical parts, but a basic unity of inseparable faculties (of mind, volitions, emotion, etc.) that sustains absolute sameness through change. Finally, I am aware that I am not capable of gradation (I cannot become two-thirds of a person).[16]

In near-death experiences, numerous people report themselves to have been disembodied. They are not aware of having bodies in any sense. Rather, they are aware of themselves as unified egos that have sensations, thoughts, and so forth. Moreover, Christians who understand the biblical teaching that God and angels are bodiless spirits also understand by direct introspection that they are like God and angels, for although they have bodies they have the same sorts of powers God and angels have. The New Testament teaching on the intermediate state (i.e., between death and the resurrection) is intelligible in light of what they know about themselves and also implies that they will, and therefore can, exist temporarily without bodies. In II Corinthians 12:1–4, Paul asserts that he may actually have been disembodied. Surely part of the grounds for Paul's willingness to consider this a real possibility were his own awareness of his nature through introspection, his recognition of his similarity to God and angels in this respect, and his knowledge of biblical teaching. All of the factors mentioned in this paragraph imply that people can conceive of themselves as existing in a disembodied state, that this provides grounds for thinking that this is a real possibility (even if it is false, though, of course, I do not think it is false), and, thus, one cannot be one's body, nor is one's body essential to one's identity.

Argument #4: Free Will, Morality, Responsibility, and Punishment

Consider the following argument:

1. If I am a physical object (e.g., a brain or a body), then I do not have free will.
2. But I do have free will.
3. Therefore, I am not a physical object.
4. I am either a physical object or a soul.
5. Therefore, I am a soul.

When I use the term *free will,* I mean what is called libertarian freedom. I can literally choose to act or refrain from choosing. No circumstances exist that are sufficient to determine my choice. My choice is up to me. I act as an agent who is the ultimate originator of my own actions. Moreover, my reasons for acting do not partially or fully cause my actions; I myself bring about my actions. Rather, my reasons are the teleological goals or purposes for the sake of which I act. If I get a drink because I am thirsty, the desire to satisfy my thirst is the end for the sake of which I myself act freely. I raise my arm *in order to* vote.

If physicalism is true, then human free will does not exist. Instead, determinism is true for at least two reasons.[17] First, one's overall brain or body state at time *t1* plus the laws of nature and other causal inputs at that time determine or fix the chances of one's overall brain or body state at the next moment of time *t2*. Second, at any time *t*, the macrophysical properties and behaviors of the physical brain or body are determined by and dependent on what is going on at the microphysical level. If I am just a physical system, there is nothing in me that has the capacity to freely choose to do something. Material systems, at least large-scale ones, change over time in deterministic fashion according to the initial conditions of the system and the laws of chemistry and physics. A pot of water will reach a certain temperature at a given time in a way determined by the amount of water, the input of heat, and the laws of heat transfer.

Now, when it comes to morality, it is hard to make sense of moral obligation and responsibility if determinism is true. They seem to presuppose freedom of the will. If I "ought" to do something, it seems to be necessary to suppose that I *can do* it, that I could have done otherwise, and that I am in control of my actions. No one would say that I ought to jump to the top of a fifty-floor building and save a baby, or that I ought to stop the American Civil War in 2014, because I do not have the ability to do either. Further, free acts seem to be teleological. We act for the sake of goals or ends. If physicalism (or mere property dualism) is true, there is no genuine teleology and thus no libertarian free acts.

It is safe to say that physicalism requires a radical revision of our commonsense notions of freedom, moral obligation, responsibility, and punishment. On the other hand, if these commonsense notions are true, physicalism is false.

The same problem besets property dualism. There are two ways for property dualists to handle human actions. First, some property dualists are *epiphenomenalists.* A person is a living physical body having a mind, the mind consisting, however, of nothing but a more or less continuous series of conscious or unconscious states and events which are the effects but never the causes of bodily activity. Put another way, when matter reaches a certain organizational complexity and structure, as is the case with the human brain, then matter produces mental states as fire produces smoke or the structure of hydrogen and oxygen in water produces wetness. The mind is to the body as smoke is to fire. Smoke is different from fire (to keep the analogy going, the physicalist would identify the smoke with the fire or the functioning of the fire), but fire causes smoke, not vice versa. The mind is a by-product of the brain and causes nothing; the mind merely "rides" on top of the events in the brain. Hence, epiphenomenalists rejects free will, since they deny that mental states cause anything.

A second way that property dualists handle human action is through a notion called *event-event causation.*[18] To understand event-event causation, consider a brick that breaks a glass. The cause in this case is not the brick itself (which is a substance), but an event, viz., the brick's being in a certain state—a state of motion. And this event (the brick's being in a state of motion) was caused by a prior event and so on. The effect is another event, viz., the glass's being in a certain state—breaking. Thus, one event—the moving of a brick—causes another event to occur—the breaking of the glass. Further, according to event-event causation, whenever one event causes another, there will be some deterministic or probabilistic law of nature that relates the two events. The first event, combined with the laws of nature, is sufficient to determine or fix the chances for the occurrence of the second event.

Agent action is an important part of an adequate libertarian account of freedom of the will. One example of agent action is this typical case: my raising my arm. When I raise my arm, I, as a substance, simply act by spontaneously exercising my active powers. *I* raise my arm; I freely and spontaneously exercise the powers within my substantial soul and simply act. No set of conditions exists within me that is sufficient to determine that I raise my arm. Moreover, this "I," this agent, is characterized by the power of active freedom, conscious awareness, the ability to think, form goals and plans, to act teleologically (for the sake of goals), and so forth. Such an agent is an immaterial substance and not a physical object. Thus, libertarian freedom is best explained by a substance dualism and not by physicalism or mere property dualism.

Unfortunately for property dualists, event-event causation is deterministic. Why? For one thing, there is no room for an agent, an ego, an "I" to intervene and contribute to one's actions. I do not produce the action of raising my arm; rather, a state of desiring to raise an arm is sufficient to produce the effect. There is no room for my own self, as opposed to the mental states within me, to act.

For another thing, all the mental states within me (my states of desiring, willing, hoping) are states that were deterministically caused (or had their chances fixed) by prior mental and physical states outside of my control, plus the relevant laws. "I" become a stream of states or events in a causal chain that merely passes through me. Each member of the chain determines that the next member occurs.

In summary, then, property dualism denies libertarian freedom, because it adopts either epiphenomenalism or event-event causation. Thus, property dualism, no less than physicalism, is false, given the truth of a libertarian account of free will, moral ability, moral responsibility, and punishment. Our commonsense notions about moral ability, responsibility, and punishment are almost self-evident. We all operate toward one another on the assumption that they are true (and these commonsense notions seem to assume libertarian free will). However, if physicalism or property dualism is true, we will have to abandon and revise our commonsense notions of moral ability, responsibility, and punishment because free will is ruled out.

Argument #5: Sameness of the Self over Time

Consider the following argument:

1. If something is a physical object composed of parts, it does not survive over time as the same object if it comes to have different parts.
2. My body and brain are physical objects composed of parts.
3. Therefore, my body and brain do not survive over time as the same object if they come to have different parts.
4. My body and brain are constantly coming to have different parts.
5. Therefore, my body and brain do not survive over time as the same object.
6. I do survive over time as the same object.
7. Therefore, I am not my body or my brain.
8. I am either a soul or a body or a brain.
9. Therefore, I am a soul.

Premise (2) is commonsensically true. Premise (4) is obviously true as well. Our bodies and brains are constantly gaining new cells and losing old ones, or at least, gaining new atoms and molecules and losing old ones. So understood, bodies and brains are in constant flux. I will assume that (8) represents the only live options for most ordinary people. This leaves premises (1) and (6).

Let's start with (1). Why should we believe that ordinary material objects composed of parts do not remain the same through part replacement?[19] To see why this makes sense, consider five scattered boards, labeled *a* through *e*, each located in a different person's back yard. Commonsensically, it doesn't seem that the boards form an object. They are just isolated boards. Now, suppose we collected those boards and put them in a pile with the boards touching each other. We would now have, let us suppose, an object called a pile or heap of boards. The heap is a weak object, indeed, and the only thing unifying it would be the spatial relationships between and among the parts *a–e*. They are in close proximity and are touching each other. Now, suppose we took board *b* away and replaced it with a new board *f* to form a new heap consisting of *a*, *c–f*. Would our new heap be the same as the original heap? Clearly not, because the heap just is the boards and their relationships to each other, and we have new boards and a new set of relationships. What if we increased the number of boards in the heap to 1000? If we now took one board away and replaced it with a new board, we would still get a new heap. The number of boards does not matter.

Now imagine that we nailed our original boards *a–e* together into a makeshift raft. In this situation, the boards are rigidly connected such that they do not move relative to each other; instead, they all move together if we pick up our raft. If we now took board *b* away and replaced it with board *f*, we would still get a new object. It may seem odd, but if we took board *b* away and later put it back, we would still have a new raft because the raft is a collection of parts and bonding relationships to each other. Thus, even though the new raft would still have the same parts (*a–e*), there would be new bonding relationships between *b* and the board or boards to which it is attached.

Now think of a cloud. From a distance, it looks like a solid, continuous object. But if you get close to it, say on a plane flight, it becomes evident that it is a very loose collection of water droplets. The boundaries are vague, and for any droplet near the "edge," it is arbitrary whether it is a genuine part of the cloud as opposed to being a droplet outside the cloud. The cloud is like a heap of boards or like a raft. If new droplets are added and some removed, it is, strictly speaking, not the same cloud.

Now, consider our bodies and brains, and assume they are mere physical objects composed of billions of parts. From our daily vantage point, they appear to be solid, continuous objects. But if we could shrink down to the level of an atom, we would see that they are in reality like a cloud—gappy, containing largely empty space, and composed of billions of atoms (molecules, cells) that stand in various bonding relations between and among those parts. If we were to take a part away and replace it, we would have a new object. The body and brain are like the cloud or our raft. Besides the parts and the relationships among them, there is nothing in the body or brain to ground its ability to remain the same through part replacement. This is the fundamental insight behind the view that the body and brain cannot remain the same if there is part alteration.[20] Since the body and brain are constantly changing parts and relationships, they are not the same from one moment to the next in a strict philosophical sense (though, for practical day-to-day purposes, we regard them as the same in a loose, popular sense).

Some have responded to this argument by claiming that mereological essentialism does not apply to at least some mereological aggregates because even though these aggregates experience part replacement, if their structure remains the same, this is sufficient to ground endurance. Unfortunately, this response does not work, because a proper metaphysical analysis of such aggregates does not provide an entity adequate to ground their literal identity through part alteration. And the aggregate's structure is one such inadequate entity.

To see this, suppose we have some mereological aggregate *W*, say a car, at some time *t*, and let "the *p*s" refer distributively to all and only the atomic simples (assuming such) that make up *W*. Now, given that the *p*s just are a specific list of simples taken distributively without regard to structure, it would seem obvious that if we have a different list of simples, the *q*s, it is not identical to the *p*s even if the two lists share all but one part in common. This same insight would be true if we took "the *p*s" and "the *q*s" collectively as referring to some sort of mereological sum. In either case, there is no entity "over and above" the parts that could serve as a ground of sameness through part alteration.

Now, *W* (a car) has different persistence conditions than, and thus is not identical, to the *p*s (all and only the ultimate parts that make up the car.) *W* could be destroyed and the *p*s (taken in either sense) could exist. Now, let *S* stand for all and only the various relations that stand between and among the *p*s. *S* is *W*'s structure. Is *W* identical to *S* and the *p*s? I don't think so. *W* has its own structure, say in comparison to some other whole *W** that is exactly similar in structure to *W*. *W* and *W** have their own exactly similar structures. Given that *S* is a universal (a type of structure that many cars could exhibit,) it is not sufficient for individuating (making particular) *W*'s specific structure. For that we need *SI*, *W*'s *structure-instance*, *W*'s token of *S*, and *SI* will consist of all and only the specific relation-instances that are instantiated between and among the *p*s. Let "the *r*s" stand for all and only the relevant relation-instances that compose *SI*. It is now obvious that *SI* is a mereological aggregate composed of the *r*s. If the *r*s undergo a change of relation-instances, it is no longer the same collection of relation-instances. Given that *SI* just is a mereological aggregate or, perhaps, a specific ordering of the *r*s, if the *r*s undergo a change of relation-instances, *SI* will cease to exist and a different structure (perhaps exactly similar to *SI*) will obtain since there is no entity to serve as a ground for *SI*'s sameness through part replacement. If *W* is the *p*s plus *SI*, it seems to follow that *W* is subject to mereological-essentialist constraints.

So much for premise (1). What about premise (6)? Why should we think we survive as the same object over time? Suppose you are approaching a brown table and in three different moments of introspection you attend to your own awarenesses or experiences of the table. At time t_1 you are five feet from the table and you experience a slight pain in your foot (P_1), a certain light brown table sensation from a specific place in the room (S_1), and a specific thought that the table seems old (T_1). A moment later at t^2 when you are three feet from the table you experience a feeling of warmth (F_1) from a heater, a different table sensation (S_2) with a different shape and slightly different shade of brown than that of *S1*, and a new thought that the table reminds you of your childhood desk (T_2). Finally, a few seconds later (T_3) you feel a desire to have the table (D_1), a new table sensation from one foot away (S_3) and a new thought that you could buy it for less than twenty-five dollars (T_3).

In this series of experiences, you are aware of different things at different moments. However, at each moment of time, you are also aware that there is a self at that time that is having those experiences and that unites them into one field of consciousness. Moreover, you are also aware that the very same self had all of the experiences at T_1, T_2, and T_3. Finally, you are aware that the self that had all the experiences is none other than you yourself. This can be pictured as follows:

$$\begin{array}{ccccc} t_1 & \rightarrow & t_2 & \rightarrow & t_3 \\ \{P_1, S_1, T_1\} & & \{F_1, S_2, T_2\} & & \{D_1, S_3, T_3\} \\ I_1 & & I_2 & & I_3 \end{array}$$

$$I_1 = I_2 = I_3 = \text{I Myself}$$

Through introspection, you are aware that you are the self that owns and unifies your experiences at each moment of time and that you are the same self that endures through time. This is pretty obvious to most people. When one hums a tune, one is simply aware of being the enduring subject that continues to exist during the process. This is basic datum of experience.

Moreover, fear of some painful event in the future or blame and punishment for some deed in the past appear to make sense only if we implicitly assume that it is literally I myself that will experience the pain or that was the doer of the past deed. If I do not remain the same through time, it is hard to make sense of these cases of fear and punishment. I would not have such fear or merit such punishment if the person in the future or past merely resembled my current self in having similar memories, psychological traits, or a body spatio-temporally continuous with mine and consisting of many of the same parts as my current body.[21]

Finally, some have argued that to realize the truth of any proposition or even entertain it as meaningful, the very same self must be aware of its different parts. For example, a sentence that expresses a proposition will consist of a subject, verb, and predicate. If one person-stage contemplated the subject, another the verb, and still another the predicate, literally no self would persist to think through and grasp the proposition as a whole.

For these and other reasons, we are warranted in believing that the I or self survives over time as the same object, even though its body or brain undergo part replacement.

6. The General Irrelevance of Neuroscience in Addressing the Ontological Nature of Consciousness and the Soul

It is widely believed that neuroscience is the proper field for discovering the nature of consciousness and the soul. But for at least five reasons, this belief is false. Before I present them, I note that below I present two ways in which empirical evidence may be relevant to deciding between physicalist and dualist views of consciousness and the self (I, ego, soul). This acknowledgment is important for many of the arguments to follow in subsequent chapters of this collection. However, given the scientism pervasive among Western intellectuals, it is important to show why the exclusivist claims regarding the empirical sciences and their hegemony in topics within philosophy of mind is an overreach.

#1: The Issue of Empirically Equivalent Theories

The neuroscientific empirical discovery of mental-physical correlations (causal or dependency relations both ways) provides no evidence whatsoever for physicalism. Indeed, as we will see below, the central issues regarding the mind—what is a thought, feeling, or belief; what is that to which my self is identical—are basically commonsense and philosophical issues about which scientific discoveries are largely irrelevant. Science is helpful in answering questions about what factors in the brain and body generally hinder or cause mental states to obtain, but science is largely silent about the nature of mental properties or states.

To see the basic irrelevance of empirical data, consider the following. We have discovered that if certain neurons (*mirror neurons*) are damaged, then one cannot feel empathy for another. How are we to understand this? To answer this question, we need to get before our minds the notion of *empirically equivalent theories*. If two or more theories are empirically equivalent, then they are consistent with all and only the same set of empirical observations. Thus, an appeal to empirical data cannot be made in favor of one of such theories over the others.

Three empirically equivalent solutions to the discovery of the function of mirror neurons come to mind: (1) strict physicalism (a feeling of empathy is identical to something physical, e.g., the firings of mirror neurons); (2) mere property dualism (a feeling of empathy is an irreducible state of consciousness *in the brain* whose obtaining depends on the firing of mirror neurons); and (3) substance dualism (a feeling of empathy is an irreducible state of consciousness *in the soul* whose obtaining depends (while embodied) on the firing of mirror neurons).

Of these three, no empirical datum can pick which is correct. This is why three Nobel Prize winners working in neuroscience and related fields—John Eccles (substance dualist), Roger Sperry (mere property dualist) and Francis Crick (strict physicalist) could hold different ontologies regarding consciousness and the self even though they all knew the same neuroscientific data. Of course, it could be argued that where the data cannot settle a question, epistemic simplicity is an appropriate tiebreaker, and so the substance dualist will insist that the arguments and evidence for substance dualism are better than those for the other two options mentioned above. But not all scientists will accept epistemic simplicity as a decisive consideration, so it would appear that there is no sure route from data to the establishment of one metaphysical view over another.

Yet for at least two reasons, it does not follow from these points that empirical data is never relevant to building a case for one metaphysical view vs. another, e.g., property or substance dualism vs. strict physicalism (everything is physical). First, we can distinguish between *strong* and *weak* empirical equivalence. Applied to issues in philosophy of mind, the former implies that it is never the case, or at least, it is not the case for a particular issue, that empirical data themselves provide any grounds at all in favor of a physicalist vs. a dualist position (e.g., as in mirror neurons), while the latter implies that there can be some issues where a physicalist and dualist view are both logically consistent with the empirical data, but the data can still contribute to a cumulative case for one view vs. the other. For example, in previous generations, extreme versions of substance dualism implied no dependency or interaction with the brain at all! Empirical studies have shown extreme substance dualism to be false. Thus, only on a case-by-case basis can one decide the relevance of empirical data to the mind-body debate.

Second, property or substance dualism can generate a research program and make predictions. For example, by arguing that information is *sui generis*, irreducible, and grounded in minds, with auxiliary assumptions, one can predict that the role information plays in the origin and development of kinds of organisms is central and uneliminable. Given certain substance dualist views of the soul, it follows that libertarian free will is true, that consciousness is a wholistic unity and not capable of being analyzed atomistically, that there should be a binding problem not suggested by physicalist views of the "self," that there should be strong evidence for near-death experiences, and so forth. Moreover, some versions of substance dualism entail a wholistic organicism, an irreducible complexity, regarding organisms' bodies.

#2: The Nature of the Central Issues in Philosophy of Mind

Upon careful reflection on the central issues in philosophy of mind, it becomes evident that the methods and findings of neuroscience are simply irrelevant in addressing and solving these questions. These questions can be grouped into four areas of reflection.

1. *Ontological Questions:* To what is a mental or physical property identical? To what is a mental or physical event identical? To what is the owner of mental properties or events identical? What is a human person? How are mental properties related to mental events? (E.g., do the latter exemplify or realize the former?) Are there (Aristotelian or Leibnizian) essences and, if so, what is the essence of a mental event or of a human person? Am I essentially a human, a person, or both?
2. *Epistemological Questions:* How do we come to have knowledge or justified beliefs about other minds and about our own minds? Is there a proper epistemic order to first-person knowledge of one's own mind and third-person knowledge of other minds? How reliable is first-person introspection and what is its nature? (E.g., is it a non-doxastic seeming or a disposition to believe?) If reliable, should first-person introspection be limited to providing knowledge about mental states or should it be extended to include knowledge about one's own ego?

3. *Semantic Questions:* What is a meaning? What is a linguistic entity and how is it related to a meaning? Is thought reducible to or a necessary condition for language use? How do the terms in our commonsense psychological vocabulary get their meaning?
4. *Methodological Questions:* How should one proceed in analyzing and resolving the first-order issues that constitute the philosophy of mind? What is the proper order between philosophy and science? Should we adopt some form of philosophical naturalism, set aside so-called first philosophy, and engage topics in philosophy of mind within a framework of our empirically best-attested theories relevant to those topics? What is the role of thought-experiments in philosophy of mind and how does the "first-person point of view" factor into generating the materials for formulating those thought experiments? Is conceivability a reliable guide to modality?

If one reflects carefully on these questions, it becomes clear that they are distinctively philosophical (and theological) questions to which neuroscience is simply irrelevant.

#3: Rebuttal to the Alleged Physicalist Implications of Localization Studies

As noted above, Nancey Murphy claims that "science has provided a massive amount of evidence suggesting that we need not postulate the existence of an entity such as a soul or mind in order to explain life and consciousness."[22] This evidence consists of the fact that "biology, neuroscience, and cognitive science have provided accounts of the dependence on physical processes of *specific* faculties once attributed to the soul."[23] Elsewhere she claims: "My argument in brief is this: all of the human capacities once attributed to the mind or soul are now being fruitfully studied as brain processes—or, more accurately, I should say, processes involving the brain, the rest of the nervous system and other bodily systems, all interacting with the socio-cultural world."[24]

A different but related objection against substance dualism (SD) claims that, when compared to the findings of neuroscience, it is explanatorily impotent. Thus, Ian Ravenscroft writes: "My claim is not that SD is definitely false; rather my principal argument will be that its lack of explanatory power relative to physicalism rules it out of contention given the current evidence."[25] I offer three responses to Murphy and Ravenscroft.

First, many substance dualists do not believe in a substantial ego primarily because it is a theoretical postulate with superior explanatory power. Rather, they take the ego to be something of which people are directly aware. Thus, belief in a substantial, simple soul is properly basic and grounded in self-awareness. Given this dualist approach, the point is that advances in our knowledge of mental-physical dependencies are simply beside the point. And the further debate about whether this dualist approach is the fundamental one for defending substance dualism is not something for which advances in scientific knowledge are relevant.

Second, in those cases where substance dualism *is* postulated as the best explanation for a range of purported facts, typically those facts are distinctively philosophical and not the scientific ones Murphy mentions. Arguments from the unity of consciousness, the possibility of disembodied survival or body switches, the best view of an agent to support libertarian agent causation, and the metaphysical implications from the use of the indexical "I" are typical of arguments offered by substance dualists, and the facts Murphy mentions are not particularly relevant for assessing these arguments.

Finally, the discovery of "the dependence on physical processes of *specific* faculties once attributed to the soul" does not provide sufficient grounds for attributing those faculties to the brain rather than to the soul. There is an important distinction between describing the nature, proper categorization, and possessor of a capacity, on one hand, and explaining what conditions are necessary for its actualization, on the other. To see this, it is important to get clear on the use of "faculty" as the term has been historically used in discussions of substances in general and the soul in particular. Recall that a faculty of some particular substance is a natural grouping of resembling capacities or potentialities possessed by that thing. For example, the various capacities to hear sounds would constitute a person's auditory faculty. Moreover, a capacity gets its identity and proper metaphysical categorization from the type of property it actualizes—its manifestational property. The nature of a capacity-to-exemplify-*F* is properly characterized by *F* itself. Thus, the capacity to reflect light is properly considered a physical, optical capacity. For property dualists, the capacities for various mental states are mental and not physical capacities. Thus, the faculties that are constituted by those capacities are mental and not physical faculties.

Now, arguably, a particular is the kind of thing it is in virtue of the actual and potential properties/faculties essential and intrinsic to it. Thus, a description of the faculties of a thing provides accurate information about the kind of particular that has those faculties. Moreover, a description of a particular's capacities/faculties is a more accurate source of information about its nature than is an analysis of the causal or functional conditions relevant for the particular to act in various ways. The latter can either be clues to the intrinsic nature of that particular or else information about some other entity that the particular relates to in exhibiting a particular causal action. Remember, there is a difference between attempting to describe, categorize, and identify a capacity's nature and possessor, and proffering an explanation of the functional or causal conditions that must be present for that capacity to be actualized.

For example, if Smith needs to use a magnet to pick up certain unreachable iron filings, information about the precise nature of the magnet and its role in Smith's action does not tell us much about the nature of Smith (except that he is dependent in his functional abilities on other things, e.g., the magnet). We surely would not conclude that the actual and potential properties of a magnet are clues to Smith's inner nature. Similarly, functional dependence on, and causal relations to, the brain are of much less value in telling us what kind of thing a human person is than a careful description of the kind-defining mental capacities, i.e., faculties, that human persons as such possess.

#4: Neuroscientific Methodology Relies on First-Person Reports

In establishing mental-physical correlations (causal or dependency relations), neuroscience must assume the commonsense dualist notion of private access. In studying brain events to establish these correlations, the relevant mental state is made aware to the subject by first-person introspection constituted by direct, private access by that subject. These neuroscientific correlations ultimately depend on first-person reports regarding what is going on in the person's consciousness, just as the dualist predicts.

But it is extremely odd and inexplicable that this would be necessary if mental states and the self were physical. If this were the case, all the facts involved in such studies would be publicly accessible, but they are not.

Moreover, as Christian philosopher and Oxford professor Richard Swinburne has noted, given the fact that the neuroscientist must rely on first-person testimony to know that a particular conscious event occurred, "if apparent testimony is to constitute evidence that conscious events occurred, the scientist must… assume that the *subject* is caused to say what they do by a *belief* that the conscious events occurred and an *intention* to *tell the truth* about their *belief*"[26] [italics mine]. The terms in italics refer to mental entities. If we do not assume this and, instead, take them as physical, then the "subject's report" is nothing but sounds caused by brain events that cause them. In this case, physical determinism obtains, and one would no more trust a subject's "report" than the sounds that come out of a parrot's mouth.

#5: Scientific Explanation Requires Necessitation

In standard scientific explanation, a state (also called an event) or a change of state is explained. For example, we may try to explain why the pressure of a gas is in a certain state or why it changed into that state. Or we might try to determine why a tornado was formed. In all these cases, one event causes another event in accordance with some law of nature. So, we have *event-event causation* (event *A* causes event *B*) or *state-state causation* (state *A* causes state *B*).

Associated with event causation is a "covering law" model of explanation, according to which some event is explained—or covered—by giving a valid deductive or inductive argument for that event. Such an argument utilizes two features: (1) a universal or statistical law of nature and (2) some initial causal conditions.

Take, for example, the following explanation for why metal rod *X* expanded:

1. All metal rods expand if heated.
2. Metal rod *X* was heated.
3. Therefore, metal rod *X* expanded.

According to the covering law model of explanation, we explain why rod *X* expanded when we subsume it under a *general law of nature* (all metal rods expand if heated) and factor the *initial conditions* into that law (rod *X* is metal and was heated). Plug the fact (stated in premise 2) into the law of nature (as stated in premise 1), and, presto, we get the conclusion which explains why it happened (conclusion 3).

Another example of the "covering law" model of explanation can be seen using the ideal gas law. You may remember the formula, $PV=nRT$, where P stands for pressure, V for volume, *n* for the number of particles (in moles), R for a constant, and T for temperature. So a covering law model of explanation for the temperature of the gas would look like this:

1. $PV=nRT$.
2. The gas in our container has P_1, V_1 and n_1.
3. Therefore, the gas in our container has T_1.

In this case, we want to explain why our gas has temperature T_1. And we explain this by citing the law of nature (expressed by the ideal gas equation in premise 1), plugging in our initial conditions (as stated in premise 2), and therefore explaining the fact (stated in the conclusion, 3) that the temperature is T_1.

Even though many believe the covering law of explanation is a completely adequate standard scientific explanation, others hold—correctly in my view—that while it may be a *necessary* condition of explanation, it is not *sufficient* by itself. They argue that it needs to be supplemented by some model that tells us *why* the universal law is true in the first place and why, given a constant volume, a certain amount of gas *must* increase in pressure if the temperature increases. We need a model to explain the necessitation of pressure increase given such and such conditions along with an increase in temperature.

Let us reflect further on the attempt to seek to explain the temperature of the gas in our container. Yes, PV=*n*RT is the relevant law of nature of ideal gases. However, one can still ask *why* this equation succeeds in describing the behavior of gases. And to answer that, scientists have developed a model that contains a mechanism that *undergirds* and further *explains* the ideal gas law.

The model, in this case, is the ideal gas model. *Gases* are taken to be collections of tiny atoms or molecules that are assumed to be point particles that engage in completely efficient elastic collisions (no loss of momentum). Moreover, *temperature* is reduced to atomic or molecular motion (some gases consisting of single atoms, others of molecules), and *pressure* is reduced to the rate at which the atoms or molecules of the gas collide with a certain area of the container wall. Thus, for example, if we keep the volume constant (as in a pressure cooker), then raise the temperature, the atoms or molecules are going to get agitated and move around much faster (this is what temperature actually is in the ideal gas theory) and this will, in turn, cause more of them to hit the container wall per second (increasing the pressure).

So the ideal gas model provides an *explanatory picture* of what is going on, including a *mechanism* (agitating the gas's atoms or molecules) that further explains the ideal gas law. And it provides necessitation in the explanation: the pressure *must* increase with temperature increase under constant conditions. This is the difference between a *model* and a *law*. All the ideal gas *law* does is to present correlations among P, V, *n* and T. The *law* is actually not an explanation at all. Rather, it is the thing to be explained. As it is, it is just a contingent correlation among the relevant factors, and the correlation is a mere brute fact without explanation. It is the ideal gas *model* that provides necessitation and explanation.

Here's another example. Neuroscientists are trying to develop laws that govern the relationship between one brain state and another brain state. For example, one proposed law goes like this: "Neurons that fire together, wire together." In other words, if a group of neurons all fire together, they tend to groove or group together, so that they seem wired together to fire simultaneously on a regular basis. So far, so good.

But now suppose that neuroscientists try to go beyond these strictly physical laws, to add new laws that correlate brain states (like C fibers firing) with *mental* states (like feeling pain). We would end up with an unruly list of hundreds of thousands, maybe millions, of brute-fact correlations between various mental and physical states. The expansion of neuroscientific theory to include these correlations would make the theory so utterly complex that the theory itself would be undermined.

As naturalist philosopher Jaegwon Kim notes, if we do not identify mental and physical properties, or mental and physical states, and, instead, leave them as two different things that are correlated with each other, then "all such correlations would have to be taken as 'brute' basic laws of the world—'brute' in the sense that they are not further explainable and must be taken to be among the fundamental laws of our total theory of the world... But such a theory of the world should strike us as intolerably complex and bloated—the very antithesis of simplicity and elegance we strive for in science."[27]

Because Kim seeks to operate within a scientific framework only, the oddness of such correlations is a problem for him. Thus, mental and physical correlations cannot be explained by neuroscience because they are brute. Not only that, but they are bizarre (they involve entities—irreducible mental entities—that sit oddly and are not at home with the rest of the physicalist ontology). And such correlations are the very antithesis of the physicalists' favorite epistemic value—simplicity.

But there is another problem—a damaging one for the claim that neuroscience provides explanations for the nature of consciousness and the self—that plagues advocates of neuroscientific adequacy. These correlations are not only odd and brute; they are also contingent. The obtaining of conscious states and the phenomenal properties that are essential to them is contingent relative to their associated "right physical situations." My claim is based on the widespread and, in my view, successful employment of a host of thought-experiments (e.g., the knowledge argument, the possibility of absent or inverted qualia, zombie worlds, and disembodied existence) that highlight the contingency in question. In regard to the employment of these thought-experiments to support the reality and contingency of qualia, Kim observes that the property dualist "case against qualia supervenience therefore is not conclusive, though it is quite substantial. Are there, then, considerations in favor of qualia supervenience? It would seem that the only positive considerations are broad metaphysical ones that might very well be accused of begging the question."[28]

Kim goes on to say that these broad metaphysical considerations amount to the question-begging assumption that physicalism *must* be true.

Second, necessitation allegedly justifies or at least attenuates the adoption of emergent properties in analyzing the obtaining of conscious states by (allegedly) "satisfying" the location problem: emergent properties are permitted as long as they are rooted in, generated by, dependent on, and determined by microphysical properties. As David Chalmers notes, without some sort of asymmetric dependency between a strongly emergent property and a physical basal property, "it is not easy to square the view with a contemporary worldview on which everything depends on what is going on in physics..."[29] Indeed. Unfortunately, the contingency of mental-physical correlations (or the very existence of mental states in the first place!) violates the physicalist worldview (as correctly depicted by Chalmers) and it makes these correlations recalcitrant to neuroscientific explanation.

7. Conclusion

Some philosophers have claimed that property and substance dualism is an obsolete relic of a bygone era. Nothing could be further from the truth. In the last few decades there has been a resurgence of property dualism and, more recently, of substance dualism. In this chapter, I have tried to show why this is happening. And I have also tried to show that philosophy, not science, is the main field of study when it comes to describing the ontological nature of consciousness and its possessor or unifier.

At the same time, I have acknowledged ways in which scientific empirical evidence can lend support for or against a viewpoint about consciousness, the soul, and related matters. Also, I have suggested that dualist views of consciousness, the soul, intelligence, and information (construed as non-physical) may fruitfully generate testable research programs.

Given these claims about philosophy, empirical sciences, and their relationship, we are in a position to understand the following claim made by a world-renowned Christian scholar. Speaking of issues in philosophy of mind, he noted that he would rather have the opinion of a scientifically informed philosopher than a philosophically uninformed scientist. Fortunately, these are not our only options. The real cash value of this chapter is the case presented that when addressing issues in philosophy of mind, it is best for philosophers and scientists to work together. This book is an attempt to do just that.[30,31]

NOTES

1. J. Gresham Machen, *The Christian View of Man* (New York: Macmillan, 1937), 159.
2. See John Cooper, *Body, Soul & Life Everlasting*, rev. ed. (Grand Rapids, MI: Eerdmans, 2000).
3. N. T. Wright, *The Resurrection of the Son of God* (Minneapolis: Fortress, 2003), 131–4, 190–206, 366–7, 424–6.
4. See Charles Taliaferro, "Emergentism and Consciousness," in *Soul, Body, and Survival*, ed. Kevin Corcoran (Ithaca, NY: Cornell University Press, 2001), 60.
5. Jaegwon Kim, "Lonely Souls: Causality and Substance Dualism," in *Soul, Body, and Survival*, 30.
6. Henry Wellman, *The Child's Theory of Mind* (Cambridge, MA: MIT Press, 1990), 50. Cf. Jesse Bering, Carlos Hernandez Blasi, and David Bjorklund, "The Development of 'Afterlife' Beliefs in Religiously and Secularly Schooled Children," *British Journal of Developmental Psychology* 23 (2005): 587–607; Paul Bloom, *Descartes's Baby: How the Science of Child Development Explains What Makes Us Human* (New York: Basic Books, 2004); Paul Bloom, "Religion Is Natural," *Developmental Science* 10 (2007): 147–51; Rochat and Striano, "Social-Cognitive Development in the First Year," *in Early Social Cognition: Understanding Others in the First Year of Life*, ed. P. Rochat (Mahwah, NJ: Lawrence Earlbaum Associates, 1999), 3–34; V. A. Kuhlmeier, P. Bloom, and K. Wynn, "Do 5-Month-Old Infants See Humans as Material Objects?" *Cognition* 94, no. 1 (2004): 95–103; Henry Wellman and A. K. Hickling, "The Minds 'I': Children's Conception of the Mind as an Active Agent," *Child Development* 65 (1994): 1564–1580; E. S. Spelke, A. T. Phillips and A. L. Woodward, "Infants' Knowledge of Object Motion and Human Action," in *Causal Cognition: A Multidisciplinary Debate*, eds. D. Sperber, D. Premack, and A.J. Premack (Oxford: Clarendon Press, 1995), 44–78; R. Saxe, R. Tzelnic, and S. Carey, "Five-Month-Old Infants Know Humans are Solid, Like Inanimate Objects," *Cognition* 101, no. 1 (2006): B1–B8; Jesse Bering and David Bjorklund, "The Natural Emergence of Reasoning about the Afterlife as a Developmental Regularity," *Developmental Psychology* 40, no. 2 (2004): 587–607. I am indebted to Brandon Rickabaugh for these sources.
7. Nancey Murphy, "Human Nature: Historical, Scientific and Religious Issues," in *Whatever Happened to the Soul?*, eds. Warren S. Brown, Nancey Murphy, and H. Newton Malony (Minneapolis, MN: Fortress Press, 1998), 18.
8. Murphy, "Human Nature," 17; cf. 13, 27. Also, see Nancey Murphy, "For Nonreductive Physicalism," in *The Blackwell Companion to Substance Dualism*, eds. Jonathan J. Loose, Angus J. L. Menuge, and J. P. Moreland (Oxford: Wiley-Blackwell: 2017), 317–327.
9. Nancey Murphy, *Bodies and Souls, or Spirited Bodies?* (Cambridge: Cambridge University Press, 2006), 56.
10. See Stewart Goetz and Charles Taliaferro, *A Brief History of the Soul* (Malden, MA: Wiley-Blackwell, 2011).
11. See J. P. Moreland, *The Soul: How We Know It's Real and Why It Matters* (Chicago: Moody Press, 2014), 74–116; J. P. Moreland, William Lane Craig, *Philosophical Foundations for a Christian Worldview*, 2nd ed. (Downers Grove, IL: InterVarsity Press, 2017), 210–250.
12. Stewart Goetz, "Modal Dualism: A Critique," in *Soul, Body and Survival*, ed. Kevin Corcoran (Ithaca, NY: Cornell University Press, 2001), 89.
13. For more on the unity of consciousness, the binding problem and split-brain phenomena, see Tim Bayne, "The Unity of Consciousness and the Split-Brain Syndrome," *The Journal of* Philosophy 105, no. 6 (2008): 277–300, and Tim Bayne and David Chalmers, "What is the Unity of Consciousness?" in *The Unity of* Consciousness, ed. Axel Cleeremans (Oxford: Oxford University Press, 2003), 23–58. For an empirical argument against physicalism that centers on some of these

considerations, see Eric LaRock, "An Empirical Case against Central State Materialism," *Philosophia Christi* 14, no. 2 (2012): 409–426.

14. William Hasker, *The Emergent Self* (Ithaca, NY: Cornell University Press, 1999), 122–46.
15. Cf. Keith Yandell, "A Defense of Dualism," *Faith and Philosophy* 12 (October 1995): 548–566, and Charles Taliaferro, "Animals, Brains, and Spirits," *Faith and Philosophy* 12 (October 1995): 567–581.
16. In normal life, I may be focusing on speaking kindly and be unaware that I am scowling. In extreme cases (multiple personalities and split brains), I may be fragmented in my functioning or incapable of consciously and simultaneously attending to all of my mental states, but the various personalities and mental states are still all mine.
17. For three reasons, quantum indeterminacy is irrelevant here: (1) The best interpretation of quantum indeterminacy may be epistemological and not ontological; (2) If quantum indeterminacy is real, events still have their chances fixed by antecedent conditions, and this is inconsistent with agent causation since on this view nothing fixes the chances of a free action; (3) Some scholars hold that determinism reigns or is at least closely approximated at the macro-level even if ontological indeterminism rules the micro-physical world.
18. Timothy O'Connor formerly argued that agent causal power could be an emergent property over a physical aggregate. See his *Persons and Causes* (NY: Oxford, 2000). Subsequently, O'Connor changed his view and opted for the idea that the agent is an emergent individual. See Timothy O'Connor and Jonathan D. Jacobs, "Emergent Individuals," *The Philosophical Quarterly* 53 (October 2003): 540–555. For a critique of O'Connor, see J. P. Moreland, *Consciousness and the Existence of God* (London: Routledge, 2008), chapter four.
19. For more on problems of material composition, see Michael Rea, ed., *Material Constitution: A Reader* (Lanham, MD: Rowman & Littlefield: 1996), and Christopher M. Brown, *Thomas Aquinas and the Ship of Theseus* (London: Continuum, 2005).
20. The view I am advancing is called mereological essentialism (from the Greek word *meros*, which means "part"). Mereological essentialism is the idea that an object's parts are essential to its identity such that it could not sustain its identity to itself if it had alternative parts. Animalists and constitutionalists deny mereological essentialism. For a brief exposition of these views, see Eric Olson, *What are We?: A Study in Personal Ontology* (Oxford: Oxford University Press, 2007), chapters two and three. In different ways, each view claims that, under certain circumstances, when parts come together to form a whole, as a primitive fact, the whole itself just is the sort of thing that can survive part alteration. In my view, this is just an assertion. The whole just is parts and various relations, and neither the parts nor the relations can sustain identity if alternatives are present. The whole is not a basic object—it is identical to its parts and relations.
21. Some claim that what unites all of one's various psychological stages into the life of one single individual is that the latter stages stand in an immanent causal relation to each other. But an immanent causal relation is one that holds between two states in the same thing. Thus, before a causal relation can be considered an immanent one, there must already be the same thing that has the two states. Because the immanent causal relation presupposes sameness of the thing in question, it cannot constitute what it is for the thing to be the same. Further, the immanent causal view confuses what it is that causes an object to endure over time with what it is for the object to remain the same.
22. Murphy, "Human Nature," 18.
23. Murphy, "Human Nature," 17.
24. Murphy, *Bodies and Souls, or Spirited Bodies?*, 56.
25. Ian Ravenscroft, "Why Reject Substance Dualism?," in *The Blackwell Companion to Substance Dualism*, 267.
26. Richard Swinburne, *Mind, Brain, and Free Will* (Oxford: Oxford University Press, 2013), 119.
27. Jaegwon Kim, *Philosophy of Mind*, 3rd ed. (Boulder, CO: Westview Press, 2011), 101.
28. Jaegwon Kim, *Philosophy of Mind*, 2nd ed. (Boulder, CO: Westview Press 2006), 233.
29. David J. Chalmers, "The Combination Problem for Panpsychism," in *Panpsychism: Contemporary Perspectives*, eds. Godehard Bruntrup and Ludwig Jaskolla (Oxford: Oxford University Press, 2017), 194.
30. I wish to thank three anonymous referees for their careful reading of, and excellent suggestions regarding, an earlier draft of this chapter.
31. For a rigorous, recent defense of substance dualism, see Brandon Rickabaugh and J. P. Moreland, *The Substance of Consciousness: A Comprehensive Defense of Contemporary Substance Dualism* (Oxford: Wiley-Blackwell, 2023).

4. Methodological Naturalism and the Mind

Robert A. Larmer

1. Introduction

The term "methodological naturalism" appears to have been first used by Edgar Brightman.[1] In 1983, Paul de Vries independently coined the term, advocating that, as a matter of method, scientists, whatever their metaphysical beliefs, should always posit a physical cause for any event taking place in the world.[2] Since then, proponents both secular and religious have enthusiastically embraced methodological naturalism as a necessary condition of scientific investigation, insisting that adopting such a method in no way commits one to any specific metaphysical position. Science, they contend, is naturalistic only on the level of its methodology, but is neutral with respect to metaphysics. Thus, given its presumed metaphysical neutrality, methodological naturalism provides a way in which science can be collegially pursued by those with differing metaphysical positions.

Science cannot, however, be so easily demarcated from metaphysics as advocates of methodological naturalism conjecture; what one thinks to be the nature of reality cannot be neatly separated from the methods one uses to investigate it.[3] Science takes as its purview of investigation all that occurs in the world. Insofar as methodological naturalism requires that all events in the world must be elucidated in terms of physical causes, that it is illegitimate ever to appeal to a non-physical cause in seeking an explanation, it precludes any possible recognition of non-physical causes by science, even if they are active.[4] If science is pursued in order to investigate the causes of events, then as an enterprise it should not mandate a methodology which prohibits, even in principle, the recognition of non-physical causes.[5]

Consequently, the adoption of methodological naturalism requires justification. This is especially true where the phenomena investigated strongly resist explanation in terms of physical causes. In such instances, espousing methodological naturalism functions not to further scientific investigation, but rather to impose an explanatory straitjacket that prevents doing justice to the data. What is known as the "hard problem of consciousness"—a term employed by those insisting on a physicalist account of the mind—is one such instance.[6]

My goal in this chapter is to demonstrate that adopting methodological naturalism as a prerequisite for engaging in philosophy of mind is unjustified. I shall argue for this conclusion in two ways. First, I shall very briefly elucidate just one of many reasons why any naturalist account of the mind faces enormous problems. My exposition can be brief, since other contributors to this volume will engage in that task in greater detail. Second, I shall examine proposed justifications of methodological naturalism and show that they fail. Once it is understood that investigation of the mind need not be shackled by insisting on a methodology that assumes the truth of physicalism, a more adequate understanding of the mind and its relation to the material world can be contemplated.

2. Naturalism and the Mind

THE OFT-REPEATED claim that methodological naturalism is metaphysically neutral is especially implausible in the context of investigating the mind. Any theory of the mind put forward by methodological naturalists will be fully naturalistic in the sense that mind will be viewed as totally dependent upon non-teleological physical processes. Theories of mind developed by methodological naturalists must, therefore, be indistinguishable from theories of mind developed by *metaphysical naturalists*. At least when it comes to investigating the mind, no principled distinction can be drawn between the practice of methodological naturalism and commitment to metaphysical naturalism.

Acceptance of methodological naturalism necessitates that one commits to treating the physical realm as causally closed. As a matter of method, one is committed to the claim that "if a physical event has a cause (occurring) at time *t*, it has a sufficient physical cause at *t*,"[7] which is to say that tracing the causal ancestry of a physical event never takes one outside the domain of the physical. This commitment requires that to the degree that one accepts the reality of irreducible mental phenomena, they must be understood as supervenient, which is to say, entirely dependent on bodily states. Understood as supervening on bodily states, "the mind cannot vary independently of the body."[8] There can be no free-floating mental states, since every mental state is anchored in a physical-neural base on which it is entirely dependent.[9] Thus, "in order to cause, or causally affect, a supervenient property, you must cause, or tinker with, its supervenience base."[10]

Treating the physical realm as causally closed requires, however, either accepting that mental states have no causal power *qua* their intentional content, e.g., a belief or purpose, or that the actions they are presumed to cause would have occurred even in their absence. Jaegwon Kim puts the argument thus:

> Suppose that a mental event, *m*, causes a physical event *p*. The closure principle says that there must also be a physical cause of *p*—an event, *p**, occurring at the same time as *m*, that is a sufficient cause of *p*. This puts us in a dilemma: Either we have to say that *m* = *p**—namely, identify the mental cause with the physical cause as a single event—or else we have to say that *p* has two distinct causes, *m* and *p**, that is, it is causally overdetermined. The first horn turns what was supposed to be a case of mental-to-physical causation into an instance of physical-to-physical causation, a result only a reductionist physicalist would welcome. Grasping the second horn of the dilemma would force us to admit that every case of mental-to-physical causation is a case of causal overdetermination, one in which a physical cause, *even if the mental cause had not occurred*, would have brought about the physical effect.[11] [emphasis added]

On the assumption that we are not prepared to accept that every case of mental to physical causation is an instance of overdetermination, either *m* or *p** must be disqualified as a cause of *p*. Methodological naturalism's acceptance of treating the physical realm as causally closed, necessitates, however, that *m* be disqualified as the cause of *p*.[12]

This commits the methodological naturalist to either denying the existence of mental states altogether or viewing them as exerting no causal influence on physical states. On such an understanding, my intention to provide a critique of methodological naturalism has absolutely nothing to do with the actions my fingers perform on the keyboard I use to type this sentence. Kim gives a very clear and definite statement of this view, writing that "qualia [i.e., the felt and experienced qualities of conscious states]... cause no effects in the physical domain... they can play no role in behavior production, and behaviors cannot be evidence for the presence or absence of qualia."[13]

Further, given a commitment to the causal closure of the physical, and acceptance that mental states supervene on brain states, it follows that mental-to-mental causation is impossible.[14] One must either accept a radical epiphenomenalism in which "mental events have no causal efficacy at all, no power to cause any event, mental or physical"[15] or reject epiphenomenalism altogether, opting for a reductionist materialism which attempts a functionalist account of the mental.[16]

Either alternative appears to undermine our reasoning capacities. Consider, for example, the following argument.

Premise One: a > b.

Premise Two: b > c.

Conclusion: a > c.

As we have just seen, on the epiphenomenalist view mental states can in no way influence one another. This means that in no way is the belief a > c the result of the beliefs that a > b, and that b > c. Abandoning epiphenomenalism in favor of simply identifying mental events as physical *tout court* yields a similar result. On the reductionist physicalist view, mental states will be viewed as identical with physical states and thus can be deemed to be in causal relation with one another. They will not, however, causally interact by virtue of their informational content, the result being that the informational content of the beliefs that a > b, and b > c has nothing to do with the belief that a > c. For the reductive materialist, "phenomenal consciousness can be regarded as an unintended side effect of the evolution of the neural systems in higher organisms... qualia as intrinsic qualities remain epiphenomenal."[17] Given such a view, our belief that a > c must be understood not to be the result of rational thought, but rather the product of physical causes whose operation is unrelated to the comprehension of logic.

That such problems result from methodological naturalism's commitment to treating the physical realm as causally closed is hard to deny. Equally hard to deny is that we typically view our actions as the result of rational thought. Such actions must be distinguished from what merely happens to us. Rationality requires agency. However, as John Bishop notes:

> Agent causal-relations do not belong to the ontology of the natural perspective. Naturalism does not essentially employ the concept of a causal relation whose first member is in the category of person or agent (or even, for that matter, in the broader category of continuant or "substance"). All natural causal relations have first members in the category of event or state of affairs... the problem is that the natural perspective *positively rejects the possibility that any natural event should be agent-caused.*[18]

Consequently, the cost of methodological naturalism's insistence that the physical realm must be viewed as physically closed is enormous, since nothing short of giving up all conception of persons as rational, purposive agents will suffice. Given the enormity of this price, the presumed justifications for accepting methodological naturalism as a prerequisite of scientific investigation must be examined. It is to this task that I now turn.

3. Proposed Justifications of Methodological Naturalism

3.1 Methodological Naturalism Reveals the Inconceivability of Non-natural Causes

PERHAPS THE poorest defense of methodological naturalism one comes across is the question-begging strategy of insisting that employing methodological naturalism reveals that any claimed need of a non-physical cause for a physical event must inevitably be self-refuting. Barbara Forrest attempts such an argument insisting that by definition the supernatural, i.e., the non-physical, is "unknowable by means of scientific inquiry,"[19] going on to insist that confirmation of any new cause "would only demonstrate that this newly verified aspect of reality had all along never been supernatural at all."[20]

Such a defense of methodological naturalism can safely be discarded. Not only is it question-begging in its circularity, it empties the terms "physical" and "non-physical" of all content. An implication of Forrest's claim is that even God, understood as the ontologically distinct creator *ex nihilo* of all other entities, would have to be conceived as a "physical entity."[21]

3.2 Questioning Methodological Naturalism Reveals Tainted Motives

Another obviously fallacious justification of methodological naturalism is the claim that it should be accepted on the grounds that those who criticize it have tainted motives. For example, criticisms of methodological naturalism made by proponents of intelligent design theory are frequently dismissed on the basis that those advancing the criticisms are "creationists."

Leaving aside the fact that most proponents of intelligent design theory are not "creationist" in the sense in which that term is usually understood, namely as referring to those who believe the earth is less than 10,000 years old, it is clear that, on pain of committing the *ad hominem* fallacy,[22] arguments are to be evaluated on their own merits, not simply dismissed by attacking the persons making them. Defending methodological naturalism requires more than ridiculing those who reject it.

3.3 Science in Principle Excludes Recognition of Non-physical Causes

One frequently meets the claim that methodological naturalism is justified inasmuch as science, by definition, cannot contemplate the existence of non-physical causes. Robert Pennock, for example, claims that methodological naturalism functions as "a scientific ground rule,"[23] that, as a matter of method science must not countenance any appeal to non-physical causes.[24] Pennock's view appears to be that methodological naturalism functions as a necessary, though not sufficient, condition of scientific investigation, serving to at least partially demarcate science from investigations which should be considered unscientific.

Such an argument for adopting methodological naturalism is inadequate on at least two counts. First, lacking reasons why science cannot legitimately consider whether evidence of non-physical causes exists, it amounts merely to an arbitrary stipulation. Simply asserting without argument that science prohibits ever recognizing the existence of non-physical causes is question-begging.

Second, it is commonly recognized by philosophers of science that demarcationist proposals are notorious for failing to provide either necessary or sufficient conditions to distinguish science from other disciplines. Larry Laudan, in his well-known article "The Demise of the Demarcation Problem," notes that determined searches for demarcation criteria have all failed.[25] Appeal to falsification as the hallmark of scientific theories fails by both including too much and excluding too much. As Laudan notes, "polywater dabblers, Rosicrucians, the-world-is-about-to-enders, primal screamers, water diviners, magicians, and astrologers all turn out to be scientific... just so long as they are prepared to indicate some observation, however improbable, which (if it came to pass) would cause them to change their minds."[26] String theory, on the other hand, would have to be dismissed as a non-scientific since it posits extremely small-length scales unfalsifiable by any foreseeable experiment.[27]

Nor do claims that scientific knowledge is unique in exhibiting growth, or that later scientific views embrace earlier ones, or that scientific claims are well-tested, succeed in clearly demarcating science from other rational modes of inquiry.[28] Growth in knowledge is hardly unique to science (we know more about the historical causes of World War II than we did formerly), not all later scientific views embrace earlier ones (it is a sobering thought that belief in Caucasian superiority was once deemed scientific), and much of scientific theory is not well-tested (it was not until recently that we found evidence that confirmed the existence of the Higgs Boson particle.) Del Ratzsch is thus correct in his observation that attempting to define science as necessarily employing methodological naturalism is "*prima facie* problematic for the simple reason that no one actually has a completely workable definition of science (nor even necessary and sufficient conditions), and that proposed definitions have been historically unstable."[29]

3.4 Non-physical Causation Implies a Chaotic Universe

Much discussion of methodological naturalism takes place in the context of questioning the scientific credentials of intelligent design theory. Advocates of accepting methodological naturalism as a prerequisite of scientific investigation frequently make the claim that taking seriously the possibility of non-physical agency implies a chaotic universe and thus methodological naturalism must be accepted. Pennock, for example, expresses concern that God, a non-physical agent, "may simply... zap anything into or out of existence... in any situation; any pattern (or lack of pattern) of data is compatible with the general hypothesis of a supernatural agent unconstrained by natural law."[30] Richard Lewontin expresses similar concern, writing that "we cannot live simultaneously in a world of natural causation and of miracles, for if one miracle can occur, there is no limit"[31] and "at every instant all physical regularities may be ruptured and a totally unforeseeable set of events may occur."[32]

There are at least two reasons for thinking this line of argument is mistaken. First, and most obviously, critics such as Pennock and Lewontin are guilty of a category mistake, inasmuch as they are arguing that adopting methodological naturalism somehow determines whether the universe is orderly. Whether the universe is in fact orderly is independent of the methodology we employ in investigating it. All that adopting methodological naturalism guarantees is that if non-physical causes of physical events exist, they can never be recognized. Science on such a view must be understood not as committed to pursuing the truth about reality, but rather as committed to pursuing the best physical explanation that can be formulated of phenomena. If immaterial minds exist, whether divine or non-divine, and act on the world, then the adoption of methodological naturalism guarantees that at best science will be incomplete, but more probably it will at important points be incorrect, since methodological naturalism will lead us to accept naturalistic explanations, no matter how implausible, over explanations in terms of rational agency.

Second, it hardly follows that immaterial agents must be conceived as necessarily irrational. One searches in vain for any theist who views himself or herself committed to the claim that God is liable at any moment to zap anything into or out of existence. What God freely wills will not be simply arbitrary or irrational, but in accordance with His rational nature. As Evan Fales, no friend of theism, notes, "it does not follow from the fact that God is a free agent that His purposes and behavior (including the occasional performance of a miracle) cannot be made intelligible or studied in systematic ways."[33]

Thus, for example, religious believers insist that if an event is to be recognized as a miracle then it must take place in a context in which it can reasonably be understood to have religious significance, which is to say it can reasonably be held to further what we take to be God's purposes, and that the regularity to which the event constitutes an exception is strongly confirmed and is known to apply to the same type of physical circumstances in which the event took place.[34]

Mutatis mutandis, it follows that recognizing human minds as immaterial causes in no way implies a chaotic universe. We routinely do find it possible to explain human actions in terms of agents' beliefs and purposes. Such teleological explanations do not of course imply determinism, but neither do they imply some sort of chaos that precludes either understanding why certain actions take place or being sure that they were not simply happenings but rather the product of agency.

3.5 Allowing for the Possibility of Non-physical Causation Is a "Science-stopper"

Advocates of methodological naturalism often claim that taking seriously the possibility of non-physical causes is a "science-stopper."[35] This charge is made on two fronts. At the psychological level it is charged that taking seriously non-physical causes will lead to scientists becoming lazy and liable to abandon the search for physical causes. At the conceptual level it is charged that explanations in terms of non-physical causes are not falsifiable and thus any acceptance of their operation puts a stop to further scientific investigation. To the degree that they are accepted they preclude the possibility of scientific investigation.

The claim that acceptance of the reality of non-physical causes is liable to make scientists intellectually lazy is not well-evidenced. Many giants of science have believed in the immateriality of the human mind, without this belief being detrimental to either the quantity or quality of their research. To take only one example, Robert Boyle (1627–1691), a pioneer of modern experimental scientific method and the founder of modern chemistry, accepted both miracles and the immateriality of the human mind, without this affecting the excellence of his scientific research.[36] It is fair to conclude that "neither science nor scientists may be vulnerable to the temptations of intellectual sloth as presumed" by advocates of methodological naturalism.[37]

The charge that because non-physical causes are not directly observable, they are not falsifiable, and thus any claim that they exist inevitably undermines the possibility of scientific progress, seems an example of special pleading. Atoms and sub-atomic particles are not directly observable, but that hardly justifies viewing claims of their existence as unscientific. So long as there is strong evidence that unobservable causes of a certain nature are required to explain observed phenomena, it is legitimate to accept their existence. If such evidence is lacking or better explained by different causes, then that goes a long way towards falsifying any claim to their existence and operation.

Further, as has already been noted, attempts to provide general criteria by which science may be demarcated from non-science are fraught with difficulty, so much so that philosophers of science have largely abandoned attempts at resolving the "demarcation problem." It will not do, therefore, to dismiss explanations in terms of non-physical causes as in principle unscientific and thus "science-stoppers." We routinely do accept explanations of certain physical events in terms of rational agents performing specific actions for reasons. Given the excellent reasons to think such agency is genuine and incompatible with the truth of physicalism, it will not do to insist that recognizing its existence is unscientific.

3.6 Methodological Naturalism Is Inductively Justified

Philosophically sophisticated advocates of methodological naturalism are increasingly inclined to defend it not as necessarily constitutive of doing science, but rather as a well-evidenced inductive generalization that should be accepted based on its fruitfulness. While accepting that methodological naturalism is not a definitional aspect of science, Christian theists Patrick McDonald and Nivaldo Tro recommend its acceptance on the basis that it is "an empirically validated methodology and as such should be honored unless and until a better framework come to the fore."[38] Likewise, metaphysical naturalists Maarten Boudry, Stefaan Blancke, and Johan Braeckman defend methodological naturalism on the basis that it functions "as a provisory and empirically grounded attitude of scientists, which is justified in virtue of the consistent success of naturalistic explanation and the lack of success of supernatural explanations in the history of science."[39] Elliot Sober expresses this line of argument well when he writes "naturalistic science has been a success… the modest defense I would offer of methodological naturalism is simply this: *if it isn't broken, don't fix it.*"[40]

Of all the defenses of methodological naturalism on offer, this is the least question-begging. Nevertheless, there seems little reason to see methodological naturalism's adoption as justified, especially given that it precludes recognizing non-physical causes even if they are operative in the world. Explanations, whether they be in terms of physical or non-physical causes, should be accepted based on fundamental explanatory virtues such as simplicity, coherence, scope, etc.

Given the observance of these virtues, there is no need to add on a blanket requirement that explanation must necessarily be in terms of physical causes. If physical causes do provide the best explanation of an event, then they will be invoked. If it turns out that a non-physical cause was posited, but further investigation provides a better explanation in terms of physical causes, then that explanation will have shown its superiority. There is, therefore, no need to accept methodological naturalism as some further requirement of scientific investigation independent of the practice of fundamental explanatory virtues.

It bears emphasis that, at the level of actual practice, methodological naturalism functions not as a provisional inductive generalization that can be abandoned in the light of evidence, but rather as an absolute embargo on ever taking seriously the possibility of non-physical causation. Consider, by way of example, Lewontin's forthright admission that "it is not that the methods and institutions of science somehow compel us to accept a material explanation of the phenomenal world, but on the contrary, that we are forced by our a priori adherence of material causes to create an apparatus of investigation and a set of concepts that produce material explanations, no matter how counterintuitive... moreover that materialism is absolute."[41]

Typically, invocations of methodological naturalism are employed polemically as discrediting devices, as "*machines de guerre*," whereby an opponent's position can be dismissed as "unscientific" and thus unworthy of being taken seriously. Laudan notes that "many of those most closely associated with the demarcation issue have... hidden (and sometimes not so hidden) agendas of various sorts"[42] and goes on to argue that "if we... stand... on the side of reason, we ought to drop terms like 'pseudo-science' and 'unscientific' from our vocabulary; they are just hollow phrases which do only emotive work for us."[43]

Given the failure of the demarcation quest to provide some sort of Maginot Line between "science" and "non-science," the real issue regarding the acceptability of an explanation should not be whether it is cast in terms of physical, non-agential causes but whether it is in accord with fundamental explanatory virtues.[44] As John Earman notes, "it does not much matter what label one sticks on a particular assertion or an enterprise; the interesting questions are whether the assertion merits belief and whether the enterprise is conducive to producing well-founded belief."[45]

A second reason to doubt the acceptance of methodological naturalism as inductively justified lies in the distinction that must be drawn between what may be termed "nomological" science and "historical" science. Paul Draper explains this distinction:

> Scientists engaged in nomological science formulate laws, models, and other interesting if-then generalizations, often testing them by experiment and prediction, and making inductive generalizations based on observable data. In historical science, on the other hand, not all causal explanations fit the covering law model and many hypotheses about the past cannot be falsified and cannot be tested by prediction or experiment. Instead, they are judged on their simplicity, their fit with general background knowledge about the world, and their ability to explain specific known facts. What all this shows is that methodological naturalism cannot be adequately defended by describing something called *the* scientific method then arguing that it cannot be applied to the supernatural. For more likely than not, the method described will be characteristic of nomological science, while appeals to the supernatural would naturally be used to answer historical questions.[46]

Once this distinction is recognized, it becomes clear that the success that nomological science has had in finding physical causes of phenomena does not necessarily provide a good inductive argument for adopting methodological naturalism in historical sciences. The fact that turtles are slow-moving and easy to catch does not provide warrant for thinking that cheetahs are slow-moving and easy to catch. Similarly, the fact that explanations in terms of physical causes have enjoyed success in nomological science does not automatically warrant the assumption that explanations in terms of physical causes will enjoy the same success in historical sciences.

4. Agency and the Historical Sciences

Explanations in terms of appeal to rational agency are typically found in historical as opposed to nomological science. Yet it is precisely in areas of historical science that any attempted justification of methodological naturalism in terms of a firmly grounded induction is most tenuous. Indeed, inductive arguments run different times in the opposite direction. In terms of explaining the origin of the universe and its apparent fine tuning, the origin of life, and the existence of consciousness, nomological science has conspicuously failed.[47]

A little over a century ago, it was possible to view the universe as eternal and the structure of living cells as relatively simple. This is no longer so; the more we know, the harder it is to avoid positing agency. In cosmology scientists have had to come to terms with our fine-tuned universe coming into existence without any physical cause.[48] In biology, although nomological science is adequate to deal with structures that manifest repetitive order (e.g., crystals), or simple complexity (e.g., mixtures of random polymers), it fails to provide any plausible explanation of the specified complexity found in DNA, RNA, and proteins.[49] The structures characterizing living things recalcitrantly resist explanations in terms of physical causes, so much so that James Tour is prepared to argue that "those who think scientists understand how prebiotic chemical mechanisms produced the first life are wholly misinformed."[50]

At the very least, it seems natural, no pun intended, to consider whether the origin of the universe and the origin of life are best explained in terms of rational agency. We can raise this question *precisely* because we daily experience ourselves as rational agents. We are conscious beings routinely familiar with the fact that our purposes and desires affect the world. The problem for the naturalist, whether methodological or ontological in outlook, is that attempted accounts of the qualitative aspects of consciousness and of their causal powers inevitably fail.

As is typical, Kim, in the introduction to his *Philosophy of Mind,* accepts that physical indiscernibility entails psychological indiscernibility, that there can be no mental difference without a physical difference, and that creatures could not be psychologically different and yet physically identical.[51] Equally typically, over three hundred pages later, he concludes that a naturalist account of the mind cannot account for the intrinsic qualities of mental states or view them as having causal power *qua* their intentional content.[52]

It is this inability of naturalism that leads Thomas Nagel to argue that "consciousness is the most conspicuous obstacle to a comprehensive naturalism that relies only on the resources of physical science."[53] He goes on to conclude that short of denying the existence of consciousness altogether—surely a self-refuting position despite the claims of eliminative materialists—we must recognize that "conscious subjects and their mental lives are inescapable components of reality not describable by the physical sciences."[54]

Given the lack of convincing justification for methodological naturalism, and given that its acceptance inevitably leads to naturalist accounts of the mind unable to account for the intrinsic qualities of mental states and their causal powers *qua* intentional states, the claim that acceptance of methodological naturalism is a prerequisite of scientific investigation of the mind cannot be sustained. As Kim notes,

> Giving up mental causation amounts to giving up our conception of ourselves as agents and cognizers. Is it

> even *possible* for us to give up the idea that we are agents who decide and act... Whether or not a stance on the mind-body problem is acceptable depends importantly, if not solely, on how successful it is in giving an account of mental causation.[55]

The price of accepting methodological naturalism as a prerequisite for investigating the mind is simply too high, inasmuch as it requires giving up our conception of ourselves as rational agents. Giving up this understanding is hardly conducive to trusting our ability to investigate the natural world. It is, therefore, unsurprising that critics such as Alvin Plantinga,[56] William Hasker,[57] Angus Menuge,[58] and Victor Reppert,[59] to name only a few, have raised concerns that naturalist accounts of the mind are subject to the charge of self-refutation.

5. If Not Methodological Naturalism, Then What?

ACCEPTANCE OF methodological naturalism skews from the very outset of any investigation what will count as evidence and what form explanations can take, inasmuch as it requires that evidence be interpreted, and theories constructed, in such a manner as to conform to a naturalist framework. Far from being free to follow the evidence where it leads, scientists accepting methodological naturalism are, if they are consistent, forced to conceive of the world as entirely uninfluenced by agential causes.[60] Ratzsch is correct to note that "if part of reality lies beyond the natural realm, then science cannot get at the truth without abandoning the naturalism it presently follows as a methodological rule of thumb."[61]

Earlier, we noted the failure of attempts to solve the demarcation problem. This failure is important inasmuch as it signals that what distinguishes science from non-science is not one prescribed method of analysis, but rather rational disciplined investigation of publicly available evidence in accordance with fundamental explanatory virtues. I suspect that one reason that methodological naturalism has been so readily adopted is that explanations in terms of physical causes work well in nomological science. What must be stressed, however, is that the acceptance of such explanations should be based on how well they explain, not because it has been prescribed in advance that the only acceptable type of explanation is naturalistic. For example, it is Occam's Razor, not methodological naturalism, which persuades me that an explanation of ocean tides in terms of the moon's gravitational force should be accepted. One does not need some prior commitment to methodological naturalism to find such an explanation convincing.

The issue, then, is not whether explanations in terms of physical causes work well in certain areas of science, but whether science requires an absolute prohibition on explanations in terms of agency. Insisting on the adoption of methodological naturalism guarantees that this issue never gets addressed. It guarantees that the existence of nonphysical causes of events in the world can never be recognized, even if they operate.

The answer to the question "If not methodological naturalism, then what?" is that the various areas of scientific investigation yield credible results to the degree that they combine disciplined investigation of publicly available evidence with loyalty to fundamental explanatory virtues. Adding on the requirement of methodological naturalism, that explanations must always be in terms of physical and not agential causes, only hinders the ability of scientists to go where the evidence may lead.

6. Should A Metaphysical Naturalist Embrace Methodological Naturalism?

I HAVE been making the case that far from being a necessary condition of scientific investigation, methodological naturalism functions as a hindrance to scientific investigation inasmuch as it prohibits ever considering explanations in terms of agential causes. Might it be argued that metaphysical naturalists are justified in adopting methodological naturalism in conducting scientific investigation? The answer, I argue, is no.

Almost invariably, metaphysical naturalists do not argue that the existence of non-physical entities is logically impossible. Rather, they argue that scientific investigation reveals that there is no need to posit non-physical causes to explain what happens in the world and thus Occam's Razor shaves away any reason to believe in non-physical causes.

Yet the existence of physical events which are best explained as produced by a non-physical cause would count against any attempt to justify metaphysical naturalism by appealing to Occam's Razor. On pain of begging the question, the metaphysical naturalist should not adopt a methodology which prohibits ever positing a non-physical explanation, since such a methodology guarantees that the requirements of Occam's Razor cannot be met. As Stephen Dilley notes:

> When joined to MN [methodological naturalism], PN [philosophical, i.e., metaphysical, naturalism] cannot be disconfirmed within science... Since MN requires that all scientific evidence be given a natural explanation, evidence can never disconfirm PN, *no matter what the evidence on hand actually is*. This is not to say that scientific evidence fails to disconfirm PN as a matter of fact, but that it cannot as a matter of principle. Empirical evidence cannot so much as murmur against PN, and no rival hypotheses, however modest, can cast a shadow on its scientific nature.[62] [emphasis in original]

Consider the following imaginary exchange between a metaphysical naturalist and a non-naturalist:

> Non-Nat: Why do you as a metaphysical naturalist adopt methodological naturalism?
>
> Meta Nat: I am a metaphysical naturalist because I do not believe that one should multiply entities needlessly. Science has had lots of success in explaining the world in terms of physical causes and there is good reason to think that success will continue. Given the probable truth of metaphysical naturalism it makes sense for me to be a methodological naturalist.
>
> Non-Nat: I disagree that there is no good evidence for non-physical causes. I propose to show you that there is evidence for such causes.
>
> Meta Nat: Such evidence cannot exist.
>
> Non-Nat: Why not?
>
> Meta Nat: Scientific investigation requires the adoption of methodological naturalism. It is illegitimate for a scientist to posit a non-physical cause for a physical event.
>
> Non-Nat: You said that your acceptance of methodological naturalism follows from the probable truth of metaphysical naturalism.
>
> Meta Nat: Yes, if you are a metaphysical naturalist, acceptance of methodological naturalism seems to follow.

Non-Nat: Remind me again why you think metaphysical naturalism is probably true.

Meta Nat: I accept Occam's Razor. If there is no evidence that non-physical entities exist, then one should not believe in them. Science demonstrates there is no need to posit immaterial entities to explain the world.

Non-Nat: Would accepting methodological naturalism ever permit one to posit a non-physical cause of an event?

Met Nat: No, otherwise it would not be methodological naturalism that one accepts.

Non-Nat: Let me make sure I am getting your position right. Your acceptance of metaphysical naturalism is based on there being no evidence for non-physical causes?

Met Nat: Yes, I think I have been clear about that.

Non-Nat: And you base your endorsement of methodological naturalism on your acceptance of metaphysical naturalism?

Met Nat: Yes, you seem to keep circling back to questions I have already answered. At what are you getting?

Non-Nat: Ironic that you mention circles. You accept metaphysical naturalism on the basis that there exists no evidence that non-physical causes exist, yet adopt a methodology which rules out the possibility of ever recognizing their existence. You are using your metaphysic to justify your acceptance of methodological naturalism, but your acceptance of methodological naturalism guarantees that even if evidence for the existence of non-physical causes exists it can never be recognized as such.

Met Nat: You may have a point, but I am still confident that we never need to look outside the physical realm for explanations.

Non-Nat: Perhaps, but it will not do to adopt a methodological requirement that rules out any possibility that your position can be challenged by contrary evidence.

7. Nomological Science and Methodological Naturalism

No one questions that explanations in terms of physical causes are routinely found for events best investigated by means of nomological science. This has led some critics of methodological naturalism, e.g. Dilley, to suggest that adoption of methodological naturalism is appropriate in the context of nomological science, but inappropriate in the context of historical science, such as the origins of the universe and its fine tuning, the origin of life, and the origin and operation of consciousness.[63] At first glance, such a proposal seems eminently sensible, inasmuch as it recognizes both the success of nomological science in providing physical explanations of lawlike events and the distinction that must be drawn between nomological science and historical science. Nevertheless, even so modest an employment of methodological naturalism has little to recommend it.[64]

This is so for two reasons, one theoretical and one practical. First, at the theoretical level, the explanations in terms of physical causes found in nomological science do not require one to employ methodological naturalism in order to find and accept them. As has already been argued, such explanations are the result of disciplined investigation of public evidence combined with loyalty to fundamental explanatory virtues. If physical causes do indeed provide the best explanations of regularly occurring phenomena, then they will be accepted without needing to add a requirement that the only legitimate form of explanation is in terms of such causes.

Second, at the practical level, the adoption of methodological naturalism in nomological science invariably bleeds over into claims of the illegitimacy of exploring explanations in terms of agency in historical science. In practice, methodological naturalism is only explicitly invoked as a rhetorical weapon whereby explanations in terms of agency can be dismissed in advance as "unscientific" and hence deemed less than intellectually respectable. Far from advancing scientific investigation, methodological naturalism, like so many of its demarcationist predecessors, typically serves as means to furthering an ideological end.

8. Conclusion

I conclude that adopting methodological naturalism as a prerequisite for engaging in scientific study of the mind is unjustified. This is so for two reasons. First, the adoption of methodological naturalism guarantees that any account of the mind and its operation will be fully naturalist. Naturalist accounts of the mind, however, require that mental states be understood as causally inert, as epiphenomena existing "in total causal isolation from the rest of the world, even from other mental events."[65] Not only does such a conception of the mind fly in the face of our daily experience of ourselves as rational purposive agents; it removes any warrant one might have for taking naturalism, whether it be metaphysical or methodological, seriously.

Second, the adoption of methodological naturalism is not in any way a requirement of scientific investigation. At best, it is a superfluous add-on to nomological science. Explanations in terms of physical causes characterize nomological science, but the acceptance of such explanations is presumably based on how well they work, not on a mandated necessity that all explanations be in terms of physical causes. At worst, methodological naturalism constitutes an explanatory straitjacket that forbids ever explaining a physical event in terms of rational agency, guaranteeing that even if such agency exists and is operative in the world it can never be scientifically recognized.

Notes

1. Edgar Brightman, "An Empirical Approach to God," *The Philosophical Review* 46, no. 2 (March 1937), 157.
2. Paul de Vries first introduced the term at a conference in 1983, later publishing the article "Naturalism in the Natural Sciences: A Christian Perspective," in *Christian Scholars' Review* 15, no. 4 (1986), 388–396.
3. E. A. Burtt, commenting on the claim that methodology need have no links to metaphysics, writes: "[T]here is no escape from metaphysics, that is, from the final implications of any proposition or set of propositions. The only way to avoid becoming a metaphysician is to say nothing... If you cannot avoid metaphysics, what kind of metaphysics are you likely to cherish when you sturdily suppose yourself to be free from the abomination. Of course, it goes without saying that in this case your metaphysics will be held uncritically because it is unconscious; moreover, it will be passed on to others far more readily than your other notions, inasmuch as it will be propagated by insinuation rather than by direct argument... The history of mind reveals pretty clearly that the thinker who decries metaphysics will actually hold metaphysical notions of three main types. For one thing, he will share the ideas of his age on ultimate questions, so far as such ideas do not run counter to his interests or awaken his criticism... In the second place, if he be a man engaged in any important inquiry, he must have a method, and *he will be under a strong and constant temptation to make a metaphysics out of his method,*

that is, to suppose the universe ultimately of such a sort that his method must be appropriate and successful... Finally, since human nature demands metaphysics for its full intellectual satisfaction, no great mind can wholly avoid playing with ultimate questions... But inasmuch, as the positivist mind has failed to school itself in careful metaphysical thinking, its ventures at such points will be apt to appear pitiful, inadequate, or even fantastic." See Burtt, *The Metaphysical Foundations of Modern Physical Science*, 2nd edition, revised (London: Routledge and Kegan Paul, 1932), 224–226. [emphasis added]

4. See, for example, Maarten Boudry, Stefaan Blancke, and Johan Braeckman, "Grist to the Mill of Anti-evolutionism: The Failed Strategy of Ruling the Supernatural Out of Science by Philosophical Fiat," *Science and Education* 21, no. 8 (Aug. 2012), 1151–1165. Erkki Vesa Rope Kojonen aptly dubs this the "truth seeking objection" to blanket methodological naturalism; see Kojonen, "Methodological Naturalism and the Truth Seeking Objection," *International Journal for the Philosophy of Religion* 81 (2017): 335–355.
5. For example, acceptance of methodological naturalism would require that religious experiences in which a subject claims to have encountered God directly be given an alternative explanation in terms of natural causes.
6. David Chalmers's "Facing up to the Problem of Consciousness," *Journal of Consciousness Studies* 2, no. 3 (1995): 200–219, appears to be the origin of this phrase.
7. Jaegwon Kim, *Philosophy of Mind*, 3rd ed. (New York: Westview, 2011), 214.
8. William Hasker, *The Emergent Self* (Ithaca, NY: Cornell University Press, 2001), 59.
9. Kim, *Philosophy of Mind*, 218.
10. Kim, *Philosophy of Mind*, 214.
11. Kim, *Philosophy of Mind*, 215.
12. Kim, *Philosophy of Mind*, 217.
13. Kim, *Philosophy of Mind*, 322.
14. Kim, *Philosophy of Mind*, 219.
15. Kim, *Philosophy of Mind*, 219.
16. Kim, *Philosophy of Mind*, 219.
17. Kim, *Philosophy of Mind*, 322.
18. John Bishop, *Natural Agency* (Cambridge: Cambridge University Press, 1989), 40.
19. Barbara Forrest, "Methodological Naturalism and Philosophical Naturalism: Clarifying the Connection," *Philo* 3, no. 2 (2000): 7–29.
20. Forrest, "Methodological Naturalism and Philosophical Naturalism," 25.
21. On drawing the distinction between material and non-material agential causes, see Hasker, *The Emergent Self*, 61–64.
22. This particular form of the *ad hominem* fallacy is sometimes known as the genetic fallacy.
23. Robert T. Pennock. "Can't Philosophers Tell the Difference between Science and Religion?: Demarcation Revisited," *Synthese* 178, no. 2 (January 2011), 184.
24. Pennock, "Can't Philosophers Tell the Difference?," 185.
25. Larry Laudan, "The Demise of the Demarcation Problem," in *Physics, Philosophy and Psychoanalysis*, eds. R. S. Cohen and L. Laudan (Dordrecht: D. Reidel, 1983), 111–112.
26. Laudan, "Demise of the Demarcation Problem," 121.
27. Lars-Gőan Johansson and Keizo Matsubara, "String Theory and General Methodology: A Mutual Evaluation," *Studies in History and Philosophy of Science Part B: Studies in History and Philosophy of Modern Physics* 42, no.3 (August 2011), 199.
28. Laudan, "The Demise of the Demarcation Problem," 122–23.
29. Del Ratzsch, "Natural Theology, Methodological Naturalism, and 'Turtles All the Way Down,'" *Faith and Philosophy* 21, no. 4 (Oct. 2004), 441.
30. Robert T. Pennock. "Naturalism, Evidence and Creationism: The Case of Phillip Johnson," in *Intelligent Design Creationism and Its Critics: Philosophical, Theological, and Scientific Perspectives*, ed. Robert T. Pennock (Cambridge: MIT Press, 2001), 89.
31. Richard Lewontin, "Introduction," in *Scientists Confront Creationism*, ed. Laurie Godfrey (New York: Norton, 1983), xxvi.
32. Lewontin, "Introduction," xxvi.
33. Evan Fales, *Divine Intervention: Metaphysical and Epistemological Puzzles* (New York: Routledge and Kegan Paul, 2010), 5.
34. Robert Larmer, *The Legitimacy of Miracle* (Lanham, MD: Lexington, 2014), 88.
35. Pennock, "Naturalism, Evidence, and Creationism," 90.
36. J. J. MacIntosh and Peter Anstey, "Robert Boyle," *Stanford Encyclopedia of Philosophy*, Winter 2018, https://plato.stanford.edu/archives/win2018/entries/boyle/.
37. Ratzsch, "Natural Theology, Methodological Naturalism, and 'Turtles All the Way Down,'" 441.

38. Patrick McDonald and Nivaldo Tro, "In Defense of Methodological Naturalism," *Christian Scholars' Review* 38, no. 2 (Winter 2009), 203.
39. Maarten Boudry, Stefaan Blancke, and Johan Braeckman, "How Not to Attack Intelligent Design Creationism: Philosophical Misconceptions About Methodological Naturalism," *Foundations of Science* 15, (2010), 227.
40. Elliot Sober, "Why Methodological Naturalism?," in *Biological Evolution: Facts and Theories: A Critical Appraisal 150 Years after The Origin of Species*, eds. G. Aulette, M. LeClerc, and R. Martinez (Rome: Gregorian Biblical Press, 2011), 375.
41. Richard Lewontin, "Billions and Billions of Demons," *New York Review of Books*, January 9, 1997, 28.
42. Laudan, "Demise of the Demarcation Problem," 119.
43. Laudan, "Demise of the Demarcation Problem," 125.
44. The failure of the demarcation quest to provide any kind of litmus test between "science" and "non-science" is widely acknowledged in the literature. For an attempt to defend the possibility of such a test see *Philosophy of Pseudoscience: Reconsidering the Demarcation Problem*, eds. Massimo Pigliucci and Maarten Boudry (Chicago: University of Chicago Press, 2013).
45. John Earman, *Hume's Abject Failure* (Oxford: Oxford University Press, 2000), 3.
46. Paul Draper, "God, Science, and Naturalism," in *The Oxford Handbook of Philosophy of Religion*, ed. William Wainwright (Oxford: Oxford University Press, 2005), 290.
47. Stephen Dilley, "Philosophical Naturalism and Methodological Naturalism," *Philosophia Christi* 12, no. 1 (2010), 133–134.
48. See, for example, Robert J. Spitzer, *New Proofs for the Existence of God* (Grand Rapids, MI: Eerdmans, 2010), 13–74.
49. Thomas Nagel notes that "no viable account, even a purely speculative one, seems to be available of how a system as staggeringly functionally complex and information rich as a self-reproducing cell, controlled by DNA, RNA, or some predecessor, could have arisen by chemical evolution alone from a dead environment," and that "recognition of the problem is not limited to the defenders of intelligent design." See Nagel, *Mind and Cosmos* (New York: Oxford University Press, 2012), 123.
50. James Tour, "Are Present Proposals on Chemical Evolutionary Mechanisms Accurately Pointing toward First Life?," in *Theistic Evolution: A Scientific, Philosophical, and Theological Critique*, eds. J. P. Moreland et al. (Wheaton, IL: Crossway, 2017), 191.
51. Kim, *Philosophy of Mind*, 10.
52. Jaegwon Kim argues for a functionalist account of intentional-cognitive states. He admits, however, that "no one has yet produced a complete functional definition or analysis of belief and… none is in sight." See his above-cited *Philosophy of Mind*, 329.
53. Nagel, *Mind and Cosmos*, 35.
54. Nagel, *Mind and Cosmos*, 41. Nagel is reluctant to give up naturalism, since he thinks the alternative to naturalism is some form of theism. Nevertheless, he admits that "an understanding of the universe as basically prone to generate life and mind will probably require a much more radical departure from the familiar forms of naturalism than I am at present able to conceive." See *Mind and Cosmos*, 127.
55. Kim, *Philosophy of Mind*, 202.
56. Alvin Plantinga, *Warrant and Proper Function* (New York: Oxford University Press, 1993).
57. Hasker, *The Emergent Self*.
58. Angus Menuge, *Agents Under Fire* (Lanham, MD: Rowman and Littlefield, 2004).
59. Victor Reppert, *C. S. Lewis's Dangerous Idea* (Downers Grove, IL: IVP, 2003).
60. Whether such consistency is in fact possible is quite a different matter. One is reminded of the remark widely attributed to Haldane, namely that "teleology is like a mistress to a biologist: he cannot live without her but he's unwilling to be seen with her in public." Sciences such as cryptology, forensics, anthropology, archaeology, etc. routinely offer explanations in terms of agency, without making any effort to show that such explanations are reducible to more fundamental explanations in terms of non-teleological physical causes.
61. Del Ratzsch, *Science and Its Limits: The Natural Sciences in Christian Perspective* (Downers Grove, IL: Inter-Varsity Press, 2000), 105.
62. Dilley, "Philosophical Naturalism and Methodological Naturalism," 129.
63. Dilley, "Philosophical Naturalism and Methodological Naturalism," 120.
64. Hugh Gauch, for example, notes that "contemporary scientific practice is far from a consistent and convincing implementation of methodological naturalism." See Hugh G. Gauch Jr., *Scientific Method in Brief* (Cambridge: Cambridge University Press, 2012), 99.
65. Kim, *Philosophy of Mind*, 201.

Unit 2: Philosophy of Mind: Dualism, Idealism, and Physicalism

5. Substance Dualism

Stewart Goetz and Charles Taliaferro

1. Introduction

Belief in the existence of the soul as an entity (a "substance" in older philosophical language) that is distinct from its physical or material (we use the terms interchangeably) body is usually referred to as "substance dualism," or simply "dualism." Most people who have thought about this belief maintain that it is commonsensical in nature and not based on any philosophical argument. Even those who deny the existence of the soul concede that the vast majority of us initially believe in its existence. For example, the experimental cognitive scientist Jesse Bering writes that human beings are believers in dualism,[1] and the psychologist Nicholas Humphrey insists that there is a human inclination to believe in dualism. Toward the end of his book *Soul Dust*, Humphrey mentions other scholars who also acknowledge this ordinary belief in dualism:

> Development psychologist Paul Bloom aptly describes human beings as "natural-born dualists." Anthropologist Alfred Gell writes: "It seems that ordinary human beings are 'natural dualists,' inclined more or less from day one, to believe in some kind of 'ghost in the machine.'... Neuropsychologist Paul Broks writes: "The separateness of body and mind is a primordial intuition.... Human beings are natural born soul makers, adept at extracting unobservable minds from the behaviour of observable bodies, including their own."[2]

One must always be aware of the distinction between an ordinary belief and a philosophical treatment or development of it. While most of us as ordinary persons initially believe in the existence of the soul, some individuals go further and philosophize about the soul's nature and its relationship with its physical body. The failure to keep in mind the distinction between an ordinary belief and a philosophical treatment of it not infrequently leads many theologians and biblical scholars to assert wrongly that the concept of the soul is a Greek idea and to argue that the Christian church mistakenly appropriated this Greek concept in its articulation of a Christian anthropology. But the concept of the soul is *not* a Greek idea. While Greeks like Plato philosophized about the nature of the soul, the idea itself is found everywhere at every time. Hence, contemporary theologians and biblical scholars who reject the idea of the soul (because it is supposedly Greek and unbiblical) in favor of a monistic Hebrew view of the self or person confuse the initial, commonsense belief in the soul with philosophical reflection about the soul's nature and its relationship with its material body.[3]

In what follows, we first present an extremely condensed overview of the history of thought about the soul as a substance in its own right, drawn from our book *A Brief History of the Soul*.[4] We then introduce the idea of property dualism as it is found in Aristotle's thought about the soul, before turning our attention to the so-called problem of causal interaction and the question whether property dualism in its contemporary non-Aristotelian, materialist form is any more successful at addressing this supposed difficulty than substance dualism. We suggest that any problem raised by causal interaction for substance dualism is also a problem for property dualism. Furthermore, we argue that responses in defense of libertarian free will to experiments of Benjamin Libet in and of themselves provide no support for the truth of substance dualism as opposed to property dualism or vice versa. Finally, we conclude with some brief thoughts about science and how evidence in support of its explanatory success is also evidence for the explanatory relevance of purposeful explanations and the occurrence of mental-to-physical causal explanations involving the soul.

2. Brief Historical Survey: Plato, Augustine, and Descartes

2.1 Plato

From what we know about philosophy in the ancient world, Plato (428/7–348/7 BCE) presents one of the most developed philosophical accounts of the soul and its nature. Plato (we assume that Socrates puts forth Plato's views) seems to think that a person is his soul. For example, in the *Phaedo* Socrates gives instructions to friends to assure another person, Crito, "that when I am dead I shall not stay, but depart and be gone."[5] As the self or "I" which survives death, the soul is essentially alive and can never perish.[6] Plato explains the soul's indestructibility by the fact that it is indissoluble.[7] It is indissoluble because it is without parts and, thus, non-composite in nature.[8] Yet, Plato also suggests that the soul has at least three parts: the appetitive, which is non-rational and experiences pleasure and pain,[9] the rational,[10] which beholds the Forms (universals) and ought to rule over the other two parts; and that which is spirited in nature[11] and has the function of supporting the rational part when it is in conflict with the appetitive. Perhaps Plato's contrasting assertions, that the soul is non-composite and that it has parts, can be reconciled by assuming he believed the parts of the soul are not separable *substances* or *entities* in their own right that make up or compose a soul, but rather are numerically distinct *powers* or *capacities* (dispositional *properties*). In other words, perhaps Plato believed the soul is simple as a substance, while complex in terms of its properties. This distinction can be illustrated by an example from the *Republic*:

> But, I [Socrates] said, I once heard a story which I believe, that Leontius the son of Aglaion, on his way from the Piraeus under the outer side of the northern wall, becoming aware of dead bodies that lay at the place of public execution at the same time felt a desire to see them and a repugnance and aversion…[12]

The philosopher David Armstrong uses this story about conflicting psychological properties or states to support the position that the self (soul) has substantive and (in principle) separable parts.[13] But the story does not provide support for this position. If the psychological properties (the desire and repugnance) were substantive and separable parts, then the loss of one or both would entail a corresponding substantial diminishment in the size of the self. To see that this is not the case, one can imagine that Socrates loses one or both psychological properties (it is not in the least an uncommon occurrence to lose a desire for this or that). Would Socrates experience a substantial loss of parts of himself? Not in the least. All of him would remain. He would undergo a change in his desires, which is a psychological change, but all of him would remain.

Our assumption that Plato believed in the distinction between simplicity as a substance and complexity in terms of properties is supported by the fact that he was aware of what is today regularly referred to as the unity of consciousness. In a discussion of knowledge and perception in the *Theaetetus*, Socrates acknowledges that "it would surely be strange that there should be a number of senses ensconced inside us, like the warriors in the Trojan horse, and all these things should not converge and meet in some single nature—a mind, or whatever it is to be called—*with* which we perceive all the objects of perception *through* the senses as instruments."[14]

Plato is aware that an account of thought and perception that likens an individual soul or mind to a group of individual thinkers and perceivers as parts of a whole is inadequate. In such an account, one soul sees the picture on the screen, another hears the phone ringing, yet another smells the aroma coming from the kitchen, and still one more feels the tap of the dog's tail on her leg, and there is no one soul that sees the picture, hears the phone ring, smells the aroma, and feels the dog's tail. This seems inadequate, because everything that is sensed is unified or bound together in one soul or consciousness. It is the unity of consciousness which leads contemporary neuroscientists to look for one spot in the brain where the seeing, hearing, smelling, and touching all come together in one unified conscious experiencer.[15]

We highlight one other aspect of Plato's view of the soul, which is that the soul, because it is itself inherently alive, gives life to its physical body[16] and, as a self-mover,[17] directly imparts movement to that body.[18] This position on life will be the default position until Descartes, and the claim about motion will become known as the problem of soul-body causal interaction.

2.2 Augustine

Aurelius Augustine (354-430 CE) was a pagan rhetorician who converted to Christianity at thirty-three years of age in the latter part of the fourth century. Like Plato, Augustine thinks of himself as a soul: "I myself am my soul."[19] Augustine believes that each of us knows what a soul is because each of us is one. Nothing can be more present to the soul than itself.[20] A soul "simply cannot not know itself, since by the very fact of knowing itself not knowing, it knows itself."[21] Augustine seems to affirm that the soul directly knows its non-bodily nature. In his work *On the Trinity*, he discusses the Delphic command to "know thyself"[22] and maintains that the whole soul knows whatever it knows. Given this is the case, when the soul knows that *it* knows, it knows all of itself. And because the whole soul knows the whole of itself, there is nothing (including a bodily nature it might have) that is hidden from itself. In line with the nature of self-knowledge, Augustine discusses the idea that the soul might go looking for itself.[23] Normally, when we go looking for something, we have an idea of what it is for which we are looking. For example, when we work on a jigsaw puzzle, we have an idea of the shape and color of the piece for which we are looking, in light of the shapes and colors of the already-assembled surrounding pieces. However, in the case of the soul, it makes no sense to think we are aware of some of its parts and their properties in light of which we can go searching for some other part with its matching properties. The fact that such an idea is not intelligible makes clear that the soul is aware that it has no substantial and separable parts because it is aware that it is simple.

Augustine believes a soul has a body. How is the former related to the latter? Like Plato, Augustine maintains that the soul is the principle of life and gives life to its body: "The soul by its presence gives life to this earth- and death-bound body. It makes of it a unified organism and maintains it as such, keeping it from disintegrating and wasting away. It provides for a proper, balanced distribution of nourishment to the body's members. It preserves the body's harmony and proportion, not only in beauty, but also in growth and reproduction."[24] If the soul imparts life to its body, how are the two related? Are they related spatially? Augustine maintains the soul is located in space, but not in the same way that a material object is located in space. A material body occupies space by diffusion. That is, as a whole it is spread out in a region of space and occupies that space by having each of its separable parts occupy smaller regions of the larger space. For Augustine, the important point is that a body as a whole is not present in its entirety in each of the regions of space that are occupied by its parts. However, a soul, while in the space occupied by its body, is not "diffused throughout the whole body, as is the blood."[25] Rather, a soul is simultaneously present in its entirety (as a whole) at every region of space occupied by its body. Why claim the soul occupies space in this way? Augustine answers:

> [Because] it is the entire soul that feels the pain of a part of the body, yet it does not feel it in the entire body. When, for instance, there is an ache in the foot, the eye looks at it, the mouth speaks of it, and the hand reaches for it. This, of course, would be impossible, if what of the soul is in these parts did not also experience a sensation in the foot; if the soul were not present, it would be unable to feel what has happened there... Hence, the entire soul is present, at one and the same time, in the single parts, and it experiences sensation as a whole, at one and the same time, in the single parts.[26]

Augustine's point is as follows. When a material object touches (and perhaps damages) a soul's body at a particular point (e.g., a toe), the soul in its entirety experiences a sensation (e.g., pain) at that point in space. And if that body were simultaneously touched at a different location (e.g., a finger), then the soul in its entirety would simultaneously experience pain at both locations. If the soul were diffused through the space that it occupies, one part of it occupying one location (e.g., the toe) and another part occupying another location (e.g., the finger), then one part of the soul would experience one pain and another part a different pain, but neither part would experience the other part's pain. But this is not how things are. One and the same soul in its entirety is simultaneously experiencing both pains, which is reflected by the assertion that *I* (in the sense of all of me) am experiencing pains in my toe and finger at the same time.

The soul's simplicity makes possible its presence in its entirety at different points in space. However, this simplicity as a substance is compatible with complexity in terms of properties. Thus, Augustine writes:

> When we come to a spiritual creature such as the soul, it is certainly found to be simple [as a substance] in comparison with the body; but apart from such a comparison it is multiple, not

> simple. The reason it is simpler than the body is that it has no mass spread out in space, but in any body it is whole in the whole and whole also in any part of the body... And yet even in the soul it is one thing to be ingenious, another to be unskillful, another to be sharp, another to have good memory; greed is one thing, fear another, joy another, sadness another; some of these things can be found in the soul without others, some more, some less; countless qualities [properties] can be found in the soul in countless ways. So it is clear that its nature is not simple but multiple.[27]

Finally, while Augustine believes in soul-body causal interaction, he does not discuss the issue philosophically. He simply asserts that the soul-body union is something marvelous and beyond human comprehension.[28]

2.3 Descartes

The influence of René Descartes (1596–1650) on philosophical thought about the soul has been so great that almost every contemporary book on the philosophy of mind begins not only with substance dualism but Descartes' version of it, to the exclusion of all other dualist accounts. One of us had a professor in graduate school who actually told his introductory philosophy class that Descartes *invented* dualism. While that statement is categorically false and laughable, what did Descartes have to say about it that was and remains so influential?

Like Plato and Augustine, Descartes maintained that he is a soul (mind): "This 'me,' that is to say, the soul by which I am what I am, is entirely distinct from body, and is even more easy to know than is the latter."[29] What he knows about himself as a soul is that he is a simple substance, "single and indivisible,"[30] yet possessing multiple faculties (properties):

> There is a great difference between mind and body, inasmuch as body is by nature always divisible, and the mind is entirely indivisible. For, as a matter of fact, when I consider the mind, that is to say, myself inasmuch as I am only a thinking thing, I cannot distinguish in myself any parts, but apprehend myself to be clearly one and entire; and although the whole mind seems to be united to the whole body, yet if a foot, or an arm, or some other part, is separated from my body, I am aware that nothing has been taken away from my mind. And the faculties of willing, feeling, conceiving, etc. cannot be properly speaking said to be its parts, for it is one and the same mind which employs itself in willing and in feeling and understanding.[31]

As a simple substance, the soul, unlike its complex body, is indivisible into parts. Yet, the soul *seems* to be united to the entirety of its body. Thus, Descartes points out that before he meditated (philosophized) about the soul, he "did not stop to consider what the soul was, or if I did stop, I imagined that it was something extremely rare and subtle like a wind, a flame, or an ether, which was spread throughout my grosser parts."[32] Again, because the soul feels pain in its toe and in its finger, it is true to say that "the soul is really joined to the whole body, and that we cannot, properly speaking, say that it exists in any one of its parts to the exclusion of the others."[33] Descartes adds, in language reminiscent of Augustine, that "I... understand mind to be coextensive with the body, the whole in the whole, and the whole in any of its parts."[34]

However, Descartes breaks with the dualist tradition when he maintains that the soul does not give life to its body. The body is a mechanism whose movements, for the most part, but not entirely, can be explained without any reference to the soul:

> And as a clock composed of wheels and counter-weights no less exactly observes the laws of nature when it is badly made, and does not show the time properly, than when it entirely satisfies the wishes of its maker... [so also] I consider the body of a man as being a sort of machine so built up and composed of nerves, muscles, veins, blood and skin, that though there were no mind [soul] in it at all, it would not cease to have the same motions as at present, exception being made of those movements which are due to the direction of the will, and in consequence depend upon the mind [as opposed to those which operate by the disposition of its organs]...[35]

Descartes believes that, because the body is a mechanism, death should not be thought of as the demise of the body as a result of the departure of the life-giving soul:

> Let us consider that death never comes to pass by reason of the soul, but only because some one of the principal parts of the body decays; and we may judge that the body of a living man differs from that of a dead man just as does a watch or other automaton (i.e. a machine that moves itself), when it is wound up and contains in itself the corporeal principle of those movements for which it is designed along with all that is requisite for its action, from the same watch or other machine when it is broken and when the principle of its movement ceases to act.[36]

In Descartes' account of the soul-body relationship, death is the irreversible breakdown of the mechanical body, and the soul leaves (ceases to be causally related to) the body because of this brokenness:

> From observing that all dead bodies are devoid of heat and consequently of movement, it has been thought that it was the absence of the soul which caused these movements and this heat to cease; and thus, without any reason, it was thought that our natural heat and all the movements of our body depend on the soul: while in fact we ought on the contrary to believe that the soul quits us on death only because this heat ceases, and the organs which serve to move the body disintegrate.[37]

Not only does Descartes hold that the soul does not give life to its body, but also he believes that the soul is not, strictly speaking, located in any of the space occupied by its physical body, because it is not located in space, period. It cannot be located in space because anything located in space is in principle divisible into parts, whereas the soul, because it is a simple substance, is indivisible. Because an experience of pain in a toe occurs in the soul and the soul is not in space, the pain does not occur in the toe. Rather, the pain is represented as being in the toe, though strictly speaking it is not there because the soul is not there. It is represented as being there because, when the body is intact and functioning properly, causal impact of a material body on the foot leads to a sequence of events ending in the brain and the pain is *projected* on to (*represented* as being in) the place of the initial causal impact:

> When I feel pain in my foot, my knowledge of physics teaches me that this sensation is communicated by means

of nerves dispersed through the foot, which, being extended like cords from there to the brain, when they are contracted in the foot, at the same time contract the inmost portions of the brain which is their extremity and place of origin, and then excite a certain movement which nature has established in order to cause the mind to be affected by a sensation of pain represented as existing in the foot. But because these nerves must pass through the tibia, the thigh, the loins, the back and the neck, in order to reach from the leg to the brain, it may happen that although their extremities which are in the foot are not affected, but only certain ones of their intervening parts (which pass by the loins or the neck), this action will excite the same movement in the brain that might have been excited there by a hurt received in the foot, in consequence of which the mind will necessarily feel in the foot the same pain as if it had received a hurt.[38]

Descartes points out that while the soul is a patient with respect to an experience of pain, it is an agent on those occasions when it chooses to move the limbs of its body. On these occasions, the soul produces an initial effect event in the brain of its body to produce a causal chain that ends with the movement of a limb. "And the whole action of the soul consists in this," Descartes writes, "that solely because it desires something, it causes the little gland to which it is closely united to move in the way requisite to produce the effect which relates to this desire.... [W]hen we desire to walk or to move our body in some special way, this desire causes the gland to thrust the spirits towards the muscles which serve to bring about this result."[39]

But how does the soul causally produce the initial brain event? Descartes concedes that he has nothing informative to say about this issue: "for soul and body [operating] together we have no notion save that of their union."[40] In other words, the causal connection between the soul and its body is a primitive or basic relationship about which nothing informative can be said:

> Though we are not in a position to understand, either by reasoning or by any comparison drawn from other things, how the mind which is incorporeal, can move the body, none the less we cannot doubt that it can, since experiences the most certain and the most evident make us at all times immediately aware of its doing so. This is one of those things which are known in and by themselves and which we obscure if we seek to explain them by way of other things.[41]

In a historical reconstruction of the "mystery" of mind-body interaction, Daniel Garber proposes that Descartes treats the evident fact of mind-body interaction as a paradigm case of explanation:

> For Descartes, mind-body interaction is the paradigm for *all* explanation, it is that in terms of which *all* other causal interaction must be understood... Mind-body interaction must be basic and intelligible in its own terms since if it were not, then *no* other kind of causal explanation would be intelligible at all; to challenge the intelligibility of mind-body interaction is to challenge the entire enterprise of causal explanation. Furthermore, we *cannot* give a simpler or more easily understood account of causal interaction than mind-body interaction because there are no more

> basic or more inherently intelligible ways of explaining the behavior of anything open to us.[42]

Garber's proposal for understanding Descartes' position on soul-body causal interaction is itself in need of interpretation. For example, Garber initially states that for Descartes soul-body interaction is the paradigm for *all* explanation, but he then goes on to claim that Descartes believed all other *causal* interaction must be understood in terms of it. It is highly doubtful that Descartes regarded all explanation as causal in nature. He surely believed in the distinction between ultimate and irreducible teleological explanation, which is explanation in terms of purposes, and causal explanation. However, this is not the place to enter into detailed analysis and defense of a particular reading of Descartes. Instead, we will use Garber's proposal to formulate our own point about soul-body causal interaction (which Descartes would have endorsed), which is that our purposefully chosen and intended performances of bodily actions require that we, as souls, are capable of causally producing purposeful movements of our physical bodies. Indeed, given that science itself consists in part in purposefully explained mental and bodily actions (e.g., much scientific research is undertaken for the purpose of curing diseases and overall improvement of the quality of human life), science requires soul-body causal interaction, so that it would be impossible to discover through scientific investigation how physical entities causally interact with other physical entities without souls causally interacting with physical entities. In this way soul-body causal interaction is primary.

3. Property Dualism

SUBSTANCE DUALISM is not the only form of dualism. Some philosophers believe that while there is no soul that is a substance in its own right which is distinct and in principle separable from its physical body, there is nevertheless an irreducible difference between what is psychological in nature and what is material or physical in nature. Thus, these philosophers believe that while human beings are fundamentally material organisms with no substantial soul, human beings yet have two irreducibly different kind of properties, psychological and material. This view is often termed "property dualism" or "dual-aspect theory." We will briefly summarize the views of the philosopher Aristotle, whom many consider to be one of the earliest proponents of this kind of dualism.

3.1 Aristotle

Aristotle (384–322 BCE) affirms the existence of the soul and maintains the soul gives life to its body.[43] The soul is the *form* or "first actuality" of a body which potentially has life.[44] To say that the soul is the form of the body is to claim something like the following: the soul is a vital (life-giving) principle which informs its body in the sense of configuring the body into the entity that it is with its organs and their purposes (e.g., the heart is for pumping blood and the eyes are for seeing). However, Aristotle maintains the soul is not a substance in its own right which does or could survive the dissolution of its body. He claims that that which exists as a substance is the soul-body composite, for example, the individual man, Socrates. Thus, it is ultimately not the soul, as opposed to the body, which thinks, chooses, experiences pleasure and pain, etc., but the soul-body composite: "Perhaps indeed it would be better not to say that the soul pities or learns or thinks but that the man does in virtue of the soul."[45] But though it is not a simple soul which pities, learns, thinks, etc., nevertheless Aristotle recognizes the unity of consciousness pointed out by substance dualists:

> Therefore… discrimination between white and sweet cannot be effected by two agencies which remain separate; both the qualities discriminated must be present to something that is one and single… What says that two things are different must be one; for sweet is different from white. Therefore what asserts this difference must be self-identical, and as what asserts, so also what thinks or perceives. That it is not possible by means of two agencies which remain separate to discriminate two objects which are separate is therefore obvious.[46]

Aristotle's point is that if one subject apprehends whiteness and another apprehends sweetness, then it would not be obvious to either subject that what it apprehends is different from what the other apprehends. However, it is obviously known that whiteness is different from sweetness, which requires the existence of a single subject that is aware both of the whiteness and sweetness and of their difference from each other.

Aristotle's thoughts about human action are particularly interesting and relevant to the issue of soul-body causal interaction, which is the topic of the next section. Aristotle maintains that "the origin of action—its efficient, not its final [teleological or purposeful] cause—is choice, and that of choice is desire and reasoning with a view to an end."[47] Given that choice and the desire that leads to it and to bodily action are mental in nature, Aristotle makes clear that there must be a causal connection between what is mental and what is bodily in nature: "so… the organ whereby desire produces [bodily] movement… is something bodily, whose investigation belongs with that of the common functions of body and soul."[48] How the mental leads to movements in the body seems to be as follows: There is the faculty or capacity of desire[49] whose operations (actualizations) are directed at what is good or believed to be good. This object of desire is an end, goal, or final cause. The subject of the desire, the individual human being, reasons about how to achieve the goal, where the goal of the reasoning process is a purpose, because "all practical thought processes have termini—they are all *for* some purpose."[50] The desire and the reasoning process causally bring about the choice to act, where that choice in turn causally produces the relevant bodily motions, which are aimed at bringing about the achievement of the purpose.

In contrast with Aristotle, most contemporary property dualists have ceased to believe in the existence of the soul. Instead, they consider psychological properties of a self to be aspects of a human body or brain. We will discuss their view in the next section in the context of the problem of causal interaction.

4. The Problem of Causal Interaction

Substance dualists are typically causal interactionists who maintain that causation occurs between a soul and its body. However, because of the difficulty of saying anything informative about how soul-body causal interaction occurs (see our earlier discussion of and comments about Descartes' view of this issue), opponents of substance dualism insist that a non-substance-dualist view of a human being must be correct. Some philosophers go so far as to insist that not only is there no soul that causally interacts with its physical body, but also that there is nothing psychological or mental in nature, period.[51] What is psychological is reducible to or eliminable in favor of what is physical in nature. For the purposes of this chapter, we will simply take for granted that such a position is a nonstarter. If we know anything about ourselves, it is that we are psychological beings who experience irreducibly real mental events.

Some philosophers who share our belief that we are psychological beings who are the subjects of irreducibly real mental events nevertheless believe there is something deeply problematic about causal, interactionist substance dualism and, therefore, reject it in favor of some form of property dualism (dual-aspect theory of events) according to which human beings are a single material substance with two irreducibly distinct types of properties, psychological and material (physical). According to these philosophers, when we make a free choice, the choice does not occur in an immaterial soul but is rather a mental event in a fundamentally material substance which is also the subject of material events. Thus, if a person chooses to write a letter, the mental event of choosing causally produces a material brain event which in turn produces other material brain and bodily events which ultimately issue causally in the movement of the individual's hand in ways teleologically suitable for carrying out the writing of the letter.

However, how is this dual-aspect account of mental-physical causal interaction explanatorily superior to a substance dualist account of writing the letter? On the former account, it still is the case that the mental aspect of a physical substance must causally produce an initial physical effect which issues in a physical chain of events. The mental event of a physical substance must produce the causal series of physical events, because otherwise we would be left with a complete physical causal story in which the mental event played no causal explanatory role. So even the property dualist must acknowledge irreducible mental-to-physical causality.

At this point, the substance dualist will want to know from the property dualist how this mental-to-physical causation occurs.[52] What can the dual-aspect theorist say that will elucidate its occurrence? It seems the dual-aspect theorist will have to maintain that a mental event of a physical substance causes a physical event of a physical substance, and that is all that can be said. Nothing more informative can be provided. But the substance dualist will want to know how this is an improvement over the substance dualist's claim that a mental event in a soul causes a physical event in its body, period. If the substance-dualist view ought to be rejected because nothing informative can be said about how the mental-to-physical causation takes place, then the property dualist view also ought to be rejected for the same reason. The explanatory dependency of purely physical effects on the mental aspect of a physical substance would be just as intellectually impenetrable as the soul-body causal interaction of the substance dualist. In short, the substance dualist has an effective *tu quoque* (you, too) response at hand when it comes to the problem of causal interaction.

A property dualist might concede that the foregoing reasoning is correct, but insist that substance dualism suffers from an additional problem that warrants its rejection. For example, Timothy O'Connor, a theistic property dualist, writes: "The fundamental problem [with a substance dualist's view of the soul and body] is that our sciences point to highly continuous processes of increasing complexity, but the two-substance account requires the supposition of abrupt discontinuity. The coming to be at a particular point in time of a *new substance* with a suite of novel psychological capacities... would be a highly discontinuous development."[53]

What does O'Connor suggest as an alternative? He proposes that at some point in the continuous biological story, psychological or conscious properties "different in kind" *emerged* from the hierarchically-structured physico-chemical properties of the brain: "These conscious states have distinctive intrinsic features, immediately apprehended by their subject, that in no way resemble the sorts of features science attributes to complex neural states."[54] A substance dualist will likely respond to O'Connor, as he responded to the property dualist above, that if substance dualism is objectionable on the grounds that it introduces discontinuity into the developmental processes in the natural world, then O'Connor's dual-aspect view is no less objectionable. After all, O'Connor's view appeals to emergent psychological or conscious properties which, in virtue of their being different in kind from physico-chemical properties, are just as discontinuous, strange, and magical as a specially created soul. Once again, the substance dualist has a *tu quoque* response to the property dualist.[55]

5. The Libet Experiments

The experimental work of Benjamin Libet[56] is often used to argue against libertarian free will and the explanatory efficacy of purposes. There is much debate about how to understand the details of Libet's experiments, especially the matter of which brain events are correlated with or represent which mental events (e.g., desires, choices, intentions). However, as we briefly explain below, regardless of the way in which this issue is resolved, it must still be the case that Libet had to presuppose the teleological explanatory efficacy of purposes and the mental events they explain (e.g., choices and intentions) in the conduct of his experimental work.

Very briefly (more complete descriptions of Libet's experiments are available elsewhere in this book), Libet used an electroencephalogram (EEG) test to show that activity in the brain's motor cortex could be detected some 300 milliseconds before a subject believed he had consciously decided to initiate movement of a finger. Libet interpreted this brain activity, and not the subject's conscious decision, as the real initiator of the movement of the finger. Another lab corroborated Libet's work using functional magnetic resonance imaging (fMRI).[57] Subjects were asked to press one of two buttons while watching a "clock" composed of a random sequence of letters appearing on the screen. They reported which letter was visible at the moment they decided to press one button or the other. The experiment allegedly made clear that a few hundred milliseconds before the subjects were aware of choosing to push one of the buttons their brains had, unbeknownst to them, already "chosen" and, thereby, determined which button they would consciously choose to push. Some[58] have generalized the results of these experiments to support the position that moments before we are aware of choosing what we will do in non-experimental daily life—a time in which we at least occasionally subjectively seem to have freedom to choose how we will behave—our brains have unbeknownst to us already "decided" and determined what we will do. We then consciously make our own decision and mistakenly believe we are initiating the course of our behavior.

As we discussed earlier in this chapter, actions are performed for purposes, where acting for a purpose (reason) is an irreducibly teleological notion. To conduct his experiments, Libet had to depend upon the reports (verbal or otherwise) of his subjects in order to learn about the time at which they were consciously aware of choosing to move their fingers.[59] Were Libet's subjects free to lie to (deceive) him? One would assume so. A lie, however, is more than communicating a falsehood. It is misrepresenting the way things are for the purpose of misleading someone else. But if one is free to choose purposefully to misrepresent what is the case, one is also free to choose purposefully to convey the truth. Libet assumed that his subjects were freely choosing to tell him the truth about when they consciously decided to move their fingers for the purpose, say, of assisting him with his experiment. And this assumption presupposed that the subjects' purposefully explained free choices to report the correct time of their conscious decisions were the ultimate or originating causes of their bodily movements involved in their reports.

Libet might respond that the subjects did not freely choose to tell the truth on each run of the experiment. They simply *deterministically intended* to tell the truth because they had no reason not to do so. If so, two further points are relevant. First, deterministically intending to tell the truth for a purpose is itself an action that is explained teleologically. So Libet had to assume the explanatory efficacy of purposes. Second, appealing to teleologically explained determined intentions to tell the truth during the experiments presupposes that the subjects made a purposefully explained choice to tell the truth before they participated in the running of the experiments. And the issues discussed in the previous paragraph about choosing to tell the truth for a purpose will have to be readdressed at this point.

Does the fact that Libet's experiments presuppose the reality of purposefully explained choices and intentions provide any reason for preferring substance over property dualism? That is, if it can be shown that the Libet experiments fail to provide any good reason for rejecting explanations of our choices in terms of irreducible purposes (goals, aims), then is it reasonable to conclude that substance dualism, as opposed to property dualism, is true? It seems to us that a defense of libertarian free will in response to the Libet experiments provides no evidence that supports a substance-dualist view of the self over a property-dualist view (or vice versa). A defense of purposeful explanations of our choices is in and of itself compatible with either substance or property dualism. Hence, if one is a substance dualist, one will hold that free choices explained purposefully are mental events in a soul, whereas if one is a dual-aspect theorist, one will maintain that free choices explained purposefully are mental aspects of a material self. If we are correct, the debate between substance and property dualists will have to be resolved, if it can be resolved, independently of the issues raised by the Libet experiments.[60]

6. Substance Dualism Today

ONE OF the most serious challenges facing substance dualism today is the widespread assumption that we currently have a problem-free concept of the physical world as disclosed in the natural sciences, but only a tangential, ill-informed concept of consciousness and mental life. A long-time opponent of substance dualism, Daniel Dennett, sees consciousness as something that should be fully accounted for in the natural sciences:

> We know how atoms are structured, how chemical elements interact, how plants and animals propagate, how microscopic pathogens thrive and spread, how continents drift, how hurricanes are born, and much, much more. We know our brains are made of the same ingredients as all the other things we've explained, and we know that we belong to an evolved lineage that can be traced back to the dawn of life. If we can explain *self-repair in bacteria* and *respiration in tadpoles* and *digestion in elephants*, why shouldn't *conscious thinking in H. sapiens* eventually divulge its secret working to the same ever-improving, self-enhancing juggernaut?[61]

What is missing in this portrait? We believe that what is missing is an appreciation of the evident primacy of consciousness (which we regard as a feature of the soul; the soul is conscious) in the practice of science itself (as well as in virtually all our activities). We would have no idea about any of the explanations Dennett references without having knowledge about and confidence in the nature and evident reliability of our consciousness and our mental lives. *Consciousness is far from some "secret working"; it is the most immediately apparent real phenomenon that makes science work.* Stan Klein castigates those who, like Dennett, treat our mental lives as secondary to knowledge gained through the natural sciences:

> According subjectivity, at best, "second class citizenship" in the study of mind is particularly ironic in virtue of the fact that subjectivity is the very thing that makes the scientific pursuit of such knowledge (actually any knowledge) possible. Timing devices, neuroimaging technologies, electroencephalographs, and a host of modern means of obtaining objective knowledge about the mind are useless absent an experiencing subject... To believe otherwise has the absurd consequence of rendering our knowledge of mind (or, more generally, reality) dependent in its entirety on the provisions of an experienced conduit stipulated either to be unworthy of study or essentially nonexistent.[62]

We are not claiming that appreciating the evident reality of consciousness *leads to* substance dualism. On the contrary, we, like most or all commonsense individuals, start with belief in the soul-body distinction. But we are maintaining that appreciating the power of the natural sciences does not lead to Dennett's or any other form of materialism which denies the unique role of the conscious soul in accounting for the nature of human and some other animal life. It is noteworthy that none of the examples of causal explanation cited by Dennett involves events such as reasoning, thinking, or conscious reflection. The fact that reasoning, thinking, conscious reflection, and so on are indispensable in practicing (or even understanding) the natural sciences is justification for insisting that substance dualism is closer to the truth in explaining the occurrence of many events in the physical world than philosophies that deny the robust role of consciousness in life. Any evidence of the "ever-improving, self-enhancing juggernaut" of the natural sciences is at the same time evidence of the mental-to-physical explanatory power of the conscious, purposeful activities of souls.

Notes

1. Jesse Bering, "The Folk Psychology of Souls," *Behavioral and Brain Sciences* 29 (2006): 453–62.
2. Nicholas Humphrey, *Soul Dust* (Princeton, NJ: Princeton University Press, 2011), 195.
3. The literature on theological and biblical anthropology has greatly increased in recent years. Good treatments of the topic can be found (in alphabetical order) in John Cooper, *Body, Soul, and Life Everlasting: Biblical Anthropology and the Monism-Dualism Debate* (Grand Rapids, MI: Eerdmans, 2000); Joshua R. Farris and Charles Taliaferro, eds., *The Ashgate Companion to Theological Anthropology* (Surrey, England: Ashgate, 2015); and Joel B. Green and Stuart L. Palmer, eds., *In Search of the Soul: Four Views of the Mind-Body Problem* (Downers Grove, IL: InterVarsity, 2005).
4. Stewart Goetz and Charles Taliaferro, *A Brief History of the Soul* (Oxford: Wiley-Blackwell, 2011).

5. Plato, *Phaedo*, trans. Hugh Tredennick, in Edith Hamilton and Huntington Cairns, eds., *The Collected Dialogues of Plato* (Princeton, NJ: Princeton University Press, 1961), 115d.
6. Plato, *Phaedo*, 105d-e.
7. Plato, *Phaedo*, 80a-b.
8. Plato, *Phaedo*, 80b and 78c.
9. Plato, *Timaeus*, trans. Benjamin Jowett, in Edith Hamilton and Huntington Cairns, eds., *The Collected Dialogues of Plato* (Princeton, NJ: Princeton University Press, 1961), 71a, 77b.
10. Plato, *Republic*, trans. Paul Shorey, in Edith Hamilton and Huntington Cairns, eds., *The Collected Dialogues of Plato* (Princeton, NJ: Princeton University Press, 1961), 440e.
11. Plato, *Republic*, 439e.
12. Plato, *Republic*, 439e.
13. David Armstrong, *The Mind-Body Problem: An Opinionated Introduction* (Boulder, CO: Westview Press, 1999), 23.
14. Plato, *Theaetetus*, trans. F. M. Cornford, in Edith Hamilton and Huntington Cairns, eds., *The Collected Dialogues of Plato* (Princeton, NJ: Princeton University Press, 1961), 184d.
15. John Searle, *The Mystery of Consciousness* (New York: New York Review, Inc., 1997), 33. Would the assumption of the existence of the soul as the simple subject of a unified consciousness warrant neuroscientists seeking a single place in the brain where the different bodily (brain) events directly causally associated with the different modes of a person's sensory and psychological life are unified (bound together)? We are not convinced the assumption of the soul's existence would warrant this investigatory project. Why couldn't there be multiple termini in the body (brain) directly causally linked with the many events involving the multiple modes of sensing and mental life of a soul? If a soul is simultaneously located in its entirety at many points in the space of its body (or brain; see our discussion of Augustine), perhaps it interacts with its body (or brain) at more than one place in the latter. If a soul is located at a single point in the space of its body (or brain), perhaps it interacts with its physical body (or brain) at multiple points, assuming causal interaction at a distance is possible. Or if a soul is not located in space at all but can causally interact with one part of its body (or brain) at one point in space, why could it not causally interact with multiple parts of its body (or brain) located in their respective different points in space? One might also wonder whether science has anything to say about whether a soul is located in space and, if it is, whether it is located at one or many points in space. We believe science has nothing to say about this metaphysical issue. Finally, if one is a physicalist, then the quest of neuroscientists for a single spot in the brain where all the events involving the different modes of sensing and mental life are unified seems justified, because the unity of consciousness, by hypotheses, is a physical feature of a body (or brain).
16. Plato, *Phaedo*, 105c-d.
17. Plato, *Laws*, trans. A. E. Taylor, in Edith Hamilton and Huntington Cairns, eds., *The Collected Dialogues of Plato* (Princeton, NJ: Princeton University Press, 1961), 896a.
18. Plato, *Timaeus*, 69c, *Laws*, 898e.
19. Aurelius Augustine, *The Greatness of the Soul*, trans. Joseph M. Colleran, in Johannes Quasten and Joseph Plumpe, eds., *The Greatness of the Soul and The Teacher* (New York: The Newman Press 1950), 30.61.
20. Aurelius Augustine, *Confessions*, trans. R. S. Pine-Coffin (New York: Penguin Books), VIII.4.9.
21. Augustine, *Confessions*, X.2.5.
22. Aurelius Augustine, *On the Trinity*, trans. Edmund Hill (New York: New City Press, 1991), X.2.6.
23. Augustine, *On the Trinity*, X.2.6.
24. Augustine, *The Greatness of the Soul*, 33.70.
25. Augustine, *The Greatness of the Soul*, 30.61.
26. Aurelius Augustine, *The Immortality of the Soul*, trans. Ludwig Schopp (New York: CIMA Publishing Company, 1947), 16.25.
27. Augustine, *On the Trinity*, VI.2.8.
28. Aurelius Augustine, *The City of God*, trans. Marcus Dods (New York: The Modern Library, 1993), XXI.10.
29. René Descartes, *Discourse on the Method*, in *The Philosophical Works of Descartes*, vol. I, trans. E. S. Haldane and G. R. T. Ross (Cambridge: Cambridge University Press, 1967), 101.
30. René Descartes, *Descartes: Philosophical Letters*, ed. and trans. Anthony Kenny (Oxford: Clarendon Press, 1970), 75.
31. René Descartes, *Meditations*, in *The Philosophical Works of Descartes*, vol. I, trans. E. S. Haldane and G. R. T. Ross (Cambridge: Cambridge University Press, 1967), 196.
32. Descartes, *Meditations*, 151.
33. René Descartes, *The Passions of the Soul*, in *The Philosophical Works of Descartes*, vol. I, trans. E. S. Haldane and G. R. T. Ross (Cambridge: Cambridge University Press, 1967), 345.

34. René Descartes, "Reply to Objections," in *The Philosophical Works of Descartes*, vol. II, trans. E. S. Haldane and G. R. T. Ross (Cambridge: Cambridge University Press, 1967), 255.
35. Descartes, *Meditations*, 195.
36. Descartes, *The Passions of the Soul*, 333.
37. Descartes, *The Passions of the Soul*, 333.
38. Descartes, *Meditations*, 196–7.
39. Descartes, *The Passions of the Soul*, 350, 351.
40. Descartes, *Descartes: Philosophical Letters*, 252.
41. Descartes, *Descartes: Philosophical Letters*, 262.
42. Daniel Garber, "Understanding Interaction: What Descartes should have told Elisabeth," *Debates in Modern Philosophy*, ed. Stewart Duncan and Antonia LoLordo (London: Routledge, 2013), 50–51.
43. Aristotle, *De Anima*, trans. Hugh Lawson-Tancred (New York: Penguin Books, 1986), 402a.
44. Aristotle, *De Anima*, 412a.
45. Aristotle, *De Anima*, 408b.
46. Aristotle, *De Anima*, trans. J. A. Smith, in *Introduction to Aristotle*, 2nd edition, ed. Richard McKeon (Chicago, IL: The University of Chicago Press, 1973), 426a.
47. Aristotle, *Nicomachean Ethics*, trans. W. D. Ross, in *Introduction to Aristotle*, 2nd edition, ed. Richard McKeon (Chicago, IL: The University of Chicago Press, 1973), 1139a.
48. Aristotle, *De Anima*, 433b.
49. Aristotle, *De Anima*, 433b.
50. Aristotle, *De Anima*, 433b, 407a.
51. E.g., Alex Rosenberg, *How History Gets Things Wrong: The Neuroscience of Our Addiction to Stories* (Cambridge, MA: MIT Press, 2018).
52. The substance dualist's question need not arise out of the supposition that the property dualist is claiming that events, *as ontologically independent or not modifications of substances*, are the *loci* of causal relations. Like a substance dualist, the property dualist can reasonably affirm that events always occur in substances, and that some events involving substances, i.e., causal events, produce other events involving substances, i.e., effect events. What the substance and property dualist disagree about is not this event-substance ontology but the nature of the substance in which mental events occur. The substance dualist believes mental events occur in souls, while the property dualist believes mental events occur in physical human bodies or only the brains of these bodies.
53. Timothy O'Connor, "Do We Have Souls?," Big Questions Online, Jan. 8, 2013, https://answptest2.dreamhosters.com/2013/01/08/have-souls/.
54. O'Connor, "Do We Have Souls?"
55. A reviewer of this chapter objects to property dualism on the grounds that "none of the physical sciences give[s] us a reason to think that purely physical objects have subjectivity… and it therefore looks *ad hoc* to claim that physical objects just do have these properties and hence have non-physical powers." Three points are appropriate in response.

    First, if physical objects are "purely" physical, then the reviewer has merely stipulated from the outset that they cannot have mental properties and be the subject of mental events. No property dualist will be intellectually moved by this stipulation.

    Second, a property dualist need not affirm property dualism on the basis of reasons provided by one or more of the physical sciences. For example, Peter van Inwagen maintains that it just seems to him in first-person experience that he is a human physical animal: "When I enter most deeply into that which I call *myself*, I *seem* to discover that I am a living animal. And, therefore, dualism seems to me to be an unnecessarily complicated theory about my nature." See "Dualism and Materialism: Athens and Jerusalem?," *Faith and Philosophy* 12, no. 4 (1995), 476. The emphases are van Inwagen's. Similarly, Eric Olson asks, "Why suppose that we are animals?" and answers, "Well, that's how it appears." See "For Animalism," in *The Blackwell Companion to Substance Dualism*, eds. Jonathan J. Loose, Angus J. L. Menuge, and J. P. Moreland (Oxford: Wiley-Blackwell, 2018), 298.

    On behalf of substance dualism, one might concede van Inwagen and Olson's assertions about what seems to be the case in terms of the self's spatiality (e.g., as we pointed out above, even substance dualists like Augustine and Descartes contend that it appears to us as if we, as souls, fill the space occupied by our physical bodies [souls are located spatially], Augustine affirming and Descartes denying that appearance here is reality), but question their implicit assumption that if the self seems to be located in space then it must seem to be physical, because seeming spatiality is sufficient for seeming physicality. Also, one might remind van Inwagen that dualism is not a theory about human nature. It is a description of the way things seem to be with respect to the self and its body. But in response to the reviewer's comment, it is nevertheless wrong to suggest as he or she does that because the physical sciences do not ascribe mental properties to physical substances that property dualists have no reason to affirm their position.

Third, there are *non-scientific*, metaphysical reasons to think that substance dualism is preferable to property dualism. For example, as we have described in our book *A Brief History of the Soul*, the simplicity of the self has played a significant role in arguing for the existence of the soul as a substance that is distinct and separable from its physical body. The skeletal structure of this argument is as follows: 1) I am essentially indivisible; 2) My body is essentially divisible; therefore, 3) I am not my body. Thus, my psychological properties are not properties of my body, assuming they are properties of me.

56. Benjamin Libet, *Mind Time: The Temporal Factor in Consciousness* (Cambridge, MA: Harvard University Press, 2004).
57. Chun Siong Soon, Marcel Brass, Hans-Hochen Heinze, and John-Dylan Haynes, "Unconscious Determinants of Free Decisions in the Human Brain," *Nature Neuroscience* 11 (2008): 543-45.
58. E.g., see Sam Harris, *Free Will* (New York: Free Press, 2012).
59. Richard Swinburne makes this point in his *Are We Bodies or Souls?* (Oxford: Oxford University Press, 2019), 133–136.
60. The reviewer mentioned in endnote 55 claims that "one reason to think that the Libet experiments do provide evidence favoring substance dualism over property dualism is if one thinks that active power is more at home in a substance dualist ontology." True. If one believes that physical objects have no active causal power which they exercise, then one will prefer substance over property dualism. But of course property dualists who are also libertarians do not believe physical objects (substances) lack active causal power. At a minimum, they believe physical objects have the active causal power to choose.
61. Daniel Dennett, *From Bacteria to Bach and Back* (New York: W. W. Norton & Co, 2017), 18.
62. Stan Klein, "A Defense of Experiential Realism," *Psychology of Consciousness: Theory, Research and Practice* 2:1 (2015): 42–43.

6. Mere Hylomorphism and Neuroscience

James D. Madden

1. Introduction

The ancient doctrine of hylomorphism has made something of a comeback in recent years among both philosophers working in the metaphysics of material composition and philosophers of mind. It is attractive to the latter, because hylomorphism promises to split the difference between materialism and more robust forms of dualism in intuitively satisfying ways. Nevertheless, hylomorphism as it has been applied to the mind-body problem is not without its critics. On the one hand, many worry that hylomorphism is really just an obfuscating version of substance dualism conveniently trading between incompatible accounts of the soul either as some sort of non-substantial, abstract entity or an individual substance-like entity.[1] On the other hand, some worry that hylomorphism is really just a "polite form of materialism" that offers nothing more intuitively satisfying than standard forms of non-reductive physicalism, except for the fact that its talk of "souls" is attractive to those concerned with squaring materialist approaches to mind and orthodox religious views.[2]

My hunch is that these criticisms are less due to a failure to understand hylomorphism on the part of its critics, and more so as a result of the way in which hylomorphism is presented by its proponents. Aristotle's original proposal for a distinction between matter and form was not intended as a doctrine narrowly construed for the philosophy of mind, and even less so to solve anything akin to the modern mind-body problem that has vexed us these last few centuries. Rather Aristotle's hylomorphism is a holistic doctrine that can only be understood as it incorporates and synthesizes elements from his philosophy of nature, metaphysics, teleological ethics, and philosophy of "mind." In what follows, I will present Aristotle's hylomorphism as it can be extracted from some of the more salient moments in his *Physics*, *Metaphysics*, *Nicomachean Ethics*, and *De Anima*. I call the position I will construct "Mere Hylomorphism," because I take it as the minimal doctrine one must accept to be a hylomorphist in continuity with Aristotle, though I am aware that other versions worthy of consideration are on offer. Moreover, I do not propose Mere Hylomorphism as a definitive interpretation of Aristotle, but simply as a plausible position that can be constructed from some of his most central texts. I will recommend Mere Hylomorphism on the grounds that it reveals hylomorphism as having intuitive appeal while clearly being neither a sly dualism, nor a polite materialism. We will also see that Mere Hylomorphism is recommended by the fact that it underwrites an integration between philosophy of mind and neuroscience.

2. Form, Matter, and Soul

Consider the desk at which I am writing this chapter. It is no doubt a composition of material parts, e.g., boards, screws, bits of glue, etc. Notice, however, that the desk is not *identical* to that collection of material parts. Though the desk is located in the very same place as that collection, and touching it would involve touching that collection (or at least some subset of that collection), they are not the *very same thing*. I realize that this is an odd claim when whispered in the ear of the person on the street. What else would a desk be but just the collection of boards, screws, glue bits, etc.? It is not as silly as it first sounds. If two things are identical, then they must persist under the same conditions. If Jim Madden and the person writing this chapter are one and the same being, then there had better not be conditions under which the one comes to be or passes away without the other doing the same. If they could exist separately, then they just are not the very same thing, assuming we are using the relevant terms in the appropriate manner.[3]

Now, if we smashed the desk with a sledgehammer such that we are left only with a pile of a broken boards, screws, etc., we would likely conclude that the desk has passed away, while the parts have clearly persisted. That is, the very same collection of parts can exist as composing a desk, as it does now, or not composing a desk after we have smashed the desk to bits. Thus, there is a *synchronic unity* that is lacked by the mere occurrence of the collection of its parts, i.e., the desk is right now a unified whole that is distinct from its parts. Moreover, if we replaced just one screw in the desk, we would not feel as though we had destroyed the rightful property of my employers. Indeed, if we did so incrementally, we might even change out every piece of this desk without ever having a sense of losing the original item. Thus, whereas we saw earlier that the parts exist without the desk, we can see here that the desk can exist without this particular set of parts (or even any member of that set of parts), supposing they are replaced incrementally. The desk has a *diachronic unity* over and above the collection of its parts.[4]

Notice that the desk enjoys, at least for our ordinary common sense, a sort of *ontological priority* over its parts. If you asked me to count the objects in this room, and I stopped and counted each part of the desk but not the desk, or I counted the parts and the desk separately, you would find my sorting quite odd. Rather, we would only expect to count the desk and other so-called "middle-sized dried goods." The most important fact is the *being of the desk*, not its parts. In other words, the desk occupies a more prominent place in our ontology than its parts. A reason for this is the *explanatory priority* the desk enjoys with respect to its parts. The proper characterization of something as a "desk foot" or a "desk drawer" is parasitic upon its being properly characterized as an actual or possible part of a desk. The nature of a "desk drawer" is derivative from the nature of desks, and the reciprocal is not the case; i.e., we do not specify what counts as a desk in terms of what counts as a drawer.

The notion that these curious facts about the desk can be generalized so as to apply to all material objects is the ancient Aristotelian doctrine of *hylomorphism*. The hylomorphist claims that all material objects (at least those that can be divided into parts) are composed of *matter* (a set of parts that have a potency to compose an object of that kind) and an additional principle of unity, the *form*, that accounts for the actuality of such a composition.[5] Notice that the form is not a part in the same sense as the items we find in the matter. We might think of the form as what answers to a definition of the object that limits the loss, gain, or rearrangement of constituents inasmuch as the object will retain its identity as a certain kind of thing.[6] As Mark Johnston puts it:

> The idea that each item will have some such canonical statement true of it might be fairly called "Hylomorphism." For it is the idea that each complex item admits of a real definition, or statement of its form, understood as a principle of unity.... For any complex item that persists over time, the relevant principle of unity will impose not only synchronic constraints but also diachronic constraints; it will not only constrain how the parts of the item have to stand at a given time, it will also constrain how successive realizations of the item stand to one another for the item in question to persist. So, for the complex item that persists in a manner that allows for loss, exchange, or rearrangement of at least some of its parts, the principle of unity will specify how much and what kind of change in material parts secures the continued existence of the item.[7]

Notice that Johnston, along with Aristotle, takes the form not merely as the real definition of an object, but as that *in* the thing that most importantly corresponds to the real definition; i.e., the real definition of the object is the "statement of its form," but the form is not the statement. The form is what grounds the real definition. Hylomorphism entails an extra-linguistic ontological commitment: something *is* the form of a unified, complex object. Certainly, there is a sense in which the matter (the composing parts) grounds the definition; e.g., there are limits to the materials that can be used to make a desk, such that we might include some material specification in the definition. Nevertheless, as Aristotle famously argues in the *Physics*, "In fact, the form is more nature than the matter is. For each thing is said to be when it actually is more than when it potentially is."[8] The definition of "desk" does not apply to the pile of sawdust left after the destruction of the desk, not least of all because the very same pile of sawdust is just as much potentially a wood carving of Abraham Lincoln as it is a desk. Thus, matter is a necessary condition for a desk, but it is not what makes the difference between what is merely potentially a desk and what actually is so.

What then is the form, if it is not the matter? What answers to the definition of desk such that its parts compose a unified object, rather than an aggregated potential for such an object? Certainly, the causal history leading up to the occurrence of a desk as a unified whole is part of the that story, but what counts as a desk-making process? Likely, it is a process leading to the occurrence of a piece of furniture suitable for the execution of certain activities—writing philosophy papers, storing files, discussing with clients, etc. Anything at the ready to serve such tasks can plausibly be thought of as a desk. There is a broad sense in which the pile of sawdust has the potential for such activities, i.e., the sawdust could be formed such that it is ready to serve in desk ways, but a desk is something *currently* suitable to these ends without further change. The point here is that the form of a desk is a direct disposition, what I call the *being-able*, that must be present in order for those parts to perform the function of a desk. It is this *being-poised-to-perform* that Aristotelians are referring to with talk of the form of an object. The form of the desk is not a discrete part, but a *ready potency* (or what Aristotle calls a *first actuality*) the parts come to have in virtue of their composing the whole. The Aristotelian notion of form is subsequently closely linked to teleology. Certainly, we can think of the form of the desk as an arrangement of the parts (or more likely a broad range of such arrangements) that constitutes its aggregate of matter as a desk, but that arrangement will always be an *arrangement for* something. What counts as a desk? A set of materially appropriate constituents arranged so as to enable certain activities. This is why Aristotle sees such an intimate relationship between the form and the "final cause" or purpose of an object.[9] The form of an object is the direct ability to (or enablement for) engaging activities that are definitive of a certain kind of thing. A form is not a part of a material being, but a readiness to do what such a thing is supposed to do as a member of its essential kind.[10]

As I said above, the hylomorphist claims that this account in terms of form-matter composition can be generalized to cover all material objects.[11] This generalization includes not just artifacts, such as desks, but living beings too. Take a cat, and let's call him Fluffy. Is Fluffy identical to his material parts? No more so than the desk. The fact that the Fluffy could have all order of unfortunate accidents that would leave us with a complete collection of unassembled cat parts *sans* cat shows that Fluffy has a synchronic unity distinct from his parts. Moreover, the fact that Fluffy is constantly exchanging matter with his environment throughout his lifespan shows that he has a diachronic unity distinct from his aggregated parts at any one moment. Like the desk, Fluffy also has an ontological precedent over his parts. Above the molecular level, Fluffy's parts have no stability of their own once separated from Fluffy.[12] The organs, tissues, cells, etc. that compose Fluffy will die and decompose once they are separated from Fluffy. The being of these parts presupposes the completed whole of the cat (or some outside causality to keep them alive artificially). If we ask why livers, muscle cells, etc. exist, it will be very difficult to answer such questions without appeal to the whole organism to which they belong. These parts exist in order to serve the existence of the organism they compose, and when separated from a living substance they immediately begin to decompose. Even at the molecular, atomic, or subatomic level, we cannot explain (to any great level of satisfaction) why they are assembled in the way they are without appealing to the functional role they play in the living organism, though they can exist outside such a composition. In this light, we can see Fluffy as enjoying an explanatory and ontological precedence over his parts.

We then account for Fluffy's unity in terms of form and matter for reasons similar to those that led us to this same conclusion with respect to artifacts such as our desk. Notice, however, that the case is even stronger for Fluffy than for the desk. Though the desk enjoys a sort of precedence over its parts, there is also a sense in which nothing really novel comes into existence with the coming to be of the first desk. The desk has no capacities that are not straightforwardly capacities of the solidity of the wood composing it. One could rest his books on a stack of wood in a pinch. Becoming a desk is a refinement of wood, but it is really just another accidental way of being wood. Fluffy, however, engages in activities and has powers that are unprecedented among the powers of his parts. Cats, along with all living things, undergo changes and develop capacities that are utterly novel compared to their parts at all levels of analysis. The desk, ultimately, is just decomposing wood (it goes through no changes that are different from the ordinary changes of the sawdust composing it), but Fluffy does "his own thing" as it were. That is not to say that there is some mystery as to how cat components give rise to cat powers, but they do amount to something new under the sun. For this reason, we might say that the coming to be of a cat is the occurrence of a new *substance*, not merely an *accident* of its parts. This distinction between a substantial being and an accidental being, for Aristotle, depends crucially on the fact that the former and not the latter has "within itself a starting-point of moving and being at rest."[13] That is, the distinctive conditions for synchronic and diachronic unity that make for a desk are all imposed by external agencies, whereas the distinctive continuity of a cat is something it possesses, to some significant degree, independently of external agency.[14]

Thus, Aristotelians have long held that a living thing is a composite not just of a form and matter, but of a *soul* and matter. Here all that is meant by "soul" is the form of a living thing, i.e., the distinctive *being-able* that marks the difference between its composing matter and the living substance that is so composed. As Aristotle most famously puts it: "the soul is the first actualization of a natural body that has life potentially."[15] That is, the soul is the ready potency of an organism that marks it as distinct from its composing parts. The soul is not a separate life-force or spirit that *actualizes* or otherwise acts on the matter as a distinct, external particular. Rather, the soul is the life of the organism in the sense that "it is the *actualization* of such a body"; i.e., it is the *being-able* that must be maintained in order to distinguish the living organism from either a dead, former member of that species or a mere aggregate of beings with the unactualized potential to be such an organism.[16] Like any other form, the soul does not *do the actualizing* of the living body (that is the work of an external, *moving cause*), but *is the actualization* of the living body.[17] To be a living body of a certain kind is to have at the ready a certain *being-able*, a readiness to perform the activities characteristic of that sort of thing.[18]

Consider the two following analogies Aristotle uses in his own explication of the notion of soul:

> It is just like this: if an instrument—for example, an ax—were a natural body, its substance would be the being for the axe, and this would be its soul.
>
> For if the eye were an animal, sight would be its soul. For that is the substance of an eye, the one in accord with the account. And the eye is matter for sight, and if this fails, it is no longer an eye....
>
> As, then, are cutting and seeing, so too is being awake and actuality, and as are sight and the capacity of the instrument, so is the soul, whereas the body is what is potentially [alive]. But just as the eye jelly and sight are an eye, so in the case the soul and the body are an animal.[19]

Following the analogy, the "soul" of the eye would be sight, and the "soul" of an ax would be its suitability for chopping. By "sight" we do not mean actually seeing something (a closed eye is still an eye) any more than suitability for chopping entails actually chopping. Sight, rather, is the *being-able-to-see* that we only find in a living eye (as opposed to a decomposing set of eye parts detached from an organism). *Being-able-to-see* is the actuality of *being-an-eye*, and it has ontological and explanatory precedence over parts of the eye; they exist and are what they are because of their role in the function of the eye. Leaving aside the analogy (eyes are not living substances, but parts of living substances, and axes are "soulless" artifacts), Aristotle would say that the soul of a cat is the *being-poised* to engage in cat actions that distinguishes Fluffy from his composing parts (and other kinds of things those parts, at finer levels of decomposition, might have composed). Actually *being* a cat is *to be able* to perform distinctively feline activities (certainly this also requires having distinctively feline parts—fur, a tail, claws, etc.—but all those parts are ordered to the definitive activities of cat-kind), so the actualization of a cat is just such a readiness, which is what Aristotle means by the soul of the cat.[20] Notice too that *being-able* is dispositional; i.e., being a cat is not a matter of actually performing the distinctively cat activities, but to have a disposition toward those activities that is lacked by non-cats, even if some individual cat never exercises such a disposition.[21]

Merely having the requisite soul is sufficient for being an organism of as certain kind, but notice that the soul, though it is a "first actuality," is likewise a potency for a certain kind of activity. Anything with a determinate disposition for definitively feline activities (whatever its stage of development) is a cat. Aristotle, however, maintains that something exists to the degree that it is actuality not mixed with potency; i.e., *being-X* is not *potentially-being-X*, but *actually-being-X*. All of an organism's activities taken in isolation are fleeting movements, giving way to yet more fleeting movements. Each activity results in yet another actuality mixed with potency for change; e.g., however magnificent Fluffy might be in any given distinctively cat-like exercise of his mouse-hunting capacity, that will only set him up for another movement of some sort. The organism comes to be "unconditionally" or is "actual being" in the fullest sense inasmuch as it finally brings its essential activity to completion. Since all of its particular doings are somehow incomplete, Aristotle concludes that the organism is only fully actualized inasmuch as it lives a certain kind of life as a "complete action." Each act, even those definitive of a certain kind of organism, is a fleeting movement, whereas living a certain type of life, flourishing, is an end in itself and possibly stable throughout the organism's entire lifespan. Fluffy's soul is ultimately not merely a disposition to pull off a particular act of cat-like hunting or reproduction, but to live a life full of these activities. The soul is then the readiness to follow a certain way of life that is desirable for its own sake for a certain kind of organism, and that manner of flourishing is determinative of the soul for that kind of organism.[22] In other words, for Aristotle, the soul is ultimately a ready tendency toward a type of distinctive happiness or flourishing that counts as the fullest being of that kind of thing played out over its entire history.[23]

Thus, the Aristotelian hylomorphist sees nature as rich with souls. All living things are "compounds" of souls and material parts. That is no trivial claim. Hylomorphism (of this sort, and there are other versions) has profound philosophical ramifications, as it rules out any kind of strict reduction of organisms to their constituents.[24] Organisms are not just collections (ultimately) of subatomic particles. Rather, nature is suffused with entities irreducible to their composing parts, living things, which cannot be understood solely in terms of their constituents, but must be seen in light of their definitive *ready disposition toward a kind of life*, their *souls*. In fact, Aristotle himself thought that the soul is the object of proper scientific study, at least for the life sciences.[25] The distinctive *being-able* of kinds of living things is what the biologist is ultimately out to understand and explain.[26]

At the same time, we must be careful not to make too much out of this talk of souls. In fact, I often wish we had a different word. When we hear "soul" today, we think of ghosts, substantial minds, our truest selves, angels, etc. That is, we think of a soul as a kind of substance that interacts with a body in the way a virus is a substance that interacts with an organism. Notice, however, that the hylomorphist means no such thing by "soul." The soul is not a substance, but the definitive *being-able* of a substance. We also tend to use the word "soul" in ways that primarily associate it with consciousness or otherwise psychological attributes. The soul is supposedly a thinking substance, or the bearer of my psychological as opposed to my merely physical properties. Notice, however, that the Aristotelian path to the soul does not run through consciousness at all. Aristotle does not introduce the notion of a soul to account for thinking things but living things in general (some of which are thinkers). The soul is posited to solve problems about the identity and unity of complex living things, and there are plenty of non-conscious things that have souls. When the Aristotelian claims that even plants have souls, that is not because she believes there are ghosts or minds inhabiting the trees. It is just to say that there is a definitive *being-able* of the tree that marks it as something over and above the aggregate of its parts. I belabor this point so as to disabuse the reader of the common tendency to think of hylomorphism as a sort of substance dualism. The soul and the organism are not separate, individuated substances. Rather, the former is the actuality of the latter, and as Aristotle puts it, "we should not inquire whether the soul and the body are one.... For, since one and being are said of things in many ways, the controlling way is actuality."[27] Of course, in some living things, the definitive *being-able* is a realized capacity for certain conscious or psychological activities, and the complete life of such organisms will include the exercise of such psychological capacities. Part of being a cat is *being-poised-to* hunt certain types of prey, respond to injury, pursue particular mates, etc., and all these definitively feline activities are hard to envision without the cat being conscious. Thus, Fluffy's soul is, in part, a *being-able* to engage in particularly feline consciousness. This is not at all to say that Fluffy's soul engages in such consciousness.[28] Rather it is Fluffy that engages in this distinctively feline consciousness. Fluffy's soul is (in part) the *being-able* to engage in such activity, but it is Fluffy, as a biological whole, that so engages, or at the very least hylomorphism is consistent with such an emergentist or holistic view of consciousness.[29]

Finally, hylomorphism extends this account to human beings. That is, a human being is not identical to the aggregate of her physical parts, because she possesses a synchronic and diachronic unity distinct from those parts, and she enjoys an ontological and explanatory precedence over those parts. Moreover, since a human being is a living thing, a substance novel as compared to its composing parts, it is a composite of matter and that special sort of form Aristotelians calls a "soul." Notice, however, that none of this is to say anything particularly interesting about human beings as such, but simply to include human beings in a generalized Aristotelian account of nature. There is nothing special, according to the Aristotelian, about our having souls, but that does not answer the question of whether there is anything special about the souls we have.

3. Is the Soul Immaterial?

THE SOUL is the *being-able* for the activities definitive of a certain kind of living being. Thus, distinctiveness of the soul of a certain kind of thing is really a question of whether such a thing is capable of engaging in activities that distinguish it from all other living beings. Is there anything unique about the human soul? Well, that question can only be answered by first querying whether there is a definitive human activity. Philosophers in the Aristotelian tradition have standardly defined humankind as the *rational animal.* There is no need to dispute that other animals have cognitive capacities (or at least the rudiments thereof). We are all aware of stock examples of non-human animals "figuring things out," "making decisions," "solving problems in novel ways," and "operating according to concepts" in some sense. These interpretations of non-human animal behavior are useful predictive stances, but also get something right about animal psychology. Animals do things that make good sense, but, as I will argue, we are distinct from them because we do things *because* they make good sense.[30]

Suppose Fluffy engages in certain stable patterns of behavior that make sense as interfaces with the world (these are clearly not accidental doings). Furthermore, suppose Fluffy can refine or adjust those behavioral patterns to fit indefinitely many specific circumstances. In such cases, we then have every reason to think of Fluffy's behavior as rational as we do in the case of human beings who engage in stable yet indefinitely adjustable manners of dealing with the world. I suspect ordinary feline hunting behaviors fit this bill rather well. It is perfectly natural to say that "Fluffy knows how to hunt a rat" and that "Fluffy was clever in hunting that rat in particular." I take it that this interplay between generality and flexibility within a rubric is the very mark of a kind of rationality. I find it then irresistible to say that Fluffy's hunting is *somehow* rational. In other words, I am quite comfortable saying that "Fluffy understands how to hunt a rat" or that "Fluffy has reasons for why he stalked that way" because Fluffy can work between the relevant generalities and particularities in a way that competently interacts with the world.[31]

Notice, however, that the conceptual or rational status of Fluffy's behavior plays no role in the motivational structures leading to Fluffy's hunting strategy on any particular day. Though Fluffy's behavior is rational, it is not rational *to Fluffy*. The rationality of the behavior does not occur to Fluffy, and subsequently Fluffy does not engage in such behavior *because it is rational.* Although we might say that Fluffy's distinctively feline *being-able* involves his being capable of doing what makes reasonable sense (both in general and in particular) for a cat, Fluffy is not motivated by his activity's rational status. This is all to say that, though Fluffy's behavior is *implicitly* rational, we have no reason to conclude that this fact is something *explicit* to Fluffy. Some philosophers put it like this: Fluffy's behavior, though it makes perfectly good sense to interpret it as rational in some broad sense, is not *explicitly normative.*[32] Fluffy does not engage in his particular hunting behavior on any given day because it conforms to a norm of rationality. Fluffy's hunting is the right thing to do, but he does not do it that way because it is the right thing to do. The fact that it conforms to such norms is irrelevant to Fluffy. The feline psychological repertoire is utterly devoid of any second-order notions that allow normative considerations to occur. Fluffy's attention is solely fixed on the particularities of the hunt, and the rational status of his activity is not borne by his conscious deliberations. Rather, Fluffy's rationality is carried by his entire bodily comportment as an animal aimed at making sense, i.e., at acting toward flourishing, in an environment. One might say Fluffy is *rational*, but not *logical* in the sense that Fluffy's behavior instantiates or conforms to certain rational norms (e.g., the standard of good hunting for feline-kind), but those norms are not topically explicit to Fluffy. He does not make these norms of felinity explicit so that he can use them to guide, evaluate, or improve his behavior.[33]

Technically, though we might say that Fluffy engages in rational behaviors, there is no need thereby to ascribe any particular logical *content* to his thinking, because Fluffy is not thinking about the rational status of his behaviors. There is no discrete "mental state" (or what have you) that carries Fluffy's rationality. Rather Fluffy's rationality just is his way of being a cat in some particular situation. When we say that "Fluffy understands how to hunt rats" we are not ascribing some special sort of psychological predicate to Fluffy. Rather, we are just identifying a distinctively feline bodily comportment present in Fluffy that makes sense in his environment. This is a rational comportment and one subject to "on the spot" adjustments (novel problem-solving) by Fluffy, but it is bodily "all the way down," as it were. That is, Fluffy's ability to engage in responsively sensitive rat-hunting-know-how as a contribution to cat-flourishing does not require our positing of anything more on the part of Fluffy than the ordinary conscious states required by the fact that Fluffy registers the presence of mice, dogs, water, potential mates, etc., and I am happy to agree that that sort of consciousness can be accommodated by broad-minded versions of naturalism. There is no reason to think there is anything interestingly non-physical or immaterial about such creatures, however interesting these creatures are in the exercise of their distinctive capacities.[34] Fluffy's rationality is an instantiated conceptuality; the embodiment of implicit reasons for acting. Fluffy doesn't *hold* those reasons, though in a sense he *is* those reasons. Being a cat just is *being-able* to act in accord with such reasons.

We should be careful here with what is meant by "embodied" or "bodily comportment." I do not mean that Fluffy's reasons for acting can be identified with or otherwise reduced to a discrete event or process within his nervous system.[35] Fluffy's rationality is identifiable with neither a discrete physical occurrence nor a discrete mental occurrence, because it is not a discrete occurrence at all. Rather, Fluffy's rationality is something that supervenes on a complicated relation among particular organisms (Fluffy and the rat serving as his prey), the interactions between the evolutionary histories of their species, the contingent circumstances of their current environment and individual histories etc., etc. Put together, all of these factors serve to make sense of what Fluffy is doing, but that rational status is possessed by no one contributor to this state of affairs. It is an attribute borne by the whole, but none of the parts. We might say that the rationality of Fluffy's hunting behavior is an indefinitely complex relationship with the entire "world" around which he "knows" his way, though that knowing is only implicit within the global circumstance.[36] Fluffy might be said to think, but he does not think about his thinking.

Is the Mere Hylomorphist a materialist or physicalist when it comes to cats and their seemingly rational behavior? Allow me the philosopher's proverbial "Yes" and "No." The "Yes" comes inasmuch as the Mere Hylomorphist does not think any sort of immaterial substance need be added to Fluffy's physiology to account for his instantiation of rational behavior. Fluffy is a material being. The "No" comes because Fluffy's rationality is not physically identifiable.

We cannot point out a single physical event, process, occurrence, etc., that is Fluffy's rationality. Thus, if you mean by "materialism" the claim that "everything is material," then Mere Hylomorphists are not materialists, because they grant that there are things, e.g., Fluffy's knowing how to catch a rat currently in his view, that are not identifiable as material things. That, however, is not to commit oneself to the claim that Fluffy's knowing his way around is identified with some immaterial entity. One can say that there is no discrete physical organ for cognition, while denying that there is a discrete non-physical organ for cognition, so long as one is also claiming that cognition is not the sort of thing carried by a discrete organ.[37] To say that Fluffy knows how to hunt is not to identify a particular process or event in Fluffy's nervous system or mind, but to say that *Fluffy participates in the world in a way that makes sense.*[38]

The human case is more complicated in some important respects, but not in all ways. Of course, just like those of Fluffy and of all living things, the vast majority of our behaviors are not explicitly, but only implicitly, rational. That is, we mostly get around the world unreflectively "doing the human thing," mediating between the stable patterns of our practical activities that have perennially served our kind well and the particular circumstances in which we must apply them. In this sense, humans, just like cats, are expressions of a rationality that supervenes on the material world in which they dwell. Though our manner of adjustment and the range of objects to which we can so adjust differs from Fluffy's (we are less attuned to smells, but we have recourse to opposed thumbs and upright posture), these powers are in no way ontologically foreign to those we find suffused throughout the kingdom of living things. We are in this way just another species of animals *being-in* the natural world, and I see no need to posit anything immaterial in humans or any other animals to account for this being.

The complicating factor is the fact that, unlike Fluffy, we are not only expressions of a rational order, but also *logical.* As Robert Brandom puts it, we do not just behave in ways that conform to norms or concepts, we are "concept mongerers."[39] Not only can our behaviors be interpreted in logical categories (and properly fall under those logical categories), *we* are logical interpreters of those behaviors. The rationality of our activities is part of the motivating structures that explain them, at least (hopefully) some of the time. That is, human beings not only do what makes sense, but *we can do what makes sense because it is what makes sense.* The fact that a certain activity is the right thing to do in some situation is often part of the reason why a human being endeavors to do it. Surely it is plausible to say "Fluffy knew that was the best way to hunt the rat," but only in the human case can we say "Smitty caught the rat that way, *because* he knew that was the best way to do it." The normative status of the activity is a relevant category of explanation of Smitty's, though not Fluffy's, activity, even when these activities are guided by similar ends and happen to conform to similar norms. Only in the human case does the normative conformity explain something.[40]

In order for the rationality of the action to play a role in the explanation of Smitty's performance, its normative status must be something that Smitty can make explicit to himself. That is, Smitty must not only be able to engage in behaviors that implicitly instantiate concepts, but he must further be able to call to mind or otherwise grasp the relevant concepts. The conceptual content must have some bearing on what he did. Human rationality entails a power to assume a second-order stance toward our doings. We are able *to ask* ourselves which norms or concepts our behaviors or possible behaviors do or do not conform to (rather than merely performing behaviors that do or do not so conform). That is not say that every action of Smitty's, or even any of his actions, are direct products of explicit conceptual analysis and normative evaluation of the alternatives. Rather, it merely entails that it makes sense to ask of Smitty "What were you doing?" or "Why did you do that?" in a way that it does not makes sense to ask that of non-human animals, because only human beings (as far as we know) are able to answer such queries by appeal to commonly held conceptual contents. The answers to these questions are things we expect to be explicitly available to Smitty upon reflection, inasmuch as he has achieved the status of being a distinctively human agent.[41] The mark of human rationality, our distinctive *being-able*, is this power to take a second-order stance that makes our reasons for action (and here we include knowing as a kind of acting) explicit. As Brandom puts it, our ability to articulate and bind ourselves to conceptual norms because they are conceptual norms distinguishes us from "merely natural creatures."[42] One might say (as Brandom definitely would not, though I believe it would be entirely at home in his position) that the power to concept-monger (the ability to think about thinking) *is* the distinctive human soul.

To possess a human soul is not only to have a tendency toward certain flourishing-constituting behaviors, but to be able to articulate and critically analyze the norms governing those behaviors in a manner which itself contributes to that flourishing. As Aristotle famously argues in the *Nicomachean Ethics*, "a human being's function is supposed to be a sort of living, and this living is supposed to be activity of the soul and actions that involve reason"; i.e., it is integral to human flourishing that we exercise our ability to make the conceptual norms governing our activities (both theoretical and practical) explicit and to place them under rational scrutiny.[43] By asking with Aristotle, "What is the good life for human beings?" we are looking for an articulation of the natural norms governing human behavior inasmuch as they are ordered to our flourishing, and attempting to achieve this articulation is a central ingredient to what Aristotle sees as that state of flourishing. In short, our ability to take a second-order stance toward the norms governing our behavior (the human soul) distinguishes us from non-human animals, while the exercise of this ability also partly constitutes our happiness.

So far, what I have presented is consistent with a polite form of materialism, but hylomorphism (in the full Aristotelian context in which I have attempted to place it in the foregoing) carries some heavy ontological consequences of an immaterialist stripe. Human rationality presupposes an ability to distance oneself from the world in which we are otherwise perfectly at home. We can remove our attention from the direct objects of our activities (practical or theoretical) in order to ask whether these manners of acting are normatively up to snuff. This distancing is only possible inasmuch as the concepts humans "monger" abstract from the particularities of the situations in which we implicitly act out their dictates. The fact that our normative bearing toward our own activities requires an abstract conceptual explicitness has the consequence that we must posit a discrete mental state on the part of human beings that we do not need to posit in an account of the cognition of living things in general. In order for Smitty to adopt the normative stance toward his theoretical and practical doings, he must have an explicit grasp of the relevant concepts or norms governing these activities. Whereas there was no need to posit an intentional object of Fluffy's thinking in addition to the rat he is hunting (and other relevant items in the vicinity), Smitty's critical-normative awareness requires that he be aware of the relevant concepts as such. For example, for Smitty to be motivated to make an inference *because* it is an instance of *modus ponens*, Smitty must have *modus ponens* as an object of his attention. In other words, distinctively human thinking is thinking about thinking that puts the status of our thinking up against explicitly held norms and concepts.[44]

Human rationality then requires us to posit a special kind of mental content on the part of humans. That is, we have among our intentional objects concepts or norms that are trans-temporal and trans-spatial. The concept of *modus ponens* is identically instantiable across an infinity of possible physical instantiations. In fact, *modus ponens* is indifferent to its physical instantiation. Almost anything, under the proper interpretation, can be an instance of *modus ponens*. In fact, understanding a logical concept or norm such as *modus ponens* is to be able to recognize its physically disparate instantiations as instantiations of the same abstract concept. Whatever else physical things are, they are not indifferent to their material instantiation, and therefore the objects of conceptual explicitness are not anything physical, even in a broad sense.[45] The objects of our distinctively human thinking are not immaterial only in the sense that a cat's knowing its way around is immaterial, i.e., there is no discrete material object or event with which it can be identified. Rather, conceptual contents are immaterial in the much stronger sense that they are not occurrences in space or time at all. Smitty's articulation of the conceptual norms grounding his thinking may be an event in space-time, but the intelligible content of that event is not. The thought expressing this articulation is therefore more than what can be taken as material in any sense. Since the content of thought is abstract, and the thought just is the articulation of that content, likewise the thought is abstract.[46] As Aristotle puts it: "That is how it is in the case of the understanding. And it is an intelligible object in just the way its intelligible objects are, since, in the case of those things that have no matter, what understands and what is understood are the same."[47]

Aristotle further claims that the activity of this understanding is itself "separable, unaffected, and unmixed… and it alone is immortal and eternal."[48] Thus, in this sense, we should conclude that the grasping of a concept is not a material power in the same way that the other vital functions of the human organism (or all other organisms) are material. There is something non-bodily about the human (normative) mode of rationality, because it involves conceptual contents indifferent to their material instantiations. That is not to say that we therefore should identify our conceptual articulations with the activities of a disembodied agent, a mind or some such, but only that human rationality involves a participation in something that is transcendent of the material world.[49] Aristotle takes the human soul as uniquely separable from matter, but not because he thinks human souls exist as discrete individuals without matter.[50] Rather, the soul is separable from matter in the sense that its exercise brings the human organism into participation with something that is truly immaterial in Aristotle's sense (eternal, unchanging, invariant, etc.). Aristotle famously defines the divine nature as an eternal act of self-reflection, i.e., thinking about thinking.[51] Notice, then, that Aristotle defines both the exercise of distinctive human rationality and the nature of divinity as self-reflective thought. Such thought makes explicit the normative grounds of one's thinking. When human beings manage to bring the true normative structure of the universe into explicit conceptual grasp, their act of thought brings them into a sort of union with Aristotle's god.[52] Once again, Aristotle takes these transcendent acts of conceptual articulation, or contemplation, as constitutive of human happiness.[53] Though the Mere Hylomorphism account of human nature is certainly not a sly form of substance dualism, it is likewise quite impolite by any materialist standards.

4. Mere Aristotelianism and Neuroscience

It is now fairly standard to claim that one can be a non-materialist about minds and still recognize the obvious successes the neurosciences have had in finding the physiological grounds in the central nervous system for psychological phenomena by arguing that those states are necessary though insufficient conditions for the corresponding psychological phenomena. We are, then, still in need of some additional non-physical principle to account for the psychological states. The Mere Hylomorphist who has argued that conceptual contents are non-physical may avail herself of just such a strategy. Be that as it may, Mere Hylomorphism is not a theory of mind, but a theory of soul, and the former and the latter are not the same thing. Something can have a soul (a unifying disposition for certain kind-defining activities) without having a mind (the ability to make conceptual contents explicit). Indeed, plants have souls and no mind, and at earlier stages of human development we all lacked minds (or at least we were far less minded than we currently are). Certain organisms' souls will include the disposition to develop a mind (in the fullest sense for human beings, and to some maybe lesser degree for other animals), though even in those cases the soul is not the mind, but the direct potency toward such development. With these points on the table, we can see a much more intimate connection between neuroscience and Mere Hylomorphism.

Consider the following remarks from Aristotle:

> These bodily parts, then, are in a way prior to the compound, but in a way not, since they cannot exist when they are separated. For it is not a finger in any and every state that is the finger of an animal, rather, a dead finger is only homonymously a finger. Some of these parts, however, are simultaneous, namely, the ones that are controlling and in which the account and the substance are first found—for example, the heart, perhaps or the brain (for it makes no difference which of them is of this sort).[54]

That is really compact, and probably unintelligible without the paragraphs leading up to it. Let's unpack it a bit. Take your run-of-the-mill body part, e.g., a finger. As we discussed earlier, for a hylomorphist, a finger is not prior (either temporally or explanatorily) to the organism as a whole (what Aristotle is here calling "the compound"), because the finger (as being a finger) depends on the compound for its being; separate the finger and it begins to disintegrate, and fingers do not antedate the organism to which they belong (at least under natural conditions). In this passage, however, Aristotle is stopping us from generalizing that claim to all of the parts of the organism. He claims that some parts are (a) "simultaneous" with the organism as a whole, because they are (b) "controlling" and (c) that in which the "account and the substance are first found." His examples are (d) maybe the heart or the brain. Let's unpack this a bit further.

As far as *simultaneous* goes, there is some quaint ancient embryology in play here, but I think in principle we can see that Aristotle is not far off the mark. We do not just pop into existence with the full complement of human organs, but come online incrementally. Whatever the first part of the organism is, that part and the organism as a whole came to be simultaneously. So, if the liver were the first part of the organism, then the liver and the organism came to be at the same instant. In a sense, the liver would be separable from the rest of the organism (the compound), because it did indeed exist separately from the rest of the organism. Now, notice the special relationship between, using my rather silly example, the liver and the soul. At that early point of development, the actualization of the organism was nothing more than the liver. In those circumstances, *the being-able* of that organism would be no more than to be a functioning liver. In this originary embryonic stage, the organism is a liver with the direct disposition to grow into a more complex organism. There is a sense, then, in which we could say at one point in development, the soul was a disposition of the liver (if the liver really were the first organ in the order of development).

Next, Aristotle starts to lay out some conditions for being parts that are simultaneous with the whole. The first he mentions is *control.* That is, some parts play a special role in directing the rest of the organism's activity and development. Aristotle is not committed to the notion that the controlling parts come online literally first in fetal development. His candidates for controlling parts are the heart or the brain. On his view of embryology, the newly formed organism goes through a vegetative (non-animal organism), non-human animal, and human stage of a development. I take it that a controlling part need only be online at the onset of the development of its distinctive stage of being, and I take it that these controlling parts will be running the show in that organism for its entire lifespan.

Aristotle then tells us that the "account and substance are first found" in these controlling parts (which I am taking as the same as simultaneous parts). In other words, as soon as the controlling parts are up and running, we have something fitting the definition of this kind of organism. The controlling parts, because they are sufficient markers of the coming to be of the organism, are themselves the places in which the soul is first (and primarily) located. That is, simultaneous parts are those that most distinctively signify the coming to be of an organism of a certain kind (the occurrence of the organism and the occurrence of such a part of an organism are simultaneous), because a simultaneous part must be the locale of the disposition for the most distinctive activities of the organism. Notice that Aristotle actually thinks that the soul can be spatially located. The soul as the substance of the organism is first found in its controlling parts. Those controlling parts maintain their governing role throughout the life of the organism, so it will always be the case that, in some sense, the soul is "there" with the controlling parts (as long as they have that role), even if in another sense Aristotle thinks that the soul somehow suffuses the body. Think of it this way: a soul is the first actuality of an entire living organism, but it acts as such by initially being the first actuality of the living controlling parts. Certainly, without the controlling parts, the organism will cease to be an integrated whole (i.e., it will die), so there is a sense in which the controlling parts have a role as the primary actuality of the compound of organs for the entire life span of the organism (i.e., it is because of its controlling parts that a certain organism is an organized unity).

Notice that the liver is certainly not a simultaneous or controlling part of an organism. It does not control the whole of the organism, and it is not the location of the distinctive activity of any animal (many species are relatively similar in their liver-function). Aristotle mentions the heart and the brain as possibilities. I think the "perhaps" is a sign that the heart is not his preferred hypothesis. I do not think it matters whether the heart or brain literally come first in embryonic development, but which of the two marks the coming to be of an organ with a governing power over the organism as a whole in its final, even if a yet unrealized, stage of development. Given what we know now, the brain is obviously the controlling part of the human organism (or any organism with a brain for that matter). There is a sense in which the living brain, as a controlling part, is the actualization of the organism. That is, to put it maybe a bit too simplistically, there is a sense in which the brain is the "place" of the soul more distinctly than any other part.[55]

Thus, as neuroscientists become more adept at producing functional neuroimages, they are not merely taking pictures of a physical object. Neuroscience is a science of the soul. Neuroscientists are rendering images of the human soul, captured in its activities, and yielding insights into our nature and the conditions of our flourishing. That is not to say that the soul, in its entirety, is something captured in any discrete image, but rather that that neuroscientist is able to render a moment in the soul's development in much the same way a single frame in a film shows us a moment in a narrative. To see why this is a fruitful insight, consider a recent study which found a decrease in symptoms of poststroke depression by direct stimulation of the brain.[56] This study is interesting to us, because the network targeted *most directly* by this intervention is associated with higher-level cognitive capacities, and not affective emotional states.[57] This is exactly what the Mere Hylomorphist would expect, since the exercise of our distinctively human powers to take a higher-order, critical stance toward our conceptual contents is integral to our flourishing.[58] The Mere Hylomorphist would see this study as confirming a connection between the human soul (as manifest in the controlling part of the human being, the central nervous system) and the Aristotelian understanding of human flourishing. Thus, once we understand the holistic nature of hylomorphism, i.e., that it is an integrated theory of material composition, biological distinctiveness, mindedness, and flourishing, we can see the cross-germinating possibilities for its relationship with neuroscience.

An additional point of contact, albeit more speculative, is the Aristotelian treatment of the *vegetative*, *sensitive*, and *rational* functions of the human soul. An organism whose definitive activities do not involve conscious interaction with its environment or affective emotional states (i.e., an organism not a subject of sensation, but still engaged in the business of metabolism and reproduction), is said by Aristotle to have a *vegetative* soul.[59] He calls the *being-able* of an organism definitively ordered to activities requiring sensation (and conscious affection) a *sensitive* soul. What Aristotle calls the *rational* soul, is distinctive to the human case, as it points to animals capable of entertaining explicit conceptual contents. Notice, however, that Aristotle takes these various soul-types as integrative within a single organism.[60] That is, a sensitive animal does not have a vegetative soul and a distinct sensitive soul, but one soul; e.g., a cat soul "contains" the *being-able* for the definitive activities of such an organism, sensitive and vegetative alike. Aristotle claims that the sensitive-affective function of the cat soul is inextricable from the vegetative function; there are no sensations without metabolism and respiration. The case is similar for the human soul; i.e., it is an integration of the vegetative, sensitive-affective, and rational dispositions. Human sensation presupposes the basic functionality of life, and human rationality, Aristotle claims, requires sensitive or emotional grist for its mill; there is no sensation without metabolism, and no rationality without sensation.[61]

Nevertheless, Aristotle argues that the exercise of human rationality can exert a sort of downward executive function on the sensitive function.[62] Indeed, Aristotle, like all Greek philosophers in the Socratic School, saw the harmonization of this two-way hierarchy as constitutive of the good life. Given what I have introduced above regarding the brain as the primary location of the soul within the organism, we should expect to see neuroscientific evidence of this integration of the three levels of functionality. Does the brain, in the interaction of these various modules, operate in such a way that the vegetative grounds the sensitive, and the sensitive grounds the rational; while the rational exerts a downward executive control on the sensitive, including the emotions, to some degree? Aristotelian speculation about the integration of the three functions of the soul not only provides a testable hypothesis for the neuroscientist, but also has important clinical implications. Common therapeutic practices such as cognitive behavioral therapy seem to presuppose something very much akin to the Aristotelian stance on the two-way hierarchy between affective states and rational control, so further investigation of the Aristotelian prediction could help us understand why such techniques have traction. Moreover, the notion that our disinterestedly rational functionality is grounded in and yet can exert control over our affective capacities may help us understand such phenomena as the apparent efficacy of placebo and contemplative states (meditation, prayer, mystical experience), especially should we find structural evidence of such integration in the nervous system.

All of this is to say that Mere Hylomorphism can propose speculative hypotheses for human flourishing based on the conceptual and phenomenological analysis of the soul as the distinctive *being-able* of the human organism, and the neurosciences can in principle confirm or disconfirm these proposals. Mere hylomorphism is uniquely situated for this partnership with neuroscience, as opposed to other philosophical stances, because of its strong affirmative stance toward embodiment (the soul is a definitive disposition of a living human organism, that can even be located primarily in the central nervous system) and its equally strong non-materialist account of human rationality and ultimate flourishing. In other words, Mere Hylomorphism can provide the clinically minded neuroscientist with a realistic and yet holistic perspective on human nature, which can generate concrete, verifiable research proposals. This exchange between Aristotelian philosophy and neuroscience may then have welcome applications for the advance of humans toward flourishing (e.g., the treatment of recalcitrant forms of depression in the case I mention above), as both disciplines are ultimately aimed at the full actualization of *human-being*.

NOTES

1. See Kevin Corcoran, "The Constitution View of Persons," in *In Search of the Soul: Four Views of the Mind-Body Problem*, ed. Joel B. Green and Stuart L. Plamer (Downers Grove, IL: Intervarsity Press, 2005), 146; William Hasker, *The Emergent Self* (Ithaca, NY: Cornell University Press, 1998), 166; William Hasker, "On Behalf of Emergent Dualism," in *In Search of the Soul*, 94–95; and Gordon Barnes, "The Paradoxes of

Hylomorphism," *The Review of Metaphysics* 56 (March 2003): 501–523. I address these criticisms, defending the specifically Thomistic version of hylomorphism at which they are aimed, in *Mind, Matter, and Nature: A Thomistic Proposal for the Philosophy of Mind* (Washington, DC: The Catholic University Press of America, 2013), 280–284.

2. Thanks to Mark Johnston for cautioning me against the temptation for "polite materialism" in my own grappling with hylomorphism.
3. Of course, this point raises the problems of contingent identity and the legion of literate works dealing with that issue. I will leave that thicket aside for our purposes and ask the reader to trust me that all that business can be appropriately smoothed out.
4. Readers of contemporary analytic metaphysics will be quick to point out that I beg the question motivating a vast corpus of contentious philosophical literature, wherein one will encounter plenty of thinkers quite happy to deny that the desk (or any other complex physical object) is in fact a singular unified entity at all. Be that as it may, I cannot wade into those deep waters here, so I will simply rely on the deliverances of common sense in what follows; i.e., I will assume that those entities to which we refer with common nouns—"desk," "ox," "river," etc.—do in fact possess some sort of unity justifying reference to them as enduring entities. As I will discuss below, I do not deny that there are different types of unity (substantial as opposed to accidental unities), which carry different levels of ontological gravity, but I will nevertheless assume that desks, oxen, and rivers are real entities in some sense. To get started on this debate, see Trenton Merricks, *Objects and Persons* (New York: Oxford University Press, 2001); Peter Van Inwagen, *Material Beings* (Ithaca, NY: Cornell University Press, 1990); and Michael Rea, ed., *Material Constitution: A Reader* (New York: Rowman and Littlefield, 1996).
5. We have to be careful with this term "form." Although Aristotle does sometimes refer to the "form or shape" of a thing, he also speaks of the form as the "account of the essence" or "paradigm." See *Physics*, trans. C. D. C. Reeve (Indianapolis, IN: Hackett Publishing), 194b25. That is, the form is not literally, or at least not in all cases, the shape of something, but, as we shall see, the principle of unity that is defined in terms of a definitive functionality. Notice also, even material "simples" that are naturally indivisible are matter-form compounds in Aristotle's sense, inasmuch as they are subject to change; mobile beings are composites of act and potency.
6. We might say that the form is that which determines what does nor does not count as a merely incremental loss or gain of parts, as opposed to a substantial change. Thanks to Andrew Jaeger for making this point.
7. Mark Johnston, *Surviving Death* (Princeton, NJ: Princeton University Press, 2010), 202–203. See also Johnston's influential statement of a similar approach to hylomorphism in his paper "Hylomorphism," *Journal of Philosophy* 103 (2006): 652–698. For an insightful and critical treatment of this way of motivating hylomorphism (the unique persistence conditions of wholes over their parts), see E. J. Lowe, "Form without Matter," *Matter and Form: Themes in Contemporary Metaphysics*, ed. David S. Oderberg (Malden, MA: Blackwell Publishing, 1999), 1–21.
8. Aristotle, *Physics*, 193b5. Further references to Aristotle will be given by the Bekker numbers (page, section, and line numbers from the Bekker edition).
9. Aristotle, *Physics*, 194a20–194b15. Of course, a desk, as an artifact, does not have a natural purpose, but a purpose imposed on its matter by our contrivance. For this reason, Aristotle would say that a desk has merely an *accidental form*, as opposed to a *substantial form*. This distinction will be further elaborated below.
10. Of course, the notion of an essential kind of an artifact, such as a desk, is at best foggy. Surely, however, a desk, though not strictly a member of an essential kind, is more so than a pile of woodchips. We might, then, say that the form is the disposition that specifies something as a member of an essential kind *to the degree that such a thing is a member of an essential kind*. Or, as Aristotle famously puts it, "Something is said to be in many ways," *Metaphysics*, trans. C. D. C. Reeve (Indianapolis, IN: Hackett Publishing, 2016), 1003b32. See also 1017a7–1017b8.
11. I find talk of a "matter-form composition" dissatisfying as this way of putting it makes the form sound like a part. I will, however, reluctantly put it this way in lieu of a better phrase.
12. Some hylomorphists are willing to push this ontological precedence of the whole organism over its parts to the most fundamental level of material composition. See Patrick Toner, "Emergent Substance," *Philosophical Studies* 155 (2011): 65–81; and Madden, *Mind, Matter, and Nature*, 235–242.
13. Aristotle, *Physics*, 192b13.
14. There is also no deeper fact of the matter as to what counts for a desk than what human beings claim to count as a desk, or at least it is entirely our practical needs that define the "final cause" of a desk. We would

do no violence to the objective order of things, if we somehow came to characterize desks and tables as the same species, instead of distinctive species within the common genus of furniture. The categorization of organisms, however, shows greater resistance to human whim. Of course, we could be wrong about our current taxonomy of organisms, but it strains credulity to claim that there is no sense in which by distinguishing a dog from a cat that we are "breaking nature at its joints" in a way that we are not when we distinguish tables from desks.

15. Aristotle, *De Anima*, trans. C. D. C. Reeve (Indianapolis, IN: Hackett Publishing Company, 2017), 412a26–27.
16. *De Anima*, 412a20. My emphasis.
17. Aristotle observes a sense in which the formal and moving causes, along with the final cause, can coincide ("amount to one"), but he does not use "cause" univocally, but analogically, when making this point. See *Physics*, 198a20-27.
18. There is a sense in which the soul or, more generally, the form, is a cause of the organism or object, but only as the "paradigm" or "account of the essence" that the moving cause is ordered to bring about (*Physics*, 194b25). That is, in order to bring about Fluffy, the activity of Mr. and Mrs. Fluffy had to be ordered toward the soul of cat.
19. *De Anima*, 412b10–132, 412b18–20 and 412b27–4134a3.
20. For an interpretation of Aristotle on soul along the lines I have followed here, see Johnathan Lear, *Aristotle: The Desire to Understand* (New York: Cambridge University Press, 1988), 96–101.
21. Something is, then, a cat as soon as an individual with a disposition toward a certain set of definitive feline activities comes to be. A very early-stage cat zygote *is* a cat, as it is on its way toward (developmentally ordered to) these activities, in a way that that, say, a very early-stage dog zygote is not; the latter is in no sense naturally on its way toward, say, distinctively feline hunting techniques, whereas the former is so disposed. Once again, there is an intrinsic connection between the notion of a *soul* (form) and the *final cause* (ultimate end of development) of an organism. As soon as something comes to be as disposed to the final cause of a certain organism, then we have an organism of that type, e.g., something with the soul of a cat. Another way to look at this point is to see that the definitively feline activities include following a certain process of fetal development. Once there is something engaged in that process (even though very few, if any, of the stages of that process have been completed), that thing is a cat. Notice also that the gametes can be distinguished from the being of the cat. Neither the sperm nor the egg that occasioned Fluffy's coming to be were *themselves* engaged in the developmental course distinctive of cats; i.e., the gametes were not on *their* way to being cats. As soon as something took of that course, something new came to be as a cat.
22. See Aristotle, *Metaphysics*, 1048a30–1048b35. Note also how this account supports Aristotle's claim that one can only be happy once having lived a complete life. See Aristotle, *Nicomachean Ethics*, trans. C. D. C. Reeve (Indianapolis, IN: Hackett Publishing, 2014), 1099b9–1101b9.
23. Obviously, this account of soul is rife with teleological connotations, but I do not see any reason to think that these should be seen as odious to anyone with broadly Darwinian sensibilities. However one understands the definitive dispositions of species of organism to have been put in place (and whether those dispositions and species are stable over long stretches of times), I don't see any serious reason to doubt that such dispositions ultimately aim toward the flourishing of such organisms.
24. It should also be clear that Aristotle's hylomorphic metaphysics cannot be easily separated from his ethical views, in particular his central claims in Books 1 and 10 of the *Nicomachean Ethics*, which themselves cannot be entirely extricated from natural theological views. For a synoptic account, see Lear, *Aristotle: The Desire to Understand*, 293–320.
25. See *Physics* 194b9–15, and *De Anima* 403a3–28.
26. I have approached the matter/form and matter/soul distinction in Aristotle primarily from his remarks in the *Physics* and *De Anima*, which is fairly typical of the scholastic tradition. C. D. C. Reeve provides an excellent introduction to a different approach to the same conclusion by Aristotle, relying on his theory of sexual reproduction and broader scientific works. See C. D. C. Reeve, *Aristotle: A Quick Immersion* (New York: Tibidabo Publishing, 2019), 13–37. Reeve's book is also an excellent overall introduction to Aristotle in general, with particular emphasis on how his views have fruitful bearing on a swath of contemporary issues. Of course, Reeve does not claim (nor do I) that Aristotle's biological and cosmological views can be adopted today wholesale.
27. *De Anima*, 412b5–8.
28. *De Anima*, 408a28–b18.

29. For a recent defense of the claim that consciousness should be understood as an engagement of a whole, not merely a part of an organism, see Maxwell Bennett and Peter Hacker, *Philosophical Foundations of Neuroscience* (Oxford: Blackwell, 2003), 68–80. This position does entail that consciousness is a biological phenomenon, but not in the sense that it can be reduced to biology mechanistically construed, nor even that consciousness is determined by, even if not identical to, its biological underpinnings. Rather, the idea is that life, *bios*, is itself something on the continuum of consciousness that enjoys a certain autonomy (a "needful freedom") from its constituents. For an elaboration of this point, and much else in this chapter, from a very helpful phenomenological perspective, see Hans Jonas, *The Phenomenon of Life: Toward a Philosophical Biology* (Evanston, IL: Northwestern University Press, 2001). See also, Evan Thompson, *Mind in Life: Biology, Phenomenology, and the Sciences of Mind* (Cambridge, MA: Harvard University Press, 2007).
30. I am basically playing on a distinction that Robert Brandom makes between *sentience* (animal awareness generally) and *sapience* (distinctively human intentionality) and employs at various places in his writings. This distinction is a key doctrine in Brandom's vast body of work, but for succinct explanation see his *Reason in Philosophy: Animating Ideas* (Cambridge, MA: Harvard University Press, 2009), 135. For a classic statement by Aristotle to the effect that the distinctive human virtue is the ability to take a second–order, critical (reason–giving) stance toward our theoretical and practical activities, see *Nicomachean Ethics*, 1102a4–1103a10
31. For a more detailed account defending the notion that higher, non–human animals should be taken as implicitly rational, see Alasdair MacIntyre, *Dependent Rational Animals: Why Humans Need the Virtues* (La Salle, IL: Open Court Publishing, 1996), 53–61.
32. The notion that the mark between non–human and human animal rationality is the ability to articulate norms is another central theme of Brandom's (though he does not often use the term "rational" to apply to non–human animals). For succinct statements by Brandom along these lines see *Reason in Philosophy*, 32–33; and Robert Brandom, *Articulating Reasons* (Cambridge, MA: Harvard University Press, 2001), 80.
33. To say Fluffy is rational (though not logical) in this sense, is really just to say that he has a cat soul in the hylomorphic scheme I sketched above. The cat soul is the *being-able* directly to engage in the distinctive behaviors of feline-kind, i.e., the ready disposition to abide the norms of being a cat. Since Fluffy is a living thing (a possessor of soul, rather than mere form), his conformity to these ontologically definitive norms is not passive, but active and differential, as he adjusts his behavior situationally so as to instantiate these norms. Fluffy is then rational inasmuch as his behavior mediates between the general and the particular in indefinitely many circumstances in ways that are ordered toward his flourishing. Whatever the means by which it came about originally, nature is such that, at least at the level at the level of organisms, it can be interpreted as "making sense," and to that degree all living things (all soul-possessing beings) are rational. The possession of a cat soul is then to be disposed toward certain behavioral norms that are themselves ordered to flourishing as a cat.
34. None of what I have argued here is to claim that there is nothing non-physical or immaterial about Fluffy's consciousness. Rather, my claim is only that there is nothing interestingly non-physical about Fluffy's consciousness. I will make the case to that effect regarding Fluffy's cognition in what follows, but I claim the same regarding qualitative consciousness (the famed "qualia" of a recent philosophy of mind). Indeed, I am happy to recognize that there is a perfectly good sense in which the qualitive aspects of consciousness (human or non-human animal alike) are not identifiable with any discrete material event or structure. That, as we shall see presently, does not license the conclusion that such phenomena must be identified with some discrete immaterial event or structure. Rather, it may be that qualia, like thoughts, just are not the sort of things that can be so discretely identified. For a similar account of sensation, even in its qualitative aspects, as accounted for as participation in a "form of life,' see also: Roger Scruton, *The Soul of the World* (Princeton, NJ: Princeton University Press, 2014), 40–43; Scruton, *Art and Imagination: A Study in the Philosophy of Mind* (South Bend, IN: St. Augustine Press, 1998), 104–106; and Anthony J. Rudd, "What It's Like and What's Really Wrong with Physicalism: A Wittgensteinian Perspective," *Journal of Consciousness Studies* 5 no. 4 1998: 454–463.
35. I am taking the notion of basic consciousness as a sort of bodily comportment from the reading of the phenomenological and psychological literature in Hubert Dreyfus and Charles Taylor, *Retrieving Realism* (Cambridge, MA: Harvard University Press, 2015).

36. For an extended account and defense of the notion that mental states are relations to worlds of involvement, see John Haugland, "Mind Embodied and Embedded," in *Having Thought: Essays in the Metaphysics of Mind* (Cambridge, MA: Harvard University Press, 1998), 207–240.
37. See Elizabeth Anscombe, "The Immortality of the Soul," in *Faith in a Hard Ground: Essays on Religion, Philosophy, and Ethics*, ed. Mary Geach and Luke Gormally (Exeter, UK: Imprint Academic, 2008), 69–83; and Ludwig Wittgenstein, *Philosophical Investigations*, trans. Anscombe, Hacker, and Schulte (Oxford: Blackwell, 2009), §146–154 and §304–306.
38. I argue that Fluffy's cognition cannot be identified with any discrete neurophysiological occurrence, but not due to the current paucity of our understanding of the complexity of the feline nervous system. Indeed, even if we had a complete, neuron-for-neuron mapping of all of Fluffy's various cognitive performances (a sort of idealized feline neuroscience), we would not thereby have identified Fluffy's cognition. Suppose we are considering Fluffy's cognition of a mouse, as an object of his hunting. Fluffy's cognition can only be understood fully in terms of a co-evolutionary history with the mouse's cognition (knowing how to evade cats). Moreover, we cannot understand Fluffy's hunting cognition without understanding his evading cognitions (it does a cat no good to be preyed upon while hunting), so now we have to add a co-evolutionary account of the cognition of, say, the hawk. Of course, with that inclusion we will need to draw the natural histories of various other animals, and even inanimate beings, into the picture. In short, a complete understanding of Fluffy's cognition requires nothing short of a complete natural history of the *umwelt* he occupies. To be a cat cognizing is to be an inheritor of a natural history involving elements far broader than anything that can be read off a mapping of the feline nervous system, which is itself only a contributor to the cognition. Even if we could produce a cat brain and stimulate it so as to be in the same state as Fluffy, such an artificial brain would not be thinking about mice or hawks unless it were appropriately related to such entities. What makes Fluffy's nervous system capable of thinking about mice and hawks is its position in a long, shared natural history with such animals. The history of that world, however, is something that is itself contingent. Thus, there really is no complete, absolute understanding available in these matters, at least in the sense in which we would expect such a final story in physics. For a treatment of the notion of the *umwelt* in cognitive explanation, see Andy Clark, *Being There: Putting Brain, Body, and World Together Again* (Cambridge, MA: The MIT Press, 1998), 24–31; for more on the importance of co-evolution in cognitive explanation, see the same work, 87–94. One could be a Mere Hylomorphist while denying the sort of extended mind thesis that Clark develops, and Clark shows no overt sympathies for hylomorphism. Once, however, we attempt to understand the importance of teleology in the Aristotelian understanding of the definitive capacities of organisms and the importance of co-evolutionary explanation for the same, it is clear that the Aristotelian should have more than a little interest in much of what Clark has to say.
39. Robert Brandom, *Making It Explicit: Reasoning, Representing, and Discursive Commitment* (Cambridge, MA: Harvard University Press, 1994), xi.
40. There is much debate over the explanatory priority of the implicit know-how we have in common with non-human animals and the explicit conceptual articulacy which seems to be a possibility only for human beings. Contemporary philosophers working broadly in the phenomenological tradition defend the priority of the former (see Dreyfus and Taylor, *Retrieving Realism*), while philosophers of the Pittsburgh School who trace their thinking back to Hegelian, Kantian, and Aristotelian sources defend the latter (see John McDowell, *Mind and World* (Cambridge, MA: Harvard University Press, 1994). For a synoptic view of the debate, see Joseph K. Schear, ed., *Mind, Reason, and Being–in–the–World: The McDowell–Dreyfus Debate* (New York: Routledge, 2013). Lee Braver's contribution to Schear's volume, "Never Mind: Thinking of Subjectivity in the Dreyfus–McDowell Debate" (143–162), comes close to my own view. For another synthetic treatment of the Hegelian and Heideggarian strands of this controversy, see Matthew Crawford's magnificent *The World Beyond Your Head: On Becoming an Individual in an Age of Distraction* (New York: Farrar, Strauss, and Giroux, 2015), 60–68.
41. Whether *homo sapiens* are the only animals with the power to make the norms governing their behavior explicit is an empirical question. Though I am unaware of compelling evidence showing that any non–human animals have such powers, if such evidence were provided, I would be quite willing to extend the anti–physicalist implications of this chapter's line of argument to such animals, including the ethical and theological consequences. For a recent presentation of the empirical evidence on animal intelligence and human

distinctiveness that I take as ultimately supportive of my stance regarding normative commitment in human cognition, see Matt J. Rossano, *Supernatural Selection: How Religion Evolved* (New York: Oxford University Press, 2010). I also have found William von Hipple, *The Social Leap: The New Evolutionary Science of Who We Are, Where We Come From, and What Makes Us Happy* (New York: Harper Publishing, 2018) helpful along these lines.

42. Brandom, *Reasons in Philosophy*, 62. See also McDowell, *Mind and World*, xix–xx.
43. Aristotle, *Nicomachean Ethics*, 1098a12.
44. Though I am using a concept from formal logic, *modus ponens*, for the purposes of this example, our material logical inferences are no less subject to distinctively human, normative scrutiny. See Brandom, *Reason in Philosophy*, 43–45.
45. This way of framing an argument for the immateriality of thought based on the logical content of distinctively human cognition owes much to James F. Ross, *Thought and World: The Hidden Necessities* (Notre Dame, IN: University of Notre Dame Press, 2008). For an extensive defense of this sort of argument, see Edward Feser, "Kripke, Ross, and the Immateriality of Thought," *American Catholic Philosophical Quarterly* 87, no. 1 (Winter 2013): 1–32.
46. This line of reasoning has been, of course, quite controversial for centuries. The most prominent recent objection is the accusation of a "content fallacy." See Robert Pasnau, "Aquinas and the Content Fallacy," *The Modern Schoolman* 75 (1998): 293–314, and Pasnau's *Thomas Aquinas on Human Nature: A Philosophical Study of "Summa Theologiae" 1a 75–89* (New York: Cambridge University Press, 2002), 45–72. I offer a reply to this criticism in "Is a Thomistic Theory of Intentionality Consistent with Physicalism?," *American Catholic Philosophical Quarterly* 91, no. 1 (2017): 1–28.
47. Aristotle, *De Anima*, 430a1–3.
48. *De Anima*, 430a17–24.
49. The question of what sort of subject underlies this process of conceptual-logical articulation is likely to arise. Specifically: what accounts for the identity of the reasoning subject throughout the process of conceptual articulation? Here, I think the hylomorphist can give the same sort of answer she gives for change in general: the constancy of the final cause as fixed by the formal aspect of the process. That is, the identity of the reasoner is provided not by an ontologically distinct, separate entity that abides during the process, but that the process of reasoning is itself aimed at a certain end that entails a certain set of related formal (logical) properties that persist throughout the process. These remarks are a painfully quick reply to a very good question from a blind review, which really deserves an essay-long response of its own.
50. Of course, later Aristotelians, most famously St. Thomas Aquinas, argue that these considerations do in fact argue effectively for the notion that the human soul can be taken as a separable individual. I have attempted to defend the plausibility of this move in "Is a Thomistic Theory of Intentionality Physicalist?" and *Mind, Matter, and Nature*, 265–273. See also Elizabeth Anscombe, "Has Mankind One Soul: An Angel Distributed Through Many Bodies?" in *Human Life, Action, and Ethics: Essays by Elizabeth Anscombe*, eds. Mary Geach and Luke Gormally (Charlottesville, VA: Philosophy Documentation Center, 2005), 17–26.
51. See Aristotle, *Metaphysics*, 1072b18–35.
52. Of course, I have done nothing in this chapter to motivate the conclusion that there is an ultimate normative structure to the universe, but this is the main concern of Aristotle's natural theology in the *Metaphysics*, and certainly that of scholastic philosophers who take those arguments as their point of departure.
53. See Aristotle, *Nicomachean Ethics*, bk. x, ch. 8.
54. Aristotle, *Metaphysics*, 1035b23–27.
55. One could make a case for the DNA as the controlling and separable part of an organism, but as DNA is neither alive nor sufficient for the organism's being alive, I doubt that a hylomorphist would take it as a candidate for the locale of the soul. That being said, there is a perfectly good sense in which DNA is the controlling part of an organism; i.e., leaving aside epigenetic influences, it does indeed set the developmental course the organism will follow for its entire lifespan, at least as a limiting factor. Notice, however, that there is likewise a perfectly good sense in which the DNA is not the controlling part of the organism; e.g., Fluffy's DNA does not determine the particularities of any actual hunting activities. Those details, which constitute Fluffy's fullest actualization as a cat, are set by his interaction with a dynamic environment, and the best candidate for who is running that show (from Fluffy's side) is his central nervous system. Thus, I believe that the hylomorphist will need to tell a complicated, multi-layered story here. On the other hand, all we need to have an organism of a certain kind is to have a living thing determined to follow a certain developmental process, and what does that determining work (even if it is insufficient for the organism to be alive) is, plausibly,

its DNA, which then implies that the DNA is a higher-order controlling part of the development of the organism. On the other hand, the dynamic later-stage activities of the organism, at least in the cat and human case, are governed directly by a brain that, though structurally limited by this higher-order controlling part, operates in an important sense independently of direct control by DNA. Thus, the soul of an organism will be located in a series of controlling parts, depending on the developmental stage of that organism and its ability to operate independently in response to the particularities of its environmental situation, as opposed to merely running its developmental program. We might then envision a coherent story involving the notion of higher-order, or developmental controlling parts, and direct, first-order controlling parts. This line of thought is, admittedly, underdeveloped and quite speculative. Elizabeth Anscombe raises possibilities for dealing with issues for the diachronic identity of the organism that are likely to arise as this account is developed. See Anscombe, "Were You a Zygote?," in *Human Life, Action, and Ethics*, 39–44.

56. J. L. Padmanabhan et al., "A Human Depression Circuit Derived from Focal Brain Lesions," *Biological Psychiatry* 86, no. 10 (2019): 749–758.
57. The "most directly" in this sentence is crucial, since in the brain, everything pushes back and forth with everything else. Everything in the brain is exquisitely interconnected. There is no completely discrete "part" of the brain responsible for higher-order thought, but various overlapping substrates to which we can assign functions.
58. In regard to the study I have cited above, it is important to note that the imaging technique being used (lesion network mapping) is not constrained to the cortex. There is no methodological reason why the lesion network mapping search should have localized the optimal treatment target to the cortex. This is important methodologically, because the cortical mapping finding was not merely due to the fact that the method only targets superficial neural tissue. In other words, the fact that the brain circuit mapping methodology is spatially unconstrained makes the case much stronger for the claim that a depression treatment targeted in higher cognitive substrates aligns with an Aristotelean hypothesis of mechanisms underlying human flourishing.
59. I have kept Aristotle's phrase "vegetative," though we would take it to include any very basic organism that lacks consciousness under this moniker, plant or otherwise. Likewise, we might now be reluctant to include all plant life as completely insensate.
60. Aristotle, *De Anima*, 435a21-435a10.
61. *De Anima*, 431b20-432a14.
62. Aristotle, *Nicomachean Ethics*, 1102b4-1103a9.

7. Of Thinkers, Thoughts, and Things: A Commonsensical Defense of Idealism

Douglas Axe

1. Introduction

My aim in this chapter is to defend one of several accounts of reality that have been called *idealism*. Most famously articulated by the early-modern Irish philosopher George Berkeley,[1] the idealism I subscribe to has been summarized by the statement: "reality consists exclusively of minds and their ideas."[2] As we will see, the terms "minds" and "ideas" need to be understood broadly in order for that summary to work, but it is at least a good starting point.

I'm motivated here not by a desire to convince professional epistemologists or philosophers of mind that Berkeley got it right. In the first place, that would require credentials I lack. In the second, the picture of reality presented by physicists of today is sufficiently unlike the picture of his day that the idealism of today should be presented differently as well.

I'm driven instead by two convictions that are amply supported by my own experience. The first of these is that the most fundamental questions are best viewed not through the formal lens of philosophy but through the familiar lens of common sense. Throughout the ages, astute thinkers have admitted as much. King Solomon's commendation of good sense and warnings about being wise in one's own eyes[3] are invitations to humble, commonsensical wisdom. John Calvin's *sensus divinitatus*—the innate sense of God in humans—is an instance of common sense that overshadows philosophical speculations about God. And, more generally, *foundationalism* is a well-established philosophical tradition that views all knowledge as resting on bedrock beliefs that are rightly accepted not because they are provable but because they are both obvious and necessary.[4]

Over two centuries ago, Scottish philosopher Thomas Reid described the foundational role of common sense, and the folly of philosophers who think that they have somehow risen above it, as follows:

> In this unequal contest betwixt Common Sense and Philosophy, the latter will always come off both with dishonour

> and loss; nor can she ever thrive till this rivalship is dropt, these encroachments given up, and a cordial friendship restored: for, in reality, Common Sense holds nothing of Philosophy, nor needs her aid. But on the other hand, Philosophy, (if I may be permitted to change the metaphor) has no other root but the principles of Common Sense; it grows out of them, and draws its nourishment from them: severed from this root, its honours wither, its sap is dried up, it dies and rots.[5]

My second motivating conviction is that the version of materialism that has become so prevalent among people of a scientific bent is highly counterproductive. This view, which I'll call *physicalism*, holds that reality consists entirely of physical *stuff*, meaning the various measurables that enter into the equations of physics (mass, energy, force, time, etc.). Some physicalists leave room for the mere existence of non-physical phenomena that emerge with natural regularity from physical stuff. Consciousness is a prime example. Physicalists either deny this phenomenon altogether or view it as a mere effect, devoid of causal power. Either way, the stuff of physics is taken to be the base reality.

Whether or not the version of idealism I have in mind precisely matches that of George Berkeley, this modern version argues against both kinds of physicalism in a way that is compelling and commonsensical—a winning combination. My aim is therefore to make this idealism more approachable to non-experts by removing the misconceptions that make it seem strange.

Berkeley preceded me in this, anticipating the grimaces and blank stares: "But, say you, it sounds very harsh to say we eat and drink ideas, and are clothed with ideas."[6] After agreeing that it does, he made it seem less so by pointing out that "we are fed and clothed with those things which we perceive immediately by our senses," and these immediate perceptions are, without exception, immaterial.

Take a freshly laundered flannel shirt, for example. What's left of it if you strip away the smell of the fabric softener, the feel of the collar against the neck and the sleeves against the arms, the hardness of the buttons against the fingernails, the familiar warmth of thick cotton, and the classic look of aged plaid? Lots of *atoms*, the physicalist would say. True enough. But when was the last time you perceived an atom *directly*, apart from immaterial sensations like these?

It can't be done. That was Berkeley's point. Once this is conceded, as it seems it must, the next point comes in the form of a follow-on question: If we only know about atoms through our conscious experience, why should we take the former to be more fundamental than the latter?

This needs unpacking, which we'll do in a moment. The point of this quick preview is simply to show that the strangeness of idealism on first encounter has a way of resolving itself on further reflection. To deny the reality of the outside world would be irreconcilably weird, but this is nothing of the kind. The idealism I'm defending places itself firmly within the broader camp of *realism*, which affirms the true existence and objective properties of the world outside our minds. It merely reinterprets the fundamental nature of that outside world.

With a bit of thought, the idealistic interpretation will be as comfortable as a flannel shirt.

2. Thinkers, Thoughts, and Things

Based on the observation that each of us perceives the outer world from the perspective of our personal inner world, I have used the terms *thinkers*, *thoughts*, and *things* to denote categories that might reasonably be thought to encompass all of reality: "The outside world consists entirely of *things* (galaxies, atoms, trees, computers, our bodies, etc.), whereas each inside world consists of the mental space in which one *thinker* has *thoughts*."[7] The outside world, then, consists of everything that lies outside all inner worlds, making it equivalent to the physical universe, by this view.

As commonsensical as this division of reality may be, there doesn't seem to be a fully satisfactory way to make sense of it. Grouping thinkers together with thoughts as *inside* stuff implies that *things* stand alone on the outside—not just in the sense of being categorically different but in the more problematic sense of being irretrievably isolated. That is, if thinkers directly interact with thoughts only,[8] such that those two categories occupy a realm completely separate from the realm of things, then what relevance could things possibly have to us thinkers? What basis could we have for believing they even exist?

In preparation for showing how idealism provides a satisfactory solution to this puzzle, I will argue not only that physicalism comes up short but also that its best-known alternative—*substance dualism*—does too. Most strongly associated with René Descartes,[9] substance dualism (henceforth, *dualism*) holds not just that the mind is distinct in substance from the brain but also that the possessor of mental states is "something quite over and above the states themselves, and is immaterial, as they are."[10] That is, dualists believe not merely that thoughts are immaterial but also that these exist only by being held in minds that are likewise immaterial.

In what follows, I won't be suggesting that dualism is as problematic as physicalism is. Instead, I see dualism as a good place for people to land when they recognize the untenability of physicalism, but idealism as the most compelling final destination for former dualists (among which I count myself).

3. *The Problem with the Physicalist Account of* ***Things***

The crux of the matter is how we view *things*, which (again) is equivalent to how we view the physical universe. In denying the existence of any non-physical realm, the physicalist is not denying the existence either of thinkers or their thoughts. Rather, the physicalist holds that thinkers and thoughts must themselves be reducible to physical things. In other words, the physicalist claims that our three categories of reality collapse into one.

The rapid ascent of artificial intelligence (AI) since the mid-twentieth century lends at least superficial plausibility to this claim. If the human brain is really just a "computer made of meat" (an expression attributed to AI pioneer Marvin Minsky[11]), then the distinction between human cognition and electronic computation is one of *instance* rather than of *kind*. In that case, computers are the purely physical thinkers that prove the plausibility of purely physical thought, thus sealing the argument in favor of physicalism.

We tend to overlook the flaw in this reasoning because we take for granted that we *ourselves* are in a position to think. Scrutinizing this assumption reveals something important, as I'll attempt to show here with my own example of critical self-reflection:

> For all the times I talk about knowing this or that, all of this "knowledge" is really predicated on the hope that I'm not profoundly confused or deceived. I have the strong impression that I have a real personal history situated within a larger world history—that I've been alive and active in this real world, experiencing real events in real time and learning from them all these years. I feel certain this isn't a false impression.

> And yet, having many times in my sleep confused dream experiences for real ones, I admit that I have no way to prove the reality of what I take to be my life history. Worse, I don't even have a way to prove my own *sanity*—that any of my thoughts or impressions are the least bit coherent. I have the strong impression that I am awake and sane in a real world populated by other people, but having frankly conceded the contrary as a possibility, I have no way to rule that undesirable alternative out, or even to assign to it a reassuringly small probability. After all, how could trustworthy math or sound reasoning bubble up from a cauldron of pure madness?
>
> Having no prospect of bootstrapping myself out of this predicament, then, I must proceed on nothing more than the *hope* that I am not deeply deluded.

In short, physicalism nullifies itself by destroying that hope. To say that neural activity in our brains is running the show when we engage in thought is to say that our own perception of our mental activity is so badly confused as to foreclose all hope of reason.

We have the strong impression that we are working with non-physical concepts when we think—drawing on knowledge and insight to shape new ideas. We labor to find the right words to express these ideas in language so that they can be recorded and communicated to others through physical media. Nevertheless, we insist that the meanings we assign to these words occupy a non-physical realm. Anyone who doubts this should try to identify the physical thing that is precisely what we mean when we use a word like *delight*, or *equal*, or *through*—such that to destroy this identified thing would be to render the corresponding word meaningless. The obvious absurdity of this proves the point.

Notice that the claim that neuronal activity is producing my impression that I am manipulating non-physical ideas is every bit as destructive to my hope of sanity as the claim that this impression is false. For example, when I apply my mental attention to prime numbers, or to justice, or to medieval European history, hoping to draw some well-reasoned conclusion about the object of my contemplation, my thought habits (attention to detail, critical scrutiny, etc.) are what guide me. Having exercised those habits, my sense of confidence that what I've arrived at is both sound and valuable is what signals success. But if this entire process is actually nothing more than a succession of delusions—sparks thrown up at me by the crackling of synapses within my skull—then I'd be a fool to place any confidence in it. To say otherwise is to say that I should have confidence in thinking that's completely delusional.

Think of it this way. If neurons are actually running the show, inducing me to have "Wow—this is a good insight!" impressions not when I actually have a good insight but whenever their circuitry forces that impression upon me, then I'm not actually a thinker at all. I'm merely being duped into *seeing* myself as a thinker. So, *the very act of supposing that my brain produces my conscious experience leaves me no option but to abandon all confidence in that experience*. And with that lost confidence goes absolutely everything—whether brains or neurons or computers or AI or science or Marvin Minsky or even my own existence. *All* of this goes up in the smoke that forces these sparks upon me.

In light of this, René Descartes's famous statement—*Cogito, ergo sum*—falls short of being a compelling argument because it doesn't distinguish between the *impression* that one is thinking and the actual *fact* that one is thinking, which remains uncertain. Since actual thinking would be required to draw his conclusion, that conclusion cannot be more certain than our *hope* that the impression is not misleading. In the end, then, the claim we're entitled to make—I *think* I think, therefore I *think* I am—is more modest.[12]

4. The Problem with the Dualist Account of ***Things***

By affirming the reality of a non-physical realm in which thinkers can have their thoughts, dualism accords with our firsthand account of our mental activity, thereby avoiding the self-defeating effect of physicalism. This is no small advantage, and yet I will argue that this advantage is more than offset by difficulties.

Even physicalism carries advantages, at least in the minds of some. Simplicity is one that has been argued.[13] If thinkers and their thoughts really could be understood as just two more examples of physical *things*, then we would indeed have a tidy picture of reality. But we now see why this doesn't work. As Albert Einstein is purported to have cautioned: "Everything should be made as simple as possible, but not simpler."[14]

For many, the other great advantage of physicalism is that, unlike dualism, it suits atheistic and agnostic worldviews. When Richard Dawkins extolled Charles Darwin as the one who "made it possible to be an intellectually fulfilled atheist,"[15] he was really extolling the theory that seemed to bring biology into the physicalist camp. Before Darwin, atheists had to admit that they had no explanation for how living things came to exist. Darwinism became a key component of physicalism, and therefore atheism, precisely because it claimed to solve this otherwise vexing problem.[16]

This raises for me the matter of how best to defend idealism to readers coming from very different starting points. To atheists and agnostics, my first point (made in the previous section) is that physicalism simply isn't a coherent option. These readers may want to skip to the next section, where I describe several deep conundrums of modern physics that cry out for resolution and for which idealism seems to be the answer. For the remainder of the present section, though, I need to lay out the problematic doctrinal implications of dualism for theists, fellow Christians in particular.

According to the book of Genesis, the physical universe came into being when God acted to bring it into being.[17] The New Testament describes God himself as a "spirit" being,[18] eternally existing as Father, Son, and Holy Spirit.[19] Humans are described as having both *flesh* (referring to the physical body) and *spirit* (referring to the immaterial self).[20] Only with the incarnation of Christ was God (specifically, God the Son) "manifested in the flesh."[21]

Those who follow these teachings therefore agree that the distinction between flesh and spirit is of profound importance. Indeed, the importance goes well beyond the mere distinction between what is physical and what is not, and yet that most basic dichotomy is clearly part of the teaching (see, for example, Luke 24:39). Being in complete agreement with the full teaching of Scripture on this matter, I am for the present purposes focusing narrowly on this basic dichotomy.

The matter I raise here is whether the divide between the realm of the physical and the realm of the non-physical can be as absolute and categorical as substance dualism implies. Did God's creation of the physical universe consist of furnishing an entirely separate category of existence, hitherto unoccupied, with stuff? Or would it be better described as the creation of something new within a category that was never empty? To say the first—that God gave a once-empty category of realness content—seems inescapably to imply that God himself was not *real* in that fundamental sense until the incarnation, and that God the Father and God the Holy Spirit will forever be unreal in that same sense. Another way to pose the question, then, is this: *do physical reality and non-physical reality have equal ontological status?* Surely the answer must be *no*, considering that the one was produced only through the other.

Consistent with this, a *yes* answer seems deeply problematic. How would the Supreme Being—the very source of existence—have broken into a realm where from eternity past he did not exist? How could there even *be* such a realm? This isn't at all like a person entering a room, where the person and the room had their locations in the same real space all along. Rather, this is a person acting in a realm *where they don't exist*. Within that realm, then, this is like *nothing* producing *something*. Indeed, if the physical realm and the spiritual realm are entirely disconnected spaces, what influence could a being who is firmly situated in one of these spaces possibly have on the contents of the other? If God's hands are spiritual and hydrogen has absolutely nothing to do with the spiritual, what levers could he conceivably have pulled to bridge that divide in order to form the stars?

On the other hand, is it possible that dualism has been unduly influenced by physicalism? Specifically, I wonder whether the physicalists were so successful in convincing people that the realm in which God was thought to reside had been *replaced* by the realm in which hydrogen resides that opponents of physicalism (substance dualists) felt they had to add God back into the picture, so that he and hydrogen could both have their domains. If so, might that response have conceded too much?

5. Conundrums that Invite a Change of Paradigm

If a large adult and a small child occupy neighboring seats on the park swings, both asking you to push them, you'll find the child much easier to push. And yet once you have both of them moving, gravity keeps them swinging back and forth at precisely the same rate, like pendulums. The only explanation for this is that gravity automatically supplies "effort" in direct proportion to what is needed to make an object move.

This equivalence between the property of a thing that gravity acts upon—*gravitational* mass—and the property of a thing that resists any change in motion—*inertial* mass—was the cause of much head scratching from the time of Isaac Newton to the time of Albert Einstein. Electricity behaves very differently. Here the tug exerted by an electric field is not proportional to the inertial mass but rather to the net charge. Why does gravity work differently? Why would the "charge" that gravity tugs upon in an object equal the inertial mass of that object when these two aspects seem so different?

This unexpected correspondence, which came to be known as the *equivalence principle,* suggested to Einstein that gravity is more intimately connected to motion than electrical or magnetic forces are. That hint led, after much brilliant work, to his general theory of relativity, which describes gravity as an aspect (curvature) of the spacetime in which motion occurs. With gravity now intimately connected to physical objects and the space in which they move, the new theory brought satisfying resolution to the mystery: gravitational mass equals inertial mass because these turn out to be two names for one physical property.

The mysterious hints pointing to idealism are more numerous and more peculiar, which makes their resolution even more satisfying, I think.

Conundrum 1: Why Does Physics Look So Much Like Math?

Chief among these hints is what Eugene Wigner, a Nobel-prize-winning physicist and mathematician, once described as "the unreasonable effectiveness of mathematics in the natural sciences."[22] We think of the physical world as being the domain of spheres made of real stuff, spinning and orbiting and occasionally colliding. The big sphere we call home is made of the same tiny spheres that make up everything else—electrons "orbiting" clusters of protons and neutrons. And then, occupying what seems like an entirely different domain, are the spheres of geometry, whose surface areas scale in exact proportion to the squares of their radii. These seem to us like *ideas*, more perfect but correspondingly less real than their physical counterparts. The substantive spheres are *things*, whereas the ethereal ones are *thoughts*, we tell ourselves.

That tidy division works only until we take an interest in the behavior of the substantive spheres. As students of classical physics, we learn that the laws governing the actions of these things are entirely (and elegantly) mathematical in form. So, the study of the world of *things* pushes us into the world of *thoughts*, and in so doing upsets our tidy categories. How can the world of stuff be more real and substantive than the world of ideas if the latter governs the former? And how did mathematical constructs gain this power of governance in the first place? With what kind of tweezers would an equation—an *idea*—compel an electron to obey it?

To grasp the problem more pointedly, consider the hypothetical laws that cosmologists are proposing in an effort to explain the origin of the universe apart from God. However confused that exercise may be, it at least has the benefit of revealing the problem with their categories. Their supposed universe-generating laws are *nothing but* math—"a giant abstract mathematical structure that encompasses all the fundamental laws of physics," as one physicist put it.[23] But how is an entirely mathematical reality supposed to beget (with a big bang) the substantive physical reality it describes? No one has expressed bewilderment at this more memorably than Stephen Hawking, who asked, "What is it that breathes fire into the equations and makes a universe for them to describe?"[24]

Conundrum 2: Why Does Matter Refuse to Be Material?

The mysteries only deepen when we progress from classical to modern physics, because here the material "stuff" of physics seems to vanish altogether! That electron we thought of as being a tiny hard sphere—a *particle*—turns out to be more accurately described by a purely mathematical construct called a "wave function." Like a parent trying to track down a wayward teenager, this wave function has only a vague notion of where the electron will be at next appearance, and like the wayward teenager, the electron seems to prefer elusiveness. When we "catch" the electron, it seems like a classical particle, but the strange behavior between catchings is most definitely math-like, not particle-like.

Furthermore, this refusal to be reduced to something unmistakably material isn't unique to electrons. When examined closely enough, all physical things are made of constituents that defy our classical preconceptions of what material stuff should be.

Conundrum 3: Why Is Physics Intrinsically Non-local and Non-reductionistic?

Physical systems, classically conceived, operate like clockwork. And since the entire universe is a physical system (one big clock), this principle should apply to everything at once. Indeed, there was a time when physicalists believed as much. In 1814, when modern physics was still in its classical Newtonian stage, Pierre-Simon Laplace famously said that we must consider the present state of the universe as the effect of its past state and the cause of the state to come.[25] The unsettling implication that the future is completely determined by the past came to be personified in *Laplace's demon*—a hypothetical being in possession of complete knowledge of the state at some past time, and therefore able to calculate all future events with mechanical precision.

At the time, no one would have thought that physics itself would bury the demon, but that's exactly what happened. Because each component in a clock is directly influenced only by those other components that touch it, one can explain the behavior of the whole clock by combining the explanations of each component. However, for reasons that lie at the very foundation of physical reality, what works for clocks turns out not to work for physical systems generally.

In the language of quantum mechanics, the term for the phenomenon that makes the universe profoundly unlike clockwork is *entanglement*. An entangled pair of particles behaves like a linked unit even if the two halves of this unit are separated by a great distance. That is, what you do to one of the entangled particles *instantly* affects the other, no matter how great the separation. Imagine the impossibility of describing how a clock works if clock parts could be paired in this way! A gear in a clock in Chicago could be more intimately linked to a gear in a clock in Tokyo than to the gears it actually touches in that same Chicago clock. And to make matters worse, the poor clockmaker in Chicago would have no idea where all these mysteriously connected faraway gears were.

The thought of this illustrates the essential relationship between local causation and mechanistic reductionism. The operation of a large mechanism can be understood as a combination of the operations of its component mechanisms *only* if the fundamental causal interactions are local. That works for clocks because the components are large enough that classical Newtonian mechanics is an adequate framework. Nevertheless, the fact that gears are made of particles that refuse to be classical in this way shows that the classical picture of causal mechanism is nowhere to be found in the mathematical regularities that physicists take to be the fundamental description of the universe. Rather, this commonsensical view is something *we bring to* our study of the universe.

Conundrum 4: How Can the True Understanding of Physics Be Rationally Incoherent?

This leads to what is surely the most uncomfortable implication of the modern understanding of physics. All of science has from the beginning been grounded on a common understanding of causal explanation that goes along these lines: to offer a physical explanation of something is to offer an orderly account of the physical causes that produced it, such that wherever and whenever these causes are present, the thing in question is sure to follow. When we think of causal *chains* we think of things happening in sequence.[26] Each effect *follows* its cause in terms of explanatory relationship (i.e., effects *are explained by* their causes), in terms of dependence (i.e., effects *depend upon* their causes), and in terms of temporal sequence (i.e., effects *are subsequent to* their causes). Any deviation from this most foundational relationship would, if taken seriously, surely wreak havoc with the entire scientific enterprise.

This is where the present understanding of quantum mechanics runs into deep trouble. As you may have guessed, entanglement is at the center of the mess. Because what we do to one of the particles in an entangled pair affects the other one immediately across any distance,[27] it is possible for the consequences of our action to be transferred *faster than the speed of light*. You may have heard that signals cannot move faster than light, but you may not have appreciated the conceptual incoherence that results from supposing otherwise.

According to Einstein's theory of relativity, the order of events is fundamentally ambiguous in this situation. That is, a faster-than-light signal can be interpreted as being received *before it was sent* with just as much legitimacy as the more sensible interpretation. Since any of the physical causes in a causal chain can be viewed as a signal (its effect being equivalent to reception of that signal) it follows that the order of cause and effect, and therefore the *relationship* between cause and effect, becomes fundamentally ambiguous once we accept the idea that actions can have instantaneous consequences over long distances. And the demonstrated reality of entanglement seems to push us to that bizarre conclusion.

It's tempting to think this problem is made less worrisome by the admittedly unusual circumstances that expose it. If the relationship between cause and effect is muddied only when physicists do experiments on entangled particles, why should we care? The answer flows from the position this relationship holds within the natural sciences. That causes precede and explain their effects is taken not just to be true but to be *fundamental*, *axiomatic*, *unassailable*—a *bedrock* truth. To admit any exception to *that* kind of truth is to remove the foundation upon which everything has been built.[28]

Interestingly, letting go of the commonsense understanding of the relationship between cause and effect leads to the very same unraveling of everything that we deduced for the physicalist view of mind in section 3. There we concluded that the very act of supposing that our brain *causes* our conscious experience leaves us no option but to abandon all confidence in that experience. We insist that our non-physical minds are the cause of our decisions and actions, which then have their effects in our bodies, including our brains. Physicalism denies that causal relationship by insisting that causation belongs exclusively to the physical realm. And this rightly leads us to reject physicalism.

Now we have seen that the results of contemporary physics, if taken to give a fundamental picture of reality, also erode any fundamental distinction between cause and effect, which leaves us in the same morass! What makes this more surprising is that it isn't so obvious that a distorted view of reality is to blame.

And yet, on closer inspection, that is indeed the culprit.

6. *Idealism as the Answer*

All of these puzzles find their solution in idealism. In fact, the puzzles themselves *point* to this solution. Further, because I see no way to situate this version of idealism within a non-theistic worldview, I won't pretend otherwise. I hope to convince you that a supreme mind, immaterial in substance and personal in nature, must lie at the center of reality. For me to use any term other than *God* to refer to this mind would be disingenuous.

As we wondered why physics looks so much like math, the idea may have occurred to you that physics *is* math. That three-word assertion turns out to be the key to understanding idealism. It needs some unpacking, though. Most theists wouldn't hesitate to affirm that God is (among other things) a mathematician, and they would see the mathematical nature of physical laws as proof of this. But they would distinguish between the laws and the stuff they govern. I've become convinced not merely that there is no compelling reason to make that distinction but, more significantly, that there are compelling reasons to drop it.

The difficulties that beset physicalism and dualism are the strengths of idealism. What use is a separate realm of *things* if all experience of them occurs among *thinkers* in their realm? *None*. Everything works seamlessly if we take things to be mathematical *thoughts* of a special kind. Specifically, physical things are aspects of an enormous and elegant mathematical construct conceived and implemented by God, in which his creatures "live and move and have their being."[29]

In a world where thinkers are the only actors and thoughts are their actions,[30] the notion that neurons are forcing deceptive impressions upon us is as silly as the notion that a thought could force itself upon us without our thinking about it. Neurons, after all, are thoughts of God—aspects of the mathematical construct that God set in motion and continuously tends. As the creator of my mind, God has absolute power over it,[31] but a mathematical idea itself (a neuron, for example) just *is*. Ideas lack the power to make anyone think about them.

The most coherent understanding of the universe in relation to its creator, then, begins with the axiom that God is the only necessary, eternal, uncreated thinker. This perfect thinker is the only one who can create other thinkers. From that position of creative omnipotence, it pleased God to create a rich, mathematical framework as a way of ordering and bounding the experience of his created thinkers (us). Having made us to exist in this way, he gives us the rich experience of his framework by presenting sensory perceptions to us in direct accordance with the state of that portion of the mathematical framework that he designed to be most intimately connected to our sensory experience—the state of our brains. Then, knowing our thoughts,[32] he works their effects out in the other direction, faithfully translating those aspects of our thoughts that should move atoms in our brains into the mathematical representations by which they become inputs to the great mathematical calculation carried out entirely by the mind of God. Like a mathematical wave, those changes to our brain state ripple outward, putting muscles in motion, which puts limbs in motion, which changes the state of the world—by changing the state of the mathematical framework that *is* the world—exactly according to his plan, which he faithfully realizes.

The first conundrum of the previous section evaporates instantly: physics looks so much like math because it *is* math. But this isn't at all to say that our human experience is reducible to math, much less that *we* are. For example, while physics (and therefore math) is certainly part of the explanation for why I sometimes feel hot, it isn't the whole explanation. Adding thermal energy to the air in the room is one way to trigger that sensation, but the sensation itself is intrinsically mental, not reducible to anything physical or mathematical. It is triggered just as reliably when my thoughts cause the kind of internal agitation that makes me want to loosen my collar and step out for some fresh air. Again, anyone who wants to think that all such events are entirely governed by physical brain states needs to reflect on the rational incoherence of that view.

The correlation between brain states and conscious experience demonstrates that much of what God brings to our awareness he brings by translating our brain states into mental perception. Much, but not all. For us to think, we must perceive our thoughts, and these, as we have seen, can't be reduced to anything physical. It follows that only *part* of what comes to our minds originates in our brains, being translated into sensory perception by God. The rest (contemplative thoughts in particular) originate in the mental realm itself. And in the other direction, *none* of the mental states that become translated into brain states are themselves brain states.

In short, our minds are the immaterial seats of our conscious experience, able, by God's action, both to affect the mathematical framework we call the physical world and to be affected by it.

Upon further reflection, the rest of the conundrums evaporate as well. To see this, we must let go of the mistaken notion that the physical world is the main purpose of the created order. Once we recognize that this mathematical framework is merely an instrument in service of the main purpose, everything becomes clear.

That crucial shift of perspective enables us to see that the world was meant to be perceived as we naturally perceive it, not as it is described by quantum physicists. A real and important drama is being played out moment to moment. We each have a real and important role in this drama. The physical world is the *set* that God created as the location for this great drama to unfold. As with any drama, the set exists to serve the higher purpose of the drama itself. The set is important, but it isn't the point.

Most physicists would have us believe that the math of quantum mechanics provides a truer picture of physical reality than common sense does. But there's an important distinction to be made between the *math* and the *picture*. The oddities of quantum mechanics are like the oddities of stage sets when viewed from angles that aren't presented to the audience. Both can be studied, but neither should be studied as though it were main thing. To do that would be to miss the picture—the backdrop for the drama.

Physical matter only defies common sense if we take it to be the base reality. Perhaps this is meant to tell us that it *isn't* the base reality—*God* is! Once we put him in his rightful place, matter becomes something he created for a purpose that's much higher than matter. We were made to perceive the physical world from the orderly perspective of classical physics, not the enigmatic perspective of modern physics. But we were also made to see ourselves—our minds, that is—as separate from the physical world. In this respect, classical physics provides a more sensible description of the physical world than modern physics does.

That said, modern physics gives us a truer glimpse of God's mathematical genius than classical physics does. He could have created a much simpler mathematical framework as the set for his great drama, but he was pleased instead to build something elaborate, overflowing with mathematical richness—richness that exposes the falsehood of physicalism! "The heavens declare the glory of God, and the sky above proclaims his handiwork."[33] Likewise, he could have populated our planet with a much simpler set of living kinds, but he was pleased to fill it with all manner of living richness. "And God saw everything that he had made, and behold, it was very good."[34] In both cases, the rich underlying structure is meant not to distract us from the great drama but rather to leave us even more in awe of its author.

7. *Idealism as Realism*

Various updated editions of Berkeley's 1710 work, *A Treatise Concerning the Principles of Human Knowledge*, summarize the arguments of his critics in comical terms. Charles Krauth, in his 1878 edition, lamented that "It is not to the credit of the metaphysicians who have combated Berkeley that so much they have written is but a prosy elaboration of the jocose misrepresentation of his views."[35] Thomas McCormack, in the preface to his 1904 edition, says of Berkeley's treatise that: "Ridicule has not been sparing of it. Argument has not been wanting. It has been laughed at, written at, talked at, shrieked at. That it has been *understood* is not so apparent."[36]

Astute philosophers, including (along with the above) Alexander Campbell Fraser,[37] have set the record straight with respect to Berkeley's thesis, such that no serious philosopher today can ignore it. Still, the visceral discomfort that motivated the ridicule is so natural that it needs to be dealt with. Krauth records that one of Berkeley's friends, master of satire Jonathan Swift, "left Berkeley standing at the door in the rain, on the ground that if his philosophy were true he could as easily enter with the door shut as open."[38] The idea here—that idealism consigns the material world to the realm of the imaginary—is both false and forgivable.

What's *perceived* to be at stake in this ongoing discomfort with Berkeley's ideas is *realism*—our innate belief that our senses inform us about a real world that has objective properties. To abandon realism would be to abandon not just science but also rationality. Visceral unease notwithstanding, idealism actually delivers more convincingly on this point than either physicalism or substance dualism do. Physicalism ascribes objective reality to things like photons and electrons with no serious account of what would lend realness to such things—of what "breathes fire into the equations."[39] Theistic dualists, for their part, ground the realness of such things in the realness of the God who created and sustains them. But the exact relationship between the supposed grounding and sustaining is a point of tension. If God granted to electrons a realness that is separable from the realness of his *conception* of them, then what does it mean for him to sustain them? If God would have to act in order for electrons to cease to exist, then "sustaining" them would seem to require no action on his part, a view that comes uncomfortably close to deism.

In contrast to this, there is the view, frequently held throughout the history of Christian thought, that God must act continually in order for any aspect of the physical universe to persist. Among the biblical passages supporting this view are those in the books of Hebrews and Colossians that, after speaking of Christ's role in creating the universe, declare that he "upholds the universe by the word of his power"[40]—that "in him all things hold together."[41] By seeing the physical universe as *nothing but* the conceptions of God being enacted by God, idealism alone accords perfectly with this doctrine. In the words of Samuel Clarke: "The course of nature, truly and properly speaking, is nothing else but the will of God producing certain effects in a continued, regular, constant, and uniform manner."[42]

In answer to Jonathan Swift, then, the jolt we experience from walking into a closed door comes from the fact that God has so embedded our mental experience into our physical bodies that physical jolts reliably bring mental jolts. This isn't at all changed by the realization that such physical jolts are, at rock bottom, mathematical conditions arising from God's faithful calculation of the present state of the universe from its prior physical state and from the translated mental states of its inhabitants. Rather, this realization solidifies our confidence in the dependability of physical law by naming God not just as the past author but also as the present enactor of this law.

That scientific investigation has revealed an unexpectedly deep and counterintuitive mathematical structure to the physical world only adds to the awe. Only by divine genius can the very same thing be commonsensical enough to be grasped adequately by children (at the level that matters most) while defying comprehensive understanding by human genius.

In the end, whatever our position in life, it is the richness of qualitative experience that makes our lives what they are, and all of this is by God's grand design. *All* of this God declared to be good. Perhaps the most beautiful aspect of idealism is that it sees this not as something delivered to us by atoms that were made by God, way back then, but as something imparted by God himself, directly and personally, moment to moment. We often live as though God were distant, but nothing could be further from the truth.

As the apostle Paul put it in a famous speech, "he is actually not far from each one of us, for 'In him we live and move and have our being.'"[43]

Notes

1. See George Berkeley, *A Treatise Concerning the Principles of Human Knowledge* (Dublin: Aaron Rhames, 1710). All subsequent references, unless otherwise noted, will be to the Open Court edition (Chicago: Open Court, 1901), available online at https://books.google.ca/books?id=Bj8_cY2iBzcC&printsec=frontcover&source=gbs_ge_summary_r&cad=0#v=onepage&q&f=false.
2. Lisa Downing, "George Berkeley," *Stanford Encyclopedia of Philosophy*, Spring 2020, https://plato.stanford.edu/archives/spr2020/entries/berkeley/.
3. See Proverbs 16:22 and 26:12.
4. Ali Hasan and Richard Fumerton, "Foundationalist Theories of Epistemic Justification," *Stanford Encyclopedia of Philosophy*, Fall 2018, https://plato.stanford.edu/archives/fall2018/entries/justep-foundational/.
5. Thomas Reid, *An Inquiry into the Human Mind on the Principles of Common Sense* [1764] (Philadelphia: J. B. Lippincott, 1878), introduction, sect. IV.
6. Berkeley, *A Treatise Concerning the Principles of Human Knowledge* (London: Penguin, 1988), 66, para. 38.
7. Douglas Axe, *Undeniable: How Biology Confirms Our Intuition That Life Is Designed* (New York: HarperOne, 2016), 237 (emphasis added).
8. I construe *thoughts* broadly to include all the stuff of direct conscious experience—ideas, emotions, sensations, intentions, etc.
9. See René Descartes, *Meditations of First Philosophy with Selections from the Objections and Replies*, ed. and trans. John Cottingham (Cambridge: Cambridge, 1996).
10. Howard Robinson, "Dualism," *Stanford Encyclopedia of Philosophy*, Fall 2020, https://plato.stanford.edu/archives/fall2020/entries/dualism/.
11. Martin Gardner, "Those Mindless Machines," *Washington Post*, May 25, 1997, https://www.washingtonpost.com/archive/opinions/1997/05/25/those-mindless-machines/228695c9-10c2-4d5d-8179-6dc4d8e04926/.
12. Descartes first makes his famous statement not in Latin but in French: "Je pense, donc je suis." This appears in his 1637 work *Discours de la Méthode Pour bien conduire sa raison, et chercher la vérité dans les sciences* (known in English as *Discourse on Method*). He later put the idea in Latin as "Cogito ergo sum" in his 1644 work *Principia Philosophiae* (*Principles of Philosophy*). Though he chose different words in his 1641 work *Meditationes de Prima Philosophia* (*Meditations on First Philosophy*), the idea seems to be the same: "hoc pronuntiatum, Ego sum, ego existo, quoties a me profertur, vel mente concipitur, necessario esse verum" (translated: "this proposition— *I am, I exist*—is necessarily true each time it is expressed by me, or conceived in my mind." [Meditation II, paragraph 3]).
13. Peter Glassen, "J. J. C. Smart, Materialism, and Occam's Razor," *Philosophy* 51, no. 197 (1976): 349–352.
14. The quotation appears to be a rephrasing of Einstein's wordier dictum: "It can scarcely be denied that the supreme goal of all theory is to make the irreducible basic elements as simple and as few as possible without having to surrender the adequate representation of a single datum of experience." See Andrew Robinson, "Did Einstein Really Say That?" *Nature* (April 30, 2018), https://www.nature.com/articles/d41586-018-05004-4.
15. Richard Dawkins, *The Blind Watchmaker* (New York: W. W. Norton & Co, 1986), 6.
16. This isn't to say that all atheists are physicalists or Darwinists. Some atheists prefer to look for other explanations for life. A notable example is Thomas Nagel, author of *Mind and Cosmos: Why the Materialist Neo-Darwinian Conception of Nature Is Almost Certainly False* (Oxford: Oxford University Press, 2012).
17. Genesis 1:1.
18. John 4:24.

19. Matthew 28:19.
20. Mark 14:38.
21. 1 Timothy 3:16.
22. Eugene P. Wigner, "The Unreasonable Effectiveness of Mathematics in the Natural Sciences" [Richard Courant Lecture in Mathematical Sciences Delivered at New York University, May 11, 1959], *Communications on Pure and Applied Mathematics* XIII (1960): 1–14, https://onlinelibrary.wiley.com/doi/10.1002/cpa.3160130102.
23. Nima Arkani-Hamed, quoted in Graham Farmelo, *The Universe Speaks in Numbers: How Modern Math Reveals Nature's Deepest Secrets* (New York: Basic Books, 2019), 255.
24. Hawking, Stephen, *A Brief History of Time* (London: Bantam, 1989), 174.
25. In French, "Nous devons donc envisager l'état présent de l›universe comme l'effet de son état antérieur, et comme la cause de celui qui va suivre." Pierre-Simon Laplace, *Essai philosophique sur les probabilités*, fifth edition, 1825 [page 3].
26. Rube Goldberg machines are humorous takes on this familiar and fundamental aspect of the physical world.
27. Gabriel Popkin, "Einstein's 'Spooky Action at a Distance' Spotted in Objects Almost Big Enough to See," *Science* (April 25, 2018), https://www.sciencemag.org/news/2018/04/einstein-s-spooky-action-distance-spotted-objects-almost-big-enough-see.
28. As a parallel, consider the place held in logic by the *law of noncontradiction*—the principle that two flatly contradictory statements cannot both be true. All of logic would evaporate if this law were not fundamentally and universally true, because one exception would make it less than what it needs to be to support the entire edifice that has been built upon it. Even non-classical logics, some of which have been explored in relation to quantum mechanics, have classical logic as their *metalogic*. After all, we still have to say either that non-classical logic applies in a given situation or that it doesn't, which is an entirely classical logical judgment.
29. Acts 17:28.
30. All human actions, by this view, are thoughts. God is the only thinker who can produce *thinkers* as well as thoughts.
31. I use "power" here in the sense of authority, not in the sense of determinacy. That is, I take human minds to be the source of human actions, but the moment-to-moment existence of a human mind, and its ability to act, to be wholly dependent on God's sustaining activity.
32. Psalms 139:2.
33. Psalms 19:1.
34. Genesis 1:31.
35. Charles P. Krauth, in George Berkeley, *A Treatise Concerning the Principles of Human Knowledge*, ed. Charles P. Krauth (Philadelphia: Lippincott, 1878), 43. The text of this edition is available at https://archive.org/details/treatiseconcerni00berkrich/mode/2up.
36. Thomas McCormack, in George Berkeley, *A Treatise Concerning the Principles of Human Knowledge*, ed. Thomas McCormack (Chicago: Open Court, 1904), editor's preface, vi.
37. Alexander Campbell Fraser, *The Works of George Berkeley* (Oxford: Clarendon Press, 1871).
38. Charles P. Krauth, in George Berkeley, *A Treatise Concerning the Principles of Human Knowledge*, Charles P. Krauth, ed. (Philadelphia: Lippincott, 1878), 42.
39. Hawking, Stephen, *A Brief History of Time* (London: Bantam, 1989), 174.
40. Hebrews 1:3 (ESV).
41. Colossians 1:17 (ESV).
42. Samuel Clarke, *The Works of Samuel Clarke, D.D.*, vol. 2 (London: Knapton, 1738), 698.
43. Acts 17:27b–28a (ESV).

8. Mind over Matter: Idealism Ascendant

Bruce L. Gordon

1. Introduction

While there are many different forms of idealism, I will defend a version of neo-Berkeleyan ontological idealism as the best account of the nature of mind and its relationship to reality. This strain of idealism is not opposed to epistemic and scientific realism. In fact, I will contend the most interesting case for ontological idealism comes from realism about modern physics, not Berkeleyan-style argumentation.[1] Various physicalisms, dualisms, and dual-aspect monisms will be discussed and criticized in the course of our examination as a prelude to articulating and defending a neo-Berkeleyan understanding of reality. Let's begin.

2. Idealism: The Very Idea[2]

Discussions relating conscious awareness to the world often ignore idealism. Idealists hold that all of reality is, in *some* sense, mental. Different senses of this claim lead to different versions of idealism. Of these differing versions, Platonic and Neoplatonic idealism,[3] German idealism,[4] process philosophy,[5] personalism,[6] and pancomputationalism[7] are beyond our present concern, though the shortcomings of panpsychism will be outlined in due course.

Berkeleyan idealism, our central focus, is *ontological* in character because it regards mental substances as foundational to all of reality: immaterial minds are the only real *substances*; everything else is an idea in a mind.[8] George Berkeley (1685–1753) held that physical things (chairs, trees, mountains, our own bodies) are really orderly collections of mind-dependent ideas produced by and reliant upon the mind of God. There is no materially substantial reality causing our perceptions, nor a substantial physical body we possess that mediates our experience of the world. Rather, physical things exist solely as ideas in the mind of God, who is the ultimate cause of the sensations and ideas that we, as finite immaterial minds, experience objectively and inter-subjectively as the physical universe. The universe thus functions (as its order, pattern, and regularity would indicate) as one mode of God's speech and communication with us.

I will defend ontological idealism of a *neo-Berkeleyan* variety. This idealism is *Berkeleyan* by being theistic and immaterialist, affirming that the phenomenological world we experience is an idea in the mind of God; it is *neo*-Berkeleyan because the arguments on which it rests have strong connections to modern science, especially relativity and quantum physics. Most accurately, my view is a *theistic* form of *quantum-informational idealism* and *conscious realism* in which the it-from-qubit transmissions defining our experiential reality[9] are communicated to us, as finite immaterial mental substances, by the mind of God, who is the only uncreated and necessarily existent immaterial Being.[10]

3. Realism about Idealism

Science tries to give us an accurate picture of what the natural world is really like, but critical interpretive judgment is needed to evaluate the extent to which a scientific theory has succeeded in capturing something fundamental about the structure, ontology, and behavior of nature. Different assessments of whether, and to what extent, a scientific theory should be interpreted realistically are often possible. At the microphysical scale, nature has a structure and behaves in a manner *described* with incredible mathematical accuracy by relativistic quantum physics. As shown in detail in my companion chapter in this volume, this mathematical description has two key metaphysical consequences: (a) the world we experience is *not* composed of material substances (material things); and (b) the regularity of reality is *not* grounded in efficient material causation.

What does this entail? The principle of sufficient reason (PSR) states that every contingent state of affairs has an explanation. It is a logico-metaphysical truth, any denial of which leads to irremediable skepticism.[11] If relativistic quantum physics entails that no material substances or causes exist to explain the world we experience, the PSR requires the objective and intersubjective nature of reality to be grounded in something immaterial.[12] The best explanation for an ordered, immaterial, experiential reality is that finite created minds experience a world provided by a transcendent uncreated mind (God) whose *actions* accomplish three things: (1) they bring coherence to the subjective and intersubjective experience of finite minds as the sole efficient cause of their perceptions of the structural and qualitative properties of the world; (2) they account for the lawlike regularity of the world; and (3) they enable the actions of morally responsible finite agents. (The last point can be explained as follows: when a morally responsible finite agent S makes a decision D that could have been otherwise and over which S had control, God actualizes the consequences of that decision in the objective (intersubjective) world experienced by other finite minds.) In short, all finite minds experience the world as a *phenomenological* reality communicated to us by God, that is, we all live and move and have our being in the mind of God (Acts 17:28), which is definitive of reality.

This ontological idealism derives from a realistic interpretation of scientific theory. A critical realistic approach to science ultimately points to neo-Berkeleyan ontological idealism as the best metaphysical picture of reality. That realism should lead to idealism in this way is ironic, for most Berkeleyan idealists have been *antirealists* about science. They should not be. Theistic idealism rises like a phoenix from the ashes that modern physics has made of material substances and efficient material causation. While God is the *vera causa* of the reality we experience, scientific exploration of phenomenological reality shows it has an intersubjectively available objective structure susceptible to proximate causal analyses that reflect the metaphysical order God transcendently imposes upon it by mental causation.[13] Insofar as our thoughts, albeit in a finite and limited way, mirror divine thought, they correspond to reality and have grasped the truth about it as *defined* by the mind of God. This picture needs closer examination in light of standard objections to Berkeleyan idealism. But first, the prospects of physicalism, dualism, and dual-aspect monism need to be evaluated.

4. Problems with Physicalism and Dualism

Let's briefly outline the most common physicalist and dualist positions and their difficulties. The discussion can't be comprehensive. Its purpose is to introduce various physicalist and dualist options and the difficulties they face, so that the virtues of theistic immaterialist idealism may be extolled by comparison.

4.1 Not Physicalism: Eliminativist, Reductivist, and Non-reductivist Physicalism Rejected

Among physicalisms, we primarily encounter eliminativist, reductivist, and non-reductivist varieties. Eliminative physicalism is the peculiar position that our commonsense understanding of the mind is completely wrong and the categories of mental states we commonly posit do not exist. For instance, eliminativists claim there is no mental state associated with having a belief. This seems self-referentially absurd: eliminativists *believe* that they have no beliefs. Of course, they will say the question has been begged by describing their assertion that beliefs don't exist as an eliminativist *belief*. The fact remains, however, that they must deny the existence of mental content with truth-conditions related to reality. This leaves them in the self-contradictory position of asserting that, while there are no truth-conditions for sentences, it's still the case that certain sentences about mental content are *false*. The only maneuver eliminativists have left is to attempt to develop and defend a deeply implausible non-truth-conditional semantics.[14] These difficulties, and the fact that eliminativism contradicts everything we may reasonably claim to *know* about ourselves, warrant its dismissal as a dead end.

This leaves reductive and non-reductive versions of physicalism, whose problems *also* afflict eliminativism, driving further nails in the eliminativist coffin. Reductive physicalism identifies the mind with the brain, reducing all mental functions to brain functions, which reduce to biochemistry, which ultimately reduces to physics.[15] By contrast, non-reductive physicalism, while claiming there is nothing about mental function that is not ultimately physical, still maintains that efforts to explain psychological and biological properties in *purely* physical terms are misguided and will inevitably miss laws, generalizations, and explanations only expressible using psychological or biological concepts.[16] Both kinds of physicalism, by affirming the causal closure of the material realm, are subject to intractable problems.

The *first* difficulty, as David Chalmers famously called it, is the *hard problem of consciousness*: how can a physical state be conscious? The fact is, it *cannot*. To think it could is simply a category mistake. There is nothing it's like to be a windmill, but there is something it's like to be a dog. Furthermore, there's nothing intrinsic to physical systems or their interactions that makes them *about* anything, but the hallmark of consciousness is *intentionality*, the fact that mental content is *about* something that consciousness *experiences*. These facts are characteristic of *conscious agency*.

The *second* problem derives from the irremediable incompleteness of physicalism, which is committed to the causal closure of the material realm. The material realm is a contingent entity without a physical explanation, which violates the aforementioned principle of sufficient reason (PSR) under causal closure, as does the absence of any explanation for its law-like behavior. As already mentioned, the PSR is a logico-metaphysical truth of which any denial leads to disastrous epistemic consequences.[17] For example, if it's possible that no sufficient reason exists for one thing happening rather than another, our current perception of reality and its accompanying memories—for instance, our perception that nature is regular—could be happening for no reason at all, so the world of our experience might be totally adventitious.[18] How could we know? Also, if a physical state of affairs can lack an explanation, then the possibility there is *no explanation* competes with every explanation for anything that occurs, undermining all of science. Denying the PSR destroys the possibility of knowledge.

Further, regarding *non-reductive* physicalism, for the *mental* to involve nothing more than the physical yet not be reducible to it requires, whether nomologically or anomalously, that macroscopic material objecthood and behavior be *reducible* to, *supervenient* upon, or *emergent* from the existence and behavior of material reality at the microscopic level, and for microphysics itself to be consistent with the existence and causality requirements of physicalism. As we will see, *none of this is the case*. Macroscopic reality is neither reducible to, supervenient upon, nor emergent from microscopic reality in any robust sense under the constraints that physicalism imposes.[19] In short, far from being able to account for consciousness, physicalism—in whatever guise—cannot even provide an adequate account of the *physical*; it is a bankrupt metaphysic.

Lastly, if physicalism is embedded in a naturalistic ontology that denies a transcendent realm exists, we must evaluate the consequences of thinking our cognitive faculties were produced by blind natural forces, which they must be if nature is all that exists and physicalism is correct. If natural selection sifting chance variations explains the origin of our faculties, what matters about them is their fitness for ensuring survival and reproduction, not their ability to represent reality as it actually is. Indeed, if natural-selection-driven conscious perception and ratiocination arose accidentally in natural history, the veridicality of our perceptions and the validity of our reasoning processes is, at best, inscrutable. As Alvin Plantinga forcefully argues,[20] undirected evolution provides no ground for supposing our cognitive faculties produce *true* beliefs, and the further removed from immediate survival our beliefs are, the less confidence we should have in their veridicality. This means that evolutionary naturalists have little warrant for believing anything (including evolutionary naturalism) to be *true*.

While this "evolutionary argument against naturalism" has provoked critical discussion,[21] it is compelling to note that computational evolutionary psychology has drawn the same conclusion. Computational experiments using evolutionary game theory demonstrate that organisms acting in accordance with the true causal structure of their environment will be out-competed and driven to extinction by organisms acting in accordance with arbitrarily-imposed species-specific fitness functions.[22] Ironically, this demonstration can only be trusted if we are *not* organisms with non-veridical fitness functions; otherwise, we have no confidence its conclusion bears any relationship to reality either. Without cognitive faculties aimed at true beliefs, all human knowledge, including science, is just a fitness-driven survival mechanism with an inscrutable connection to reality. However, the probability that properly functioning cognitive faculties are reliable guides to truth is high *on the assumption of theism*, especially Christian theism: God not only brings about our existence, he wants us to *know* and have a relationship with him, and intends we *understand* the world well enough to be its stewards. Christianity gives us access to the world and a basis for thinking that science can lead us to the truth; physicalist naturalism takes this away.

4.2 And Not Dualism Either

What about dualism? Since consciousness, intentionality, and personal agency are self-evident from first-person experience and completely intractable on a physicalist basis, it's not surprising that philosophers of mind not seduced by scientific materialism focus on other ways of understanding these things. *Dualism* takes mental and physical phenomena to be *equally fundamental* and denies that either is assimilable to the other. This raises questions about how mental and physical states *influence* and *interact* with each other given their fundamentally different natures. Insofar as a dualist affirms the dependence of the physical realm on an *immaterial* spiritual realm, which theists certainly do, these questions have corresponding formulations on a universal scale: How does God, as an *essentially immaterial* being, create and interact with an *essentially material* reality? Theistic dualism thus confronts an *interaction problem* on both the universal (God-world) and particular (mind-body) levels, whereas naturalistic dualism only confronts it on the particular level. On the other hand, without God, naturalistic dualism has no possibility of explaining the existence of a contingent universe, and it is prevented from *reducing* the mind-body problem to one of God-world interaction as Leibnizian psychophysical parallelism with pre-established harmony or Malebranchian occasionalism would do. With these things in mind, let's discuss different models of dualism, beginning with the more anemic and progressing to the more robust, that is, starting with *property dualism*, and moving to *hylomorphic dualism*, and finally *substance dualism*.[23]

4.2.1 Property Dualism

Property dualism maintains that conscious experiences involve having mental properties *not* entailed by the physical properties of those having them, even though these mental properties supervene (are dependent) on the physical properties. Individual consciousness is thus an *emergent property* transcending the physical features of the individual—the experiential properties characteristic of consciousness are *ontologically distinct* from the physical features on which they depend for their existence.[24] There are at least two issues of concern here. First, is the connection between physical states and mental properties lawlike or anomalous? An anomalous relationship would render property dualism evidentially inscrutable and explanatorily vacuous—without reliable correlations between specific physical states and mental properties, no evidence regarding their connection would be sustainable, and nothing of any explanatory value would be generable. Multiple realizability is also germane to this issue: the same conscious abilities can, at times, be correlated with different neurophysiological channels or processes. Abilities of certain kinds may correlate with specific brain functions, but a damaged brain can sometimes "rewire" so that tasks formerly dependent on a damaged part of the brain are processed on a different neurophysiological basis. Insofar as psychophysical regularities exist, however, the best we could hope to establish is an *objective correlation* between certain conscious capacities and certain brain functions in living organisms. But *correlation* is not *causation*,[25] so the status of these correlated mental properties is unclear. What *are* they? How does the brain, which is physical, get *correlated* with something in a different ontological category so that when the subvenient properties are absent in a living body, so are any supervenient mental properties? And how does mental causation, which we know from experience, get correlated with changes in the subvenient neurophysiology, especially if the physical realm is taken to be causally *closed*? We're left with an epiphenomenalism our mental life belies. This brings us to the second issue of concern, namely, the unity of our mental life.

Our self-understanding is mentally constituted; it consists in the *conjoined experience* of a variety of mental properties. But what *unifies* the experience of these properties into a *self* so their confluence is not merely adventitious and empty of significance? Clearly, whatever the *self* is, it's not physical, but rather associated with a nexus of mental properties, each of which has an independent ontological status in property dualism. Is the self, therefore, nothing more than a *bundle* of sensations and thoughts, as Hume advocated?[26] No, nor could it be. Nothing would hold the bundle of mental properties together in this view, nor explain how it could function as a *subject* for the actions attributed to it. Some kind of *substance* must unify these mental properties, explain their coherence with each other, and ground their function as a morally responsible *agent*.[27] So property dualism yields to a *substantial* dualism, which presents two candidates for consideration: *hylomorphic substance dualism* and *robust substance dualism*.

4.2.2 Hylomorphic Substance Dualism

In speaking of hylomorphic dualism, pure Aristotelianism should be distinguished from Aquinas's modification of it via Augustinian neo-Platonic substance dualism. Aristotle characterizes matter (*hylē*)[28] as what individuates a substance and form (*morphē*)[29] as what identifies it. For Aristotle, the human soul is the *form* of the body—it makes us human—but the *matter* of our bodies accounts for our individuality (numerical distinctness) as human beings. In Aristotelian metaphysics, humans have an organic-sentient-rational soul that functions as a *principle of actuality* that activates every human function and capacity. It is our form-matter *composite* that *makes* an individual human substance. There are no substances if either member of this composite is absent. Specifically, since forms do not exist apart from the matter in which they are instantiated, *the soul does not exist apart from the body* in Aristotelian conception. Human beings do *not* possess an immortal soul in Aristotelian metaphysics. This will not do for traditional Christian conceptions of the human person, so Aquinas affirms, with Augustine, that the human soul retains its own being after the dissolution of the body. *Thomistic dualism* thus asserts that a *substantial form* exists that survives the dissolution of the human body, while emphasizing the substantial *unity* of human persons as integrated form-matter composites. Disembodied souls therefore *cannot* be complete persons in Thomistic anthropology. The soul is only *potentially*, not actually, the form of the body, and is deficient since it *cannot* be conscious in a way that would require bodily organs. The soul has a *rational* identity according to Aquinas, but it has lost its *sensory* capacities. This also explains Aquinas's view that human beings survive the death of the body, but animals do not. *Rational souls* can exist without sensory input, but *sentient souls* cease to exist without it.[30]

Thomistic dualism faces both conceptual and evidential challenges.[31] Conceptually, we can ask whether the idea of a substantial form is metaphysically coherent and adequately explains the needed interaction between the body and the soul. Evidentially, we can ask whether the diminished status of disembodied substantial forms fits with the evidence from near-death experiences.

Conceptually, consider the rational subject in the hylomorphic scheme. As a hylomorphic composite, neither the soul nor the body is the thinking subject, but rather their *union*. This means the rational soul as the *form* of the body is not by itself the rational subject. The soul is sub-rational and only *actualizes* a rational subject in *union* with the body. In order to have a *subsistent* rational form, however, Aquinas must maintain that the soul *alone* is the thinking subject, privileging it in a proto-Cartesian manner as a *res cogitans* distinct from an unthinking material body. But in what sense, then, is it the *form* of the body, as opposed to a separate *substance* that animates the body? Anything subsistent is capable of independent existence, i.e., a *substance*. *Forms*, however, are *not* substances, but rather mental abstractions that function as a *principle* in the analysis of substances. Aquinas cannot consistently maintain that the rational soul is *independent* of matter within an Aristotelian hylomorphic framework. His view is either incoherent or it defaults to a proto-Cartesian substance dualism. If we opt for proto-Cartesianism and coherence, however, it won't do to say that the problem of body-soul interaction is resolved through *formal* rather than *efficient* causation, since the causal relationship is now between *independent substances*, not through an inseparable relationship between the form itself and that of which it is the form.

Evidentially, near-death experiences (NDE) pose problem for Thomistic dualism. Such experiences indicate that conscious souls separated from bodies declared clinically dead, far from having a *mere* rational existence without sensory content, often have *perceptual* experiences more intense than when embodied. This evidence starts with individuals documented to be brain-dead for a significant period of time who have revived to offer first-person accounts of out-of-body experiences containing early-stage *perceptual* details that have been independently corroborated. These *perceptual* experiences indicate retention of sensate and rational consciousness, not just rational consciousness as Thomism implies.[32] This is further corroborated by NDEs in blind subjects who were able to *see* for the first time when their consciousness was decoupled from the information-processing channels of their sightless bodies.[33] Beyond this, it is common for first-person reports to detail heightened conscious awareness, lucidity, and even new cognitive capacities as consciousness is decoupled from the constraints imposed by neurophenomenological correlations.[34] Lastly, many NDE reports detail direct awareness of the mental states of other conscious agents, both brain-linked and decoupled, in which sensory and linguistic communication channels are bypassed and thought-expression is non-local, unmediated, and unambiguous.[35]

One might wonder about the value of mental operations being tied to brain states in the first place if, as NDEs apparently indicate, the mind can function *better* when decoupled from the brain. NDE research and Christian theological anthropology suggests two observations here. First, the inhibition of mental function in the brain-linked state is understandable in Christian theology as one of the noetic effects of sin, manifested as a functional impediment in addition to moral and epistemic distortion. Secondly, there is considerable evidence in the NDE literature that decoupled minds or souls are not so much *dis*embodied as *differently* bodied. Many NDEs report a translucent body transparent to the decoupled self and invisible to those still neurophysiologically constrained. Finally, a metaphysical argument can be made that finite consciousness *requires* perceived location and perceived embodiment for action, which is only to say that there is no such thing as a "view from nowhere" for finite epistemic agents. This all makes perfect sense from the standpoint of an idealist Christian metaphysics in which embodiment, the intermediate state, and eschatological resurrection are *not* substantial changes, but changes in our divinely mediated perceptual environment and cognitive capacities.

4.2.3 Robust Substance Dualism

Hylomorphic dualism defaults to *robust* substance dualism, within which we may distinguish between *simple substance dualism* and *compound substance dualism*. Simple substance dualists maintain that while human persons *have* bodies, they are *identical* to their minds or souls. It is Cartesianism: a person is a *res cogitans*, an unextended, indivisible, thinking substance that is an immaterial simple. Compound substance dualists hold instead that human persons are soul-body composites. The body is an *accidental* proper part of each person, but the soul or mind is *essential* to that person's identity. Both kinds of dualism hold that persons can exist without their bodies, but not without their souls.

A conceptual problem arises for the compound dualist that's very similar to the problem afflicting hylomorphic dualism: *What* is the thinking subject? Is it just the soul, or is it body-soul composite? If the soul sustains personal identity *apart* from the body and the composite constitutes the person, then *both* the soul *and* the body-soul composite are thinking subjects. This means every act of thought has *two* subjects—the soul *and* the compound substance.[36] To say the least, affirming two subjects for every person was never the intention of the compound dualist, so what's left? The compound dualist could say it's not the soul that thinks, but the compound substance. As with Aristotelian hylomorphism, this makes the soul sub-rational, undermining the central motivation for dualism: the recognition that distinctive modal properties of unextended, indivisible thinking substances versus extended, divisible, unthinking matter allow the self to exist apart from the body.

Another approach might be to institute a division of labor: the soul's immaterial substance is the seat of *abstract rationality*, the body's material substance is the seat of *sensory experience*, and their compound makes the person complete. This would be a "Cartesian Thomism," stripped of hylomorphism. I suspect no Thomist would consent to it; moreover, the construction would entail the same diminished view of disembodied consciousness as standard Thomism and run into the same evidential difficulties with NDEs. The final alternative would default to simple substance dualism, maintaining that persons are simple substances, identical to their immaterial souls, who animate the unthinking matter of their bodies. To be a dualist, then, it would seem that simple substance dualism is best.

Regardless of the dualism adopted, the *interaction* between the soul and the body must be addressed. How does the mind cause behavior in the body and how does the body affect thoughts in the mind? This interaction problem is often presented as an insuperable difficulty. Elliott Sober gives a standard version of the objection:

> If the mind is immaterial, then it does not take up space. But if it lacks spatial location, how can it be causally connected to the body? When two events are causally connected, we normally expect there to be a physical signal that passes from one to the other. How can a physical signal emerge from or lead to the mind if the mind is no place at all?[37]

But why assume there must be a physical signal transferring cause and effect? Sober's argument—the standard objection of all physicalists—supposes a transfer of some physical magnitude from cause to effect is necessary for a causal interaction to take place. Since the dualist *denies* this, it's hard to see how the objection doesn't just beg the question. The interactionist objection assumes a transfer theory of causation, but there are any number of different theories of causation—transfer, nomological, powers, counterfactual, agentive, direct, and so on—and the dualist with respect to mind-body interactions insists on *direct* causation that has no need for an intervening mechanism. Nonetheless, while demanding a *physical mechanism* begs the question, there's no principled reason why causally relevant properties, or a counterfactual nomological description of mental-physical interactions, or both, might not still be articulated. A definitive rejection of interactive substance dualism requires non-dualists to have a convincing argument that direct causation of the relevant sort is *impossible*.

Even if direct mental-physical causation is coherent, however, it's still not clear that the *conservation of energy* isn't a problem. Granting that there's no physical mechanism of transference, it still seems that the production of a physical effect *absent* a physical cause might involve a *detectable* violation of the conservation of energy. Direct mental causation of physical effects would then be a constant source of such violations in dualist metaphysics.

Several responses are possible. First, the law of energy conservation holds between *physical* things, and in dualist interactionism, the mind is *not* a physical thing, so there's no reason to think the law applies and that mental causality would involve a detectable violation of energy conservation. Secondly, the principle of energy conservation has two known *exceptions* in physics: gravitational field energy in general relativity and non-local correlations in quantum physics. There's no meaningful local expression for gravitational stress-energy in general relativity and thus no meaningful energy conservation either.[38] Exceptions to this rule are possible in regions of space-time that are asymptotically flat, but no such regions exist in the observable universe, so the point is moot. Furthermore, since the best *global* models of spacetime are *not* asymptotically flat, energy conservation doesn't realistically apply to the universe as a whole.[39] Inflationary cosmology exploits the failure of energy conservation in general relativity to enable the inflaton field to be an unlimited supply of energy, inflating any spacetime region to whatever needed size while retaining constant energy density.[40] So one of the two main pillars of modern physics—the theory of relativity—entails the *falsity* of energy conservation.

What about quantum theory, the other pillar of modern physics? It entails lawlike correlations between quantum systems at spacelike separation that cannot, on pain of experimental contradiction, be explained apart from instantaneous non-local action-at-a-distance without *any* transfer of energy.[41] Quantum physics entails these correlations while demonstrating they cannot have a *physical* explanation despite the fact that they *must*—as a matter of metaphysical necessity—have an explanation.[42] Such correlations provide strong evidence that causation does *not* need mediation or mechanism or to involve a detectable energy exchange. Furthermore, given that immaterial mental substances are *not* spatially located and given that instantaneous physically unmediated correlations exist in physics, the energy conservation objection in particular, and the so-called interaction problem in general, have no compelling force for the substance dualist.[43]

So why not be a substance dualist? If substantial matter exists, simple substance dualism is not only tenable, but to be preferred over all varieties of physicalism, and over property and hylomorphic dualism as well. But other difficulties with substance dualism have to be faced. The first is the *origin* of material substance, if it exists, and the second is that its existence seems *precluded* by modern physics.

Regarding its origin, there is a dualism inherent in theism between the Creator as uncreated, necessarily existent, immaterial substance, and his Creation as contingently existing created substance. But what is the nature of created substance? If it's not just the simple immaterial substances of finite created minds, but also materially substantial bodies of varying sorts, then what is *essentially immaterial* must give rise to what is *essentially material.* We've argued there's no problem of interaction or energy conservation between immaterial and material substances if each exists, but this *doesn't* show that generation of one by the other is *possible*. Mass and energy are interconvertible, whatever they are, but immaterial substance is *neither matter nor energy*. Indeed if, as seems true, it's intrinsically *impossible* for material substance to generate immaterial consciousness, why think it's possible—going in the other direction—for something essentially immaterial to generate something *intrinsically foreign* to its nature? It's no good to invoke divine omnipotence if we're asking for something *metaphysically impossible*. Logical and metaphysical impossibility are constraints on the intelligibility of God's omnipotence. The substance dualist lacks a resolution to this problem. Perhaps we should reason as follows instead. If material substances exist, the principle of sufficient reason entails a necessary being to explain their contingent existence. But any necessary being is entirely immaterial in nature. Since it's metaphysically impossible for the essentially immaterial to generate the substantially material, and since reality *must* have an explanation, material substances *cannot exist* and *reality must be essentially immaterial.*

Is there any evidence for the immateriality of reality? *Yes*. As already mentioned, quantum physics is inconsistent with the existence of material substances and efficient material causality, making theistic quantum idealism the best explanation for the existence and coherence of our experiential reality.[44] Created reality consists of finite minds as simple immaterial substances created by God and the universe as a phenomenological construct communicated to these minds by God. In short, we have the soul of simple substance dualism, minus the materially substantial universe (inclusive of substantial bodies) that makes it substance dualism. A residue of hylomorphism is preserved in ontological idealism too. Immaterial consciousness exemplifies universal *forms* distinctive of its kind—human consciousness differs from canine consciousness, for example—and an individual consciousnesses is differentiated by its *haecceity*, its "thisness." Each immaterial consciousness possesses a primitive numerical *thisness* of substance that individuates it as a particular. A hylomorphic residue further pervades the *perceptions* of immaterial consciousness, but rather than an ontic form-matter distinction, there is a phenomenological *quantitative structure* versus *qualitative content* distinction. The simple immaterial substance of finite consciousness experiences divinely provided reality (including bodily phenomenology) *holenmerically*, providing a unified picture of human persons that accords well with lived experience and various theological and biblical desiderata.[45] Theistic ontological idealism thus preserves the tenable insights of hylomorphic and substance dualism while jettisoning the problematic aspects of their metaphysics, all within an orthodox Christian framework.

5. Problems with Dual-Aspect Monism and Agentive Cosmopsychism

An approach to the mind-body relationship we have not yet considered is *dual-aspect monism*, a view primarily associated with Baruch Spinoza (1632–1677). Spinoza argued that the *whole* of reality is a *single substance* that undergirds *both* physical and mental phenomena. Viewed under one aspect, this reality is *God*; viewed under the other aspect, it is *Nature*. This *one* substance is neither spirit nor matter but something possessing the properties of both and accessible to human understanding in *either* mode. Human beings are manifestations of this absolute substance, so souls and bodies are *not* distinct substances, but aspects or modes of being abstracted from the unified substance that *is* fundamental reality. As such, individual human beings are associated with particular bodies and their souls do *not* survive the death of the body except as something absorbed into the mind of God, that is, by becoming one with the spiritual aspect of absolute substance.[46]

Spinoza's thought deeply influenced Hegel's attempt to elide the reality of the subject-object distinction and his conception of Absolute Spirit. Spinoza's influence also pervades Bertrand Russell's neutral monism, Albert Einstein's view of nature, Peter Strawson's understanding of the human person, Alfred North Whitehead's process metaphysics, the view of God in process theology, the recent resurgence of naturalistic panpsychism, and Bernardo Kastrup's ontic idealist cosmopsychism, to name a few prominent intellectual progeny. One manifestation of Spinozist pantheism currently in vogue is cosmopsychism. This view—and panpsychist approaches in general—try to steer between the Scylla of physicalism and the Charybdis of dualism while resolving issues of consciousness and fine tuning in the universe without invoking theism.[47] Cosmopsychists claim that the general nature of consciousness, and facts about organic consciousness in particular, are explained by "consciousness-involving facts" about the universe as a whole. For these consciousness-involving facts about the universe to explain the fine tuning of the universe for life *and* the origin and cognitive fine tuning of consciousness in accordance with norms of rationality, *agentive cosmopsychism*—the thesis that the universe itself has the capacity to recognize and respond to reasons or facts about value—is needed.[48]

Taken as a serious hypothesis, agentive cosmopsychism manifests a variety of deficiencies in comparison with theism. The most obvious is that the existence of a universe with conscious properties is contingent and, by the principle of sufficient reason, requires a necessary being to explain it. God, as the necessary being, would have to bring the contingent cosmopsychic universe into existence as something separate from himself. A universe with inherently contingent properties cannot itself be necessary, so theism follows straightforwardly and moots any need for agentive cosmopsychism.[49] So consciousness-involving facts about the universe are not needed to explain the existence of cosmological fine tuning and consciousness within the universe, for God is their proper explanation. Theism is also the more parsimonious and plausible hypothesis here because it does not require the personification of obviously inanimate aspects of nature.

Theism arguably follows from agentive cosmopsychism another way. Agentive cosmopsychism postulates the universe can respond to reasons or facts about *value*, but what *grounds* these facts about what is valuable? Agentive cosmopsychism implies that value is objectively *recognized by* but *not intrinsic to* cosmic consciousness. To claim that cosmic consciousness is *necessarily* good and the standard of goodness would make cosmic consciousness into God, which is precisely what agentive cosmopsychists want to avoid. Yet values pursued by the universe that are objective, but *not* grounded in God, seem to commit the agentive cosmopsychist to moral Platonism of the kind advocated by Erik Wielenberg.[50] Wielenberg's position is problematic, however, and the proper grounding of objective values, moral and otherwise, arguably requires theism.[51] Insofar as the objective ground for the values recognized by the universe in agentive cosmopsychism leads to a moral argument for God's existence, agentive cosmopsychism is yet again a gratuitous hypothesis that defaults to theism.

We seem to have subverted the motivations for cosmopsychism, but let's return to the question of whether the conscious universe could *itself* be a necessary being. If so, concerns about agentive cosmopsychism arising from the principle of sufficient reason would be allayed. Philip Goff proposes that a necessarily existent conscious universe in a timeless, non-spatial, non-material, agentive phase gives rise to the spatiotemporal, physical phase in which we find ourselves (where these phase sortals of the universe are to be understood in the same manner in which, say, "adulthood" is a phase sortal of a person).[52] This proposal faces several difficulties.

First, if any form of the ontological argument is sound—and I'm inclined to think it is—the issue is moot: God exists and the features of the universe are explained without the metaphysical gymnastics Goff's speculative construction represents. *Second*, asserting that the conscious universe exists necessarily doesn't make it so. The goal is to present the universe as a necessary being that accomplishes things usually attributed to God without actually being God, obviating arguments that certain features of the universe require a theistic explanation. But it's hard to see how a conscious universe gerrymandered to obviate theism could be something that *necessarily* exists. Mere non-spatiotemporality and non-materiality does not imply non-contingency; such properties are merely necessary conditions, not sufficient conditions, for necessary existence. For example, in simple substance dualism, the soul is an immaterial simple without spatiotemporal location, but it is not thereby a necessarily existent entity. So the conscious universe of agentive cosmopsychism is *not* a necessary being, and quite arguably, *if* there is a metaphysically substantial, necessarily existent being, that being must have the *other* perfections attributed to God, and therefore *be* God.[53] *Third*, since Goff postulates the universe itself as the primary necessarily existent substance, we're owed an account of how its nature allows it to begin as something *entirely* immaterial and non-spatiotemporal, described exclusively by consciousness-involving facts, and *become* a mixture of consciousness-involving and matter-involving facts. How do purely consciousness-involving facts generate matter-involving facts in the *same* substance given irreducible categorial differences between mental and physical properties in that substance? How is *intrinsic* change possible in a *necessary* being? Metaphysical impossibility looms large, reminiscent of earlier worries about how essentially immaterial substance could generate essentially material substance.

Earlier resolution pointed us toward ontological idealism; perhaps it should do so again. Would a naturalistic ontic idealist cosmopsychism like Bernardo Kastrup's remain a possibility?[54] No. It still suffers from the explanatory deficiencies related to contingency that plague dual-aspect agentive cosmopsychism. But it's a step closer to the theistic idealism we will discuss next. *Finally*, insofar as agentive cosmopsychism and ontic idealist cosmopsychism are

Spinozist regarding our fate at death, so that individual organic consciousness is absorbed back into universal consciousness, they are even more strongly contradicted by near-death experiences than hylomorphic dualism.[55]

6. Against Parity: The Explanatory Superiority of Idealism

THE CONCEPT of a materially substantial universe is largely taken for granted and regarded as common sense by most people. Arguments that the actual nature of things is otherwise get dismissed as intellectual curiosities. This attitude arises from the perception that a substantial material reality provides a natural explanation for: (1) why the world develops independently of our volitions and desires; (2) why we all seem to inhabit the same universe; (3) why there are observed correlations between brain activity and our inner life; and (4) why we can be wrong about the things we take ourselves to perceive. A mind-independent reality does a good job of explaining these facts. Let me suggest that a properly articulated theistic ontological idealism can do a better one. We've seen indications of this, but let's make a more detailed case.[56]

Stanford theoretical physicist and cosmologist Andrei Linde, offers a caveat about the foundations of science:

> According to standard materialistic doctrine, consciousness, like space-time before the invention of general relativity, plays a secondary, subservient role, being considered just a function of matter and a tool for the description of the truly existing material world. But let us remember that our knowledge of the world begins not with matter but with perceptions. I know for sure that my pain exists, my "green" exists, and my "sweet" exists. I do not need any proof of their existence, because these events are a part of me; everything else is a theory. Later we find out that our perceptions obey some laws, which can be most conveniently formulated if we assume that there is some underlying reality beyond our perceptions. This model of material world obeying laws of physics is so successful that soon we forget about our starting point and say that matter is the only reality, and perceptions are only helpful for its description.[57]

Linde's point is that our perceptions provide the *content* of our experience. While we have a clear and distinct awareness of this content and we know our perceptions are *not* the product of our own mind, nonetheless, what is *not* given in our experience is *the underlying ontic nature* of the objects we perceive. This needs investigation, not presumption.

One investigative path to theistic ontological idealism is provided by the philosopher-theologian George Berkeley (1685–1753). Berkeley argued that since an object presents itself to us as a collection of sensible qualities, and sensible qualities are ideas, the objects of our experience are nothing more than a collection of ideas. Since ideas are mind-dependent, he reasoned, it follows that the entire physical world of our experience is mind-dependent and does not exist apart from being perceived. This led him to the central tenet of his ontology: *esse est percipi aut percipere*, that is, to be is to be perceived or to be a perceiver. This means that nothing exists that is not itself a mind unless perceived by some mind. Of course, Berkeley did not think that the objects of our perception were the product of our own minds. He found it obvious, as should we, that we are not consciously producing the images, smells, sounds, tastes, and textures we encounter via our senses, nor do we have any reason to believe that we are dreaming or hallucinating and producing those experiences for ourselves unconsciously. If our minds are not producing these experiences, however, it follows from Berkeley's argument that the source must be some *other* mind. That mind, he contended, could be nothing less than divine:

> To me it is evident... that sensible things cannot exist otherwise than in a mind or spirit. Whence I conclude, not that they have no real existence, but that, seeing they depend not only on my thought, and have an existence distinct from being perceived by me, there must be some other mind wherein they exist. As sure, therefore, as the sensible world really exists, so sure is there an infinite omnipresent Spirit, who contains and supports it.[58]

Now, this reasoning is hardly unassailable, for even though its monistic immaterialism is more parsimonious than a dualism of spirit and matter, nothing would dictate that the bundle of sensible qualities constituting the objects of our perception have their causal origin in *direct communication* from God as opposed to being *mediated* by material substances created by God. Thus far, Berkeley's argument is insufficient for this conclusion. Nonetheless, our discussion of the problems with property dualism makes it clear that the bundle of sensible qualities must *originate from something substantial* and, as perceived, *inhere in a substantial self*. The *natures* of these originative and receptive substances, however, cannot be inferred simply from the observation that our percepts come to us as bundles of sensory qualities; these natures *still* need to be investigated.

To this point, investigation has revealed two factors that would favor theistic ontological *idealism* over theistic ontological *dualism*. The first is its parsimony: why postulate the existence of material substances mediating the world to us when God exists and can communicate the idea of it directly to our minds in an intersubjectively coordinated way? The second is the removal of a mystery smacking of metaphysical impossibility: it is clear, from its contingent nature, that whatever the material world really is, the principle of sufficient reason requires its existence to be explained by a necessarily existing and therefore immaterial being. But how does immaterial substance cause material substance (spacetime and mass-energy) when the latter is *completely foreign to its nature*? God's omnipotence does not extend to the metaphysically impossible, so for theistic ontological dualism to be tenable, an explanation of how this is possible is needed.[59] Theistic ontological idealism does not face this problem: God, as uncreated immaterial substance, brings finite immaterial minds into existence and, by direct divine communication to these souls, provides the reality they experience as "physical." These two facets of idealism's explanatory superiority over both dualism and (especially) physicalism are significant and, by themselves, arguably tip the scale in its favor. But further considerations break any semblance of parity and secure the superiority of theistic ontological idealism.

No symmetry exists between *physicalism* and *idealism* in respect of the physicalist's claim that everything can be explained by matter and its properties as opposed to the idealist's assertion that everything can be explained by mind and its properties. The hard problem of consciousness, an artifact of physicalism, has *no* solution because it rests on the category mistake of thinking the mental can either be reduced to or produced by the physical. Nothing can make a purely *material* thing—whether considered ontically or functionally—into a subject that experiences mental content and exhibits intentionality, with awareness that is *about* something, so that there is something that it is like to be that thing. There's nothing it is like, from the inside, to be something purely physical like a rock, but this quite clearly characterizes our mental life.[60] No corresponding "hard problem of matter" exists for the idealist, however. It is *not* the case that ontological idealism is incapable of explaining how *mental states alone* could explain what matter is without the existence of material substances, or have an observed correlation with brain processes without the existence of a material brain—something we will say more about momentarily. Idealism is clearly superior to physicalism for this reason, in addition to the others already mentioned.

Let's examine further the perceptions that constitute our direct experience of the world and consider the nature of the objects of our awareness. Our subjective and intersubjective life-world or *lebenswelt*, as Edmund Husserl called it,[61] is quantitatively and qualitatively *concrete*, arising from the particularity of our experience. This level of concrete particularity is the foundation of what we consider most *real*; every abstraction from it takes us farther away from the reality we actually experience.[62] Moving beyond the concrete mental perceptions that are the intentional objects of direct consciousness to the abstract concept of a material substance, then abstracting from this to the properties of material substances, so that we can further abstract to nomological relationships among physical properties as explanatory of the behavior of both material substances and our own states of consciousness, is a progressive movement to *greater and greater levels of abstraction* from our immediate conscious experience. It should be clear from this that mind and matter *do not exist at the same level of explanatory abstraction.*[63] Mind is the foundation within which and out of which abstractions are made, *but matter itself is an abstraction of mind.* Thus there is not, nor could there be, explanatory parity between matter and mind. Mind holds explanatory priority and matter is derivative; the explanatory relationship between them is asymmetric. This significantly erodes the epistemic and metaphysical basis for physicalism, the physical side of dualism, and the physical mode of dual-aspect monism.

Even so, if we turn to the abstractions of science, we find that fundamental physics, as the most basic science, lends further support to theistic ontological idealism. As mentioned earlier, the mathematical description of the world in relativistic quantum physics has two primary consequences: (a) the world we experience is *not* composed of material substances (material things); and (b) the regularity of reality is *not* grounded in efficient material causation.[64] In short, not only foundational philosophical reflection, but foundational scientific investigation, points to the explanatory superiority of theistic ontological idealist metaphysics.

What of neuroscience? Given everything that's been said, it's clear that the human brain and its activity are known only insofar as the brain is observed and examined in acts of *perception.* The phenomenology of our brain states, which we've discovered are correlated with the information-processing channels for our perceptually embodied experiences, is subject to empirical study just like every other aspect of experiential reality. That neurophenomenological analysis reveals different brain states to be correlated with different aspects of cognitive and bodily functionality or malfunctionality in brain-linked consciousness (as opposed to brain-decoupled near-death experiences) is no more surprising or unintelligible on a theistic ontological idealist basis than the fact that our perceptual experience and ability to function is affected, say, by whether our leg is healthy or broken. I've sketched the details of an idealist research program in neuroscience elsewhere,[65] but the existence of correlations between observed brain activity and our inner life poses no problem for ontological idealism.

Lastly, let's deal with a standard objection to idealism: if the reality we experience is merely phenomenological, if it is *constituted* by perception, how can we ever be said to be mistaken about what we perceive, or to have an illusion, or to be subject to hallucinations? Furthermore, if God is the source of our perceptions and our perceptions are wrong, is it not the case then that God is deceiving us?

Such objections misunderstand the *nomological structure* of the world in theistic ontological idealism. As Andrei Linde observed about our perceptions, we have discovered that they obey some laws, which means, especially in physics, that they are subject to quantitative mathematical descriptions. The qualitative structure of our experience is also regular: when our perceptual faculties are functioning normally in a normal environment, we predictably have certain contextually appropriate visual, auditory, olfactory, gustatory, and tactile experiences. These regularities of experience enable us to have rational expectations about the future and to interact with each other successfully on the basis of a shared reality about which we have similar expectations. As we investigate our perceptual world, *we discover new things about it*; this tells us that *our perceptual reality does not come to us pre-interpreted in every way*, even though the *kinds* of experiences we can have are correlated with the kind and limitations of the perceptual faculties we perceive ourselves to have. We thereby discover our perceptual reality has a *lawful* structure, and in so doing, we come to understand that some direct perceptions are inaccurate in a lawful way.

Just because we *see* something, then, doesn't mean we have a *proper understanding* of what we see. For example, a stick that *appears* bent in water is not really bent in the way that it appears. We come to understand that its appearance is inaccurate (illusory) when we discover its apparent bentness is the result of the lawful behavior of light as a phenomenon when passing through media of different densities.[66] Similarly, we discover that certain pharmaceuticals or certain aberrations in the neurophenomenology of brain function are regularly correlated with the experience of hallucinations. Those affected by such things have perceptions that others in the same cognitive environment not similarly affected do *not* have because the cognitive faculties of those so affected can be *seen* not to be functioning normally. These are all things we discover to be part of the lawful structure of perceptual reality, so all of these things are perfectly intelligible and explainable within an idealist metaphysic. Those who assert otherwise have misunderstood how reality has a regularity both *structured for* and *perceived by* finite created agents if reality has the nature that theistic ontological idealists ascribe to it.

7. An Ideal End

We have sojourned into the mental nature of reality, basking in the warmth of a neo-Berkeleyan immaterialism brought to life, in part, by the glow of critical realism in modern physics. We have sojourned as well among the constructions of physicalists and dualists and dual-aspect monists, taking note of absurd features and searching for answers, before returning home to theistic ontological idealism and its more satisfying resolution of our questions. And being at home, we can reflect expansively on the virtues of home, and ultimately, on the Master of the house, in whom we live and move and have our being. This is perhaps the most satisfying thing of all, for if theistic idealism is true, not only is *all* of our experience immediate evidence that God exists, but unbelief is a performative contradiction, for the very breath that denies God is provided by God as the one who is before all things, and in all things, and in whom all things hold together.[67]

Notes

1. See my companion essay in this volume, "Consciousness and Quantum Information," for a development of this point.
2. A helpful account of the historical development of different streams of idealism is J. Dunham, I. H. Grant and S. Watson, *Idealism: The History of a Philosophy* (Kingston: McGill-Queen's University Press, 2011). See also P. Guyer, "Idealism," *Stanford Encyclopedia of Philosophy*, 2015, https://plato.stanford.edu/entries/idealism/. For a shorter introduction, see my article "Idealism," in *Dictionary of Christianity and Science*, eds. Paul Copan, Tremper Longman III, Christopher L. Reese, and Michael G. Strauss (Grand Rapids, MI: Zondervan, 2017), 372–373.
3. Plato's (428–348 BCE) theory of ideas (or forms) maintains that the material world is a pale shadow of the absolute reality constituted by eternal, unchanging, ideal forms that are grasped by the mind. Participation in these forms gives identity to everything in the world of our experience and understanding these forms is the purpose of all knowledge. While Platonic forms can only be apprehended by minds, the Platonic theory is not often called "idealism" today because the forms, while not material, are also not mind-dependent. The Neoplatonism of Gregory of Nyssa (*c*. 335–*c*. 395 CE) and Augustine (354–430 CE), which turned Platonic forms into ideas in the mind of God that differentiate and give intelligibility to the world of our experience—an interpretation of the *logos* doctrine (John 1:1–3; Colossians 1:16–17) as the divine reason infusing reality—is accurately classified as a form of idealism. It strongly influenced later Christian forms of idealism, including those of John Scotus Eriugena (815–877), George Berkeley (1685–1753), and Jonathan Edwards (1703–1758). Arguably, variations of this metaphysic also undergird the primary role accorded to information among theistically oriented philosophers, scientists, and mathematicians pursuing intelligent design research in physics, cosmology, and biology.
4. German idealism begins with Immanuel Kant (1724–1804). It is best classified as *epistemological*: the existence of something non-mental is conceded, but everything that can be known about this mind-independent reality is permeated by the formative structures of the mind. Our perceptions of the world, Kant maintained, are organized by space and time as modes of human cognition and by innate categories of the understanding (quantity, quality, relation, modality) that structure our perception and conception of what is given to us in experience. The self as a transcendental unity of consciousness that precedes and grounds experience—what Kant called the *transcendental unity of apperception*—is the source of these modes of cognition and categories of understanding and it applies them to our "raw" experience of the world. Thus, we never experience reality-in-itself (noumenal reality) but only reality as it *appears* to us (phenomenal reality) through the innate structuring of the human mind.

   In thinking about ontological versus epistemological idealism as such, we observe that ontological idealism *is opposed to both dualism and materialism*, whereas epistemological idealism *need not be opposed to either*. G. W. F. Hegel (1770–1831) rejected Kant's transcendental idealism, denying a difference between what is given in experience and the categories that structure it—in short, denying any *real* distinction between subject and object—and asserting instead that everything (including all finite consciousness) exists in interrelationship with everything else as part of *one* evolving conscious substance conceived as Absolute Spirit. This Absolute Spirit is a perfectly interrelated, all-inclusive thinking whole that is in the process of actualizing and fulfilling the transient existence of all finite things. Thus, in contrast to Berkeleyan and Kantian idealism, which recognize a plurality of mental subjects, Hegelian idealism is monistic, even pantheistic, maintaining that everything that exists is a form of self-actualizing Absolute Spirit. There are some obvious similarities between Hegelian idealism and Spinoza's pantheistic dual-aspect monism.

Spinoza's dual-aspect monism and related ideas will be examined in sect. 5. For more on German idealism, see: N. Boyle, L. Disley, and K. Ameriks, eds., *The Impact of Idealism: The Legacy of Post-Kantian German Thought. Volume 1: Philosophy and Natural Sciences* (Cambridge: Cambridge University Press, 2013); H. Kim and S. Hoeltzel, eds., *Kant, Fichte, and the Legacy of Transcendental Idealism* (Lanham: Lexington Books, 2014); and M. Rohlf, "Immanuel Kant," *Stanford Encyclopedia of Philosophy*, 2010, http://plato.stanford.edu/ entries /kant/. For examinations of Spinoza's influence on Hegel, see: Efraim Shmueli, "Hegel's Interpretation of Spinoza's Concept of Substance," *International Journal for Philosophy of Religion* 1, no. 3 (1970): 176–191; Pierre Macherey, *Hegel or Spinoza*, trans. Susan M. Ruddick (Minneapolis: University of Minnesota Press, 2011); and Gregor Modor, *Hegel and Spinoza: Substance and Negativity* (Evanston: Northwestern University Press, 2017).

5. Alfred North Whitehead's (1861–1947) *process philosophy* and its theological outworking is idealist in that it offers a broadly neo-Hegelian conception of nature's historical progression in which all things have an irreducibly mental aspect and are being drawn by a divine "lure" toward actualization and fulfillment. Process philosophy is based on the premise that being is dynamic and its dynamic nature should be the primary focus of any comprehensive philosophical account of reality and our place within it. Process metaphysicians reject substance metaphysics and argue that the traditional mind-body problem dissolves if all basic constituents of reality are short-lived processes of information transfer (actual occasions) that have "mental" and "physical" poles of emphasis dependent on the context of analysis. Process metaphysics, and panpsychism as well, both owe much to Spinoza's dual-aspect monism (see sect. 5 below). In any case, the idealist component of process metaphysics emerges in that the mental pole of every event is always present and influences dynamic development. But the central problem with describing processes as a series of *events* apart from an underlying view of substances or sequences of substances in and by which such events occur, is that castles are being built in the air. This kind of incoherence is common to all views that seek to eliminate notions of substance from their metaphysics: action requires an actor; processes require participants; and becoming requires being. For the *locus classicus* of process philosophy, see A. N. Whitehead, *Process and Reality* [1929] (New York: Free Press, 1979). For standard works in process theology, see J. Cobb and D. R. Griffin, *Process Theology: An Introductory Exposition* (Philadelphia: Westminster Press, 1976); P. Teilhard de Chardin, *The Future of Man* (New York: Image Books, 1959 [2004]); J. F. Haught, *God After Darwin: A Theology of Evolution* (Boulder, CO: Westview Press, 2001); J. A. Jungerman, *World in Process: Creativity and Interconnection in the New Physics* (Buffalo, NY: SUNY Press, 2000).
6. Personalism, in its various guises, emphasizes the centrality of the person to a proper understanding of the world, especially in the humanities and human sciences, and it focuses on the person as the ultimate locus of explanation in every branch of philosophy—ontological, epistemological, and axiological. The varied character of personalist approaches has led to tensions among idealist, phenomenological, existentialist, and Thomistic factions within this philosophical school. See T. D. Williams, "Personalism," *Stanford Encyclopedia of Philosophy*, 2013, http://plato.stanford. edu /entries/personalism/.
7. See endnote 10 for some comments on pancomputationalism. As for panpsychism in general, panpsychist views are getting a widespread hearing these days as metaphysicians and philosophers of mind grapple with the explanatory demands of consciousness and order in the universe. In panpsychist thought, organic forms of consciousness that we pre-theoretically associate with humans and animals are not fundamental but rather grounded in a more basic form of consciousness that permeates nature. This *constitutive panpsychism* has bottom-up and top-down versions. The bottom-up form, *micropsychism*, holds that everything, including all the facts about human consciousness, can be grounded in consciousness-involving facts at the micro-level—macrophenomenal truths are (wholly or partially) grounded in microphenomenal truths. Just as atoms combine to give rise to physical objects, "psychic" atoms of some sort combine to generate complex forms of consciousness. This poses a significant *combination problem*: How do the "experiences" of fundamental physical entities such as quarks and photons combine, for example, to yield human conscious experience? The proposal, if not regarded as a category mistake, is intractable at best and subject to wild speculation, as is its top-down version. From the top-down perspective, known as *cosmopsychism*, the claim is made instead that all facts about consciousness in general and organic consciousness in particular can be grounded in consciousness-involving facts concerning the universe as a *whole*. The conscious universe subsumes all its conscious components as part

of its larger consciousness. This approach is readily assimilable to various forms of pantheism, especially those conceived in a Spinozist vein. For a helpful survey, see Philip Goff, "Panpsychism," *Stanford Encyclopedia of Philosophy*, 2017, https://plato.stanford.edu/entries/panpsychism/. For critical exploration, see G. Brüntrup and L. Jaskolla, eds., *Panpsychism: Contemporary Perspectives* (Oxford: Oxford University Press, 2017). Finally, see also T. Nagel. *Mind and Cosmos: Why the Materialist Neo-Darwinian Conception of Nature is Almost Certainly False* (Oxford: Oxford University Press, 2012).

8. R. M. Adams, "Berkeley's 'Notion' of Spiritual Substance," *Archiv für Geschichte der Philosophie* 55 (1973): 47–69; R. M. Adams, "Idealism Vindicated," in *Persons: Human and Divine*, eds. P. van Inwagen and D. Zimmerman (New York: Oxford University Press, 2007), 35–54; S. H. Daniel, "Berkeley's Christian Neoplatonism, Archetypes, and Divine Ideas," *Journal of the History of Philosophy* 39, no. 2, (April 2001): 239–258; Lisa Downing, "Berkeley," *Stanford Encyclopedia of Philosophy*, 2011, https://plato.stanford.edu/entries/berkeley/. S. Cowan and J. S. Spiegel, eds., *Idealism and Christian Philosophy* [vol. 2 of *Idealism and Christianity*] (New York: Bloomsbury Academic, 2016); J. R. Farris, M. Hamilton, and J. S. Spiegel, eds., *Idealism and Christian Theology* [vol. 1 of *Idealism and Christianity*] (New York: Bloomsbury Academic, 2016); H. Robinson, *Perception* (New York: Routledge, 1994); J. Foster, *The Case for Idealism* (London: Routledge & Kegan Paul, 1982); J. Foster, *The Nature of Perception* (Oxford: Oxford University Press, 2000); J. Foster, *A World for Us: The Case for Phenomenalistic Idealism* (Oxford: Oxford University Press, 2008); T. Goldschmidt and K. L. Pearce, eds., *Idealism: New Essays in Metaphysics* (Oxford: Oxford University Press, 2017).

9. John Wheeler proposed that everything in the universe—that is, every *it*—is ultimately derived from quantum information expressed as quantum bits (*qubits*). I defend a version of this thesis in my companion essay in this volume, arguing that our minds experience the world as *it-from-qubit* communications from the mind of God.

10. In the contemporary context, the neo-Berkeleyan idealism I defend involves *informational idealism*, *quantum idealism*, and *conscious realism*. Informational idealism holds that information *precedes* both experience and matter/energy. This is correct as far as it goes, but it is ultimately untenable if the attempt is made to turn information itself into a new kind of foundational entity. It's true that our experience is structured both quantitatively and qualitatively by information that is not generated by us, and the *awareness* of which provides the content for our experience. It is also true that what we call *matter* or *energy* is structured by that which *informs* it and can be mentally *abstracted* from it. This has led some to advocate a kind of *ontic pancomputationalism*: the universe is made by abstract, ungrounded information processing—computation precedes and constitutes matter, energy, and mind. This is yet another effort to construct a dual-aspect theory in the spirit of Spinoza. See: E. Fredkin, "An Introduction to Digital Philosophy," *International Journal of Theoretical Physics* 42, no. 2 (2003): 189–247; L. Floridi, "Trends in the Philosophy of Information," in P. Adriaans and J. van Benthem, eds., *Handbook of the Philosophy of Science, Volume 8: Philosophy of Information* (Amsterdam: Elsevier, 2008), 113–131; G. Piccinini, "Computation in Physical Systems," *Stanford Encyclopedia of Philosophy*, 2017, https://plato.stanford.edu/entries/computation-physicalsystems/; and M. Tegmark, *Our Mathematical Universe: My Quest for the Ultimate Nature of Reality* (New York: Vintage Books, 2014). As Bernardo Kastrup observes in *The Idea of the World: A Multi-disciplinary Argument for the Mental Nature of Reality* (Washington, DC: Iff Books, 2019), 25ff., ontic pancomputationalism suffers from a Lewis-Carroll-style Cheshire-cat problem: How can the grin of the cat exist without the cat? Information requires either a physical (matter/energy) or a mental substrate, otherwise, it is like spin without a top, waves without water, or a dance without a dancer. Clearly, pancomputationalism is not a tenable form of informational idealism and further reflection on the metaphysics of information is needed. In this regard, William Dembski gestures in the direction of grounding the metaphysics of information in immaterial substances (minds) in W. A. Dembski, *Being as Communion: A Metaphysics of Information* (Burlington: Ashgate Publishing Company, 2014), but does not commit to embracing ontological idealism. Closely connected to this issue, of course, is the well-established conclusion that quantum physics is *not* compatible with local realism about material substances and that reality, even non-locally, lacks a definitive character apart from our *experience* of it. The idealist implications of this fact have been extended by recent experiments separating quantum properties from anything like a physical substrate in a manner precisely analogous to Carroll's separation of the grin from the Cheshire cat. The absence

of a *material substrate* in such cases requires the existence of an *immaterial mental substrate* that grounds a kind of *quantum idealism*. I explore the relevant literature and press this argument in my companion essay in this volume, "Consciousness and Quantum Information," though I have been making variants of this argument for a number of years (see my other essays referenced there). All of this indicates that *consciousness* is the fundamental ground of reality—it is not what needs to be explained, but rather *what does the explaining*; it is the *explanans*, not the *explanandum*. This leads to *conscious realism*: consciousness is fundamental to the universe and immaterial mental substance is the fundamental ground of reality. Immaterial consciousness is the subject of experience providing the substrate for the information structures characterizing physical reality and communication among minds. In other words, consciousness (immaterial agency) is the foundational reality and starting point for *everything* else, immaterial substances are the *fundamental constituents* of reality, and the phenomenological reality of our experience (the universe) is ultimately *personal and relational*, not impersonal.

11. I argue these points about the PSR in my companion essay in this volume. For a more extended argument, see my essay "How Does the Intelligibility of Nature Point to Design?," in William Dembski, Casey Luskin, and Joseph Holden, eds., *The Comprehensive Guide to Science and Faith: Exploring the Ultimate Questions about Life and the Cosmos* (Eugene, OR: Harvest House, 2021). For a more thorough discussion, see Alexander R. Pruss, *The Principle of Sufficient Reason: A Reassessment* (Cambridge: Cambridge University Press, 2006), and Alexander R. Pruss and Joshua L. Rasmussen, *Necessary Existence* (Oxford: Oxford University Press, 2018); for a more popular discussion, see Joshua Rasmussen, *How Reason Can Lead to God: A Philosopher's Bridge to Faith* (Downers Grove, IL: IVP Academic, 2019).
12. This principle of sufficient reason (PSR) would also apply if material substances and causes *did* exist to serve in a proximate explanatory role, for the existence of such things would, in such case, also be a contingent fact in need of explanation. But explanations cannot rest on contingent facts all the way down. At some point, a necessary being must provide the ultimate ground for all contingent explanations, and no *material* entity is non-contingent, so the necessary being in question must be *immaterial*. In short, even if you're committed to the existence of material substances, you can't avoid God.
13. To say that God is the *vera causa* of something is to say that he is the *true cause* of the phenomenon, as opposed to the apparent and proximate cause. God alone has real power to actualize things. The regular causal antecedents that we observe allow us, in divine providence, to reliably anticipate what will happen next, but they lack the intrinsic power to produce it. This power comes from God as the sole efficient cause of everything in that segment of the universe not subject to the influence of rational creatures with genuine moral (libertarian) freedom.
14. P. Boghossian, "The Status of Content," *Philosophical Review* 99 (1990): 157–84; P. Boghossian, "The Status of Content Revisited," *Pacific Philosophical Quarterly* 71 (1991): 264–278; M. Devitt and G. Rey, "Transcending Transcendentalism," *Pacific Philosophical Quarterly* 72 (1991): 87–100; for a comprehensive overview of eliminativism that is hopelessly optimistic, see W. Ramsey, "Eliminative Materialism," *Stanford Encyclopedia of Philosophy*, 2019, https://plato.stanford.edu/entries/materialism-eliminative/.
15. Reductive physicalism was defended prominently by Jaegwon Kim (1934–2019), among others. See: J. Kim, *Mind in a Physical World: An Essay on the Mind-Body Problem and Mental Causation* (Cambridge: MIT Press, 1998); J. Kim, *Supervenience and Mind: Selected Philosophical Essays* (Cambridge: Cambridge University Press, 1993); and J. Kim, *Philosophy of Mind*, 3rd ed. (New York: Routledge, 2018).
16. Non-reductive physicalism is represented variously by Donald Davidson's anomalous monism in D. Davidson, "Mental Events" (1970) reprinted in D. Davidson, *Essays on Actions and Events* (Oxford: Oxford University Press, 1980), 207–224, by John Searle's biological naturalism in J. Searle, *The Rediscovery of the Mind* (Cambridge: MIT Press, 1992), and by many others. It has, unfortunately, become trendy in Christian circles as well and is championed by Christian philosophers like Peter van Inwagen, Lynne Rudder Baker, Nancey Murphy, Kevin Corcoran, and others, though these Christian philosophers do not hold to a strong causal closure principle where divine action is concerned.
17. I elaborate upon and defend these and related claims in sect. 4 ("The Principle of Sufficient Reason and the Nullification of Natural Necessity") of my companion essay in this volume.
18. That is to say, these things may be happening by chance, for no intrinsic reason related to considerations that would connect them to other things in our experience, so our experience may actually be a chance collection of unrelated events that we believe (also

by chance, and mistakenly) to have a relationship to each other.

19. I make this clear in my companion essay in this volume in sect. 3 ("From Microscopic to Macroscopic"); see also B. L. Gordon, "The Incompatibility of Physicalism with Physics," in *Christian Physicalism? Philosophical-Theological Criticisms*, eds. J. R. Farris and R. K. Loftin (New York: Lexington Books, 2018), 371–402.
20. The *locus classicus* for Plantinga's argument is chapter 12 of his *Warrant and Proper Function* (Oxford: Oxford University Press, 1993).
21. See J. Beilby, ed., *Naturalism Defeated? Essays on Plantinga's Evolutionary Argument Against Naturalism* (Ithaca, NY: Cornell University Press, 2002).
22. See J. Mark, B. Marion, and D. Hoffman, "Natural Selection and Veridical Perceptions," *Journal of Theoretical Biology* 266 (2010): 504–15. Note that a "fitness function" is a mathematical expression that characterizes, in terms of the performance of something relative to its alternatives, how close that thing is to achieving a certain goal, for example, survival. What Hoffman et al. showed was that organisms functioning on the basis of accurate (veridical) representations of an objective environment can be outcompeted and driven to extinction by organisms with arbitrary fitness functions tuned to environmental utility (usefulness) rather than veridicality. This means that useful perceptions readily diverge from and replace true perceptions as organisms struggle to survive. See also C. Prakash et al., "Fitness Beats Truth in the Evolution of Perception" (http://cogsci.uci.edu/~ddhoff/ FitnessBeatsTruth_apa_PBR, forthcoming), and Donald D. Hoffman, *The Case Against Reality: Why Evolution Hid the Truth from Our Eyes* (New York: W. W. Norton & Company, 2019).
23. A sympathetic critical survey of the literature on dualism may be found in H. Robinson, "Dualism," *Stanford Encyclopedia of Philosophy*, 2016, https://plato.stanford.edu/entries/dualism/. Robinson adds a fourth dualism, *predicate dualism*, which asserts that mental predicates are not reducible to physical predicates, but it's really just a form of non-reductive physicalism (anomalous monism).
24. Should this emergence be understood *causally*, so that mental properties *arise from* physical properties of a certain kind in some mysterious way? While it is hard to see why this should be so, it is certainly the intent of those who wish to affirm the causal closure of the physical realm (philosophical naturalism) but recognize the distinctive character of the mental and its irreducibility to the physical. Unless mental properties are then called some additional species of *physical* property, however, as long as causal closure is maintained, it's difficult to see how this view avoids an *epiphenomenalism* that renders the mental causally otiose. A good discussion of property dualism and an attempt to incorporate it within naturalistic metaphysics is provided by David J. Chalmers, *Consciousness: In Search of a Fundamental Theory* (Oxford: Oxford University Press, 1996), 123–171.
25. There is no coherent theory of how material substance could *cause* that which is immaterial, nor of how that which is immaterial could *cause* that which is material, but as we will discuss in sect. 4.2.3, there are some reasonable things that can be said about how immaterial *consciousness* could affect material outcomes.
26. David Hume, *A Treatise of Human Nature* [1739–40], edited by D. F. Norton and M. J. Norton (Oxford: Clarendon Press, 2007) Book I, Part IV, Section VI.
27. Neo-Humeans, such as Derek Parfit in *Reasons and Persons* (Oxford: Clarendon Press, 1984) and Barry Dainton in *The Phenomenal Self* (Oxford: Oxford University Press, 2008), modify Hume by *rejecting* the claim that mental states of awareness can have an identity apart from the bundle to which they belong (which Humeans maintained because they thought the mind did not perceive any connection among the objects of awareness), inventing instead *co-consciousness* relations that hold among the elements of the bundle and are also objects of awareness. Nonetheless, they deny that co-consciousness relations, which are intended to function as the nexus of the bundle, require substantial entities. But this is incoherent too. The relational nexus constituted by co-consciousness, like first-order mental properties, is an object of awareness, and simultaneous objects of *awareness* do not exist apart from a subject that is aware, for *without the subject there is no awareness*, that is, there is nothing that *is* conscious. The conscious subject is not reducible to or replaceable by mere properties and relations. Neither monadic properties nor polyadic relations can function as *agents*. Consciousness is no more instantiated as a property without a subject that *is* conscious than ocean waves exist without water or live performances without performers. Both Humeans and neo-Humeans are attempting to build castles in the air with assertions that—proceeding from themselves as subjects—are performative contradictions.
28. The "*y*" in "*hylē*" represents the Greek letter *upsilon*, which was traditionally rendered "y" in English transliterations of Greek words. The "y" in the English word

"hylomorphic" reflects this traditional transliteration. It should be noted, however, that it has become increasingly common to represent the *upsilon* not with a "y" but with a "u"; thus, the Greek word for "matter" is now frequently rendered as *hulē*, which gives the English reader a clearer idea of the original Greek vowel sound. In this particular word, the accent is on the first syllable, the "*u*" is long, and the "*ē*" represents the Greek "long e" vowel (*ēta*), and is pronounced like the "ey" in "they."

29. The accent on *morphē* is on the last syllable, and the final "*ē*" is pronounced like the "ey" in "they."
30. Aquinas's discussion of these points can be found in his *Summa Theologica*, I, 75–89, most especially 75–79, and in his *Summa Contra Gentiles*, II, 49–51. If the concept of a substantial form makes sense, it is not clear to me why, if animals have *memories* and even *dreams*, their souls could not also continue to exist after death without sensory input. They may have no *further* sensory input, but why wouldn't their memories of sensory input and the possibility of dreaming be sufficient to sustain their substantial identities? Of course, this is all predicated on the intelligibility of a *substantial form*, which is problematic for reasons about to be discussed, and further contradicted by the evidence of near-death experiences (NDEs) in which consciousness decoupled from the body experiences heightened *sensory* input.
31. In discussing the tenability of hylomorphic and substance dualism, I have benefitted from reading the interactive polemics about these and related topics on the internet blogs of Bill Vallicella (https://maverick philosopher.typepad.com/maverick_philosopher/) and Ed Feser (http://edwardfeser.blogspot.com/).
32. This also suggests that the sensate souls of non-rational animals may survive separation from their bodies. Of course, in the context of ontological idealism, this would hardly be surprising, for *the experience of embodiment* for *any* finite conscious being is itself merely phenomenological.
33. Kenneth Ring and Sharon Cooper, *Mindsight: Near-Death and Out-of-Body Experiences in the Blind*, 2nd ed. (Bloomington, IN: iUniverse, Inc., 2008).
34. See Pim van Lommel, *Consciousness Beyond Life: The Science of the Near-Death Experience* (New York: HarperOne, 2010); Pim van Lommel, "Near-Death Experiences: The Experience of the Self as Real and Not as an Illusion," *Annals of the New York Academy of Sciences* 1234, no. 1 (2011): 19–28; and Pim van Lommel, "Non-local Consciousness: A Concept Based on Scientific Research on Near-Death Experiences During Cardiac Arrest," *Journal of Consciousness Studies* 20, nos. 1-2 (2013): 7–48. See also: J. M. Holden, B. Greyson, B. James, and D. James, eds., *The Handbook of Near-Death Experiences: Thirty Years of Investigation* (Santa Barbara: Praeger, 2009); J. Long and P. Perry, *Evidence of the Afterlife: The Science of Near-Death Experiences* (New York: HarperOne, 2010); C. Fracasso and H. Friedman, "Near-Death Experiences and the Possibility of Disembodied Consciousness," *NeuroQuantology* 9, no. 1 (2011): 41–53; T. Rivas, A. Dirven, and R. H. Smit, *The Self Does Not Die: Verified Paranormal Phenomena from Near-Death Experiences* (Durham: International Association for Near-Death Studies, 2016); and J. C. Hagan, ed., *The Science of Near-Death Experiences* (Columbia: University of Missouri Press, 2017).
35. See again Ring and Cooper, *Mindsight*; van Lommel, *Consciousness Beyond Life*; and van Lommel, "Non-local Consciousness." Also see L. P. Fenwick, "Non-local Effects in the Process of Dying: Can Quantum Mechanics Help?" *NeuroQuantology* 8, no. 2 (2010): 155–163, and L. Shan, "Consciousness as an Entity with Entangled States: Correlating the Measurement Problem with Non-Local Consciousness," *NeuroQuantology* 16, no. 7 (2018): 70–78.
36. This point is made by Eric T. Olson in his essay, "A Compound of Two Substances," in Kevin Corcoran, ed., *Soul, Body, and Survival: Essays on the Metaphysics of Human Persons* (Ithaca, NY: Cornell University Press, 2001), 75.
37. Elliott Sober, *Philosophy of Biology*, 2nd edition (Boulder, CO: Westview Press, 2000), 24.
38. Robert Wald, *General Relativity* (Chicago: University of Chicago Press, 1984), 70n6.
39. Carl Hoefer, "Energy Conservation in GTR," *Studies in the History and Philosophy of Modern Physics* 31, no. 2 (2000): 187–199.
40. G. Börner, *The Early Universe: Facts or Fiction?* (New York: Springer-Verlag, 1988), 298; J. Peacocke, *Cosmological Physics* (Cambridge: Cambridge University Press, 1999), 26.
41. These are, of course, the well-known Bell correlations. John Bell's original papers are J. S. Bell, "On the Einstein-Podolsky-Rosen Paradox" (1964) and "On the Problem of Hidden Variables in Quantum Mechanics" (1966), reprinted in J. S. Bell, *Speakable and Unspeakable in Quantum Mechanics* (Cambridge: Cambridge University Press, 1987), 14–21 and 1–13 (respectively). There has been much discussion and extension of these results in the literature. I've examined this

literature and analyzed its significance in a number of places, most conveniently (and non-technically) in my companion essay in this volume, "Consciousness and Quantum Information" (see sect. 2).

42. See B. L. Gordon, "The Necessity of Sufficiency: The Argument from the Incompleteness of Nature," in J. L. Walls, J. Dougherty, and T. Dougherty, eds., *Two Dozen (or so) Arguments for God: The Plantinga Project* (Oxford: Oxford University Press, 2018), 417–445, and also sect. 4 ("The Principle of Sufficient Reason and the Nullification of Natural Necessity") in my companion essay in this volume.

43. Robin Collins provides an excellent discussion of these points and a strong defense of substance dualism in R. Collins, "Modern Physics and the Energy-Conservation Objection to Mind-Body Dualism," *American Philosophical Quarterly* 45, no. 1 (2008): 31–42, and also in R. Collins, "The Energy of the Soul" and "A Scientific Case for the Soul," in Stewart Goetz and Mark C. Baker, eds., *The Soul Hypothesis: Investigations into the Existence of the Soul* (New York: Continuum, 2011), 123–137 and 222–246 (respectively).

44. Again, see my companion essay "Consciousness and Quantum Information," and references therein, for a development of this argument.

45. The holenmeric manner in which the immaterial soul experiences reality is worth reflecting upon briefly. In a substance dualist metaphysics of the person, it is usually maintained that the immaterial soul is a mereological simple (not composed of separable parts) that is holenmerically present throughout the body, which is to say, it is fully present to the body as a whole and wholly present in each part of the body. While this is somewhat mysterious in dualist metaphysics, in ontic idealism it is much clearer. For the ontic idealist, the immaterial soul is an immaterial simple as well, and the very fact that the body is *phenomenological* rather than substantial guarantees that the soul, as the conscious subject, is fully present to the experience of the body as a whole and wholly present in each part of the body experienced, for the body is an experiential *mode* of the soul. As experienced through the mode of the body, that subset of the world mediated to us by divine consciousness is *also* wholly present in the immaterial soul. This is what is meant by saying that finite consciousness, as a simple immaterial substance, experiences divinely provided reality, inclusive of bodily phenomenology, holenmerically. In the same manner, in ontic idealism, God is holenmerically present in all of creation, for the phenomenological reality we perceive individually and collectively exists wholly in God's mind, a fact that constitutes and explicates his omnipresence. Thus it is that we live and move and have our being in God himself (Acts 17:28).

46. For the development of these themes, see Baruch Spinoza: *Ethics*, I, Proposition XXIX; *Ethics*, III, Proposition II; *Ethics*, V, Propositions XXX–XLII; *Short Treatise on God, Man, and His Well-Being*, I, Chapters 8–9; *Short Treatise on God, Man, and His Well-Being*, II, Appendix II. For a helpful, if overly sympathetic overview of Spinoza's thought, see Steven Nadler, "Baruch Spinoza," *Stanford Encyclopedia of Philosophy*, 2020, https://plato.stanford.edu/entries/spinoza/. Bernardo Kastrup's idealist cosmopsychism is a modern variant of this metaphysic and treats death similarly (see the references in note 54).

47. See my brief discussion of varieties of panpsychism in note 7 above. Regarding the supposed virtues of constitutive cosmopsychism, particularly agentive cosmopsychism, see: Philip Goff, *Consciousness and Fundamental Reality* (Oxford: Oxford University Press, 2017); Philip Goff, "Conscious Thought and the Cognitive Fine-Tuning Problem," *Philosophical Quarterly* 68, no. 270 (2018): 98–122; and Philip Goff, "Did the Universe Design Itself?," *International Journal for Philosophy of Religion* 85 (2019): 99–122.

48. See especially P. Goff, "Did the Universe Design Itself?," 108.

49. See the discussion in the main text below.

50. Erik J. Wielenberg, *Robust Ethics: The Metaphysics and Epistemology of Godless Normative Realism* (Oxford: Oxford University Press, 2014).

51. Making this argument is beyond the scope of our present concerns. See William Lane Craig, "Erik Wielenberg's Metaphysics of Morals," *Philosophia Christi* 20, no. 2 (2018): 333–338. See also the exchange between David Baggett and Erik Wielenberg on the moral argument for God's existence in Gregory Bassham, ed., *C. S. Lewis's Christian Apologetics: Pro and Con* (Leiden: Brill, 2015), 121–169.

52. P. Goff, "Did the Universe Design Itself?," 120.

53. For an extended discussion of this issue, see: Alexander R. Pruss and Joshua L. Rasmussen, *Necessary Existence* (Oxford: Oxford University Press, 2018), especially ch. 8; Alexander R. Pruss, *Infinity, Causation, and Paradox* (Oxford: Oxford University Press, 2018), especially ch. 9; Joshua Rasmussen, "From a Necessary Being to God," *International Journal of the Philosophy of Religion* 66 (2009): 1–13; and Joshua Rasmussen, "Could God Fail to Exist?," *European Journal for Philosophy of*

Religion 8, no. 3 (2016): 159–177. For a popular discussion of these and related themes, see Joshua Rasmussen, *How Reason Can Lead to God: A Philosopher's Bridge to Faith* (Downers Grove, IL: IVP Academic, 2019).

54. Bernardo Kastrup, *The Idea of the World: A Multi-disciplinary Argument for the Mental Nature of Reality* (Washington, DC: Iff Books, 2019), and Bernardo Kastrup, "The Universe in Consciousness," *Journal of Consciousness Studies* 25, nos. 5/6 (2018): 125–155.
55. See the discussion at the end of sect. 4.2.2.
56. See note 8 above for an introduction to the contemporary literature on Berkeleyan immaterialism, phenomenalistic idealism, and its place in Christian theology and metaphysics. While Bernardo Kastrup's ontological idealism is not developed from a theistic standpoint (see footnote 54), his exposition and defense of some key concepts is both clear and useful, so I draw on it selectively in the discussion that follows.
57. Andrei Linde, "Universe, Life, Consciousness," Physics and Cosmology Group, *Science and Spiritual Quest* program, CTNS, University of California, Berkeley, 1998, 12. Available at: http://www.andrei-linde.com/articles/universe-life-consciousness-pdf.
58. George Berkeley, *Three Dialogues between Hylas and Philonous* [1713] (Chicago: Open Court, 1969), 64. For those who haven't noticed the allusion in Berkeley's title, "Hylas" is the character who advocates for matter (*hylē*), whereas "Philonous" loves mind (*philo-* ["loving"] + *nous* ["mind"]). Berkeley first presented this argument in *The Principles of Human Knowledge* [1710] (Indianapolis: Bobbs-Merrill, 1957), sect. 29, 146–156. For a good general introduction to Berkeley's philosophy and how it fits into Christian theology, see James S. Spiegel, "Idealism and the Reasonableness of Theistic Belief," in *Idealism and Christian Philosophy*, 11–28, and James S. Spiegel, "The Theological Orthodoxy of Berkeley's Immaterialism," in *Idealism and Christian Theology*, 9–33.
59. If, for reason of doxastic inertia, one were inclined to insist that the existence of material substances is a properly basic belief, then given that God is necessary to explain the existence of material reality (whatever its nature), it would follow that God, as an immaterial being, is *somehow* able to bring a substantial material reality into existence, even if *we* don't know how. Now, if it *had* to be the case that material substances existed, this would undoubtedly be true. But given the genuine metaphysical puzzlement the wholly immaterial generation of a substantially material reality creates, surely it counts in *favor* of ontic idealism and *against* substance dualism that this puzzlement *does not exist* in an idealist metaphysic. When one adds to this consideration what is made clear in my companion essay in this volume, namely that fundamental physics irremediably problematizes the idea of substantial matter, one sees that the possibility of generating material substantiality from immateriality *doesn't need to be explained*, because material substances *do not exist*. At this point, not just the explanatory superiority of ontic idealism, but the genuine need of it, is manifest.
60. See sect. 4.1 and references therein for further discussion of problems with physicalism.
61. Edmund Husserl, *The Crisis of the European Sciences and Transcendental Phenomenology: An Introduction to Phenomenological Philosophy* [1936] (Evanston: Northwestern University Press, 1970).
62. Maurice Merleau-Ponty, *The Primacy of Perception: And Other Essays on Phenomenological Psychology, the Philosophy of Art, History and Politics, 1947–61* (Evanston: Northwestern University Press, 1964).
63. See B. Kastrup, *The Idea of the World*, ch. 2, especially 29–38.
64. As also noted earlier, the evidential basis for these claims and their implications are defended in my companion essay in this volume, "Consciousness and Quantum Information," and other writings of mine referenced there.
65. See Bruce L. Gordon, "Idealism and Science: The Quantum-Theoretic and Neuroscientific Foundations of Reality," in *The Routledge Handbook of Idealism and Immaterialism*, eds. J. R. Farris and B. P. Göcke (London: Routledge, 2022), 536–575, especially the extended discussion in sect. 5 ("Neuroscience, Consciousness, and Theistic Quantum Idealism," 548–558. See also the sections on neuroscience, mind, and consciousness in my companion essay in this volume, "Consciousness and Quantum Information."
66. To clarify this situation further, we have a visual perception that the stick is bent and a tactile perception (if we feel the stick under water) that the stick is straight. Which perception should take precedence, since both are provided to us by God? Is God deceiving us in one case and not the other? Or is it perhaps the case that there's no real fact of the matter? Of course not, and indeed, no more so than if there were a materially substantial reality *created* by God that catalyzed perceptions of this nature. To understand why, consider that through a persistent investigation of our perceptual reality, we eventually arrive at a law of refraction characteristic of the phenomenon we call

light as it passes through media of different densities. Once this law has been discovered, it becomes clear that our tactile perception is correct and the bentness of the stick is an optical illusion. So how, you then ask, is the stick conceptualized by God in communicating the idea of it to us? It is, quite simply, conceptualized within the nomological perceptual structure he has designed and implemented: the stick is a perceptual object that illusorily appears bent because of the laws of refraction, an illusion that is confirmed by our feeling the stick under water and realizing it is not bent. The nomological structure of perceptual reality—which God faithfully maintains, but which he could have made to be otherwise because there is no necessity in it—reveals that he is not a deceiver. Rather, we initially misinterpreted our visual reality, but on investigation discovered that we were experiencing an optical illusion explained by the refractive laws of light that were found to govern our perception of the world. As I am trying to make clear, being an ontic idealist does *not* mean that our perceptual reality comes to us already interpreted any more than it would if the reality behind it were one that is materially substantial. Where the investigation of nature is concerned, as the biblical proverb states, "it is the glory of God to conceal things, but the glory of kings is to search things out" (Proverbs 25:2, NRSV).

67. Colossians 1:16–17. We have not, in the course of our discussion, found a place to examine the Christian orthodoxy of theistic ontological idealism, or the light it can shed on the doctrine of God, the doctrine of creation, the doctrine of providence, the doctrine of man, the doctrine of Christ, the doctrine of the Holy Spirit, and personal and corporate eschatology. This must await another occasion. In the meantime, a good introduction to some of the resources that theistic idealism brings to bear on philosophical and systematic theology can be found in the previously cited volume, *Idealism and Christian Theology* (New York: Bloomsbury Academic, 2016).

9. The Simple Theory of Personal Identity and the Life Scientific

Jonathan J. Loose

1. Introduction

Processes of scientific reasoning and observation are rightly held in high esteem. The aim of this chapter is to discover what characteristics a human person must have if he or she is to engage in these processes. It turns out that the necessary characteristics have to do with conscious experience and with the human subject and that, in order to have them, conscious experience and the human subject must be understood in ways that are inconsistent with materialism (or physicalism). If this is correct, then to take the view that human persons are wholly material things is not to esteem science, but to undermine it.

Recently Daniel Dennett was interviewed as part of the BBC series *The Life Scientific*, which celebrates the achievements of great scientists of our time. The presenter warmly commended him, suggesting that Dennett "thinks like a philosopher but that hasn't stopped him making a serious contribution to science." Going further, he suggested that Dennet's philosophical approach may even have improved that contribution since the sciences emerged from natural philosophy in the first place.[1] While any affirmation of the compatibility of science and philosophy is encouraging, Dennett's way of achieving this is not. A staunch defender of naturalism and materialism, Dennett thinks that anything incompatible with those views cannot be real. We must eliminate the self as a concrete entity and excise phenomenal first-person experience from consciousness. To explain consciousness and the self by denying them is not only deeply unsatisfying, it seems as obviously false as any position could be.

Despite these startling eliminations it may seem that materialism esteems science by ensuring that nothing is beyond its explanatory grasp and so perhaps, after all, it is a reasonable commitment within a "life scientific." However, this is a profound mistake. Materialism undermines science, since it renders the observations and reasoning on which scientific activity depends unjustified and the aspirations and accomplishments that scientists cherish incoherent. If we wish to esteem and celebrate the life scientific, then we should reject materialism by expanding our ontology to include persons as essentially immaterial entities not composed of parts.

The incompatibility of materialism and the life scientific is made clear by a consideration of personal identity, of what we human persons are essentially. Personal identity can be studied "at a time" by asking what characteristics persons possess at any given moment and using this to evaluate the sort of thing we might be (e.g., organisms or minds). It can also be studied "over time" if we ask what criterion determines whether or not a person at time A and a person at time B are identical (in the numerical sense of being one and the same person).

Views of personal identity are typically considered either complex or simple.[2] Complex theorists hold that a person is identical to an entity composed of parts (whether physical, such as an organism, or psychological, such as a set of memories) and persists through time in virtue of whatever psycho/physical continuities sustain that entity (e.g., an ongoing life process or a chain of memories). In contrast, simple theorists hold that personal identity consists not in psycho/physical continuities, but in some "further fact." Most proponents of the simple theory argue that the "further fact" in question is that persons essentially are, or involve, simple, immaterial substances.

Considering personal identity "at a time," I will argue that human persons experience a unified consciousness at any given moment that cannot be accommodated by a complex view of the self and, furthermore, that we cannot simply eliminate the self in response. This leaves a simple view of personal identity as the only way to accommodate those aspects of observation and reasoning that are central to scientific activity and depend on the unity of consciousness. Considering personal identity "over time," I will bring out a significant mistake made by complex theorists.

This again leaves a simple view of personal identity as the only sufficient way to underpin aspects of the observation and reasoning central to scientific activity that depend—in this case—on the identity of a person over time. This includes the coherence of the scientist's aspirations and achievements. Scientific aspirations only make sense if the person who has them can persist into the future to become a scientist (or a greater scientist), and a person can only be celebrated for his or her "life scientific" if he or she is the one who previously made the significant discoveries being attributed to him or her.

2. Science and First-Person Experience

Before turning to arguments about personal identity, it is important to recognize that they require us to reflect on first-person experience. This is a very important point. We experience ourselves as mental subjects; subjects of experience who may be in various mental states. These mental states are phenomenal (there is a something-that-it-is-like to be in a particular mental state), private (our access to our own mental states "from the inside" is privileged in the sense that nobody else has that access, and through having it we are the ultimate authorities on our own experience), and intentional (our mental states are about things other than themselves). Reflection on first-person experience is a process radically different from the scientific activities that have propelled knowledge forward and transformed our lives over the last five centuries.

In light of this it can seem incredible that we might learn something fundamental about what it is that we are through activities radically different from those employed by the physical sciences, and which seem naively simple by comparison. Richard Swinburne highlights this incredulity when he considers Descartes's argument from what he (Descartes) can conceive to the conclusion that he is entirely distinct from his body. Swinburne writes:

> All the great discoveries about the nature of the world made by science in

the last five hundred years have been the result of scientists doing experiments and making observations to test theories formulated using technical terms and difficult mathematics. So how could we reach such a big conclusion about human nature as Descartes purports to have reached by a short argument which relies on no results of experiments and uses no sophisticated mathematics?[3]

It seems in keeping with our esteem for science that we should somehow understand human nature purely on the basis of sophisticated analysis of publicly available data, such as an image from an fMRI scanner, or an algorithm that impresses us with the human-like behavior it produces in a robot. Such things *feel* right in a scientifically driven world; they have the experimental and mathematical credentials that reflection on first-person experience seems so sorely to lack. Such a reaction to the prospect of deriving substantive conclusions from reflection on experience is understandable, but it is also mistaken. There is no reason, beyond this felt lack of fit, to rule phenomenal experience out as a datum to be considered, and since conscious awareness is first-personal rather than third-personal we need to take a different approach to it than we take to publicly observable phenomena. Again, as Swinburne points out:

> Almost all of what science of the past five hundred years has taught us about the world concerns publicly observable objects including our own bodies and the predictable ways in which they behave. What Descartes does is to draw attention to something totally different from the publicly observable, our own conscious awareness, something about which we can be more certain than about anything else, and merely asks us to face up to what that involves.[4]

It is important to note Swinburne's final comment here that our conscious awareness is that about which we can be more certain than anything else. It is not simply that we *can* include conscious experience as a datum; rather, we *must* do so, and given our certainty about it we must take the consequences of doing so very seriously, even if the processes required involve nothing of the science and mathematics that we typically make use of when describing the physical world.

Charles Taliaferro helpfully emphasizes that we cannot even appeal to science and mathematics without referring to the mental lives through which we engage in them, and that to attempt such an appeal, as Dennett does, is "self-undermining and confused."[5] As Taliaferro explains:

> It is self-undermining to the extent that Dennett cannot presume to have any clearer understanding of nonmental physical phenomena than he does of concepts, reasons and reasoning, grasping entailment relations, reliance on experience and observations that go into the practice of the sciences, and the kind of reasoning that goes on at philosophy conferences.... To appeal to *physics*, *chemistry*, and *physiology* is (if it means anything at all) to appeal to what persons as scientists practice with their theories and observations, their conceiving of and intentionally undertaking experiments and making predictions, their recording data and drawing inferences.[6] [emphasis in the original]

In the end, a spirit of inquiry will lead us to consider *all* of the facts relevant to a given question, and when we draw out the implications of what may seem to be trivial claims about first-person experience for our understanding of ontology and human nature, we realize that their consequences are, in fact, "astounding."[7]

3. Personal Identity at a Time: The Unity of Consciousness

Having emphasized the legitimacy and importance of reflection on conscious awareness within our inquiry we turn to consider personal identity, beginning with personal identity *at* a time.

As I write I can see a garden with flowers and trees set around a grass lawn. I can see a bird flying back and forth across the lawn and I hear it calling as it goes. I also notice the smell of bread baking as lunch is being prepared in the kitchen next door and—of course—all the while I'm thinking about philosophy and the theme of this chapter. Despite this variety of experiences my consciousness exhibits a profound unity. It seems most accurate to say that I am having a single, unified experience somehow involving each of these elements. It turns out that if we characterize carefully this unity of my consciousness "at a time," we reveal important consequences for personal identity.

First, consider the structure of consciousness. There are many specific types of conscious state. There are states associated with our senses and our bodies, as well as with memory, imagination, thought, and will. These particular states occur in the context of broader background states that give the conscious field its overall "tone" and determine the range of experiences that can be had. For example, the tone of conscious experience when asleep and dreaming differs sharply from that experienced when fully awake, and dreams can involve particular (sometimes bizarre) conscious experiences that cannot be had when fully awake. These background and particular conscious states *together* determine the conscious field at a given time. What we need to consider is what we mean when we say that these states are "together." That way we can understand better the nature of the conscious field as a whole.

Tim Bayne, the leading thinker in this area, notes three different unity relations, each of which implies a different account of the unity of consciousness.[8] These are subject, representational, and phenomenal unity. First, specific and background states may be united as experiences of the same *subject*. My various experiences of seeing the garden, smelling the bread baking, and thinking about philosophy are united because they are all *mine*. Second, conscious states can also be united because their content is integrated. At a given moment particular experiences of a bird-shaped visual image and a calling sound are united as a single experience of a particular bird flying across the garden. This unity is *representational* because it has to do with consistency of content, and this is something that can apply to multiple thoughts as well as multiple sensory experiences. The final type of unity is one Bayne takes to be deeper and more primitive than the others. This is *phenomenal* unity (sometimes called "co-consciousness"): there is not only "something that it is like" to experience specific conscious states (such as hearing the birds in the garden *or* smelling the bread baking *or* thinking about philosophy), but also "something that it is like" to experience the entire conscious field as a whole, including each of the particular states and background states in the context of all the others (such as hearing the birds in the garden *while* smelling bread baking *while* thinking about philosophy). It is also the case that each of the particular experiences is different because it is had as an aspect of the conscious field as a whole. Indeed, this is why we often describe consciousness at a given time as a *field*; it is because the experiences unite together as a whole that possesses its own distinctive quality. Bayne writes, "The multiplicity of objects and relations that we experience at any one point in time are not experienced in isolation from each other; instead, our experiences of them occur as components, aspects, or elements of more inclusive states of consciousness. It is this fact—however it is to be understood—that the notion of phenomenal unity attempts to capture."[9]

Of the three different possible unity relations, Bayne considers phenomenal unity to be the heart of the matter: there is something that it is like to be a subject at a particular time and this is the subject's total conscious state. Importantly, this is a distinct state that subsumes the various background and particular states involved. (One strength of this view is that phenomenal unity seems to be preserved through disorders of consciousness, even if representational unity is not.[10]) Bayne thus offers the unity thesis:

Necessarily, for any conscious subject of experience (S) and any time (t), the simultaneous conscious states that S has at t will be subsumed by a single conscious state—the subject's total conscious state.[11]

The central insight to draw from this thesis is that the unity of the conscious field is *holistic* rather than atomistic. Unification does not merely involve the clustering together (aggregation) of particular experiences that exist independently of one another, but the production of a single, totalizing conscious state that has particular experiences as aspects. Importantly, Bayne argues that we should accept this claim because we recognize its truth in our own experience. We experience ourselves as subjects of a single, totalizing conscious state of this type. Phenomenal unity seems to be essential to consciousness. This important observation is by no means novel. As Bayne acknowledges, John Searle had written that "all of the conscious experiences at any given point in an agent's life come as part of one unified conscious field," and "Kant himself gestures at this idea in referring to the 'one single experience in which all perceptions are represented.'"[12]

What does a holistic understanding of the unity of consciousness mean for personal identity? For Bayne, it means we must reject the complex theorists' view that personal identity consists in biological and psychological continuity, and thus accept that we cannot be organisms or minds (where a "mind" is a complex of psychological states, such as memories). We will consider complex theories of personal identity in more detail below. What is worth noting here is the standard approach to answering such questions that Bayne employs. This approach involves considering the consequences of certain hypothetical, non-paradigmatic cases of personal identity. In paradigmatic cases of personal identity (such as when I recognize I am the same person as the person who was sitting by the garden earlier), various types of unity and continuity (e.g., biological, psychological) point to the same intuitively correct judgments about who is who. However, this does not entail that personal identity consists in (i.e., "just is") all these types of unity and continuity. In order to tease out a robust answer to the question of what it is that human persons essentially are, it is important to try to develop hypothetical cases that disrupt the relationship between a particular type of continuity and our intuitions about personal identity. For example, if the claim is that I am an organism, this can be tested by trying to develop a case in which the homeodynamic processes that maintain an organism in existence come apart from intuitions about personal identity, perhaps by showing that these processes continue when it is clear to us (by intuition) that the person no longer exists. If we succeed, and if the cases we design are logically possible, and if our intuitions are reliable, then it becomes clear that personal identity cannot consist in that type of continuity (or unity).

Bayne considers a hypothetical, non-paradigmatic case to show that complex views leave open the possibility that *multiple* streams of consciousness could be associated with the *same* organism or mind.[13] This is a problem, because the organism or mind would thus be unable to play the roles that a self should play, including functioning as the owner of the individual's conscious states, the thing to which "I-thoughts" refer and the provider of the individual's unique, first-person perspective.[14] This strategy of defending the logical possibility of scenarios that place intuitions about personal identity in conflict with certain types of unity (or continuity) is the strategy of the simple personal identity theorist.

If we turn to a simple view of personal identity, what is the further fact beyond biological or psychological continuity in which personal identity might consist? Bayne is concerned that this fact should take due account of the unity of consciousness, and he is attracted to the possibility that the self consists in the conscious field itself. As we have seen, the conscious field is phenomenally unified. According to Bayne we should construe this unity holistically, such that it is a simple state (not composed of parts) that subsumes particular conscious experiences as aspects. This unified conscious field could then provide the sort of "further fact" that the simple theorist is looking for. However, despite defending this "naive phenomenalism" at some length, Bayne ultimately rejects it, for the good reason that it cannot account for the unity of selves across temporal gaps in conscious experience; a person cannot be unconscious on this view without failing to exist. He thus considers the possibility that personal identity consists in the substrate that produces the conscious field,[15] but ultimately rejects this line of thinking, too (substrate phenomenalism is discussed below).

4. Virtual Phenomenalism or Substance Dualism?

The account of the self that Bayne eventually defends involves joining Dennett in taking a sharp turn away from any form of concrete existence of the self. Persons turn out to be virtual, intentional objects that only exist in order to unify the conscious field. In other words, I am a *fiction* generated by the cognitive architecture that underlies the conscious field. This fiction is generated in order to provide a way to unify the field as a representation of the world. Bayne writes, "The cognitive architecture underlying your stream of consciousness represents that stream as had by a single self—the virtual object that is brought into being by *de se* representation."[16] (Note that *de se* representation is representation of the self to the self.)

The self can unify the conscious field because, among other things, it functions as the owner of all of the particular conscious experiences involved. Appropriating and modifying Dennett's notion of the self as a center of *narrative* gravity, Bayne describes it as a center of *phenomenal* gravity and labels his theory "virtual phenomenalism." Crucially, in virtual phenomenalism, the self is nothing over and above its representation in the conscious field. It is an intentional object rather than a concrete particular. This turn to a virtual self is held to be the central discovery that enables us to understand ourselves: "We can now see where other approaches to the self go wrong: they assume that there must be some 'real' entity that plays the role of the self."[17]

Bayne argues that to eliminate the self as a concrete particular is not to explain it away. This is because we can still *talk* about virtual objects and can thus refer to the self as the owner of experiences, referent of I-thoughts, and provider of a first-person perspective, among other things. However, virtual phenomenalism remains an astonishing position. On this view persons possess the same ontological status as fictional characters. To illustrate this, consider the Belgian detective Hercule Poirot and his creator, the author Agatha Christie. It is obvious that the ways in which Poirot and Christie exist contrast sharply. Christie's literary career no doubt included hope for success, the exercise of exceptional creativity, and much effort (including wrestling with the logic of plots and characters within each book), as well as a subsequent sense of pride to be the person responsible for these literary achievements. Poirot's sleuthing career could be described in similar terms, but his hopes, brilliance, and achievements can only be referred to in a fictional, deflationary sense. This is not only because the relevant events never happened, but because a fictional Poirot does not exist as an agent to enact any actual events anyway. The astounding claim of virtual phenomenalism is that Christie and Poirot exist in a very similar way.[18]

We should also not assume that since virtual phenomenalism enables us to *talk* about selves, our use of "I sentences" is unproblematic. We must not forget that "I" no longer refers to a concretely existing entity, and this has significant consequences. Theories that eliminate whole classes of properties or substances also render whole classes of statements uniformly and systematically false.[19] Consider J. L. Mackie's belief that there are no objective moral values. Since moral claims express beliefs about objective moral values, Mackie's moral eliminativism has the profoundly counter-intuitive consequence that all moral claims are uniformly and systematically false.[20] In the same way, consider the virtual phenomenalist's belief that there are no concrete human selves. Since "I" sentences implicitly express beliefs about concretely existing selves (e.g., we take "I am tired" to mean "There is an x such that x is tired, and x is me"), all "I" sentences as they are typically understood by competent language users are uniformly and systematically false. It would be hard to overstate just how profoundly implausible this position is. Sinatra may triumphantly sing, "I did it my way," but on virtual phenomenalism this statement could not be more wrong. Sinatra did not do things in anyone's way, let alone his own, because there was no "Sinatra" behind the name—indeed, there are no concrete selves at all.

This comparison between fictional and real selves also highlights the inconsistency within virtual phenomenalism. Christie's Poirot is a representation of a Belgian detective. This is unproblematic because the concrete existence of the thing being represented (a Belgian detective) does not depend on the existence of the representation (Poirot). However, the central notion of Bayne's approach is that the self is created by processes of *de se* representation: the representation of the self to the self. This is problematic since, on this view, the existence of the thing being represented (the self) *does* depend on the existence of the representation. The thing being represented *just is* the representation. The view is thus circular and vacuous. It seems to be a way to try and pull the self up into existence by its own bootstraps in the absence of any independent entity with which it might be identified.

A final problem with virtual phenomenalism is that it lacks the phenomenological support that Bayne relies on in other areas. Bayne justifies his holistic understanding of the unity of consciousness on the grounds that we recognize it in our experience. This reasoning is inconsistent with the eliminative turn to virtual phenomenalism, since we also recognize in our experience that we exist independently of the conscious field and that we are thus not aspects of it. Introspectively we are aware of the unity of consciousness, and we are also aware of our own existence as concrete particulars distinct from our unified conscious experience. As I am writing I am aware of seeing the garden, smelling the bread baking, and thinking about philosophy, but I am also aware of myself as the one having these experiences. "I" am the one doing the perceiving and thinking. In fact, the "substrate phenomenalism" that Bayne considers and rejects pushes in this direction. He rejects it because it is not a conceptual truth that a substrate must produce one conscious field rather than many. In other words, it is not necessarily the case that a substrate must produce only one conscious field and, as we have seen, Bayne argues that something with multiple conscious streams is a poor candidate for the self and he is looking for a substantive self that necessarily could not sustain more than one. However, even if the notion of a substrate does not necessarily entail a single conscious field, it does not follow that the entities that in fact are the subjects of conscious experience are not limited in this way. Bayne has assumed that there could be an x such that x is a substrate and x is compatible with multiple conscious fields. However, in order to hold that there could *not* be a substrate that is *incompatible* with multiple conscious fields he must show that for all x, if x is a substrate, then x is compatible with multiple conscious fields. He has not shown this.

This is important because a much more consistent approach for Bayne would be to accept that some type of substrate view is likely correct. Such a view would hold that the owner of the conscious field exists independently of the field, and this claim is true to our experience; it receives the same phenomenological support as the holistic understanding of the unity of consciousness that Bayne rightly favors. His acceptance of the phenomenological evidence for unified consciousness means that he should also accept the phenomenological evidence for a self that is a substrate. This substrate would be the subject of the (single) stream of consciousness and would exist independently of it.

This line of thought surely leads to substance dualism, which holds that the individual is an immaterial substance not composed of parts (a simple). The simplicity of the self accounts well for the phenomenological evidence that the unity of consciousness is holistic, since as a simple substance I am a fitting subject and ground for a simple, holistic conscious field. Substance dualism also entails that I am what I experience myself to be: a concrete individual existing independently of the conscious field of which I am the subject. Furthermore, if this immaterial substance is both simple and identical to the subject of experience (as substance dualists typically claim), then it may after all be conceptually necessary for the self to be the subject of only *one* conscious field at any time. Since this is not a phenomenalist view, the conscious field in question would necessarily be the consciousness of the same individual.[21] This could be precisely the further fact that Bayne, having rejected complex accounts of personal identity, is seeking, and it is the view adopted by many other simple theorists.

That substance dualism provides the only rescue from the elimination of the self is a strong argument in favor of its truth. Given substance dualism, why adopt an eliminative view instead? The obvious answer is that substance dualism is inconsistent with naturalism and so for the naturalist the cost of accepting it is simply too high. The naturalist needs some other answer. Since Bayne has rejected the complex biological and psychological views that represent the naturalist mainstream and has found naive and substrate phenomenalism wanting, only an eliminativist view is left given the resources of naturalism.[22] Yet, given Bayne's sensitivity to phenomenological considerations, it seems he would do better to affirm that we are the concrete particulars that we experience ourselves to be, and thus to let go of naturalism.[23]

5. Identity at a Time and "The Life Scientific"

If unified consciousness depends on substance dualism, then those aspects of mental life that require unified consciousness depend on it, too. This includes some of the phenomena central to scientific investigation.

Many of the beliefs we hold on the basis of our visual experience are only justified because consciousness is unified. For example, as I look out across the garden I see the different elements within it, such as the lawn, flower bed, and trees. I believe that the entire lawn is green and that the flower bed is in front of the trees. I am justified in holding these beliefs because I co-experience these elements as aspects of a single, totalizing, conscious field. This enables me to apprehend directly not only the different elements of the scene, but also the relationships between these elements, which are aspects of the whole. I apprehend directly the relative positions of the trees and flowers and also the lawn in its entirety, and I thus "see" these relationships directly. It is the unity of consciousness that enables this seeing and thereby justifies these types of perceptually grounded beliefs. Systematic observations of natural phenomena leading to beliefs of this type underpin scientific theory development. Consider, for example, Sir Arthur Eddington's famed observation of a total solar eclipse during the Armistice of November 1918. The ability to observe the apparent position of certain stars relative to the Sun and to compare this with observations of the stars when the Sun was not in view enabled a measure of the degree to which light is deflected by the Sun's gravity. Einstein's general theory of relativity predicted that this deflection would be double that predicted by Newton's theory, and so these perceptually grounded beliefs about the apparent positions of the stars, justified through co-experiencing elements as aspects of a single conscious field, drove the revolution that was the replacement of Newton's formulation of the laws of gravity.[24]

The types of relation that we apprehend directly within a unified conscious field can be spatial, but they may also be logical. The unity of consciousness is therefore also necessary for the exercise of reason. We can consider two different aspects of reason: our passive recognition of self-evident logical relationships between statements and our active reflection on extended arguments to determine their validity and soundness. C. S. Lewis referred to the recognition of a logical relationship as *intellectus* and reflection on extended arguments as *ratio*.[25] The exercise of *intellectus* will depend on the unity of consciousness at a time because it requires holding in mind and correctly apprehending multiple statements simultaneously. We may say that we cannot help but "see" (directly apprehend) that certain logical relationships hold. Swinburne refers to these self-evident logical relationships as "mini-entailments."[26]

Intellectus is not an active process but a passive one. It does not involve us as agents searching for logical relationships but rather as patients who cannot help but "see" them. Just as my co-consciousness of the flower bed and the trees is all that is required for me to perceive the spatial relationships between them (their relative positions), so my co-consciousness of two statements may be all that is required for me to apprehend that they are inconsistent. It is self-evidently clear that certain statements cannot be simultaneously true, or that a further statement is entailed by their truth. It is critical to the authority of reason and the justification of beliefs produced by it that awareness of these mini-entailments results from such "seeing." This is because *intellectus* is a case of intentional causation. In intentional causation the content of one or more statements—what they are about—causes us to apprehend the content of another. As Lewis notes, "the cause-effect and the ground-consequent must apply simultaneously to the same series of mental acts."[27] This suggests that the ontology of reason is itself incompatible with naturalism. However, the main point here is that the unity of consciousness at a time is required if we are to apprehend directly relations between things of which we are co-conscious, and that our ability to reason depends on such apprehension.

The life scientific is based on strong convictions about the importance of observation and reason for constructing authoritative accounts of the physical world. For example, a scientist may have a theory that predicts that a flame will burn red, but then observe that it burns green. This important observation will prompt a reasoned argument in the form of *modus tollens*: if the theory is correct, then the flame will burn red; the flame does not burn red, and so the theory as it stands is incorrect and must be revised. However, the logical relations involved can only get a grip on us as a result of the unity of consciousness at a time and this unity strongly suggests that human persons—including those who pursue science—are not physical beings essentially. If this is true, then it simply does not follow that to esteem science involves making a commitment to physicalism. In fact, a commitment to the importance of observation and argument leads to the conclusion that physicalism is false.

6. Personal Identity over Time

Turning to personal identity over time (and noting that the arguments here are more difficult), recall that the complex view of personal identity is that it consists in psycho/physical continuity, while the simple view is that it consists in some "further fact."

Complex theorists are particularly impressed by the fact that in most if not all real situations in which we intuit that a person at time A is the same person as a person at time B, there are psycho/physical continuities that hold between "them." So it is perhaps unsurprising that complex theorists come to believe that these continuities not only accompany but also constitute personal identity. Contrastingly, simple theorists also have clear reasons for their opposing view. First, they offer hypothetical cases in which personal identity comes apart from the psycho/physical continuities that, according to complex theorists, constitute it. If these cases are logically possible, and if our intuitions about personal identity are reliable, then personal identity cannot consist in psycho/physical continuities, and the complex view must be false. Second, they argue that their preferred simple view does not entail contradictions, despite the complex theorists' claims to the contrary. Finally, simple theorists can also argue that they are the only ones who understand the term "personal identity" in the sense that is implicit in the paradigm examples through which it is learned. The complex view may offer criteria consistent with these paradigm examples, but they are not the criteria that the examples impress upon us when we encounter them. Taken together these reasons form a strong case. Swinburne writes: "My feeling about how this debate is going these days is that we are getting our point across."[28] We will consider these three reasons in turn.

Regarding the first reason, consider the observation that psycho/physical continuities always accompany personal identity over time and so these must *constitute* personal identity. Simple theorists respond that the presence of such continuities in paradigm cases is not unexpected on the simple view and thus not a reason to prefer the opposing complex view. Tight links over time between memories, character, and a particular brain are most simply explained by assuming that they are all properties of the same person. Hence these continuities are to be expected where there is personal identity over time, because they are good *evidence* of it. However, personal identity need not be *constituted* by such continuities.[29] Indeed, the simple theorist argues that there can be logically possible cases (albeit hypothetical ones) in which biological and psychological continuities are independent of personal identity. If there is agreement that these situations are logically possible—as there often is—then it is agreed that there can be personal identity without psycho/physical continuity, and thus the complex view is false.

Many such thought-experiments are employed in discussions of personal identity in order to challenge complex views. For example, Bayne uses one himself, rejecting the animalist view (that a person is identical to a particular organism) through reflection on van Inwagen's example of Cerberus, the dog of Greek mythology, an organism with two heads and two brains, neither of which is responsible for the homeodynamic processes that sustain the life of the organism. Neither brain can be identified with the organism (since neither is responsible for the life-process), yet each sustains an independent conscious stream. Bayne argues that the organism cannot be identified with the self in this case, since "I thoughts" will not make sense. For example, if one conscious stream in Cerberus includes an experience of pain while the other does not, the organism cannot coherently have the thought "I am in pain" or "I am not in pain."

Swinburne offers a thought-experiment to demonstrate the inadequacy of psycho/physical continuity as an account of personal identity over time. It can be summarized as follows: Due to a brain disease, I have my right brain hemisphere replaced with one taken from a clone with slightly different memories and character traits to my own. The result of the operation is a man who bears some psycho/physical similarity both to me and to the clone. A couple of years later the same procedure is required for my left hemisphere. With both hemispheres replaced, the resulting man is materially and psychologically very different from the man I was before the operations. The question is whether this resulting person is the same person as me. (The situation is intended to render the answer to this question wholly ambiguous, and it can be manipulated to ensure this is the case, since the frequency of operations and/or the proportion of the brain replaced in each operation can be increased or decreased in order to ensure that it is wholly unclear whether or not I have survived.) The question is whether I have or have not survived the operation. However, even though we have all of the relevant facts about biological and psychological continuity, we do not know the answer. We must conclude, therefore, that full knowledge of all the psycho/physical facts does not determine whether I have survived or not, and so we must also conclude that personal identity is not constituted by such facts. We can accept that usually psycho/physical facts provide good *evidence* of personal identity, but we must conclude from our very unusual case that personal identity can come apart from these facts and so it is not *constituted* by them. Arguments such as this provide good reason to reject a complex view.

We now come to the second reason why simple theorists reject the arguments of complex theorists. Complex theorists argue that the simple view is logically impossible because it entails a contradiction. The complex theorist tries to persuade people of the logical impossibility of statements about personal identity that are independent of strong psycho/physical continuities. The following statement, inspired by Locke and offered by Swinburne, presents just such an argument made by an advocate of the complex view. The statement is followed by a claim about the necessary truth of an ethical principle and then seeks to show that this results in a contradiction:

> [The statement] "Socrates is the same person as the mayor of Queenborough, but has none of the same brain, memory or character as the mayor," together with what they may claim to be a necessary truth, "no one should be punished for any act which they cannot remember doing," entails "both {the mayor should be punished for any immoral act of Socrates} and not-{the mayor should be punished for any immoral act of Socrates}."[30]

If sound, this argument would show that *without* psycho/physical continuity, questions of personal identity entail contradictions, and so we cannot do without it. The complex theory must therefore be true. The problem for the complex theorist is that the soundness of this or any other similar argument is typically hard to establish. After all, why should we agree that the argument is sound, since the ethical premise (that it is necessarily true that no one should be punished for any act which they cannot remember doing) hardly seems compelling? For example, drunk drivers who do not recall hitting a pedestrian can still be found guilty of vehicular homicide because it was a foreseeable consequence of their actions. The overarching point here is that it is difficult to find any sound argument that will demonstrate that the simple view leads to a logical contradiction. By pointing out and demonstrating this difficulty, simple theorists thus attempt to show that the simple theory is not *a priori* metaphysically impossible, and they go on to argue that it is entailed by situations that all agree are logically possible, in which case it is the complex view that must be false.

Third, a further claim that the simple theorist can make is that simple and complex theorists come to understand personal identity differently because they acquire different senses of personal-identity-relevant language. Consider sentences of the type, "Jon is the person who..." (a more familiar abbreviation of a clearer statement of identity: "Jon is the same person as the person who"). Such sentences come to be understood by first understanding various paradigm examples (e.g., "Jim is the person you saw last week"; "You are the person who had a headache just now."). We observe that those paradigm examples are also examples of people with strongly continuous bodies, memory, and characters. What is important to notice is that these observations are consistent with two possible understandings of personal identity: first, that the observed continuities are what personal identity *consists in* (the complex theorist's view); second, that they are *evidence* of an underlying personal identity that consists in a further fact (the simple theorist's view).

Given the two possible understandings referred to above, the simple theorist's argument is not that the complex theorist lacks a coherent understanding of "personal identity" as it is typically used in paradigmatic cases, but rather that the complex theorist's understanding of personal identity is not the one that is implicit in some of the paradigm examples from which we learn about it. Simple theorists focus on those paradigm examples involving one's *own* identity, and they argue that those examples lead everyone to understand personal identity (e.g., phrases such as "is the same person as") from the *inside*. Consider the question, "Are you *the person who was* complaining of a headache just now?" Simple theorists claim that in answering this question, "the person who…" quite clearly refers not to psycho/physical continuity, but to an ongoing, underlying identity with which each person is directly acquainted. It is "something that can only be described as a direct awareness of [the self] as a continuing subject of experience."[31] So the complex theorist may offer a *coherent* analysis of "personal identity" as typically used, but it is not the one that is *implicit* in many of the paradigm examples through which we learn about it.[32]

These different linguistic conceptions help to explain why there are two groups of theorists and why disputes arise between them about what is entailed by personal identity statements. This also creates the need for proponents of the simple view to produce the thought experiments discussed above that entail the falsity of the complex view, in order to show to those who understand personal identity language in a different sense that their sense, while legitimate, is not the one implicit in the examples from which it was learned.

It is also worth noting that if it is the simple understanding of personal identity that is implicit in the examples from which we learn about it, then the simple view is also the commonsense view and the complex theory is, in contrast, a revisionist position. If we abide by an approach in which common sense is deemed innocent until proven guilty, then the burden of proof rests on the complex theorist to show that there is a need for an alternative. To prefigure the conclusion of this chapter, it is my argument not only that the complex theorist fails to do this, but that there are good reasons, having to do with esteeming science, to hold firmly to the simple theorist's commonsense assumption.

There are good reasons to adopt the simple view. First, there is not a successful argument to show that it is logically impossible, and it is entailed by situations that all agree to be logically possible. This entails that the complex view is false. Second, even if the complex view of personal identity is accepted as a legitimate inference from the paradigm examples of personal identity through which we learn to talk and think about it, it is an inferior inference compared to the simple view. These two reasons lead us to conclude that personal identity does not consist in psycho/physical continuities but in some further fact.

It is highly likely that psychological and physical continuity exhaust the possible things in which personal identity might consist, given materialism, unless personal identity is to be eliminated. Given that psychological and physical continuity fail to account for personal identity, and that the self cannot be eliminated, it seems that it is materialism that should be rejected. The simple view is, of course, consistent with the belief that a human person consists of a body and a soul, with the latter being the essential part that constitutes personal identity.

7. Identity over Time and "The Life Scientific"

The identity of persons over time is necessary not least because it enables us to have prudential concern for our futures and responsibility for our past actions. These concerns are important for scientists as much as for anyone else. For example, it is only coherent for a young scientist to hope that one day she will be celebrated for her great discoveries if the person who will work towards those discoveries will in fact be her and not some other person who will succeed her. Later on, when the discoveries have been made, it will only be coherent to celebrate and reward her for them if she is the same person who made them. (In a similar vein, we may wonder how it is that eminent, eliminativist philosophers of mind justify cashing their royalty checks for popular books written years ago that, on the basis of their eliminative tendencies, they did not write.)

Personal identity over time has relevance not only for prudential concern and responsibility, but also for scientific activity itself. We must return to matters of observation and reason. The scientific method requires not only accepting the testimony of others with respect to the disconfirmation or verification of their findings, but—crucially—it also requires first-person epistemic access to the disconfirmation or verification of one's own findings. I must be able to test a hypothesis and have first-person epistemic access to the results of that test. To illustrate, consider Moreland's discussion of the need for "fulfilment structures"—the verification or disconfirmation of thoughts or concepts through a sequence of relevant first-person experiences. He gives the everyday example of having a query about whether the chair on the other side of the room is scratched. Fulfilment of the query involves having a series of perceptual experiences as one approaches the chair and inspects it. These must be experiences for the same individual if they are to fulfil the query such that the individual concerned has first-person epistemic access to the truth about the chair and whether or not it is scratched.[33] If it is not the same person throughout the process, then the self at the end of the sequence verifies the scratch for someone else.

Turning to reason, we focus now not on Lewis's *intellectus,* but on *ratio*; reasoning that requires one to consider premises together in order to see what follows logically. The evaluation of the validity of a logical argument is a temporally extended practice central to scientific theory. It can be compared to comprehending a sentence. If a different person comprehends each word in a sentence, then there is no person who comprehends the whole and its meaning is lost. In the same way, in order to grasp a statement as the conclusion of a syllogism, or series of syllogisms, the person who considers the premises individually must be the same individual who considers them together and draws the conclusion from them in order to know and be justified in holding that the conclusion is or is not valid.[34] As A. C. Ewing wrote:

> To realise the truth of any proposition or even entertain it as something meaningful the same being must be aware of its different constituents. To be aware of the validity of an argument the same being must entertain premises and conclusion; to compare two things the same being must, at least in memory, be aware of them simultaneously; and since all these processes take some time the continuous existence of literally the same entity is required.[35]

It is clear that personal identity over time, which is sustained only by a simple view of identity inconsistent with materialism, is critical for the observational and rational processes on which scientific activity relies.

8. Conclusion

In order to esteem the processes of observation and reason involved in scientific activity, it is important to inquire about the characteristics that a human person must have in order to engage in them, and what kind of thing a human person must be in order to possess those characteristics. It turns out that both observation and reason rely on the holistic unity of consciousness and the persistence of human persons through time. These in turn cannot be grounded in the mainstream complex accounts of personal identity at or over time that are consistent with materialism but require a simple view that is inconsistent with it.

If one aspires to a life scientific, or hopes one day to be recognized or even celebrated for one's own scientific achievements then, once again, the identity of the person over time is a matter of concern. Here again the failure of the complex view of personal identity points clearly towards persons who exist both at and over time as combinations of immaterial substances—souls—and bodies, with the soul as the essential part.

In short, if one esteems the life scientific, one should not be a materialist.

Notes

1. Daniel Dennett, "Daniel Dennett on the Evolution of the Human Brain," interview by Jim Al-Khalili, *The Life Scientific*, BBC Radio 4, April 4, 2017, https://www.bbc.co.uk/programmes/b08kv3y4.
2. Derek Parfit, "Personal Identity and Rationality," *Synthese* 53, no. 2 (Nov. 1982), 227. To explore this distinction in detail, see Georg Gasser and Matthias Stefan, eds., *Personal Identity: Complex or Simple?* (Cambridge: Cambridge University Press, 2012).
3. Richard Swinburne, "Cartesian Substance Dualism," in *The Blackwell Companion to Substance Dualism*, eds. Jonathan J. Loose, Angus J. L. Menuge, and J. P. Moreland (Oxford: Blackwell, 2018), 133.
4. Swinburne, "Cartesian Substance Dualism," 133.
5. Charles Taliaferro, "Substance Dualism: A Defense," in *The Blackwell Companion to Substance Dualism*, ed. Jonathan J. Loose, Angus J. L. Menuge, and J. P. Moreland (Oxford: Blackwell, 2018), 46.
6. Taliaferro, "Substance Dualism," 46-47.
7. Swinburne, "Cartesian Substance Dualism," 133.
8. Tim Bayne, *The Unity of Consciousness* (Oxford: Oxford University Press, 2012).
9. Bayne, *Unity of Consciousness*, 11.
10. While an integrative agnosia may lead to a breakdown of representational unity as someone fails to experience objects as unified wholes, this does not entail a loss of phenomenal unity as there will still be a state that captures the entirety of what it is like to be the subject at that moment.
11. Bayne, *Unity of Consciousness*, 16.
12. Bayne, *Unity of Consciousness*, 15–16. The internal references given by Bayne to Searle and Kant have not been reproduced.
13. The main thought experiment considered is van Inwagen's example of the two-headed creature, Cerberus. See Peter van Inwagen, *Material Beings* (Ithaca: Cornell University Press, 1995), 191.
14. If the self is identified with an organism (or mind) with two streams of consciousness, then one stream of consciousness may include pain experiences and the other not. In that case, is the organism (or mind) in pain? Nor is it clear in this case that the organism (or mind) is the thing to which "I thoughts" refer. To what does an "I thought" in just one of the streams refer? The role of providing a first-person perspective is also problematic since each stream will have its own. Hence, a single self cannot have multiple streams of consciousness.
15. Bayne considers a view that identifies the self with the entity that produces the conscious field, but he then also rejects this since it is not logically necessary that the substrate would produce just one unified conscious field.
16. Bayne, *The Unity of Consciousness*, 289.
17. Bayne, *The Unity of Consciousness*, 290.
18. Bayne argues we should not draw this comparison, since all selves are virtual and so there simply are no "real" selves to compare with virtual ones. However, this seems to beg the question, given the existence of other views that take the self to be a concrete particular.
19. We might ask what unicorns eat for breakfast. Whatever answer we give will be false because there are no unicorns!
20. J. L. Mackie, *Ethics: Inventing Right and Wrong* (New York: Penguin, 1977).
21. Bayne's recognition of the importance of phenomenal unity for the unity of consciousness is helpful, but ultimately it is subject unity that resides at its heart. The unity of consciousness is grounded in the unity of

the human subject. This only becomes clear when one accepts that personal identity consists not in the conscious field itself, but in a separate substance.

22. For a detailed discussion of the unity of consciousness and naturalism focused on Bayne's work, see J. P. Moreland, "Substance Dualism and the Unity of Consciousness," in *The Blackwell Companion to Substance Dualism*, eds. Jonathan J. Loose, Angus J. L. Menuge, and J. P. Moreland (Oxford: Blackwell, 2018), 184–207.
23. For a more detailed discussion of the relationship between holistic unity, virtual phenomenalism, and naturalism, see the previously cited Moreland, "Substance Dualism and the Unity of Consciousness."
24. See, for example, Daniel Kennefick, *No Shadow of a Doubt: The 1919 Eclipse That Confirmed Einstein's Theory of Relativity* (Princeton, NJ: Princeton University Press, 2019).
25. Stewart Goetz, *C. S. Lewis* (Oxford: Wiley-Blackwell, 2018), 37.
26. "*s1* mini-entails *s2* iff anyone who asserts *s1* is thereby (in virtue of the rules for the correct use of language) committed to *s2*. 'British mail boxes are red' mini-entails 'British mail boxes are coloured'…" Richard Swinburne, *Mind, Brain, and Free Will* (Oxford: Oxford University Press, 2013), 18.
27. Goetz, *C. S. Lewis*, 37; see also C. S. Lewis, *Miracles: A Preliminary Study* (London: HarperCollins, 2016), 23–24.
28. Richard Swinburne, "How to Determine Which Is the True Theory of Personal Identity," in *Personal Identity: Complex or Simple?*, eds. Georg Gasser and Matthias Stefan (Cambridge: Cambridge University Press, 2012), 115.
29. Swinburne adds a further reason to take psycho/physical continuities as evidence of personal identity: To be confident in the large amount of knowledge that we acquire from others we each need to trust our memories, including our "personal" memories. I must trust my memories "from the inside" that various things were done and experienced by me; so I must take these memories as evidence of personal identity over time. These are memories of the actions and experiences of a person with a brain strongly continuous with my brain. Hence, continuity of brain is also evidence of personal identity over time. See Swinburne, "How to Determine Which Is the True Theory," 107.
30. Swinburne, "How to Determine Which Is the True Theory," 113.
31. Swinburne, "How to Determine Which Is the True Theory," 114.
32. Furthermore, as an anonymous reviewer of this chapter noted, we typically have no access to our neurophysiological states and we are aware of significant discontinuity in our psychological states through sleep, distraction, the flux of experience, etc. It is thus very difficult to see how a person would come to grasp the idea of identity over time in this way, even if examples of personal identity are correlated with the continuities on which the complex account relies.
33. J. P. Moreland, *The Recalcitrant Imago Dei: Human Persons and the Failure of Naturalism* (London: SCM, 2009), 133–4.
34. We are, of course, very familiar with computers working through complex logical problems in steps and presenting the results to us. This may lead a physicalist to argue that the computer is evaluating a logical argument effectively over time with only a persistent physical substrate and not a persistent perceiving subject or a common "I" in existence. However, when we say that a computer is "evaluating a logical argument," the scare-quotes are critical, since a computer does not apprehend logical relationships directly. Rather, it is programmed (or, if machine learning is used, it is taught) to follow logical rules, and so its results are only justified in so far as the intentional mind of the programmer or teacher has perceived the logical relationships that must be implemented in order for it to provide the results in which we are interested. The computer lacks intrinsic intentionality and does not perceive logical relations between the content of different statements and report self-evident conclusions that are thereby justified; rather it follows rules that apply to the structure of different statements that originate in some form of guidance from a human mind that *does* possess intentionality and *does* perceive logical relationships between the content of different statements. The fact that the computer works in steps over time with no persisting "I" does not mean that logical inferences can be justified without a persisting "I"; rather it reflects the fact that the computer is incapable of producing any outputs that reflect a direct apprehension of logical relationships and are thereby justified. The computer is a tool formed and used by a reasoning mind and the justification (or otherwise) of its outputs is thus not intrinsic to the machine, but found in that reasoning mind.
35. A. C. Ewing, *Value and Reality* (London: George Allen and Unwin, 1973), 84.

10. Mirror Neurons, Consciousness, and the Bearer Question

Mihretu P. Guta

1. Introduction

In this chapter, I aim to examine the two central features that are said to characterize a class of brain cells known as *mirror neurons* (MNs). These features are:

F1. MNs respond or fire when someone reaches for an object in a goal-oriented manner.

F2. MNs respond or fire when someone observes another person reach out for an object.

Neuroscientists tell us that (F1) is an instance of *action execution* and; (F2) is an instance of *action observation* or *action understanding*.[1] Following Giacomo Rizzolatti and Laila Craighero,[2] I too call (F1) and (F2) *the functional properties* of MNs. Such a characterization of (F1) and (F2) is justified given that each is said to involve a response to relevant external stimuli.[3] MNs are said to play a key role in mirroring the mind of another person by way of simulating his or her mental states.[4] As we shall see, MNs are also implicated in a host of other mental phenomena that underlie human behavior.[5]

However, the theory of MNs does not enjoy a free ride. It has critics as well. Cognitive neuroscientist Gregory Hickok is an excellent example. In *The Myth of Mirror Neurons* he makes a book-length argument against (F1) and (F2).[6] He claims that the widely embraced properties of MNs are based on shaky empirical evidence. For example, Hickok disputes an overwhelmingly accepted view that there is direct evidence in monkeys that MNs support action understanding. Hickok claims that MNs are not needed for action understanding, since the two functional properties of MNs, namely "action execution" and "action understanding" dissociate in humans. For example, a person can understand a certain task without being able to execute it. In short, Hickok raises empirical objections against empirical evidence that the proponents of MNs offer in defense of their theory.

Contemporary controversies over MNs seem to take two forms. On the one hand, defenders of the theory argue that the functional properties of MNs provide us with deep insights into the neural mechanisms of diverse mental phenomena that underlie human behavior. On the other hand, critics of MNs (such as Hickok) seriously question whether the theory has an adequate empirical justification. Other critics of the theory take a narrowly focused approach in their examination of MNs. For example, they focus on the role of MNs in empathy, action-mirroring, and mind-reading.[7] Recently, the origin and function of MNs have also been a subject of intense controversy.[8] But a closer look at these controversies gives us very little (if any) insight into some key metaphysical issues that must be taken into consideration in discussions involving MNs.

Here three things come to mind:

a. the question of neural correlates of the functional properties of MNs;
b. the question of a causal profile of the functional properties of MNs;
c. the question of whether or not the very existence of the functional properties of MNs requires consciousness and its bearer (i.e., the self or person).

Current scientific controversies over the nature of the functional properties of MNs swing back and forth between questions (a) and (b). The controversies aim at settling questions related to correlation or causation or a mixture of both. But what seems to have slipped through the cracks in current discussions concerning MNs is question (c). On the part of defenders and critics of MNs alike, there seems to be hardly any noticeable realization that success in tackling (a) and (b) is squarely contingent on success in tackling (c).

Dealing with (a), (b), and (c) brings up problems at three different levels:

a. *the easy problem of the functional properties* of MNs;[9]
b. *the hard problem of the functional properties* of MNs;[10]
c. *the hardest problem of the functional properties* of MNs.[11]

In this chapter I attempt to show why the contemporary controversies over (a) and (b) are symptoms of a much bigger issue that involves the problem of an irreducible consciousness and its bearer, a problem for which appeals either to a particular region in the brain or to the nervous system as a whole seem to be incapable of providing any satisfactory solution. So there is a sense in which attributing the functional properties of MNs (as many neuroscientists and psychologists do), to a certain brain region(s) leads to a problem of misidentification. This is a situation whereby one mistakenly identifies an object or a property of one sort with an object or a property of some other sort. An example of the misidentification in question would be identifying mental states such as beliefs and desires with the brain states or the firing of neurons in the brain. What I call *the problem of misidentification* arises when the functional properties of MNs are identified with a certain region(s) of the brain. The solution for this problem, I will argue, lies in how we tackle the issue of consciousness and its bearer (point (c) above). Ultimately, the issues besetting the functional properties of MNs turn out to be metaphysical in nature.

In this chapter, it is not my intention to argue for or against the actual existence of MNs. In fact, for present purposes, I choose to remain agnostic about this particular matter. I will only be defending the following claim: *regardless of whether or not mirror neurons actually exist, the functional properties associated with them necessarily require consciousness and its bearer*. As we shall see, the discussion I will advance regarding this matter has a close affinity with M. R. Bennett and P. M. S. Hacker's well-known critique of attributing psychological attributes to the brain.[12] But in this chapter, my main goal is to make my own distinctive case in defense of the above claim. In doing so, I hope to make some contribution to our understanding of some of the central issues that beset the mind-body problem.

This chapter is divided into five sections. In section 1, I introduce the key issues to be discussed. In section 2, I will present a brief exposition of MNs.[13] In section 3, I will discuss the three levels of difficulty associated with the functional properties of MNs. In section 4, I will discuss the problem of misidentifying the role of MNs and suggest a way to tackle it. Finally, in section 5, I will conclude this chapter by claiming that the most promising solution for the problem of misidentification is to ground the functional properties of mirror neurons squarely in consciousness and its bearer.

2. Mirror Neurons

2.1 Mirror Neurons in a Monkey Brain

THE DISCOVERY of mirror neurons has its roots in the work of a team of neuroscientists in Parma, Italy.[14] MNs were discovered originally in the macaque monkey inferior premotor cortex, particularly in the region of the brain called F5 (frontal lobe area). The Parma team conducted a series of experiments to understand a neural mechanism for visuomotor activities. The experiments were done by using microelectrodes inserted into the cortex of the monkey. This method allowed the recording of the electrical activity of single neurons while a trained monkey was engaging in various motor movements. These motor activities include a monkey retrieving objects of different sizes and shapes from a testing box after a delay of a few seconds upon the presentation of a given stimulus. The Parma researchers reported that when the monkey engaged in goal-oriented motor acts, neuronal discharge took place in area F5. When the monkey watched experimenters engaging in motor movement such as picking up the food or placing it inside the testing box, neurons in area F5 were also activated. Many such experiments are said to have shown that the functional properties of MNs, namely, *action execution* and *action observation*, are interrelated in that they feed into each other sensorimotor information.[15]

Neurons in area F5 also have a property of selective firing pattern. The selective neuronal firing was observed in an experimental setting, in which a piece of food was placed for the monkey to pick up from a different spatial location. The monkey extended its arm and reached for the food and brought it to the mouth. The monkey also engaged in other similar motor acts such as grasping objects. In all of these cases, neurons' selective firing orientation was documented. For example, neurons that fired for the monkey's *precision grip* (which involves the index finger and thumb) were distinct from those that fired for the monkey's *finger prehension* (e.g., when grasping middle-sized objects). Similarly, neurons that fired for the monkey's *whole hand prehension* (e.g., a whole apple) were distinct from those that fired both for the precision grip and the finger prehension.[16]

Such a selective firing pattern of MNs is said to show that distinct motor acts are accompanied by distinct neuronal discharges. If this is the case, then there has to be a strong correlation between selective neuronal firings and the relevant specific goal-oriented motor acts and observations. In other words, no random non-goal-oriented movements give rise to selective neuronal discharge. This is one reason why researchers describe MNs as: "precision grip neurons," "finger prehension neurons," "whole hand prehension neurons," "manipulating neurons," and so on.[17] The term "mirror neurons" was coined to refer to a class of brain cells known as visuomotor neurons, which are implicated in facilitating *action execution* (e.g., when one grabs a certain object) and *action observation* (e.g., when one observes other people perform similar actions). Unlike the other brain nerve cells, mirror neurons are said to be involved, inter alia, in mirroring or representing the intentions of individuals.

Despite much focus being on area F5 (which is taken to be a hub of MNs), recent experimental works showed that such neurons also exist in other regions of the brain. For example, neurons in the cortex of the superior temporal sulcus display the properties of F5. They fire when one observes another person's motor activities such as "walking, turning the head, bending the torso, and moving the arms."[18] Some neurons in the same region also respond to goal-oriented hand movements.[19] There is also another brain region that is strongly associated with MNs. This is the rostral part of the inferior parietal lobule (IPL) called PF, which is responsible for sensory information, among other things.[20] More recently, Luca Bonini pointed out that the mirror-neuron network extends beyond the ventral premotor cortex and IPL to other cortical and subcortical areas of the brain such as basal ganglia.[21]

2.2 A Mirror-Neuron System in the Human Brain?[22]

In 1996 the Parma team asked: Does such a system exist in humans?[23] Answering this question did not appear to be as straightforward as it was in the case of the macaque monkeys. This is because almost no experiments involving humans had been done in which single neurons were recorded.[24] Thus, most of the evidence for MNs in humans remained indirect.[25] Given the lack of direct evidence in humans for MNs, it is common to use a looser expression, *mirror-neuron system*.[26] One of the reasons for using the term "mirror-neuron system" in the case of humans has to do with the empirical evidence that indicates that the functional properties of the brain cells in question are mediated by multiple regions of the brain. Characterizing the function of these brain cells in terms of a "system" is intended to imply the coordination of a complex network of multiple brain areas. If such a system exists in humans, then it ought to display functional properties similar to those of the MNs discovered in the macaque monkeys.[27]

The human mirror-neuron system is believed to be in the Broca's area (the area of speech production), which is taken to be the equivalent of the monkey's area F5. Other areas implicated in the mirror-neuron system include the inferior parietal lobule (IPL), the ventral premotor cortex, and the caudal part of the inferior frontal gyrus.[28] This shows that the mirror-neuron system is implicated in multiple brain areas corresponding to its multifaceted roles. For example, Hickok listed over forty instances of behavior linked to MNs. These instances include "lip reading," "drug abuse," "political attitude," and "mother-infant attachment."[29] The mirror-neuron system is also associated with empathy, that is, sharing another person's experience by taking that person's perspective.[30] Furthermore, MNs are also implicated in imitations, associative learning, social interaction, intention understanding, language and speech evolution, and many other activities.[31] Dysfunction in the mirror-neuron system is implicated in disorders such as autism, which affects, *inter alia*, the ability for intention understanding and social cognition.[32] Furthermore, MNs are praised as being "neurons that shaped human civilization" as well as "tiny miracles."[33]

The data with regards to the human mirror-neuron system comes primarily from neurophysiological and brain-imaging studies. For example, the Parma team in 1996 reported that studies from both transcranial magnetic stimulation (TMS) and positron emission tomography (PET) provided two strong pieces of evidence for a mirror-neuron system in humans. TMS is a non-invasive brain stimulation method that applies magnetic fields to the skull resulting in action potential or neuronal firing. This method can also be used to initiate involuntary muscle movements such as a finger twitch. TMS was used to stimulate the motor cortex of normal human subjects while they observed an experimenter grasping objects. It was reported that increased motor evoked potentials were recorded from the hand and arm muscles. The Parma researchers also pointed out that the motor evoked potential increased only on those muscles that were selected when the subjects made the observed motor acts actively.[34]

Similarly, by using a PET scan, it was possible to localize areas of the brain involved in the representations of motor acts. In this case, neuronal activity could be measured, albeit indirectly. It was reported that regional cerebral blood flow was measured while human subjects were observing objects as well as actions performed by other people.[35] Other techniques, such as functional magnetic resonance imaging (fMRI), electroencephalogram (EEG), and magnetoencephalography (MEG), have all been used to record brain activity while human subjects were involved in either action observation or action execution or both.[36]

3. The Three-Level Problem of the Functional Properties of MNs

Earlier (see the introduction above) I ranked the questions on correlation, causation, and consciousness and its bearer as "easy," "hard," and "hardest." The "easy" problem (the question of neural correlates of the functional properties of MNs) is surely a challenging scientific problem but is nonetheless easier than the "hard" problem (the question of the causal profiles of MNs) and the "hardest" problem (the question of whether the functional properties of MNs require consciousness and its bearer.) In this section I will deal with each of these problems in turn.

From here on, I will be using the terms "MNs" and "mirror-neuron system" interchangeably. I will also be using the expression "functional properties" as an umbrella term to refer to all mental phenomena associated with MNs or the mirror-neuron system.

3.1 The Easy Problem of the Functional Properties of MNs

The functional properties of MNs, i.e., action execution (F1) and action observation (F2), are said to be rooted in neural mechanisms. But how are we supposed to understand the link between the functional properties of MNs and their neural bases? As I put the question earlier, what are the *neural correlates of the functional properties of MNs*? At the heart of this question lies what philosophers of mind call *psychoneural correlations*. These are correlations between mental phenomena and brain states or processes.[37] This in turn involves questions about the neural correlates of consciousness (NCC). David Chalmers defines an NCC as a minimal neural system that is directly associated with states of consciousness.[38] We will take up the issue of consciousness in section 4 below.

Here my aim is twofold. First, I want to discuss what sort of progress that MN theorists can reasonably claim to have made in exploring the neural correlates of the functional properties of MNs. Second, I want to identify questions that would fall outside the purview of MNs theorists' investigations of the neural correlates of the functional properties of MNs. Taken together, these two things reveal some serious (and possibly inherent) limitations that beset the project of establishing the neural correlates of the functional properties of MNs. This is to say that, while there are things that we can know by way of investigating the neural correlates of the functional properties of MNs, there are also things about which such investigations simply remain silent.

But what exactly do we mean by psychoneural correlations? Here I begin with Jaegwon Kim's characterization of the notion of psychoneural correlation since it resonates well with most MNs theorists' views discussed in section 2 above: "*Mind-Brain Correlation Thesis*: For each type M of mental event that occurs to an organism *o*, there exists a brain state of kind B (M's "neural correlate" or "substrate") such that M occurs to *o* at time *t* if and only if B occurs to *o* at *t*."[39]

By adopting Kim's *Mind-Brain Correlation Thesis,* we can state the neural correlates of the functional properties of MNs as follows:

> *Functional Properties-MNs Correlation Thesis*: For each type M of mental functional property that occurs to a person *P*, there exists a brain state of Kind B such that M occurs to *P* at time *t* if and only if B occurs to *P* at *t*.

What do we make of the *Mind-Brain Correlation Thesis* and the *Functional Properties-MNs Correlation Thesis*? Notice that despite some minor differences in wordings, the two psychocorrelation theses are semantically equivalent. Thus, by the logical principle of *salva veritate* (substitutability of equivalent expressions), the *Mind-Brain Correlation Thesis* can be replaced with the *Functional Properties-MNs Correlation Thesis*. For example, given the *Mind-Brain Correlation Thesis*, a neural correlate is both necessary and sufficient for the occurrence of a certain mental event to an organism.[40] Similarly, given the *Functional Properties-MNs Correlation Thesis*, a neural correlate is both necessary and sufficient for the occurrence of a certain mental functional property to a person. Psychocorrelations are also said to behave in a lawlike manner, that is, non-coincidentally. More specifically, for every activation of neural fibers, it is said that there is an accompanying mental phenomenon, say, experiencing a toothache. That means, as Kim argues, that even the smallest change in one's mental life cannot take place unless it is rooted in some specific change in one's brain state. In light of such considerations, mental states are said to be supervenient on brain states. But figuring out the details of this supervenience is said to be a matter of scientific investigation.[41]

Given Kim's characterization of psychocorrelations, brain states or activity can be said to have an ontological primacy over mental states in the sense that the latter are totally dependent on the former both for their existence and function. But what precisely does the alleged ontological primacy entail? Putting aside the details, for now, here we can focus on four key claims:

i. *the necessary and sufficient claim*: neural correlates are both necessary and sufficient for the occurrence of mental states;
ii. *the state dependency claim*: mental states are totally dependent on brain activities;
iii. *the lawlike claim*: psychocorrelations are lawlike;
iv. *the psychoneural identity claim*: mental states and events are identical with brain activities or processes (i.e., the conjunction of (i)–(iii) is said to pave the way for psychoneural identity).

These claims are heavily debated among philosophers of mind. For now, there is no need to get into a detailed discussion of these controversies. Here I raise only some of the core problems that underlie them.

As it stands, claim (i), the necessary and sufficient claim, is too strong. For reasons we shall see in the remaining part of this section, if there is a sense in which mental states can be shown to operate without being strictly controlled by brain activity, then (i) would be undermined. Moreover, David Chalmers argues that one could grant the initial possibility that the neural state, *N*, is necessary and sufficient for the corresponding mental state. However, for Chalmers, such requirement is arguably too strong. This is because, as Chalmers further points out, it could well be the case that there is more than one neural correlate of a given mental or conscious state. Chalmers imagines the possibility of two neural systems, which I will (departing from Chalmers's own lettering system, for the sake of clarity) call N_a and N_b. Suppose that a certain state of N_a suffices as a neural correlate for being in pain. Similarly, suppose that a certain state of N_b suffices as a neural correlate for being in pain. In such scenarios, Chalmers argues that the states attributed to N_a and N_b are not always correlated. But Chalmers argues that the supposed two systems, N_a and N_b (including their corresponding states) can be said to be neural correlates of a mental state. In this case, Chalmers argues that neural activity in N_a is not necessary and sufficient for a mental state such as pain (as it is not necessary). The same thing is true of N_b. That means that, as Chalmers argues, if both N_a and N_b are considered to be neural correlates of mental state, we cannot thereby also require a neural correlate of mental state to be necessary and sufficient. But if neither N_a nor N_b is *necessary* as a neural correlate for mental state, that would undermine the central claim of (i), since (i) requires correlates to be not only sufficient but also necessary for a mental phenomenon.[42]

Chalmers's argument as briefly laid out above has a direct implication for (ii), the state dependency claim. Remember that (ii) is a claim about the ontological dependency of the mental on the physical. Here mental or psychophysical causation is called into question, since (ii) denies mental causes in the physical domain. In defending (ii), physicalist philosophers of mind (who reject interactive dualism), generally follow a Kimean sort of strategy. In this case, they are committed to the *causal closure of the physical domain* while rejecting what they call *causal overdetermination*.[43] The causal closure of the physical domain is the claim that every physical event that occurs at a particular time has a sufficient physical cause. In this case, no independent mental cause(s) is said to bring about a physical event. On the other hand, causal overdetermination is a situation whereby a certain observed event is said to be caused by multiple causes when a single cause would suffice to bring about an event in question. Here a "double assassination" example is often invoked. Suppose that two independent members of a firing squad hit a certain person, simultaneously resulting in the victim's death. In this case, a single shot would have been sufficient in bringing about the victim's death. But as it stands, the victim's death is said to have been overdetermined, since it involved two independent bullets hitting the victim. Likewise, if a mental cause is responsible for bringing about a physical event along with a physical cause, then the physical event in question is said to be causally overdetermined.

But the Kimeans' denial of the causal efficacy of the mental forces them to embrace *epiphenomenalism*, which denies mental causation. Elsewhere I have argued why epiphenomenalism directly conflicts with the phenomenal features of mental states, which are grounded in first-person direct awareness of one's own subjective experiences.[44] I also argued why, among other things, the causal closure of the physical domain principle fails to accommodate irreducible phenomenal features of the mental.[45] Moreover, as Lowe argues, the causal closure principle fails to eradicate the causal input of the mental from the physical domain. This is because, even if every physical effect P_2 has a sufficient physical cause P_1, it could be that P_1 operates through a mental intermediary M. So although P_1 is sufficient for P_2, so is M and because M does not occur at the same time as P_1, it does not overdetermine P_2 either. Lowe also argues that the causal input of the mental in the physical domain does not necessarily constitute systematic causal overdetermination.[46] For these and similar other reasons, claim (ii) remains highly questionable, to say the least.

Like claims (i) and (ii), claim (iii), the lawlike claim, also faces problems. Psychocorrelations are claimed to be lawlike, which are said to be scientifically investigated and explained. In this case, as Richard Swinburne argues, there has to be a scientific explanation for a one-many simultaneous mind-brain correlation. Here the phrase "one-many" refers to the sort of relationship that is said to exist between a physical state(s) and a corresponding mental state. This means that for each kind of mental event, there are one or more brain events that are said to underlie it. For example, whenever a certain brain event occurs (e.g., a firing of neurons), the mental event occurs (e.g., a feeling of headache) simultaneously, and the converse is also the case. But for reasons we shall see, Swinburne argues that one-many psychocorrelation lacks any detailed supporting evidence. Of course, neuroscientists could say that in light of a growing experimental work based on neuroimaging techniques, one-many psychocorrelation is gaining support. However, as we shall see, such optimism accomplishes very little by way of settling the sorts of problems Swinburne is pointing out. Here the main issue has to do with an inherent difference (which I take to be ontological) that exists between mental properties and physical properties for which no satisfactory scientific explanation seems to be available.

On Swinburne's view, even granting that one-many correlations between mental events and brain events are possible, settling the nature of the correlations in question still falls outside of the purview of science. Swinburne claims that science cannot explain why there are these correlations as opposed to other correlations. In other words, science leaves us with a problem of radical contingency. There is no good reason why some neural events are correlated with pain and not pleasure and others with pleasure and not pain, for example. So it is conceivable that there is an inverted world in which tea and cookies are torture and burns are pleasant, for example. Swinburne argues that to give a scientific explanation as to why some events occurred, one would have to also show that, given the initial conditions or states of affairs, the laws of nature are such that events had to occur. But in this case, he argues at length, the sort of scientific explanation that could unpack the nature of lawlike psychocorrelations is far from obvious.[47]

In Kim's view, claim (i)–(iii) together pave the way for claim (iv). Recall that (iv) is the claim about psychoneural identity, according to which mental states and events are identical with brain activities and processes. The identity in question comes in the form of both *token to token identity* and *type to type identity*. To see a type/token distinction, consider the following words: yellow, yellow, yellow, red. How many words are in the sequence? We can give two equally acceptable answers. That is, there is a sense in which it is correct to say that there are *four* words. There is another sense in which it is also correct to say that there are *two* words. We can say this because we have two word-types and four word-tokens.

A type is something that can "appear" or be instantiated in different places, whereas a token is a particular instance of a type.

In light of the foregoing type/token distinction, an example of token-to-token identity would be a reduction of a certain mental state and process that occurs at a particular time to a certain physical state and process in the brain. In this case, type-to-type identity would be a reduction of a type of mental state to a type of physical state. Psychoneural identity also entails a reduction of mind to brain. Put this way, to say that *pain* is a mental state is to say that pain is identical to a physical state such as brain activity or neuronal firing. Notice that the nature of reduction entailed by (iv) is ontological in nature in that it involves a reduction of one kind of entity, say a mental property (e.g., being in pain) to some other kind of entity, in this case, a physical property (e.g., neuronal activity). Following Brian McLaughlin, we can schematize psychoneural reduction as follows:[48]

> *The Identity Principle*: For every type of state of mental property, *M*, there is some type of physical property, *P* such that *M* = *P*.

Here the equality sign (=) stands for strict numerical identity, which can be spelled out in terms of Leibniz's law of the *indiscernibility of identicals*. According to this law, for all *x* and for all *y*, if *x* is identical to *y*, then for any property, *P*, *x* is said to have *P* if and only if *y* is said to have *P*, and vice versa. If there is anything that is true of *x* but not true of *y*, then the supposed strict identity relation would fail to hold between them. Formally put: $\forall x\, \forall y\, [x = y \rightarrow \forall P\, (Px \leftrightarrow Py)]$.[49]

As I see it, to establish psychoneural identity (claim iv above), one has to overcome at least two obstacles. First, it should be shown how psychoneural identity follows from psychoneural correlation. Since correlation entails neither identity nor as we shall see, causation, it is entirely unclear what motivates a Kimean assumption in (iv).[50] Second, it should be shown that one can do away with the phenomenology of mental states or *qualia*. If someone is experiencing a headache, then there is a "what it is like" to have such a headache, which includes a qualitative feeling of pain. In short, the question is whether it is even possible in principle to give a purely physical analysis of one's subjective experience of pain or any other relevant first-person-based subjective data.[51]

Qualia pose a serious problem for (iv), as Kim himself acknowledges.[52] If qualia characterize mental states but not physical states, then by Leibniz's law of the *indiscernibility of identicals*, it follows that strict identity relation does not hold between mental states and brain states. Unless Kimeans show us how the Leibnizian *Identity Principle* accommodates the two obstacles discussed in the previous paragraph, we are within our epistemic right to reject (iv).

To sum up: We have seen why the Kimean construal of psychoneural correlations faces significant hurdles. Moreover, since the *Functional Properties-MNs Correlation Thesis* is semantically equivalent to Kim's *Brain-Mind Thesis*, it also faces the same problems raised above. This does not mean that nothing good can come out of psychoneural correlations considered in general. But as it stands, the *Functional Properties-MNs Correlation Thesis* fails to show us what sort of progress MN theorists can reasonably be said to have made (also continue to make) in exploring the neural correlates of the functional properties of MNs. Similarly, we cannot use the *Functional Properties-MNs Correlation Thesis* to identify questions that fall outside the purview of the empirical investigations carried out on the neural correlates of the functional properties of MNs. On each of these fronts, we need to get some clarity to be able to unravel the limitations that beset the project of establishing the neural correlates of the functional properties of MNs. Although there are things we can know by way of investigating the neural correlates of the functional properties of MNs, there are also things concerning which such investigations simply remain silent.

In light of such considerations, we need to introduce a new principle that would allow us to rectify the deficiencies of the *Functional Properties-MNs Correlation Thesis* while at the same time showing us what MN theorists can and cannot achieve based on psychocorrelations. I shall call the new principle *The Mirror-Neuron Activation Pattern Principle.*

> *Mirror-Neuron Activation Pattern Principle:* For any given neuronal activation pattern *P* there is correlation-specific neuronal firing *N* coupled with a mental functional property *M* such that *M* occurs to a person *X* at time *t* (CSN_F) or correlation-unspecified neuronal firing *N* coupled with any mental functional property *M* such that *M* occurs to a person *X* at any time *t* ($CUSN_F$).

Notice that the *definiens*[53] of the *Mirror-neuron Activation Pattern Principle* is marked by a two-place logical connective called disjunction, "or." A symbolic notation for disjunction looks like "v." The entire disjunction in the definiens above can be expressed as "CSN_F v $CUSN_F$" and its component parts are called *disjunct*. For example, "CSN_F" is a disjunct. Likewise, "$CUSN_F$" is a disjunct. A disjunction connective "or" has two senses, an exclusive sense and an inclusive sense. In the former sense, for any given two things (options) presented, one can only have one of the things, not both, whereas in the case of the latter, one can have one of the two things or both. If I am told I can go to France or Brazil for an upcoming conference (where both conferences take place on the same days), then I have to choose one of the countries, since I cannot go to both; this is an example of an exclusive "or." In contrast, if I am told I can have either coffee or cookies after lunch, I can choose coffee over cookies, but I can also decide to have them both. This is an inclusive sense of "or." The application to the definition above is this: it is possible that in some cases a neural activation pattern is correlation-specific and in others correlation-unspecified, so it could be both. In light of this, I leave the "or" in the *definiens* of the *Mirror-Neuron Activation Pattern Principle* ambiguous. This is because, as we shall see, the disjunction in the definiens reflects functional ambiguity that characterizes neuronal firing.

If the mirror-neuron system exists, then a host of cognitive capacities attributed to such a system seem to have been discovered in light of CSN_F. For example, as discussed in section 2.2 above, a mirror-neuron system is implicated in empathy, imitations, associative learning, social interaction, intention understanding, language and speech evolution, and other things. Moreover, dysfunction in the mirror-neuron system is implicated in disorders such as autism, aphasia, and limb apraxia.[54] In these and similar other disorders, the functional properties of MNs (F1 and F2 in the introduction above) are said to be affected. For example, autistic people are said to lack an ability for, *inter alia*, intention understanding, communication, and social interaction skills. Aphasiac people suffer from an inability to produce language due to brain injury to the Broca's area in the frontal lobe. There is also aphasia related to language perception which results from brain damage to Wernicke's area in the temporal lobe. Similarly, people with limb apraxia are said to struggle from an inability to perform actions and in some cases to recognize actions done by others. As we saw in section 2.2, neuroimaging techniques are the main methods used in the study of mirror-neuron systems. In this case, neuronal firing is the main source of data for MN theorists in studying CSN_F. MN theorists can certainly claim to have made notable progress in answering empirical questions concerning the functions of the mirror-neuron system which serves as a physical substrate, for mental or cognitive capacities mentioned earlier. As Bennett and Hacker rightly remark:

> Empirical questions about the nervous system are the province of neuroscience. It is its business to establish matters of fact concerning neural structures and operations. It is the task of *cognitive* neuroscience to explain the neural conditions that make perceptual, cognitive, cogitative, affective, and volitional functions possible. Such explanatory theories are confirmed or infirmed by experimental investigations.[55]

Bennett's and Hacker's remarks reflect the central task that MN theorists undertake in terms of dealing with CSN_F. As I see it, the central steps that underlie the MN theorists' work can be summed up as follows:

OBSERVATION → DATA → ANALYSIS → INTERPRETATION → CONCLUSION

These steps are by no means exhaustive. I only use them here for illustrative purposes. The *arrows* in between the terms simply mean, "one thing leads to the other." Each of these steps proves to be useful provided that mistakes are not made at the level of observation. If an observation is faulty, then the subsequent steps would not give us an accurate result. In this case, the "conclusion" could end up being inconsistent with what the "data" actually implies concerning any given matter under investigation. But for now, I leave such issues aside.

Notice that the above steps are confined to CSN_F. In this case, whatever progress MN theorists claim to be making (in unpacking neural correlates of the functional properties of MNs) remains incomplete. The reason for this is that progress made based on CSN_F says nothing whatsoever concerning the other disjunct in the definiens above, which is $CUSN_F$. As we recall, $CUSN_F$ concerns correlation-unspecified neuronal firing which can be coupled with any mental functional property that occurs to a person at any time. For example, as I write this paper, it is virtually impossible to answer specific questions concerning which parts of my brain are implicated in allowing me to process my thoughts. Surely we have general answers that we can generate as to which parts of my brain may be more involved—say, the prefrontal cortex, which is implicated in facilitating capacities such as reasoning or reflection. But this says nothing about the contributions of other parts of the brain that we have not targeted in our study. There is a difference between experimentally controlled brain imaging studies and the reports we hear thereafter on the one hand, and a general lack of knowledge about the rest of the brain's role on the other hand. This is what $CUSN_F$ implies. Given $CUSN_F$, neuronal firing is a neurophysiological process that takes place at all times. Even when the brain is said to be in a *resting state*, neuronal firings continue to happen, although at a much lower intensity.[56] The only time neurons stop firing for good is when a person dies.[57]

But when MNs theorists carry out their investigations, for all sorts of practical reasons they cannot study neuronal signals that take place throughout the entire brain. So they have to simply ignore neuronal firings that are not the immediate target of their study. This is precisely what $CUSN_F$ implies. If so, MN theorists' investigations of the neural correlates of the functional properties of MNs inevitably face the problem of *underdetermination* (i.e., inability to cover adequate ground in investigating and explaining something). This problem in turn shows us why there are things that inevitably fall outside the purview of the investigations MN theorists carry out concerning a neural correlate of the functional properties of MNs. The problem of underdetermination could be the main source of functional ambiguity that besets the entire disjunction stated in the definiens, that is, CSN_F v $CUSN_F$. By "functional ambiguity" I refer to more than one neural structure facilitating neuronal firings concerning certain corresponding mental phenomena. In this case, it remains unclear how to go about settling how all these neuronal firings are related and what specific contributions they are making to a person's mental states.

At this point, we may wonder whether a problem that is said to arise from functional ambiguity is one that can never be solved, or is merely a problem that is related to our current empirical capabilities. Could it be that if we dramatically increase our ability to measure the brain at a fine-grained level, while a person experiences mental states, that we could clarify exactly the degree to which all of the neural activity in the entire brain is correlated with a particular mental state? In answering these sorts of questions, we could easily be tempted to embrace one of the following two extreme positions. First, if we say that a problem that arises from a functional ambiguity can never be solved, then we are assuming a strong burden of proof that proves to be hard to justify in any obvious way. Second, if we say that a problem that arises from a functional ambiguity can be solved only if we significantly improve our existing empirical capabilities, then we are implicitly assuming that a solution for the problem in question is strictly empirical. But this is hardly the case, for the reasons we have already seen and also the other reasons we will see.

The inadequacy of experimental work in explaining the nature of neural correlates of the functional properties of MNs has to do in part with a lack of engagement with the domain of philosophy. As Bennett and Hacker put it:

> Conceptual questions (concerning, for example, the concept of mind or memory, thought or imagination), the description of the logical relations between concepts (such as between the concepts of perception and sensation, or the concepts of consciousness and self-consciousness), and the examination of the structural relationships between distinct conceptual fields (such as between the psychological and the neural, or the mental and the behavioral) are the proper province of philosophy.[58]

I do not follow Bennett and Hacker in deeming the role of philosophy as being nothing more than clarifying concepts. This is a misguided conception of philosophy. As Timothy Williamson insightfully asked: "Where is the need for philosophers' clarifications supposed to come from, if all they do is clarify?"[59] However, despite their questionable conception of the role of philosophy, Bennett and Hacker are right in claiming that neuroscientists cannot successfully do their work alone. They have to collaborate with philosophers, who primarily depend on the *a priori* method. In section 4 below, we will see why this is indeed the case.

In light of the foregoing discussion, it could be said that the MN theorists' main challenge in settling the neural correlates of the functional properties of MNs is practical in nature. Such challenges can be said to require only practical solutions which can be achieved as neuroscientists continue gathering more facts about brain functions. So it could be said that here MN theorists only face *the easy problem of the functional properties of MNs*. Although this could well be the case, it would be foolish of MN theorists to think that the neural correlates of the functional properties of MNs require only solving practical problems.

3.2 The Hard Problem of the Functional Properties of MNs

One of the things psychocorrelations seem to imply is causation. In this case, the causal relationship could be said to hold between neural activity and the corresponding functional properties of MNs. We could easily be tempted to link up correlation with causation because of the sort of questions correlation forces us to ask. For example, as Kim remarks, if we come across a systematic correlation between any given properties or event types, we find ourselves needing an explanation of the correlation. Such explanation or interpretation of the correlation is supposed to answer questions such as: Why do the properties *F* and *G* correlate? Why is it that an event of type *F* occurs just when an event of type *G* occurs?[60] Such questions remind us of David Hume's famous account of how causal relations can easily be assumed where they do not exist or at least, in scenarios in which establishing their existence proves to be extremely difficult. Hume's account is described as the *Regularity View of Causation*. Stathis Psillos sums it up as follows:[61]

c causes *e* iff [if and only if]

a. *c* is spatiotemporally contiguous to *e*;

b. *e* succeeds *c* in time;

c. all events of type *C* (i.e., events that are like *c*) are regularly followed by (or are constantly conjoined with) events of type *E* (i.e., events like *e*).

Here there is no need to get into messy contemporary controversies that beset Hume's view of causation.[62] But one thing we can certainly say is that Hume seems to deny any necessary connection existing between any given two or more events. On his account, we have no reason to think that there is a necessary causal link between *c* and *e* as set forth in points (a) through (c) above. The best we can say in this case is that some sort of correlation seems to hold between *c* and *e*. But ruling out causal relation entirely seems a hard thing to do. This is because denying causal relations seems to conflict with our experience of the world around us, which seems to be causally infused. We find it hard to swallow Hume's view that any causal relation between *c* and *e* is nothing more than our own projection, for which no justification can be given.

By applying Hume's causal analysis to neural correlates of the functional properties of MNs, we can encounter similar complexities. Let *B* stand for the population of neurons in the Broca's area. Similarly, let *LP* stands for language production. We can put the causal analysis schema as follows:

B causes *LP* iff

a. *B* is spatiotemporally contiguous to *LP*;

b. LP succeeds *B* in time;

c. all events of type *B* (i.e., events that are like *B*) are regularly followed by (or are constantly conjoined with) events of type *LP* (i.e., events like *LP*).

As in the case of the Humean causal analysis, here too, we cannot simply infer a causal link between *B* and *LP*. By extension, the same thing applies to all other functional properties of MNs. All we can say here is that there is some sort of correlation between *B* and *LP*. It would be hard to make sense of *B* and *LP* from a causal standpoint. Imagine, for example, the ability to produce language being literally caused by a relevant population of neurons in the Broca's area. How can nerve cells bring about a non-physical cognitive capacity, in this case, an ability to produce language with all its rich complexities?

The most common reason cognitive neuroscientists have for assuming the presence of causal relation in cases like *B* and *LP* comes from lesion studies. For example, let's suppose that John, while playing football, suffers from a serious head injury in an area of the brain that is strongly implicated in language ability, say a Broca's area. Furthermore, suppose that the following day, John loses his ability to speak altogether. In this case, there indeed seems to be a strong jusification for causal inference regarding why John has lost the ability to speak. But from a Humean point of view, we cannot apply this situation to *B* and *LP* without illegitimately conflating correlation with causation. The mere sequence of events does not give us any clear evidence of the presence of causal relation between *B* and *LP*. The best we can do here is investigate what I earlier stated as CSN_F, according to which for any given neuronal activation pattern *P* there is a correlation specific neuronal firing *N* coupled with a mental functional property *M* that occurs to a person, *X* at a time *t*. Yet it seems that some sort of causal relation must be involved between the neural substrate and the corresponding mental phenomena. Again, there is no guarantee that we can get causality in this regard. As neuroscientist Gareth Jones nicely sums up:

> The simple act of finding neural correlates for certain behaviors or attitudes provides few, if any, insights into *causative factors*. Even if a certain brain structure were strongly associated with... [some sort of subjective] experience, this says nothing about whether the structure generates that experience. Simply because brain region "R" is active when behavior "B" is undertaken does not mean that changes in "R" cause "B" to take place. The opposite, in fact, could be the case, in that when an individual displays behavior "B," brain region "R" is modified, and if this occurs sufficiently often, there are significant changes to "R." Yet again, the interplay between "R" and "B" may be so close that the only tenable conclusion is that there is no definitive causative factor—the one feeds on the other.[63]

Jones's remarks here resonate well with the Humean sort of causal analysis. But the question remains: Where can we locate causality if not in psychocorrelation? This is *the hard problem of the functional properties of MNs*. We need to keep looking for an answer to this problem.

3.3 The Hardest Problem of the Functional Properties of MNs

It now seems clear that settling the problem of the functional properties of MNs goes beyond tackling issues related to pyschocorrelation, identity, and causation. Although each one of these things presents different levels of difficulties, as the hitherto discussion shows, we have yet to encounter the hardest problem of the functional properties of MNs. Here the issues we face are even more explicitly metaphysical in nature. In this regard, the problem of the functional properties of MNs comes down to answering a two-pronged question, namely: what is the role of consciousness and its bearer? Put differently, do the functional properties of MNs require consciousness and its bearer for their existence?

We need to keep in mind that each prong of this question has its corresponding questions, concerning the neural correlates of consciousness and the neural correlates of the bearer of consciousness. Working through these sets of questions will be even more demanding given that the issues of identity and causation we discussed earlier are directly linked to consciousness and its bearer. All the major problems we have looked at so far concerning the functional properties of MNs congregate around the question of consciousness and its bearer. It is for this reason that we call this stage *the hardest problem of the functional properties of MNs*.

However, for present purposes, there is no need to resolve all of these issues at once. Without attempting to undertake such a task, we can show how and why tackling the question of consciousness and its bearer as it relates to the functional properties of MNs paves the way for dealing with other issues, such as identity and causation (as they relate to the functional properties of MNs). Dealing with these questions ultimately sheds some light on the central question, i.e., where the functional properties of MNs should be located—in the brain or in something different? We can even call this the location problem of the functional properties of MNs.

4. The Misidentification Problem and Its Remedy

The discussions in mirror-neuron literature do not indicate that there is an awareness on the part of the MN theorists that the functional properties of MNs raise deeper ontological issues. Contemporary MN-related controversies over psychocorrelations, identity, and causation (as discussed in sections 3.1 and 3.2 above) are symptoms of rather a much bigger issue that involves the problem of an irreducible consciousness and its bearer. In this case, neither appeal to a particular region in the brain nor to the nervous system as a whole seems to be capable of providing us with any satisfactory solution. So there is a sense in which attributing the functional properties of MNs (as many neuroscientists and psychologists do) to a certain brain region housing a mirror-neuron system leads to the *problem of misidentification*. This is a situation whereby one mistakenly identifies an object or a property of one sort with an object or a property of some other sort. The problem of misidentification arises when the functional properties of MNs are identified with or located in a certain region or regions of the brain.

A ready-made solution for the location problem of the functional properties of MNs has been and still is *the brain*. In a way, this is not surprising. If the functional properties of MNs are not located in the brain, where can they be said to be located? Things seem to fall easily in place by invoking the brain as not only a physical substrate but also a bearer of mental phenomena. In this regard, it is not uncommon to see in neuroscientific literature phrases such as "the brain represents," "the brain knows," "the brain unifies experiences," and the like in accounting for psychological properties.[64] But have MN theorists shown that the brain is a subject of experience? Questioning the very intelligibility of ascribing psychological properties to the brain, Bennett and Hacker remark:

> We pose questions and search for answers, using... our language—in terms of which we represent things. But do we know what it is for a *brain* to see or hear, for a brain to have experiences, to know or believe something? Do we have any conception of what it would be for *a brain* to make a decision? Do we grasp what it is for a brain (let alone for a neuron) *to reason* (no

> matter whether inductively or deductively), to *estimate probabilities*, to *present arguments*, to *interpret data* and to *form hypotheses* on the basis of its interpretations?[65]

Here Bennett's and Hacker's central point is that any attempts one makes to produce answers for these sorts of questions will bear no fruit. Attributing psychological properties to the brain emanates from deep conceptual confusion which results in misidentification of the right sort of subject or object of such properties with the wrong sort of subject or object, which in this case is, *the brain*. In this regard, Bennett and Hacker argue for an alternative way to answer the sorts of questions they raised in the passage quoted above. As they remark:

> We can observe whether a person sees something or other—we look at his behaviour and ask him questions. But what would it be to observe whether a brain sees something—as opposed to observing the brain of *a person* who sees something? We recognize when a person asks a question and when another answers it. But do we have any conception of what it would be for a brain to ask a question or answer one? These are human attributes of human beings.[66]

Bennett's and Hacker's remarks here show how a misidentification mistake pointed out earlier can be rectified. In this case, psychological properties must be located in a human being, not in a part of a human being such as *a brain*. Bennett and Hacker call the attribution of the properties of the whole to a part of the whole a *mereological fallacy*.[67] Bennett's and Hacker's remarks in the two passages quoted above resonate well with what I have already implied in the discussions I advanced in sections 2 and 3 above. Attributing the functional properties of the MNs to the mirror-neuron system faces insurmountable problems. It makes no sense to attribute empathy, imitation, learning, social skills, intention understanding, language perception, and other higher-level cognitive capacities and properties to the brain substrate. Properly speaking, such things belong only to subjects of experience or persons like us. And this insight inevitably brings us to the question of consciousness and its bearer.

4.1 Consciousness and Its Bearer

The literature on mirror neurons is shockingly silent on both consciousness and its bearer (which I take to be a self). In my view, this is a profound mistake. Whatever progress one hopes to make in terms of investigating the nature of the functional properties of MNs, the appropriate backdrop to undertake such a project should be *consciousness* and *its bearer*. But why is this so? My answer is that the functional properties of MNs are inextricably linked with subjectivity, which underlies the very nature of irreducible consciousness. By "irreducible consciousness," I mean that consciousness is a property that strongly resists an ontological property reduction. An example of a reduction in question would be to consider consciousness as nothing but a brain process or a neural state. That said, consciousness itself cannot exist without a bearer. For now, we can be indifferent as to whether consciousness and its bearer are physical or not. But one thing we cannot afford to grant is that both consciousness and its bearer are somehow identifiable with the brain. This is because, as we already saw, the brain proposal is doomed to fail from the start. But those who still think that the brain needs to be taken as a source of the functional properties of MNs must first meet the burden of proof for the brain-related problems discussed earlier.

I therefore remain highly skeptical of any forthcoming proposal to establish the brain as a subject of the functional properties of MNs.

Central to an irreducible consciousness is subjectivity or phenomenology. Thomas Nagel's famously expressed it as what it is like to be in a certain mental state, say experiencing a headache.[68] But subjectivity does not exist in thin air all on its own. This is to say that subjectivity necessarily requires a subject. Or a conscious being. Thus, consciousness and its bearer fall or stand together. If this is true, as I think it is, then mental phenomena in general, and the functional properties of MNs in particular, cannot be divorced from consciousness and its bearer. Of course, both consciousness and its bearer each have neural correlates. But that does not give us any license to reduce them to the brain substrate. As E. J. Lowe argues, the self (the bearer of consciousness) is neither identical to the body taken as a whole nor with any specific part of the body such as the brain. Understood this way, the self is distinct both from the brain or the body taken as a whole. Lowe also argues that the bearer of mental properties is not the brain but the self. In saying this, however, Lowe is not ignoring the sense in which the self can also be said to be a physical thing. For example, the self is said to be a physical thing in virtue of having a physical body.[69] By the same token, David Chalmers also argues that an irreducible consciousness, despite having a physical substrate, is not itself physical. For Chalmers, the primary data for the subjectivity of consciousness comes from first-person data as opposed to third-person data.[70]

If we put together Lowe's view of the self and Chalmers's view of consciousness, we get a promising roadmap towards rectifying the misidentification problem mentioned earlier. In this case, the bearer of mental phenomena to which consciousness plays a central role is the self. By extension, the functional properties of MNs are also located in the bearer of consciousness. Moreover, the problem of causality discussed earlier also can be tackled by locating consciousness and its bearer. Causation in the context of the present discussion is something that can be said to take place as a result of an intentional act of a conscious being. If such observations are on the right track, then it is a serious mistake for contemporary MN theorists to think that any progress they make in MN investigation comes solely from a deep study of the mirror-neuron system. Instead, in light of the foregoing discussions, the sole focus on the mirror-neuron system only ends up giving us an incomplete picture. The issue we are dealing with is not only empirical but also strongly metaphysical in nature.

4.2 The Dangers of Misidentification

In this chapter, I have tried to make a case for two things. First, I have argued that mirror cognition cannot be properly understood without considering the problem of misidentification of the source of this ability. Second, I have argued that it is critically important to consider consciousness and its bearer to tackle the misidentification problem. Although these points have focused on mirror neurons, failing to heed them would have wider negative implications for the discipline of cognitive neuroscience taken in general. In this regard, we can think of at least two main dangers of misidentification.

The first danger has to do with undermining *scientific progress*. As a professor of cardiovascular physiology, Denis Noble, remarks, "the first step to scientific progress is to ask the right questions. If we are conceptually confused, we will ask the wrong questions."[71] Noble's remarks imply that asking wrong questions hinders a proper understanding of the issues we investigate in science in general and in neuroscience in particular. By contrast, asking the right questions is the bedrock of scientific progress. But to ask the right questions, as Bennett and Hacker argue, working neuroscientists cannot afford to ignore the role of philosophy in their discipline. What this means is that, as Bennett and Hacker show through numerous examples, neuroscience without philosophy cannot make genuine progress in properly grasping neuroscientific issues such as sensation, perception, memory, emotion, volition and voluntary movement, cognitive capacities, and consciousness.[72]

The second danger has to do with the acceptance of *reductionism* as a paradigm of scientific methodology.[73] The sort of reductionism invoked here portrays neuroscience as self-sufficient. Taken this way, neuroscience is said to have all the conceptual resources that allow it to tackle issues related to the mind-body problem. Again, this is hardly the case. The reductionism in question can neither establish the self-sufficiency of neuroscience nor can it render obsolete the centrality of philosophy or metaphysics for neuroscience. Embracing such reductionism in neuroscience can only come at the cost of minimizing the positive role of philosophy. But as Lowe argues, progress in non-empirical disciplines such as mathematics and logic can be and has been made. Likewise, Lowe claims that progress is achievable in philosophy despite its being a non-empirical discipline. Philosophy by its very nature is concerned with the most fundamental questions that take a center stage in human thought. In light of this, following Lowe, we can advocate for the conception of philosophy that puts metaphysics at the center of philosophy and similarly, for the conception of philosophy that puts ontology (study of being or existence) at the center of metaphysics. Once metaphysics and ontology take their appropriate places in philosophy, addressing ontological questions comes down to investigating not our own portrait of reality but reality taken "as it is in itself." Lowe claims that no special science can succeed in addressing its own ontological questions independently of the "science of being" or ontology. The reason for this, he says, is that reality is one and truth is indivisible.[74]

The key lesson we can glean from the two dangers briefly discussed in this section can be summed up as follows: neuroscience is not and cannot be self-sufficient; it needs to team up with philosophy. We can, of course, apply this finding directly to the mirror-neuron discussion presented in this chapter. In this case, both cognitive mirroring and its neural correlates can exist. But to explain the ability of cognitive mirroring, we need consciousness and its bearer. Consequently, metaphysics cannot be avoided.

4.3 Potential Objection, and Reply

Against everything I have defended in this chapter, an objection could be raised that I have underestimated the role of the brain in sustaining mental phenomena in general and the functional properties of MNs in particular. Such an objection appeals to empirical works in cognitive neuroscience. So the objector might say that I am simply providing philosophical solutions to problems that should be dealt with within an empirical domain.

A lot could be said in response to this objection. Here I only want to point out that empirical research without a proper conceptual foundation will only give rise to a lot of confusion rather than the desired outcome.[75] Elsewhere I have also argued that empirical and non-empirical works can and should complement each other.[76] Balanced discussion on human nature must leave room for an *a priori* input as well as an *a posteriori* input. If philosophical as well as scientific approaches fail to properly utilize an *a priori* (rational or philosophical) input and an *a posteriori* (empirical) input, no sustainable progress can be made in our effort to shed light on the metaphysical complexities that underlie the notion of human nature.

For example, consider an answer to the following question: *To what is the self or person identical*? This question concerns the nature of the self.

The contemporary discussions over the nature of the self are largely dominated by strong physicalism, according to which everything has to be physical and physically explainable. An overwhelming majority of professional neuroscientists are also committed to a physicalist view of the self.[77] However, a survey of the literature to date will reveal that our commonsense notions of "self"—which can be given significant intellectual justification—resist a purely physical explanation. Yet the mainstream supporters of strong physicalism have continued to push their argument against the notion of the self by taking approaches ranging from a total denial or elimination of the self to various forms of reductionism.[78] In this case, philosophers who embrace a physicalist view of the self hope that neuroscience or empirical research will, at some point, settle any recalcitrant questions concerning the nature of the self. Similarly, neuroscientists who embrace a physicalist view of the self think that empirical research on the brain and its function is the ultimate powerhouse to establish the answer for the question: *To what is the self or person identical*?

Yet seeking a proper answer to this metaphysical or ontological question cannot be established in isolation from other fundamental metaphysical issues that presuppose the self. Here I have in mind such things as the *nature of mental causation*, *consciousness*, *human agency*, *the first-person perspective*, and *the unity of the self*, just to mention a few. The question remains: Does it take only empirical research on the human brain to settle such matters? If the foregoing discussion in this chapter is any guide, no affirmative answer is available for this question. As Moreland argues, the brain is not a suitably unified mental subject. No amount of detailed knowledge of the brain and its function seems to give us insights into our own basic awareness of ourselves as selves. If this is right, which I believe it is, our knowledge of the unity of the self is not rooted in the brain.

As Moreland further argues, we are aware of our own self or center of consciousness as being distinct not only from our bodies but also from the mental experiences that we have. Hence, it seems plausible to say that we know ourselves as simple, uncomposed selves that have bodies and a conscious mental life. If this is right, then whatever experiences we have, it is not the case that our experiences exist without a bearer of those experiences. From this, it follows that we are not identical with our experiences but are enduring mental substances that have the experiences.[79] In a recent paper, Moreland remarks:

> Neuroscientific empirical data are metaphysically neutral and the empirical study of the brain and goings-on within it are blind with respect to the existence and nature of consciousness and the self. And in moments of honesty, most (if not all) neuroscientists admit this is the case.[80]

Moreland's claims here further strengthen the central arguments presented and defended in this chapter. In light of such considerations, the objection considered in this section does not have anything useful to say concerning the sorts of issues raised.

5. Conclusion

IN THIS chapter, I have aimed to examine the two central properties—action execution and action observation—which I have called the functional properties of mirror neurons. I have argued that attributing the functional properties of mirror neurons, as many neuroscientists do, to mirror-neuron systems suffers from the problem of misidentification. I have suggested how the problem of misidentification can be avoided. I have proposed that the most promising solution for the problem of misidentification is to ground the functional properties of mirror neurons squarely in consciousness and its bearer. If the solution proposed here is embraced, I believe that it would immensely enhance our understanding of the nature of the mind-body problem.[81]

NOTES

1. See, e.g., L. Bonini, "The Extended Mirror Neuron Network: Anatomy, Origin, and Functions," *The Neuroscientist* 23, no. 1 (2017): 56–67; J. M. Kilner and R. N. Lemon, "What We Know Currently about Mirror Neurons," *Current Biology* 23, no. 23 (Dec. 2, 2013): R1057–R1062; M. Fabbri-Destro and G. Rizzolatti, "Mirror Neurons and Mirror Systems in Monkeys and Humans," *Physiology* 23, no. 3 (June 2008): 171–179, doi:10.1152/physiol.00004.2008.
2. G. Rizzolatti and L. Craighero, "The Mirror-Neuron System," *Annual Review of Neuroscience* 27 (2004): 169–92.
3. Here my characterization of functional properties shares some similarities with the functionalist view of the mind. However, it would be a mistake to identify the former with the latter. According to the functionalist view of the mind, mental states lack any intrinsic nature other than the input-output role that they play in a system, in this case, the biological system. But the characterization of MNs need not consider mental properties as being devoid of intrinsic nature. In fact, defenders of MN theory do not have in mind a functionalist sense regarding the roles that MNs are said to play.
4. Marco Iacoboni, *Mirroring People: The New Science of How We Connect with Others* (New York: Farrar, Straus and Giroux, 2008); see also Iacoboni, "Imitation, Empathy, and Mirror Neurons," *Annual Review of Psychology* 60 (2009): 665, and V. Gallese and A. Goldman, "Mirror Neurons and the Simulation Theory of Mind-Reading," *Trends in Cognitive Sciences* 2, no. 12 (1998): 493–501.
5. For an extensive list of the roles MNs are said to play, see Richard Cook et al., "Mirror Neurons: From Origin to Function," *Behavioral and Brain Sciences* 37, no. 2 (April 2014): 178, and Gregory S. Hickok, *The Myth of Mirror Neurons: The Real Neuroscience of Communication and Cognition* (New York: W. W. Norton & Company, 2014), 24–25.
6. Hickok, *The Myth of Mirror Neurons*; see also Hickok's "Eight Problems for the Mirror Neuron Theory of Action Understanding in Monkeys and Humans," *Journal of Cognitive Neuroscience* 21, no. 7 (July 2009): 1229–1243.
7. For the role of mirror neurons in empathy, see Claus Lamm and Jasminka Majdandžić, "The Role of Shared Neural Activations, Mirror Neurons, and Morality in Empathy—A Critical Comment," *Neuroscience Research* 90 (Jan. 2015): 15–24; for their role in action-mirroring, see P. Jacob, "A Philosopher's Reflections on the Discovery of Mirror Neurons," *Topics in Cognitive Science* 1 (2009): 570–595; for their role in mind-reading, see S. Spaulding, "Mirror Neurons are Not Evidence for the Simulation Theory," *Synthese* 189 (2012): 515–534. Note that mind-reading capacity comes in two varieties. According to the "theory-theory" view, the mind-reading capacity of the mental states of other people is rooted in commonsense-based predictions of their behavior. In contrast, the "simulation theory" states that mind-reading capacity is rooted in taking the perspective of the other person, thereby matching it with one's own to infer the other person's mental state. For details, see Gallese and Goldman, "Mirror Neurons and the Simulation Theory."
8. See Cook et al., "Mirror Neurons: From Origin to Function"; Kilner and Lemon, "What We Know Currently about Mirror Neurons."
9. For detailed discussion of this point, see section 3.1 below.
10. For detailed discussion of this point, see section 3.2 below. Note that in my characterization of points (a) and (b) I have adopted David Chalmers's phrases, "the easy

problem" and "the hard problem"; however, my use of these phrases does not directly coincide with Chalmers's. See D. J. Chalmers, *The Conscious Mind* (Oxford: Oxford University Press, 1996), and again Chalmers, *The Character of Consciousness* (Oxford: Oxford University Press, 2010).

11. For detailed discussion of this point, see section 3.3 below.
12. See M. R. Bennett and P. M. S. Hacker, *Philosophical Foundations of Neuroscience* (Malden, MA: Blackwell Publishing, 2003).
13. For a critical but also informative account of the history of the discovery of MNs, see Hickok, *The Myth of Mirror Neurons*, chs. 1–3.
14. See M. Gentilucci et al., "Functional Organization of Inferior Area 6 in the Macaque Monkey: I. Somatotopy and the Control of Proximal Movements," *Experimental Brain Research* 71, no. 3 (1988): 475–490, and Rizzolatti et al., "Functional Organization of Inferior Area 6 in the Macaque Monkey: II. Area F5 and the Control of Distal Movements," *Experimental Brain Research* 71, no. 3 (1988): 491–507.
15. V. Gallese et al., "Action Recognition in the Premotor Cortex," *Brain* 119 (1996), 600.
16. See G. Di Pellegrino et. al., "Understanding Motor Events: A Neurophysiological Study," *Experimental Brain Research* 91 (1992): 176–180, and Gallese et al., "Action Recognition in the Premotor Cortex."
17. Di Pellegrino et al., "Understanding Motor Events"; Gallese et al., "Action Recognition."
18. Rizzolatti and Craighero, "The Mirror-Neuron System," 171.
19. Rizzolatti and Craighero, "The Mirror-Neuron System," 171.
20. See G. Rizzolatti, L. Fogassi and V. Gallese, "Neuorophysiological Mechanisms Underlying the Understanding and Imitation of Action," *Nature Reviews Neuroscience* 2 (2001): 661–670. See also Rizzolatti and Craighero, "The Mirror Neuron System," 170–174.
21. L. Bonini, "The Extended Mirror Neuron Network: Anatomy, Origin, and Functions," *The Neuroscientist* 23, no. 1 (2017): 56–67.
22. Gallese et al., "Action Recognition," 606; Rizzolatti and Craighero, "The Mirror-Neuron System"; Rizzolatti, "The Mirror Neuron System and Its Function in Humans," *Anatomy and Embryology* 210 (2005): 419–421.
23. Gallese et al., "Action Recognition," 606.
24. One study that did conduct single-neuron recording in humans was done in 2010. See R. Mukamel et al., "Single-Neuron Responses in Humans during Execution and Observation of Actions," *Current Biology* 20, no. 8 (Apr. 2010): 750–6.
25. Rizzolatti and Craighero, "The Mirror-Neuron System."
26. The phrase "mirror neurons" was introduced in the literature officially in 1996, in Gallese et al.'s above-cited work, "Action Recognition in the Premotor Cortex," 595.
27. Rizzolatti and Craighero, "The Mirror-Neuron System," 174.
28. M. Fabbri-Destro and G. Rizzolatti, "Mirror Neurons and Mirror Systems in Monkeys and Humans," *Physiology* 23 (2008): 173ff.
29. Hickok, *The Myth of Mirror Neurons*, 24–25.
30. This is taking a second-person perspective, in contrast to the first-person and the third-person perspectives. See my discussion on the difference between these three perspectives in M. P. Guta, "Consciousness, First-Person Perspective, and Neuroimaging," *Journal of Consciousness Studies* 22, nos. 11–12 (2015): 218–245.
31. See Iacoboni, "Imitation, Empathy, and Mirror Neurons," and Rizzolatti and Craighero, "The Mirror-Neuron System."
32. For details, see M. Iacoboni and M. Dapretto, "The Mirror Neuron System and the Consequences of its Dysfunction," *Nature Reviews Neuroscience* 7 (Dec. 2006): 942–951.
33. For the first phrase, see V. S. Ramachandran, "The Neurons That Shaped Civilization," TED talk, Jan. 4, 2010, https://www.youtube.com/watch?v=t0pwKzTRG5E; for the second phrase, see Marco Iacoboni, *Mirroring People*.
34. Gallese et al., "Action Recognition," 606.
35. Gallese et al., "Action Recognition," 606–607; see also Scott Grafton et al., "Localization of Grasp Representations in Humans by Positron Emission Tomography," *Experimental Brain Research* 112 (1996): 103–111, and G. Rizzolatti et al., "Localization of Grasp Representations in Humans by PET: 1. Observation versus Execution," *Experimental Brain Research* 111, no. 2 (Sept. 1996): 246–252.
36. For details, see Iacoboni and Dapretto, "The Mirror Neuron System and the Consequences of Its Dysfunction"; Rizzolatti, "The Mirror Neuron System and its Function in Humans"; Rizzolatti and Craighero, "The Mirror-Neuron System."
37. J. Kim, *Philosophy of Mind*, 2nd ed. (Cambridge: Westview Press, 2006), 81.
38. Chalmers, *The Character of Consciousness*, 44. See also Kim, *Philosophy of Mind*, chs. 3 and 4; M. Owen,

"Neural Correlates of Consciousness and the Nature of the Mind," in *Consciousness and the Ontology of Properties*, ed. Mihretu P. Guta (New York: Routledge, 2019), 241–260; M. Owen and M. P. Guta, "Physically Sufficient Neural Mechanisms of Consciousness," *Frontiers in Systems Neuroscience* 13, article 24 (July 2019): 1–14; T. Metzinger, ed., *Neural Correlates of Consciousness: Empirical and Conceptual Questions* (Cambridge, MA: MIT Press, 2000).

39. J. Kim, *Philosophy of Mind*, 3rd ed. (Boulder, CO: Westview Press, 2011), 92.
40. Kim, *Philosophy of Mind*, 3rd ed., 92.
41. Kim, *Philosophy of Mind*, 3rd ed., 92–93.
42. For details, see Chalmers, *The Character of Consciousness*, 71ff.
43. See, e.g., J. L. Bermudez and A. Cahen, "Mental Causation and Counterfactual," in *Consciousness and the Ontology of Properties*, ed. Guta, 155–173; D. Papineau, *Philosophical Naturalism* (Oxford: Blackwell, 1993) and *Thinking about Consciousness* (Oxford: Oxford University Press, 2002); B. Loewer, in "An Argument for Strong Supervenience," in *Supervenience: New Essays*, eds. Elias E. Savellos and Ümit D. Yalçin (Cambridge: Cambridge University Press, 1995), 218–225. See also S. C. Gibb, "Mental Causation and Double Prevention," in *Mental Causation and Ontology*, eds. S. C. Gibb, E. J. Lowe, and R. D. Ingthorsson (Oxford: Oxford University Press, 2013), 193–214, and Gibb's "Introduction" to that book, 1–17.
44. See Guta, "Consciousness, First-Person Perspective, and Neuroimaging"; see also M. P. Guta, "Frank Jackson's Location Problem and Argument from the Self," *Philosophia Christi* 13, no. 1 (2011): 35–58; M. P. Guta, "The Non-Causal Account of the Spontaneous Emergence of Phenomenal Consciousness," in *Consciousness and the Ontology of Properties*, 126–152; E. J. Lowe, *Personal Agency: The Metaphysics of Mind and Action* (Cambridge: Cambridge University Press, 2008), ch. 4.
45. Guta, "Frank Jackson's Location Problem"; see also J. P. Moreland, *Consciousness and the Existence of God* (London: Routledge, 2008).
46. For details, see Lowe, *Personal Agency*, chs. 1–5. See also R. Swinburne, "Mental Causation Is Really Mental Causation," in *Consciousness and the Ontology of Properties*, 174–186; Gibb, "Mental Causation and Double Prevention"; P. Menzies, "Mental Causation in the Physical World," in *Mental Causation and Ontology*, 58–87.
47. For details, see R. Swinburne, *The Evolution of the Soul*, rev. ed. (Oxford: Oxford University Press, 1997), 183–196. See also Swinburne's "Mental Causation Is Really Mental Causation," cited above.
48. See B. P. McLaughlin, "Type Materialism for Phenomenal Consciousness," in *The Blackwell Companion to Consciousness*, eds. Max Velmans and Susan Schneider (Oxford: Blackwell, 2007), 435. However, here I do not adopt McLaughlin's exact wordings, which have to do with phenomenal consciousness.
49. This law is contrasted with another controversial law known as "Identity of Indiscernible." This law tries to establish strict identity relation between objects based solely on the exact similarity relation which holds between them. That is, if two objects *a* and *b* are exactly similar in the properties that they share, then *a* and *b* are identical. But this conclusion is hardly convincing, since any given two objects despite sharing exactly similar properties can remain numerically distinct. Hence, no strict identity can hold between them.
50. See, e.g., Guta, *Consciousness and the Ontology of Properties*, 145–150.
51. See for details, T. Nagel, "What It Is Like to Be a Bat," *The Philosophical Review* 83, no. 4 (1974), 435–450 (cf. Guta, "Consciousness, First-Person Perspective, and Neuroimaging"); Chalmers, *The Character of Consciousness*; Chalmers, *The Unconscious Mind*; R. C. Koons and G. Bealer, *The Waning of Materialism* (Oxford: Oxford University Press, 2010); Moreland, *Consciousness and the Existence of God* (London: Routledge, 2008); Lowe, *Personal Agency*.
52. See, e.g., J. Kim, *Physicalism, or Something Near Enough* (New Jersey: Princeton University Press, 2005). Furthermore, the 1950s and 1960s mind-body identity theory—e.g., J. J. C. Smart, "Sensations and Brain Processes," *Philosophical Review* 68 (1959): 141–156; Herbert Feigl, "The 'Mental' and the 'Physical,'" *Minnesota Studies in the Philosophy of Science* 2 (1958): 370–497; U. T. Place, "Is Consciousness a Brain Process?," *British Journal of Psychology* 47 (1956): 44–50—which paved the way for continued controversies, quickly fell out of favor, mainly due to Hilary Putnam's multiple realization problem and Donald Davidson's anomalous monism objections: see J. Kim, *Mind in a Physical World* (Cambridge: MIT Press, 1998), ch. 1, and H. Taylor, "A Powerful New Anomalous Monism," in *Consciousness and the Ontology of Properties*, ed. Guta, 39–52. According to the multiple realizations view, a given mental state can be realized by a different physical substrate—neural, electronic, and the like. This view echoes the functionalist view of the mind, which concerns itself with the input and output roles that

mental states play in a physical system. Functionalism is indifferent regarding the intrinsic nature of mental states. Anomalous monism states that mentality cannot be subsumed under strict, scientifically predictable, exceptionless laws, despite the fact that token mental events are identical to token physical events. Each of these views challenges the Kimean sort of strict reductive approach to mentality. Such objections forced most (if not all) materialist analytic philosophers to recognize the difficulty of lumping mental properties with physical properties (see Kim, *Physicalism, or Something Near Enough* (New Jersey: Princeton University Press, 2005), and J. Searle, *The Rediscovery of the Mind* (Cambridge: MIT Press, 1992).

53. In logic, a definition has two parts, the *definiendum* and the *definiens*. The former is the term that is to be explained and the latter is the explanation.
54. Here the list can be expanded to other brain disorders and mental illnesses; see B. J. Gibb, *The Rough Guide to The Brain* (London: Rough Guides, 2012), chs. 8–10.
55. Bennett and Hacker, *Philosophical Foundations of Neuroscience*, 1.
56. See Gibb, *Rough Guide to the Brain*, 103–109.
57. See Gibb, *Rough Guide to the Brain*, 103–109.
58. Bennett and Hacker, *Philosophical Foundations of Neuroscience*, 1.
59. T. Williamson, *Doing Philosophy: From Common Curiosity to Logical Reasoning* (Oxford: Oxford University Press, 2018), 41.
60. Kim, *Philosophy of Mind*, 2nd ed., 83.
61. S. Psillos, *Causation & Explanation* (Montreal and Kingston: McGill-Queen's University Press, 2002), 19.
62. See, e.g., Helen Beebee, *Hume on Causation* (London: Routledge, 2006).
63. D. Gareth Jones, "Peering into People's Brains: Neuroscience's Intrusion into Our Inner Sanctum," *Perspectives on Science and Christian Faith* 62, no. 2 (June 2010): 125.
64. See, e.g., Bennett and Hacker, *Philosophical Foundations of Neuroscience*, ch. 3.
65. Bennett and Hacker, *Philosophical Foundations of Neuroscience*, 70.
66. Bennett and Hacker, *Philosophical Foundations of Neuroscience*, 70.
67. For details, see Bennett and Hacker, *Philosophical Foundations of Neuroscience*, ch. 3.
68. T. Nagel, *The View from Nowhere* (Oxford: Oxford University Press, 1986), and "What It Is Like to Be a Bat"; see also Moreland, *Consciousness and the Existence of God*, chs. 1–2.
69. Lowe, *Personal Agency*; see also Guta, "Frank Jackson's Location Problem."
70. For details, see Chalmers's books *The Character of Consciousness* and *The Conscious Mind*.
71. Denis Noble, "Foreword," in Bennett and Hacker, *Philosophical Foundations of Neuroscience*, xiv.
72. For details, see Bennett and Hacker, *Philosophical Foundations of Neuroscience*.
73. Reductionism comes in different forms both in science and in philosophy. Due to space limitations, I won't initiate a detailed discussion. For an excellent discussion on issues related to method and reductionism, see Bennett and Hacker, *Philosophical Foundations of Neuroscience*, part IV and appendices. See also Tuomas E. Tahko's recent *Unity of Science* (Cambridge: Cambridge University Press, 2021).
74. E. J. Lowe, *The Four-Category Ontology: A Metaphysical Foundation for Natural Science* (Oxford: Oxford University Press, 2006), ch. 1. See also M. P. Guta, "Metaphysics, Natural Science and Theological Claims: E. J. Lowe's Approach," *TheoLogica* 5, no. 2 (2021): 129–160, https://doi.org/10.14428/thl.v5i2.62723.
75. See, e.g., Bennett and Hacker, *Philosophical Foundations of Neuroscience*.
76. M. P. Guta, "Introduction," in *Consciousness and the Ontology of Properties*, ed. Guta, 1–12.
77. See, e.g., the revised second edition of Bennett's and Hacker's book, *Philosophical Foundations of Neuroscience* (Hoboken, NJ: Wiley-Blackwell, 2022).
78. See e.g., Frank Jackson, *From Metaphysics to Ethics: A Defense of Conceptual Analysis* (Oxford: Clarendon Press, 1998); D. Papineau, *Philosophical Naturalism* (Oxford: Blackwell, 1993); P. M. Churchland, *Matter and Consciousness*, rev. ed. (Massachusetts: MIT Press, 1988).
79. For more on this from Moreland, see *Consciousness and the Existence of God*, and also J. P. Moreland, "Substance Dualism: The Best Account of the Unity of Consciousness," in *Consciousness and the Ontology of Properties*, ed. Guta, 85–106. See also Lowe, *Personal Agency*.
80. J. P. Moreland, "The Fundamental Limitations of Cognitive Neuroscience for Stating and Solving the Ubiquitous Metaphysical Issues in Philosophy of Mind," *Philosophia Christi* 20, no. 1 (2018): 44.
81. The main claims presented in this chapter were presented at the Weekend Seminars I gave on the Philosophy of Neuroscience in 2016 and 2019 at Biola University. I thank my graduate students as

well as other audience members who participated in the weekend seminars and asked great questions that sharpened my thinking on the issues discussed in this chapter. I also presented the content of this chapter at the 75th Anniversary of the American Scientific Affiliation Conference on "Brain-Mind—Faith" which took place in 2016 at Azusa Pacific University. I thank the audience for the interactions I had. Finally, I thank the three anonymous reviewers for their excellent feedback.

11. In What Sense Is Consciousness a Property?

Mihretu P. Guta

This is an online-only chapter. You can access it at MindingtheBrain.org.

12. Subject Unity and Subject Consciousness

Joshua R. Farris

1. Introduction

IMAGINE FOR a moment that you experience a jolt to your brain upon accidentally touching a light socket while changing out a light bulb. Prior to this experience you encounter all the elements in your visual field of awareness, e.g., the ceiling, the ladder, the light coming out of the ceiling, and the light bulbs (both the one that has lost the capacity to give light and the new one that has that capacity). Upon experiencing the jolt to your brain, you also experience a temporary split in your experience.

When we consider this thought-experiment, we realize that there are several things going on. First, the experience of the shock is unified. But what does "unified" mean here? Philosophers talk a lot about different kinds of unity that exist in the world. There are certainly unities that are structural, like the unity of the parts of wood that are placed in a specified arrangement to make a chair, or the unity of the particles arranged in a crystalline structure, or the unity of atoms that is found in a molecule of H_2O. Consciousness, however, involves (or is) a different type of unity, a unique type of unity (and one connected with the nature of personhood) that we will look at in a bit more detail in what follows.

The nature of the unity found in the experience of the electric shock is characterized by some things that are not readily apparent in other types of unities in the world. First, it seems to presume a feature of our nature that does not map onto various garden-variety unities that we can directly investigate in a third-person way—a feature that we can't get at by observation through the five senses. Second, there is something in the rather jarring experience of an electric jolt which causes a split in perceptual experience of some of the data in one's purview. Third, there is the deeper metaphysical issue undergirding the phenomenological experience of these things, i.e.: What is it that is experiencing the felt quality of this unified experience? Is that which experiences the unified experience itself a unity?

In short, what I have just described is often attended by a common problem in philosophy and science and a quality or event for which it is difficult to provide an accounting. The problem is often called the "binding problem" of consciousness, which is related to this positive feature, event, or quality, namely the "unity of consciousness." As one might imagine, the present set of concerns has implications for what we take to be implicit in the nature and practice of science. Consciousness is central to the world and has several implications for what appears to be implicit in the world scientists study—namely its mental character. And this is important, as William Dembski has pointed out in his work *Intelligent Design: The Bridge between Science and Theology*, because mental events and mental actions provide an important bridge between scientific investigation and the metaphysics necessary to explaining it.[1]

In what follows, I explore the binding problem and its ancillary feature, the unity of consciousness. By considering the features characteristic of one's unified consciousness, we can better appreciate the relationship between the mind, as a subject, and the body. My thesis is rather simple. Going beyond the view espoused by the likes of Roderick Chisholm—the view that one particular version of the mind-body relationship, i.e., psychophysical or mind-body dualism, is the only option on offer—I argue that there are some positive features that stand out in favor of not only substance dualism, but a clear and transparent version of substance dualism, namely a variant of what I have elsewhere called neo-Cartesian substance dualism. Neo-Cartesian substance dualism is the view that I am a soul-body compound (i.e., substance dualism), and that what it is that makes me *me* is, strictly speaking, my soul (which I take as nearly synonymous with the terms mind, spirit, and immaterial substance), an immaterial substance relevantly distinguished from the material component (i.e., the body). It is this relationship that is phenomenologically implicit and metaphysically necessary in the unity of consciousness.

While some are reticent to furnish the connection between phenomenal unity and agent unity, I operate under the guiding principle that phenomenology (i.e., the study of the items of experience) is not only an experiential, epistemic lens (i.e., the appearances are real) but also has implications concerning the nature of selves. In this way, I am working in the introspective tradition of Franz Brentano, Roderick Chisholm, and J. P. Moreland, among others. The relevant sort of unity that reflects the unity of consciousness is something that is characterized by what it is like to be conscious, i.e., the feel of the experience itself. In the introspective tradition, there are additional features that seem to be present to the self, namely: the presentation selves have, the transparent ownership, and the implicit nature of the self, given self-presentation in consciousness.

The nature of the person as an immaterial substance is revealed uniquely in our consciousness. This is something that all dualists agree upon in contrast to materialists. There is something about the nature of consciousness that is uniquely captured in the immaterial substance. This is why materialism is insufficient as a basis for consciousness, according to dualists.

But there is another feature within consciousness that distinctively points us in the direction of Descartes's notion that the subject of that consciousness is simple, which means that it is indivisible. I argue in what follows for what I call neo-Cartesianism because it builds on Descartes's original idea, but where it moves past Descartes is significant. There exists a distinctive feature that we might call "particularity." Every individual person has a particularity that is indivisible and non-complex. This particularity, I believe, entails a form of creationism, in its original theological sense, i.e., the doctrine that souls are created by God directly.[2] I will not in this chapter argue directly for the origins of selves or souls as created entities, but it is a part of the package when we consider a neo-Cartesian view of selves as souls.

Consider that when we think of persons, it's not the way they look or speak, or the mannerisms they display, that we want. What we want is *them*. If that fundamental fact about them is simple and indivisible, then it is also not something complex, generalized, or a product of regular lawlike processes. In other words, it is not something that can be duplicated in the way that material bodies can (in theory) be duplicated. Instead, there is something about them that makes them *them*. This "something," I argue, is present in each individual consciousness, and it is neither duplicated nor generated. And the best explanation for this "something," I argue, is dualism.

However, not all versions of dualism can satisfactorily explain this "something." At least two versions of dualism fail do so. In one version, souls are generalizable and come about through a complex, regular, lawlike process (i.e., in the way that bodies of different types come into existence). In another version, the fact of the existence of the aforementioned "something" is unaccounted for, and no content of the "something" is provided. These dualisms, I will argue below (as I do elsewhere), contradict the experience each individual has of himself or herself. This means that they are inadequate theories of persons and their consciousness.

In a famous article written by philosopher Max Black, the author brings out an important insight into our consideration of certain "substances," namely persons. According to Black, it is conceivable that material objects could be perfect duplicates that have all the same properties. For example, take a cheeseburger. Once you describe all the properties (e.g., hamburger, bun, cheese) of the cheeseburger, you have, arguably, provided a sufficient description of it. There is no further relevant fact to be ascertained (incidental facts, such as the location of the cheeseburger, being irrelevant to its description). But, says Black, the difference when it comes to persons is significant. Properties that are, in principle, exemplified by all minds do not sufficiently describe individual personal minds. For example, the fact that I have private access to the contents of my own consciousness reveals that I am a substance that has a capacity that is not, in principle, accessible in the way that the properties of material objects are accessible by minds in general. Knowing the properties of minds in general is not sufficient to determine what is in the mind of any individual person. A sufficient description of personal consciousness (i.e., of the soul or mind of an individual) cannot be obtained by the methods used to obtain a sufficient description of hamburgers.

I will argue, from a systematic analysis of the unity of phenomenal consciousness, that we have a phenomenal unity that is not found in the natural world, a *soul*, and that this soul has a primitive, particular feature that makes me *me* and you *you*. This feature is not accommodated in all dualist options and it is certainly not accommodated in materialism with naturalistic processes.

In this chapter, I will explore these issues and the reasons favoring my brand of neo-Cartesianism. In section 2, I will consider the naturalist, or physicalist, competitors, and the challenge from the unity of consciousness, which will take us some way toward adequately describing a view of self, the subject of consciousness implicit in phenomenological unity. In section 3, I will consider some of the substance dualist options on offer, which will aid in clarifying why it is that neo-Cartesianism, or something near it, is phenomenologically given and metaphysically necessary.

2. Binding Problems, the Unity of Consciousness, and Naturalist Agents

2.1 Self-Presenting Properties in Phenomenal Unity

I begin with three points about the phenomenal unity of human consciousness.

First, I am aware of my phenomenal unity (i.e., field of consciousness), in light of the fact that I am a singular subject of that consciousness.

Second, the phenomenal field is neither (a) a set of parts transmitted or a structural part, nor (b) a holistic entity that stands on top of the collected bits. Point (a) rules out the body, brain, and central nervous system; point (b) rules out the whole animal.

Third, and following from the first two points, the phenomenal field is (or presumes) a transparent subject, a *soul*, as a *sui generis* (novel) substance.

In what follows, I argue from the broad phenomenological unity of consciousness against naturalism. By "naturalism" I mean the view that asserts that all things, including human beings, are explained from the bottom up, beginning with low-level physical particles that compose larger mereological aggregates that are themselves purely material in nature. Another name used by philosophers for this view is physicalism or materialism. I believe that no purely material view of consciousness is capable of supplying solutions to the binding problem(s) of neural items in phenomenal experience. In fact, there exists a phenomenal unity of what it is like to experience the world, a unity that requires something of a peculiar substance (i.e., an immaterial substance that is characteristically distinct from material things or events).

When we consider the binding problem, we are concerned with how varying and disparate physical items can be unified in one field of awareness. Considering this topic will take us into three different aspects of the unity of consciousness. The first aspect is what philosophers call objectual phenomenal unity. The second aspect is often called the subject of phenomenal unity. The third aspect is that which underlies and unifies different states, parts, or features in one felt experience. These three aspects are related when considering the nature of consciousness.

The first aspect, objectual phenomenal unity, posits the existence of an object that experiences features together as one thing. Consider the analogy above where we see all the features of experience held together by the person, e.g., the features of the light bulb, the ladder, and the light glowing. The second aspect, subjective phenomenal unity, is related, but highlights the subjective aspect of perception by a singular subject—namely the perceptual state that I have of the various features in one unified field of awareness. I have these experiences of the ladder, the light bulb, and the glow from the light bulb projecting. The third aspect, which brings the first two together, is the felt quality of standing simultaneous with the turning of the light bulb in the socket and the appearance of light when the bulb connects with a power source.

These three ways of articulating the unity of consciousness help us to see the inter-relationship between the nature of the subject of experience as a whole substance and the phenomenal nature of consciousness. Bringing these together provides a context for deeper reflection on the metaphysics necessary to consciousness. Some*thing* is required that unifies the bits of experience for which naturalism has no options. Let us consider the basic problem for naturalism, or material agents.

2.2 The Basic Problem for Naturalism

As was stated above, naturalism understands everything as a mereological hierarchy (a complex arrangement of atomistic parts). But as we have seen, if our phenomenal experiences grant us knowledge of the nature of consciousness, then consciousness is not atomistic. In other words, it is not built from the ground up. It is not a mereological hierarchy of parts and there exists no mechanism that is requisite for consciousness.

Let's consider a phenomenal unity of consciousness argument advanced by William Hasker:

1. I am aware of my field of consciousness and the components of it as a singular subject all at once.
2. A functionally integrated whole system can experience unity, but a set of parts cannot.
3. Therefore, my conscious items are an integrated whole rather than a set of parts (i.e., atomism is false).
4. The brain, nervous system, and body are not more than a set of physical parts arranged (i.e., holism is false).
5. Therefore, the brain, nervous system, and body lack the capacity to function as a whole; it must function as a system of parts.
6. Hence, the subject, the self, is not one of these physical things (or the parts composing them).
7. The subject is something relevantly different like a soul, an immaterial substance, a mental substance.
8. Therefore, the subject is a soul, or contains the soul as a part.[3]

Going back to our original analogy, it is easy to illustrate the features described in this version of the unity of consciousness argument. There are different facts that need some explanation. Consider the following features: I experience the light bulb surrounded by the other material parts of the base, the ceiling, the circuitry, the holder, many of which are necessary to the event of a bulb lighting, as well as the ladder for which I am standing. All of the facts about the physical objects could exist separately and they could exist in some proximal relation to the other, but that would fail to explain the facts we have just described. The physical objects themselves are not identical to the subject that unifies them, the subject that perceives them together in one event, and the subject that experiences what it is like to be a perceiver of that event.

The phrase "functionally integrated whole system" is quite a mouthful, so let's unpack that briefly before considering it in more detail below. What Hasker means is that there exists an inseparable part that is a phenomenal unity of all the parts in the neurological system. These parts function together as a whole, not as discrete parts that are loosely united. Instead, there exists a novel unity that is unlike the parts that make up the physical system. This is important, because the argument shows us that what is going on in physical systems alone is not what we see going on in consciousness. Something else is required, which he takes to be a soul. It is clear that the various parts or objects relate to one another in one's united field perception along with the additional feature of what it is like to experience all of them together in one's unified field.

These features of the unity of consciousness exist in a striking contrast to the sort of unity found in merely physical objects. It is here that materialism (which, as I explained above, I will use as a synonym for naturalism) cannot account for consciousness. Materialists lack the resources in the natural world to explain consciousness. We have different types of unities that we can point to in the natural world, and we can observe and even examine the molecular structure underlying them, but the nature of consciousness cannot be explained in terms of such things. To take the most obvious example, the type of unity describing conscious experience is privately available to the subject and exhibits a felt character—something that is not empirically detectable from an objective frame, i.e., a third-person point of view. The two explanations for unity are radically different in nature.

Let us try to gain a clearer picture of why it is that materialists, as opposed to dualists (who accept the notion of "soul"), lack the resources to explain the unity of consciousness. There exist two prominent options employed by materialists when advancing an account of phenomenal unity: atomism and holism. But both are problematic when considering phenomenal unity. Atomism, as we described it earlier, is the view that consciousness is built from the ground up, from bits of matter as the building blocks of one large unified structure. Holism affirms that consciousness is not merely the additive product of these external (and separable) parts, but a new feature resulting from the specific arrangement of these parts into a novel unity. Let's now compare these two options.

The basic idea advanced by holism is that the sum of the parts is greater than the mere total of the parts. An example would be a tornado. Though its parts (particles of air) come together due to the purely natural power of convection, the power exhibited by the tornado, once formed, is something new, beyond what would be predicted merely from looking at its constituent parts. Atomism is like a mud pie arising from particles of dirt and moisture that are swirling around and achieve an accidental unity. The parts of the mud pie are separable, and though those parts have powers, the mud pie is really not an integrated whole but merely a conglomeration of those parts with their separate powers. Holism is slightly better off, because it can show that the structural unity illustrated in a tornado is more than the sum of its parts, yet it is still the sum of its parts arranged in a specified way.

There are obvious problems for naturalist contenders. First, how could the mechanisms we have just described (atomist or holist) give rise to the unity descriptive of consciousness? Second, how could the experience of *qualia* (i.e., the subjective properties or features of experience) arise? What it is like to be conscious is important as a phenomenological launching point for considering the relation between consciousness unity and agent unity. There is something about what it is like to be in the state of experiencing the phenomenal unity that unveils what it is to be me.

This phenomenal unity, upon reflection, reveals something altogether different from the parts in the physical world, e.g., the parts that are atomistically related in the light bulb event or related in a holistic way in the case of a tornado. That the properties of conscious experience are wholly different from those of material complexes is clear. This poses a problem for materialism.

There is a more sophisticated version of materialism that attempts to stave off the objections from the unity of consciousness in Timothy O'Connor's emergent individualism. On his view there exists a novel feature of consciousness that he has argued does not fall prey to the challenges consuming standard versions of materialism. O'Connor describes his view as one that involves a novel subject that emerges from the parts, but is not identical to the parts. Advancing this thesis helps materialists because it recognizes the novelty of consciousness as a unique feature of persons that is unlike other material unities, but the question remains whether, on materialist grounds, nature has the resources to generate the distinctive features of consciousness. On close examination, it appears that O'Connor's view fails to account for these features, as I will show in the next section.

2.3 Likeness, Phenomenal Unity, and Materialism

Timothy O'Connor has defended a version of materialism that takes it that human persons are biological objects that emerge at some complex level of biological development. It is important to point out that O'Connor's materialism is a sophisticated kind of "non-reductive materialism" (or "non-reductive physicalism"). For non-reductive materialists, the human person is not identical to the underlying parts that compose the human body and brain, and they reject a reductionist view of the human person that explains the person in terms of such parts. Further, such materialists reject functionalism as insufficient for explaining the person, because it is not the lower parts that realize some new higher-order state in which we have consciousness. They agree that there is something about the nature of consciousness as a unity that defies reductive explanation to the underlying parts interacting. In one important place O'Connor develops his view with Jonathan Jacobs. They define their view as follows:

> I am indeed a biological organism, but some of my mental states are instantiations of simple or non-structural properties. A property is "non-structural" if and only if its instantiation does not even partly consist in the instantiation of a plurality of more basic properties by the entity or its parts.... Emergent features are as basic as electric charge now appears to be, just more restricted in the circumstances of their manifestation. Further, having such emergent states is, in general, a causal consequence of having the requisite type of intrinsic and functional complexity. The emergent state is a "causal consequence" of the object's having this complexity in the following way: in addition to having local influence in a manner familiar from physical theories, fundamental particles and systems also naturally tend (in any context) towards the generation of the emergent state. Their doing so, however, is not detectable in contexts lacking the requisite macro-complexity, because each such tending is, on its own, incomplete. It takes the right threshold of complexity for tendings, present in each micro-particle, to achieve their characteristic effect jointly, the generation of a special type of holistic state.[4]

Let's unpack this briefly to understand how it overlaps with other physicalist views and how it differs from Hasker's brand of emergentism (emergent dualism). In contrast to the dualism of Hasker and other dualists (particularly substance dualists), O'Connor and Jacobs affirm a version of monism (i.e., the view that we, human persons, are one kind of thing, not two kinds of things) with a distinct higher-order set of properties non-reducible to the lower-level parts that give rise to them at some level of biological complexity. While there is some debate as to whether this is strictly speaking a version of materialism or physicalism, the authors are clear that human persons are material, biological organisms, rather than souls. We are bodies, even if sophisticated ones, that are not rooted in some underlying ontology distinct from the material. What is difficult to ascertain on the O'Connor-Jacobs view is what this vague "person" is that is not strictly speaking identical to the parts underlying it or the interaction of those parts.

Mental states, or consciousness properties, are non-structured parts or properties that emerge from the underlying material structure, yet non-identical to the structure. They are *sui generis* properties not predictable from that which underlies them. How can O'Connor and Jacobs explain this? Apparently, the parts that give rise to mental properties have "tendencies," although this is controversial and undetected prior to the emergence of these mental properties. Consciousness, i.e., mental properties, are the "causal consequent" of the right sort of material complexity at some level within the mereological hierarchy. Further, and it is important to highlight this in contrast to Hasker's version of emergentism, O'Connor and Jacobs state that there is a "threshold of complexity" that is "present in each micro-particle" and amounts to "the generation of a special type of holistic state."

There is here an overlap with different versions of materialism, in that the states exist in the local micro-parts as a unique holistic state. It is here that Hasker is at pains to point out why this is problematic given one important feature in the unity of consciousness. It lacks a feature that characterizes the intrinsic unity of the subject (as a phenomenal field) and requires what Hasker says is a distinct type of "thisness" that unifies the phenomenal features in the field of awareness. For this reason, Hasker advocates for a version of emergent dualism rather than the O'Connor-Jacobs emergent individualism.

William Hasker defends a version of emergentism similar in many ways to the emergent individualism just described. However, he is clear that what is required is a novel substance and not simply a feature, property, or power. For Hasker, the fact of first-person phenomenal consciousness and libertarian free will depend for their instantiation on a novel substance not confused with the material substance or the material parts that underlie it and are necessary for the logical consequence of it. While Hasker defends what he understands to be a version of substance dualism, it is not a version of Cartesian substance dualism (or is only a very deviant one at best), for Cartesian substance dualism denies an essential or internal relation between the body and the soul, which depends for its origins on Divine creationism because there is no necessary relation between the body and the soul. Rather, for Hasker there is an essential relation between the two substances because the soul is the logical consequent of a sufficiently complex brain and central nervous system. His view can be summarized as follows:

a. Persons are novel substances (i.e., basic mental substances).[5]
b. Persons have novel powers and properties distinct from their bodies.
c. Material has endowed potency.
d. Material, in certain arrangements, has designed ends.
e. Persons are non-reducible to and non-constituted by their brains or bodies and the interaction of the material parts.
f. Persons originate from their bodies and are sustained by them.[6]

From this, it is clear that it is distinct from emergent individualism in that the novel substance is not a novel feature that somehow exists in the parts conjoined as some high-level, unique holistic state. It is on this point that one must press the materialist, in general, and the emergent individualist, in particular (with its emphasis on the conscious properties existing locally as a holistic state).

Hasker grants that O'Connor's emergent individualism offers the most satisfying materialist account of consciousness, but finds that the account falls short. He presents his critique in his recent article, "Do My Quarks Enjoy

Beethoven?" His answer is that, no, they do not. Hasker answers in this way because he has good reason to believe that, given the nature of materialism, we cannot locate the experience in the material bits. In other words, the mental substance (i.e., a property-bearer, a particular that is the bearer of consciousness) exhibits a novel property that is not metaphysically permissible on materialism.

Hasker lays out the reasons why materialists cannot account for the unity of consciousness: consciousness lacks the character of aggregates, and consciousness requires non-composed substances. Hasker advances his critique based on the quality of phenomenal consciousness, i.e., what it is like to experience consciousness. What is it like to experience a phenomenal unity and why is it the case that material aggregates (conceived either atomistically or holistically) cannot or do not make sense of the unity? Hasker explains the problem in what follows:

> Now, we are supposing, unlike Brentano in this passage, that the whole of the conscious experience exists at each point not instead of, but rather along with, the data-processing function carried on by the different parts of the brain in all their marvelous complexity. It will nevertheless still be true that there is this "quite immeasurable complication" of subjective, conscious experience at each and every point in the brain, in spite of the fact that the various points do not receive the information from the different processing units of the brain through any known causal process, nor is there any physical process occurring at those points that would correspond to the complexity of the conscious experience. At this point I merely ask the reader to consider: does this not strain one's credulity to the breaking point, and beyond?[7]

I use this quote from Hasker because it reveals what he takes to be problematic in the emergent-individualism, but below I will show that what he needs is something akin to Brentano, which he denies. Hasker makes clear that the whole of conscious experience, on his view, exists along with the data-processing points found in the brain, but that is distinct from two claims that must be the case on the materialist version advocated by the likes of O'Connor. First, it is not the case, as it is in materialism, that the conscious experience exists in each bit of neural matter. Second, it is not the case, as it is in materialism, that the conscious experience is measurable. While the firing of neurons is measurable, the nature of conscious experience is not. What this means is that consciousness is not identical either to the parts (hence not atomism) or to the material whole (hence not holism). The novelty of consciousness requires a different explanation.

Support for Hasker's position may be found in the following passage from J. P. Moreland:

> Accordingly, a physicalist may claim that such a unified awareness of the entire room by means of one's visual field consists in a number of different physical parts of the brain each terminating a different wavelength, each of which is aware only of part–not the whole–of the complex fact (the entire room). But this cannot account for the single, unitary awareness of the entire visual field. There is a what-it-is-like to have the whole visual field. If we terminate our search for an explanation for this with a holistic phenomenal field, then two problems arise. First, it is hard to see how a myriad of atomistic parts

> could give rise to a single, nonatomistic, holistic field; we are owed an account of this within the constraints of subject physicalism.
>
> Second, a basic datum of our experience is not simply this or that item of awareness of the room, but that I *have* and *am not identical to* the totalizing state. In the history of philosophy, classical substances have served to unify things in this way, and Hasker and Erik LaRock believe this ontology provides the best answer for how we could have a totalizing, unified field of consciousness. The very same substantial soul is aware of the desk to the left, the podium at the center, and, indeed, each and every distinguishable aspect of the room. But no single part of the brain correspondingly activated as a terminus for the entire visual field. Only a single, uncomposed mental substance can adequately account for the unity of one's visual field, or, indeed, the unity of consciousness in general.[8]

The foregoing discussion should make it clear that the nature of O'Connor's emergent individual is insufficient to describe subjective conscious experience. For O'Connor, and any materialist metaphysic, what is required is that there exists a property-instantiation of a physical whole, i.e., that which is composed of necessary and inseparable parts. However, Hasker presses the point that this is not possible on O'Connor's view, despite O'Connor's belief that there exists some novel feature of the whole system. Hasker employs what he calls the LPI (i.e., the local property instantiation) principle to show that all materialist alternatives are lacking. LPI is the principle that there is no individual experience of a thing that exists in the underlying parts. In other words, the experiences are not properties that are predicable of the material bits. And herein lies the problem for materialism. Experience would need to exist in the ordered arrangement of the parts to serve as an explanation of subjective conscious experience. The problem is—it doesn't.

So for Hasker, experiences of consciousness are not existent in the material composite or the parts composing it. Instead, they exist elsewhere. For this reason, Hasker puts forward substance dualism, and his preferred version of it, namely emergent dualism. This is the view that the mind is non-reducible to the parts or the parts combined and is a *sui generis* substance—a thisness, i.e., a phenomenal unity (of which self-consciousness and free will are the basic powers).

O'Connor has responded by claiming that the problem raised is a causal one, but as we have shown above, the problem is less about causality than about the location of experiences. Hasker responds, rightly in my understanding: "The problem does not lie in the pushes and pulls but rather in the complexity of the machine, the fact that it is made up of many distinct parts, coupled with the fact that a complex state of consciousness cannot exist distributed among the parts of a complex object."[9] Hasker elsewhere repeats that the problem concerns where the experiences occur.[10]

This points us in another direction regarding the subject necessary for the unity of consciousness. By taking seriously our introspective awareness (within the introspective tradition as described earlier), we discern something about consciousness that is not present on Hasker's view. We find something that is distinct and requires an object distinct from Hasker's phenomenal emergence. When I survey the contents of my own mind regarding the situation of changing out the light bulb, the transparency of my mental items becomes clear. All the elements within my purview are present to me in a way that I can direct my attention to them and reflect on them internally. This transparency can be extended to cover the nature of the subject of experience. When we consider the items of our conscious experience, the "I" is necessarily present in a way that points us beyond a phenomenal conscious unity to an inseparable entity that owns these experiences.

But more needs to be said about this substance that Hasker only partially handles. Let's consider what it is that needs explaining concerning the substance.

2.4 The Base for Unity of Consciousness

Consider the following argument, distilled from the writings of E. J. Lowe:

1. Phenomenal properties inhere in something, and are unified by something.
2. Phenomenal properties are explicable and clear.
3. Neither concatenation of perceptions nor phenomenal qualia are an adequate basis for the self's unity.
4. Therefore, the subject or agent of that consciousness is probably a metaphysically simple substance.

For Lowe, phenomenal unity, as described by Hasker, does not adequately describe the subject of conscious experience. Phenomenal unity points us to the metaphysical fact of a substance as the base for the properties of conscious experience.

There are two inter-related arguments advanced by Lowe in favor of substance dualism. The first is based on the ownership of ideas, experiences, and qualia; the second is based on the individuation of the conscious items (i.e., features internal to the conscious perspective of subjects that may or may not represent external objects).

2.4.1 Ownership Argument

The ownership argument was implicit in what was said above. On the view of properties already mentioned, properties themselves are instantiated in a thing, part, or substance. Properties are exemplified by many things (e.g., the color blue and the experience of blue exist in multiple things). A different kind of property is one that is dependent on descriptive properties of wholes (like a rich property, which is a complex property descriptive of a thing). Thoughts, experience-items, or qualia are properties, and they all presume an ownership by a thing. That thing is a subject of experience or a subject of thoughts. The experience of the sky is an experience a subject has of something (namely the sky), and that experience implies intentionality of the subject, i.e., directedness of the subject toward something (in this case the sky). But for a subject to have intentionality or directedness to an external object (like a sky) it must have ownership of the experience and of the relation to that object (in this case the sky).

Intuitively, you cannot have a thought without a thinker thinking the thought. And thoughts are not free-floating things; they are properties that are communicated by persons (i.e., subjects of experience) and belong to persons. When there is a thought, it is had by a subject of experience that can think about it, communicate it, and connect it intellectually to other thoughts in a unified way. Thoughts are therefore always me-thoughts or I-thoughts. I own my own thoughts as you own your thoughts. This notion of ownership is related to the notion of individuation, which will be discussed next.

2.4.2 Individuation Argument

Thoughts are individuated by something. When a person conveys his or her experience of the blue sky, there is a specified grammatical logic that is implicit and necessary for identifying the thought about the blue sky. The blue-sky experience, as we can call it, is had by a thing that is experiencing the blue sky. The first-person perspective (FPP), which is one's conscious perspective or angle, is a necessary condition to the experience, and the "I" is the thing that binds or unifies the elements as one field. The self postulated by materialism just won't do as the thing that owns or individuates phenomenal consciousness, because that self is composed of underlying separable parts and lacks the unity descriptive of a distinct substance. Further, the self postulated by "obscure dualisms" won't do, either, because they lack a clear and transparent agent for which we can predicate the properties and parts, both separable and inseparable, of a subject of *that* conscious experience. The most obvious example of an "obscure dualism" is Timothy O'Connor's emergent individualism, which confuses the nature of material unities with the nature of a novel substance. We will address these "obscure dualisms," using the example of emergent dualism, in more depth below, when we consider the nature of the self or subject of consciousness in finer detail.

Taking introspection seriously means that when we consider the implications of ownership and of individuation, we do so from the phenomena themselves. Following the introspective tradition means taking seriously what some (like Roderick Chisholm) call self-presenting properties. Self-presenting properties are those properties that are apparent or transparent to the subject of that consciousness. They are those properties of the "I" that are accessible to someone in his or her immediate conscious experience. They attend every thought and experience had by persons. So, going back to our analogy of the light bulb, when I experience the whole event of changing out a light bulb, it is the "I" that is present to the whole experience. It is not a mere brute phenomenal experience, but an experience *I* have. From this we can see that ownership and individuation are two of the self-presenting properties.

2.5 Summarizing the Argument to This Point

I will now summarize the argument against materialism. There are two ways of making sense of material unities of conscious experience. One is atomism and the second is holism. I will consider each in turn.

Atomism is the view that what is experienceable in consciousness is owned by each one of the physical parts of the organism (or the brain, or the neural system), parts that constitute the larger whole. Yet there is no obvious way to make sense of the physical or neural parts as owning the experiences. If they could own experiences, then, since they are separable and could exist quite apart from the human body, fragments of consciousness could exist everywhere, which does not seem plausible. Further, the notion of adding together fragmentary experiences, like adding blocks to a building, fails to describe the basic unity of experience as something prior to the parts that the subject of experience has experience about.

Then there is the option of holism. Holism takes it that conscious experience is the structural product of the underlying parts—a kind of add-on to the whole system in question. However, this inaccurately portrays consciousness. In consciousness there is a unified field of awareness. And as we have seen, this unified field of awareness appears to be owned by a *something*, a something that experiences what it is like to have that unified field of awareness about something. Experiential qualia are not something built up by a parts-to-whole relation; instead, what we have in conscious experience is an apprehension of a whole with all its parts.

From these objections to atomism and holism, we can see the outline of an argument for an alternative to those positions, namely, substance dualism. Immediately, the following reasons present themselves for substance dualism. First, the soul or mind, as an immaterial substance, provides an explanation for the binding problem (i.e., a subject of experience is a spiritual substance as defined above). Second, it is the commonsense view that furnishes a necessary logical framework for the unity of consciousness. Third, it explains the fact of ownership. Fourth, it explains the fact of individuation. Fifth, it accounts for objectual phenomenal unity, subject phenomenal unity, and subsumptive phenomenal unity.

However, there are many kinds of substance dualism: Hasker's emergent dualism, Lowe's non-Cartesian emergent dualism, Thomistic dualism, and Cartesianism (or neo-Cartesian) dualism. Which of these dualisms best accounts for what we can learn from introspection? I argue that neo-Cartesianism is the best option on offer. Going back to our original illustration, and asking, "What is it that is experiencing that jolt from the light socket?," I will argue (using the lens of introspection of our phenomenal experiences) that the subject of the experience is a *substance*, a unified, simple substance, the nature of which is clear, and I will thus rule out a variety of substance dualisms that obscure the nature of the subject.

3. Arguing for Substance Dualism and the Transparent Metaphysical Primitive "I"

In what follows, I continue the argument for substance dualism as the best explanation for a subject's conscious experience of the world. But it is not enough to accept substance dualism, as I will show. For when we consider the self-conscious character of the substance that our introspection reveals, we can affirm a transparent subject of consciousness[11] that is not readily affirmable on all substance dualism views. Under the next two headings, I will set forth the arguments for this conclusion.

3.1 Phenomenal Unity, Divisibility, and the First-Person Perspective (i.e., Self-consciousness)

1. Empirical evidence for the divisibility of perception does not support the divisibility of self-consciousness (or of its basic properties).
2. Phenomenal unity is divisible in some special cases (e.g., in the power of mental perception), but self-consciousness (along with its basic powers) is not.
3. Therefore, a phenomenal unity, by itself, does not adequately account for or capture self-consciousness (the deep problem for Humeans).
4. Therefore, Hasker's phenomenal unity view, though valid as far as it goes, is an inadequate account of self-consciousness; something more is required (i.e., following Kant, a transcendental unity of apperception requires a transcendent unified substantial self).

Ironically, "obscure dualisms" (such as Hasker's) share with materialism an inability to account for the transparent self of consciousness present in the unity of phenomenal consciousness, and this inability renders them unacceptable. I will in what follows explicitly advance an argument in favor of my neo-Cartesian dualism, which affirms a transparent subject of consciousness that is a fitting owner and individuator of the thoughts in one's unified field of phenomenal awareness.

Let us return to Hasker's "emergent self" view to see why it is an inadequate basis for phenomenal consciousness, after which we can move on to consider some sturdier, more promising substance dualist options.

Hasker accepts that my being shocked by the light bulb might mean, not only that I have a split perception or split phenomenal field, but also that I experience a split self. This follows quite intuitively from his view that selves are emergent from a sufficient parcel of matter; if there is a splitting, and if two parcels can sustain a self, then it makes sense that there could ensue two selves. However, this interpretation departs radically from the traditional assumption of many dualists, originally codified in Descartes's understanding of the simple self (as soul), when he says: "I can't conceive of half a soul."[12] Hasker thinks that emergent selves comport better with recent scientific findings (regarding split perception) than does Descartes's original dictum.

Even if this is so, Hasker still encounters the problem of the inexplicability of the self. He defines the self as the "center of consciousness," the basic power that unites the bits in one's perceptual field. This is all well and good as far as it goes. Most dualists agree with this. Yet this is where his view appears to be subject (no pun intended) to the problem with Humeanism. The self, as an immaterial subject of consciousness, is not a bundle of various parts, but a unified substance that does the work of binding the bits together. When I am aware of the items in my purview of conscious experience, I am aware of those items along with the self or "I" to which they are attached. This "I" is one that *I* am aware of upon introspection. I am not an elusive "I" that is missed upon reflection, but the "I" that is present to them to whom I properly ascribe ownership. And he concedes this point in multiple places.[13] Yet when pressed about the individual essence of the person, he insists that there is no primitive essence of persons. But his insistence is not logical, for if the "sense of self" presupposes an objective ground for the metaphysically simple self, then this favors not emergent selves (as described by Hasker) but Cartesian selves.[14]

We have reasons for preferring a more refined version of metaphysical simplicity. E. J. Lowe has helpfully stated:

> [I]t is strongly arguable that the only adequate criterion of identity for mental states and events will be one which makes reference to their subjects... [P]art of what makes an experience of mine numerically distinct from a qualitatively indistinguishable experience of yours is the very fact that it is mine as opposed to yours.[15]

Lowe advances an emergent view that originates from the underlying material base—a view much like that of Hasker. Lowe, however, is convinced of the existence of a novel substance not by phenomenal unity, but by the basicality of thoughts that find their ownership and individuation in a basic substance that is the proper referent. Lowe is clear that thoughts cannot find their proper referent in body parts or the body as a whole. There must be a novel substance which emerges, and coordinates with, and is functionally dependent on, some parcel of matter. But rather than concede to Cartesianism, with implicit indivisibility and potential disembodiment, he is inclined to affirm a distinct emergentist account.[16] What I have shown elsewhere is that this ground and criterion for identity reflected in phenomenal consciousness is a metaphysically primitive mind or "I," of which we are aware upon analysis of the phenomenological contents of our consciousness.

There are other metaphysical reasons preferring the metaphysically primitive "I." Elsewhere, I wrote:

> Consider the fact of one's qualitative experience of tasting coriander. It is

> true there are certain physical facts about coriander that have something to do with how coriander tastes, but there remains an unaccounted fact when I taste coriander quite apart from Hasker tasting coriander. There is a fact of tasting coriander that is distinct from the fact of Hasker tasting coriander that physical facts, properties, could not explain. Even if you have two qualitatively identical experiences of coriander, they are numerically distinct because of the subject of experience and the fact that experiences have different subjects.[17]

In addition to the metaphysical problems just discussed, Hasker's view faces a scientific problem—in particular, an empirical challenge from commissurotomy cases. Hasker in multiple places affirms the possibility of emergent selves from material manipulation, which is confirmed by the possibility of a dissolution of the self upon the dissolution of the body. This points, then, not to a primitive simplicity (i.e., indivisibility) but a simplicity of the self that is coordinate with a complexity of experiential items. Unfortunately for Hasker, the neuroscientific data from commissurotomy cases lends more support to a basic kind of simplicity, where the self and its basic powers (e.g., self-consciousness, volition) cannot be split, and therefore cannot emerge by stages. Empirical neuroscience thus presents an empirical case for primitive simplicity. In what follows, I will present Hasker's interpretation of commissurotomy data, and show the defects of that interpretation.

Hasker believes that the evidence from commissurotomy cases (as well as from multiple personality cases) provides a significant challenge to Cartesians and favors his "emergent dualist" view precisely because it, unlike Cartesianism, can make sense of the fact that tinkering with the brain can cause the emergence of two centers of consciousness from one. He cites two test cases that favor his view, from the work of Roger Sperry and W. J. Wilkes, where the patients have functional differences between their hands (based on the distinct hemispheres).

He summarizes his findings as follows:

> In each case, there is a conflict between the two cerebral hemispheres, each apparently operating on a different conception of how the assigned task is to be accomplished. Furthermore, a strong impression is created that we have here two *centers of consciousness*, each seeking to pursue its own agenda. This conclusion is not irresistible; it could be that in each case one of the hemispheres is not conscious but is instead proceeding "automatically," as one may perform many familiar actions without conscious attention. However, this does not seem at all a natural reading of the situation. The tasks involved are not familiar, routinized procedures like brushing one's teeth or walking along a familiar route. Rather, they involve novel, interesting tasks that receive their point precisely from the special instructions given by the experimenter. The most plausible reading of the situation, surely, is that both hemispheres are somehow conscious, and each is attempting to perform the assigned task in its own way. I submit that any theory about the mind that forces one to deny this incurs a significant empirical burden, by forcing one to reject the most plausible way of understanding the observed facts in cases such as these.[18]

On his interpretation, then, the data supports something like his emergent dualism, with a greater elasticity regarding the center of consciousness than what we find in Cartesianism with its indivisible self.

Hasker provides further support for his emergent dualism from the fact of multiple personality disorder. Somewhat speculatively, he argues that while the science of multiple personality disorder is not settled and is immensely controversial, the findings of commissurotomy cases, together with multiple personality cases, *suggestively* rule out Cartesianism and favor his emergent dualism,[19] which lacks the metaphysical simplicity of the indivisibility of self.

It is not clear that Hasker has provided a compelling argument. Regarding commissurotomy cases commonly cited in the literature, the data from which he argues could be accounted for along the lines of perception, or from competing phenomenal fields (e.g., schizophrenia cases, or cases in which one is disposed to suppress different personalities), which should be distinguished from other facets and powers of consciousness. Regarding cases of multiple personality disorder, this data too could be accommodated by an interpretation along the lines of habitual conditioning of the self to occupy different personas[20] through suppression of other personas (e.g., as when someone takes on one personality at work and a different one at home). Furthermore, the functional integration of environment, neural functioning, and the soul could have so many layers of explanation that it permits the possibility of a deeply subjective experience of different modes of conscious perceptual experience of the self in different environments. In other words, the data on multiple personality disorders does not warrant rejecting Cartesianism.

Neurosurgeon Michael Egnor takes a different line, and confirms my suspicion that one should distinguish between "perception" and perceptual powers from consciousness and other more basic powers of the person. Egnor summarizes his argument in the following:

> This understanding of the mind that emerges from studies of split-brain patients—that some powers of mind such as perception and movement can be split, while others, such as unitary sense of self, reason, intellect, and will cannot—is clearly inconsistent with *materialism*. If mind is caused by brain, without remainder, a materialist would naturally expect commissurotomy to split the mind as well as brain, and indeed the claim for split consciousness as a consequence of commissurotomy has been a dominant theme in the neuroscience literature of commissurotomy. Yet several more recent studies of split-brain patients make it clear that the most surprising outcome of commissurotomy is *how little the mind is split when the brain is split*. Dividing the brain has no impact on the sense of self or abstract reasoning. Only *perception* splits, and even that is difficult to discern without meticulous testing. The patient's daily experience, his sense of self, his exercise of powers of reason and will, remain unitary, despite disconnection of the two halves or even several lobes of his brain. Materialism is inconsistent with unity of mind in split-brain patients.[21]

Michael Egnor agrees with other dualists that unity of consciousness yields dualism of some sort, and he agrees that commissurotomy data supports such a view against materialism. He does not agree with Hasker's interpretation of the data, and he cites several reports and studies that show or suggest the indivisibility of the self.

The reports of commissurotomy, Egnor believes, reflect a divided perception or phenomenal field with a unified agent persisting. In fact, through studies of commissurotomy and hemispherectomy the conclusion is confirmed that there is a transcendent, unified, conscious self that persists even when perceptual fields are divided as they correspond to localized operations of parts of the brain.[22] Egnor's interpretation fits quite well with Cartesianism, where the self is a bounded unity and one's consciousness about self and the basic powers of self are left unscathed.

In addition to drawing the conclusion that selves exist with a deeper unity than that allowed by Hasker, Egnor draws another conclusion that is consistent with the phenomenology of the metaphysically primitive self. He argues:

> A plausible inference that can be drawn from this dichotomy is that intellectual powers of the mind are not related to the brain in the same way perception and locomotion are related. An inference that intellectual powers of mind are in some sense immaterial, as compared with the materially caused powers of perception and locomotion, is not unreasonable here.[23]

The unity of the basic "intellectual powers" lends itself to a view of the self not as tightly bound to the brain as in Hasker's view.[24]

In Egnor's account, Hasker's "emergent self" view encounters a significant challenge. On Hasker's view, one would expect the possibility that the self, as a center of consciousness, could in fact split or become two emergent selves (i.e., could undergo fission). But the experimental data supports the distinction between the metaphysically simple self, the sense of self, and the basic powers of minds or souls in contrast to other lower powers of perception. One may split and the other cannot.

There is a final point to be made against Hasker. Following Kant, it seems right that for there to be a transcendental unity of apperception (as Kant puts it[25]), there must be a transcendent unified substantial self, and it's just not clear that Hasker can offer that, at least not as the self presents to one's own self-consciousness. For a metaphysically unified self that is present and prior (at least logically) to the bits in a unified field, we must look elsewhere. So much for Hasker's emergent dualism. Let us consider some of the other options in the dualist literature.

3.2 Phenomenal Individuation and the View of E. J. Lowe

It is here that we need to consider the contents of phenomenal experiences more carefully. When we do this, we find that some versions of dualism obscure the nature of the "I" present to one's phenomenal experience. Here is a formal argument to that effect.

1. Phenomenal consciousness is both individuated and owned by something (i.e., the agent or subject unity) that is clear and accessible by introspection.
2. The agent or subject unity found in "obscure dualisms" (i.e., dualisms that undermine the transparency thesis and confuse or render ambiguous the relationship between bodily and soulish substances) is not clear and accessible by introspection.
3. Therefore, "obscure dualisms" won't do to explain the ownership and individuality of phenomenal consciousness.

In what follows, I will discuss the above conclusion in reference to the position of E. J. Lowe.

I will begin by presenting what I have elsewhere developed as the "perfect duplicate" problem.[26] The problem, briefly summarized, is this: A self or soul could not emerge from underlying material bits in a regular lawlike process. If that were the case, all selves, or souls, provided they arose from the same sorts of material, could (in a probabilistic sense) be identical. But they are not; the fact is that *I* exist, and it is this "I" and not another "I" that is present to me whenever I encounter the world.

Here we have what I will call a "particularity problem," and it is a problem found in the work of Lowe. On Lowe's view, the ontological "I" is inextricable from the first-person perspective (FPP) that makes sense of the I-thoughts. But what he does not mark out clearly in his corpus is what it is that is distinctive about the "I" in question. His failure to do so illustrates the liability produced by departing from Cartesianism for a brand of emergent non-Cartesianism. The "I" is a subject, and for that reason it is a part of the mind-body whole that is not only inseparable but also characterized by thoughts and experiences that are owned by the soul as substance. It is a novel, unique particular that goes beyond the general properties of other substances that individuate their thoughts; it is a thing of which I am directly aware upon introspection of my own I-thoughts. Without possession of the metaphysics of the particular "I," we are left with an obscure "I" that defies explanation.

The problem with Lowe's non-Cartesian substance dualism is that he never spells out what that substance or subject of experience is; he never goes beyond asserting what is metaphysically necessary to make sense of the agent of consciousness. What we need is a metaphysical individuator that furnishes the ground for the epistemic benefits of Lowe's I-thoughts. The problem may be highlighted by what Geoffrey Madell has developed in his recent work, *The Essence of the Self.*

An Uninformative and Insufficient Cartesian Self: Geoffrey Madell

Geoffrey Madell defends a Cartesian view of the self in his *The Essence of the Self: In Defense of the Simple View of Personal Identity.*[27] Like Lowe, he contends for a simple view of personal identity. He argues that the soul is a metaphysically simple substance that carries along personal identity, and he calls this the Cartesian view. Madell is a substance dualist in that he sees the mind and body as distinct property-bearers. He does not see any monistic alternative, let alone materialism, as successfully accounting for the distinct property-bearers.[28] This is related to his overarching point when we take seriously the mind, self-reference, and the first-person perspective in contrast to properties predicated of the objects (e.g., body and soul) as contingent factors of the objective order.

In many ways, his discussion reduces to a common, albeit interesting discussion of the gap between the subjective order and the objective order, which has had a long and venerable history in philosophy. Madell states: "There is a huge tension between the way we think of objects in the world, 'elements of the objective order of things,' and the way we seem compelled to think of persons. This book is an attempt to explore that tension."[29] For Madell, there is a primitive fact of the self or soul that is metaphysically simple, but the explanation of that fact is somewhat elusive. Without exploring and analyzing Madell's fascinating discussion of all the particulars (for that would be impossible in this context), it will suffice to show what it is he in fact believes to be the case and suggestively point in a direction that resolves the problem he finds irresolvable. At a minimum, this will show how his position is distinct from the sort of Cartesian view I defend.

His fundamental concern is summarized in the following passage:

> There is, however, no denying that many people will see grounds for rejecting outright the account of the self which seems to be emerging from what I have said, and that for a fundamental reason. To suggest, as I appear to have done, that there are no criteria for identity of the self over time, and no criteria which have to be satisfied for a state of consciousness to be mine at any one time, leaves one with a sort of free floating 'I'. On one hand, every attempt to establish criteria for the identity of the self, to tie it logically to some such condition as the continuity of the body or of psychological continuity, or its identity to the notion of origin, seems to break down. But to accept this is to give credence to the idea of the self as an entity which, purely as a matter of chance, alights on a certain set of properties in history but might equally have alighted on any other set. This presents a dilemma of awesome proportions, and we must eventually confront it.[30]

To help clarify what he sees as the "dilemma," he quotes the words of Michael Bitbol:

> Why do I live now, in this special period of history? Why am I me, born in this family, in this place of the world? I was taught that there were many other possibilities: being any person, at any time, or even just *not being at all*. And yet here I am, in front of you. Me, not you, here, not there, now, not then...What is the reason, if any, of this inescapable singularity? Does the fact that we all live through this mystery alleviate it in any way?[31]

There is something instructive and insightful in Madell's discussion about the uniqueness of consciousness and subjects of experience. There is something mysterious about persons that requires the information from phenomenology and defies exhaustive examination from analytic philosophical methods. But further, there is something in his discussion that helps us to see that there is something about the uniqueness of persons for which the Cartesian is better suited to account, namely the radically distinct properties of the mind in contrast to the body.

It does not appear to be true that there are no criteria for informatively or sufficiently referring to persons by our designations in the world. It is on this point that I think his Cartesianism has moved in the wrong direction. If, in fact, our designators are informative or sufficient, then the criterial issue is met.

His view explicitly does not provide an informative designation to the soul because, on his view, it is precisely this sort of designation that is irresolvable. There is an irresolvable tension between the self perceived as object and the self perceived as subject, between "A is F" and "I am F", and between "persons are souls" and "I am soul."[32] The soul, as a designation descriptive of general properties, lacks the content necessary to identify it. When we consider this self as it presents itself in our consciousness, the tension between the conceptions is striking. This may be the reason why it is that so many, even substance dualists, are less inclined to believe that there is any content accessible in the mind. By doing this, they settle for an epistemic individuator (i.e., the feature or property that distinguishes one thing from another) from the first-person perspective (i.e., what we know, not what actually refers), but they have no designator that is transparent and accessible (this is where content, an informative or sufficient designator, is necessary).

Yet, as we have shown, not only is an individuator metaphysically necessary; it is accessible through introspection upon what it means to be an "I" when I use it in reference to my*self*. This "I" that I use is a rigid designator, but it is also sufficiently informative to distinguish one self from another self. Assuming the intentional and transparent structure above (namely the *cogito* where the conscious perspective of the individual presumes an "I" to which I have access), there appears to be something of which each individual is aware or can be aware. I grant that some contingencies cannot be explained apart from a personal explanation (i.e., a Creator God assigning this soul to this body at this time). But the more fundamental metaphysical criterion can be given an analytic description that departs from Madell and other views that do not give informative accounts.

I provide here a brief technical discussion, presenting Madell's analytic, systematic argument in his own words, followed by my rebuttal of the argument. Madell states:

> No account of personal identity or of the nature of the self can possibly be right unless it can make sense of the aspects of the issue which I have tried to emphasise. I list them here and make a few comments about each.
>
> 1. The fact that there is always a gap between "X is *f* and *g*" and "*I* am *f* and *g*." This gap makes it difficult to see how there could be logically constitutive criteria of personal identity through time, since there must always be a gap between a claim such as "a range of experiences X is linked together in a specified way" and the assertion "those experiences are mine." I have suggested earlier that it is a mistake to suppose that this gap is innocuous. I call this the Criterial Gap issue.
> 2. The fact that no account of conscious perspective or conscious awareness in general can be adequate, since it leaves quite unclear how one can grasp the truth that one of these conscious perspectives is one's own… I call this the Uniqueness Issue.
> 3. The fact that no description of the world expressed without indexicals or token-reflexives can include the *contingent* truth that some tiny segment of that reality is *me*, GM. There is no such unaccounted-for contingent truth which is left out of a complete description of the physical world. All refrigerators monitor their own temperature, and a complete description of all refrigerators in the world and their self-monitoring capacity will leave nothing out. But something is missed from such a description if it purports to include a description of all the conscious beings in the world… I call [this] the Contingent Truth Issue.
> 4. The failure to acknowledge this contingent truth, or that one conscious perspective is uniquely mine, means that the self-ascription of experiences becomes impossible to understand… I call this the Self-Ascription Issue.[33]

In another paper, I critically engaged Madell and offered a way of thinking about the self as content*ful* rather than lacking content, thus meeting the problem of criteria and undercutting the apparent dilemma. What follows is a reworking of my argument.[34] When offering a sufficient designator, we can close the gaps between the self as object and the self as subject:

1. Criterial Gap Issue: "A is F" when the *user* ascribes to himself this property.[35]
2. Uniqueness Issue: No account of conscious perspective from a purely objective frame apart from the user.[36]
3. Contingent Truth Issue: All truths, held by the person, about the world are contingent upon the user using them rightly as it pertains to him/herself.
4. Self-Ascription Issue. There is no problem of self-ascription when the user uses the terms to refer to himself or herself, if the user has an adequate self-grasp.[37] Otherwise, the gap is simply one between *de dicto* and *de re*.[38]

I do have access to myself, and when I use the word "I" properly to describe myself as the subject of *this consciousness,* I use the word appropriately, in a way that closes the gap and provides a sufficient designator.

4. Conclusion

We are left, then, with neo-Cartesianism as the best explanation for the phenomenal unity of consciousness. Neo-Cartesianism, unlike other substance dualist alternatives, supplies us with a metaphysical individuator that we can informatively or sufficiently designate as the "I"—an inseparable part of the mind-body compound and a part that is a primitive metaphysical simple—of the thoughts and experiences under analysis. The dualist alternatives that affirm a metaphysically simple self (as entailing indivisibility) often do so in a way that leaves the inseparable part without sufficient designation. And, as we have seen, the neuroscientific evidence provides some support for the fact that there is a metaphysically non-divisible self (with the basic higher-order powers of first-person perception). Obscure dualisms, such as those offered by the authors we have discussed, fail to supply a metaphysical individuator implicit in the first-person perspective as spelled out through the lens of phenomenal consciousness. For these reasons, it is appropriate to maintain neo-Cartesianism.

Notes

1. See William A. Dembski, *Intelligent Design: The Bridge between Science and Theology* (Downers Grove, IL: InterVarsity Press, 1999). See also Alvin Plantinga, *Where the Conflict Really Lies: Science, Religion and Naturalism* (Oxford: Oxford University Press, 2011), especially 225–248.
2. For this original meaning, and some historical background on the doctrine, see *The Oxford Dictionary of the Christian Church*, 3rd ed., edited by F. L. Cross and E. A. Livingstone (Oxford: Oxford University Press, 1997), 429–430.
3. A similar argument, yet modified for our purposes here, is found in William Hasker, "Persons and the Unity of Consciousness," *The Waning of Materialism* (Oxford: Oxford University Press, 2010), 185. The soul is a designator for an immaterial substance and is taken to be synonymous with a mind or self.
4. Timothy O'Connor and Jonathan D. Jacobs, "Emergent Individuals," *The Philosophical Quarterly* 53, no. 213 (October 2003): 541–542.
5. See specifically William Hasker, "Persons as Substances," in *Soul, Body and Survival: Essays on the Metaphysics of Human Persons* (Ithaca: Cornell University Press, 2001), 107–119.
6. Hasker, "Persons as Substances," 107–119.
7. William Hasker, "Do My Quarks Enjoy Beethoven?," in *Neuroscience and the Soul: The Human Person in*

Philosophy, Science, and Theology (Grand Rapids, MI: Eerdmans, 2012), 39.

8. J. P. Moreland, "Substance Dualism and the Unity of Consciousness," in *The Blackwell Companion to Substance Dualism*, eds. Jonathan J. Loose, Angus J. L. Menuge, and J. P. Moreland (Oxford: Wiley-Blackwell, 2018), 189–190.
9. Hasker, "Do My Quarks Enjoy Beethoven?," 29.
10. William Hasker, "A Rejoinder to O'Connor," in *Neuroscience and the Soul*, 48.
11. The subject of consciousness is "transparent" when the subject is aware that he is a subject. Beyond that minimal definition, I add the further point that the subject S has a contentful designation internal to S's conscious states of subject S rather than a contentless S.
12. René Descartes, *Discourse on Method* and *Meditations on First Philosophy*, trans. Donald A. Cress (Indianapolis, IN: Hackett, 1980), 50. This is just one example of many in Descartes that we can point to where he says something very similar, if not identical, regarding the self, soul, or mind. He often uses the terms interchangeably. It is important to note that he often takes the referent to be present to one's conscious awareness where the *cogito* brings together the epistemic and metaphysic. If there is an intentional structure that secures this knowledge about self as a soul-substance (a view consistent with Brentano as pointed out above in the summary of Hasker's discussion), then it brings together the distinction that is oft made between epistemology and metaphysics. This is important for my argument below. This depends on his clarity and distinction as part of his original conceivability argument, which has and continues to be discussed at length in the history of analytic philosophy.
13. Hasker confirms this in a series of articles discussing the nature of the self as a metaphysically primitive self with a haecceity (i.e., a fundamental feature that informatively describes the self). See my "Souls, Emergent and Created: Why Mere Emergent Dualism is Insufficient," *Philosophia Christi* 20, no. 1 (2018): 83–92, followed by his "Emergent Dualism and Emergent Creationism: A Response to Joshua Farris," *Philosophia Christi* 20, no. 1 (2018): 93–97. For my later rejoinder, see "A Response to Hasker's 'Emergent Dualism and Emergent Creationism,'" at https://www.epsociety.org/userfiles/Farris%20(A%20Response%20to%20Hasker-Web).pdf, 1–6, and for his response to that rejoinder, see "Souls Without Thisnesses," https://www.epsociety.org/userfiles/Hasker%20rejoinder%20to%20Farris-2-3.pdf, 1–4.
14. Hasker, "Emergent Dualism and Emergent Creationism," 93-97, see specifically 95-96.
15. E. J. Lowe, "The Probable Simplicity of Personal Identity," in *Personal Identity: Complex or Simple?*, eds. Georg Gasser and Matthias Stefan (Cambridge: Cambridge University Press, 2012), 149.
16. E. J. Lowe, "Why My Body Is Not Me: The Unity Argument for Emergentist Self-Body Dualism," in *Contemporary Dualism: A Defense*, eds. Andrea Lavazza and Howard Robinson (New York: Routledge, 2014), 245–266.
17. Farris, "Response to Hasker's "Emergent Dualism and Emergent Creationism," 5.
18. William Hasker, "Persons and the Unity of Consciousness," in *The Waning of Materialism*, eds. Robert C. Koons and George Bealer (Oxford: Oxford University Press, 2010), 177.
19. Hasker, "Persons and the Unity of Consciousness," in 175–191.
20. By personas I am referring to something like masks or personalities. These may point to non-rigid designators and not the rigid designator that sufficiently informs the subject. In the same way that one can represent or take on different personalities, one singular person could do so at a deeper level that is dependent on perspective or even phenomenal experience, but this doesn't rule out the fact of a singular person (who, in some transcendent way, may have a higher-order perspective, but this need not deter us). Joseph LaPorte, "Rigid Designators," *Stanford Encyclopedia of Philosophy*, Winter 2022, https://plato.stanford.edu/archives/win2022/entries/rigid-designators/>. A designator is a technical term in this context. There are commonly rigid designators and non-rigid designators. The former applies to specific names that are fixed in meaning and apply in all possible worlds. The latter are common terms, titles, or names that are contingent and can be applied contingently (e.g., "police officer" can refer to different persons at different times, as can "the president"). There is a further distinction to rigid designators, and that is whether a specific name is informed or sufficiently informed, which aids in distinguishing self as a generable or a property that is a universal and a sufficient designator of a particular, a person.
21. See Michael Egnor, "Neuroscience and Dualism," in the present volume.
22. Egnor, "Neuroscience and Dualism."
23. Egnor, "Neuroscience and Dualism."
24. Egnor takes a Thomistic approach to dualism, because of the functional integration of the self with embodied

neurology. However, it is not clear from cases of commissurotomy that Thomism is the best option. I believe that Cartesianism, especially one that provides a functionally integrated view of mind and brain, is a better option, because the self, the sense of self, and the ability to reason abstractly are left unharmed by assaults on the brain. Thomism, in contrast, affirms that the soul, immaterial substance, or immaterial principle is somehow bound up with the material. Egnor's view is that the data seems to confirm the fact of a metaphysically simple substance, which a Cartesian would affirm. That said, the Cartesian would affirm a more robust distinction between the Thomist material substance (as on Egnor's account) and the material substance as it is ontologically. More needs be said here and developed in a separate place; here I merely signal the seeming advantage of a Cartesian account over a Thomist one.

25. This is a common view of Kant and frequently discussed in the history of philosophy. See Immanuel Kant, *Critique of Pure Reason*, trans. N. K. Smith (New York: St. Martin's, 1965), 335–376, especially 376.
26. If the phenomenal self presented here is correct, then it is incompatible with the sort of emergentist views that explain the origins of the self from lawful mechanistic processes. For an argument along these lines, see Joshua R. Farris, "A Novelty Without Particularity: The Problem from 'Perfect Duplicates,'" *Philosophy, Theology and the Sciences* 7, no. 1 (2020), 70–89. Also see Farris, "Souls, Emergent and Created," *Philosophia Christi* 20, no. 1 (2018): 83–92.
27. Geoffrey Madell, *The Essence of the Self: In Defense of the Simple View of Personal Identity* (New York: Routledge, 2014).
28. Madell, *Essence of the Self*, 1–12, 139.
29. Madell, *Essence of the Self*, 10.
30. Madell, *Essence of the Self*, 10–11.
31. Michel Bitbol, as quoted in Nicholas Humphrey, *Soul Dust* (London: Quercus, 2011), 151–152.
32. Madell, *Essence of the Self*, 1-12, 122.
33. Madell, *Essence of the Self*, 9-10.
34. See Joshua R. Farris, "Creationist-Dualism," in *The Origin of the Soul: A Conversation*, eds. Joshua R. Farris and Joanna Leidenhag (New York: Routledge, forthcoming 2023).
35. Richard Swinburne, *Are We Bodies or Souls?* (Oxford: Oxford University Press, 2019), 86–115. This is consistent with Swinburne's understanding of thisness. His work here builds on his other works, including *Mind, Brain, and Free Will* (Oxford: Oxford University Press, 2013), 141–74, and *The Evolution of the Soul* (Oxford: Oxford University Press, 1997), 333–345.
36. If one is a Berkeleyan idealist, then this knowledge would be mitigated by God's external knowledge that is communicated to created minds because all phenomena (including bodies) are explained by God's action where the contingency is an occasion of his action.
37. The sufficient grasp is had by the person having it, but, again, the feature is not universalizable or generalizable between persons, and that is what grounds the fact of individual persons.
38. See my *The Creation of Self* (forthcoming 2023) and my "Creationist-Dualism," cited above.

Unit 3: Neuroscience and Psychology

13. Neuroscience and Dualism

Michael Egnor

1. Introduction

Scientific inquiry based on a mechanistic view of nature—at least on a scale larger than the atom—has revealed much about the natural world, from the intricate biochemistry of the cell to the relativistic structure of the universe. This mechanistic view of nature posits efficient causes and truncated material causes—the interactions of matter understood, according to the laws of physics and the properties of matter, to be the only causes at work.

Neuroscience, as it extends to the study of the relation of the mind to the brain, seems an exception to mechanistic science, because the phenomenology of thought seems to have no point of contact with matter and mechanism. First-person experience—sensations, memories, beliefs, desires, and such—cannot be explained or even described in the language of matter and forces. Many of the great neuroscientists of the twentieth century—Sherrington,[1] Penfield,[2] Libet,[3] Sperry,[4] and Eccles,[5] to name a few—have been dualists or idealists of various sorts and have rejected the materialist theories of the relation between the mind and the brain.

In the twentieth century neuroscience was motivated largely by identity theory, a form of reductive materialism that claims that mental states are identical with (or can be reduced to) brain states. While this theory has fallen out of favor, neuroscientists in the twenty-first century generally assume alternative forms of materialism. They generally favor either functionalism,[6] the non-reductive materialist view that the mind is what the brain does, or eliminative materialism, a metaphysical perspective championed most notably by philosophers Paul and Patricia Churchland.[7] Eliminative materialism is based on two claims: (1) Our everyday beliefs about our mental states—that we have feelings, beliefs, desires etc.—are "folk psychology," are incorrect, and will be abandoned by the scientific community, in the same way that the historical belief that combustion is caused by the release of phlogiston was abandoned; (2) Mental states will, in time, be fully explained by the methods of neuroscience working in a purely materialist framework.[8] The "eliminative" qualifier in eliminative materialism means that all mental states will be ultimately explained materialistically; non-materialist perspectives, such as dualism or idealism, will be shown to have no explanatory power in scientific investigation.

Whether twenty-first-century neuroscientists work from the perspective of reductive, non-reductive, or eliminative materialism, most neuroscience research today is predicated on a materialist perspective. In *The Astonishing Hypothesis: The Scientific Search for The Soul,* Nobel laureate Francis Crick, the co-discoverer of the structure of DNA and pioneering molecular geneticist-turned-neuroscientist, wrote, "The Astonishing Hypothesis is that 'You,' your joys and your sorrows, your memories and your ambitions, your sense of identity and free will, are in fact no more than the behaviour of a vast assembly of nerve cells and their associated molecules. As Lewis Carroll's Alice might have phrased it: 'You're nothing but a pack of neurons.'"[9]

Some materialist neuroscientists insist that discarding belief in the immaterial mind and soul is not only good science, but will liberate mankind from ancient superstitions. Julien Musolino, a cognitive scientist and director at the Psycholinguistics Laboratory at Rutgers University, insists that neuroscience has disproven the existence of the soul, and that discarding belief in the soul will lead mankind to ever greater flourishing:

> Because they inevitably distort our perception of reality, soul beliefs corrupt our thinking on matters that are deeply important to all of us. By subverting the intuitions that underpin our criminal-justice system (i.e., influencing how we intuitively reason about moral responsibility), poisoning the debate over abortion, and muddling the question over whether people should have the right to die with dignity, soul beliefs, far from the blessing they are commonly portrayed to be, actually stand in the way of a healthier and more humane society. This is the soul's dark secret, the hidden truth.[10]

Some materialist neuroscientists and philosophers have tried to explain the mind as a product of Darwinian evolution. Philosopher Daniel Dennett, in *From Bacteria to Bach and Back: The Evolution of Minds*, provides a wholly materialist account for evolution of the mind. Dennett argues that mindless natural selection acted on the early human ability to transmit "memes"—shared ideas—to evolve the ability of humans to form new memes at will. This evolution via memes is, in Dennett's estimation, the (naturalistic) creation of the mind. Dennett proposes that the mind evolved by competition between memes, just as the body evolved by competition between organisms. Dennetts admits his story of the evolution of mind is counterintuitive, and entails what he terms "hazards to comfortable thinking," such as, in Dennett's words:

> Darwin's strange inversion of reasoning... reason without reasoners... competence without comprehension... feral neurons... words striving to reproduce... consciousness as user illusion...
>
> Yet these materialist theories of mind elide a fundamental question: Do the results of modern neuroscience *genuinely* support the view that the mind can be explained wholly on the basis of matter and its interactions? Is materialism an adequate framework for neuroscience?[11]

Modern neuroscientists and philosophers of mind nearly always begin with implicit (and at times explicit) materialist predicates for neuroscience, and then invoke the results of neuroscience to buttress materialism. Much of this volume is devoted to the philosophical critique of materialism as an explanation for the mind. And, indeed, the recourse in neuroscience to materialism as an explanation for the mind illustrates the logical fallacy of circular reasoning. My purpose here is to show that materialism is also *scientifically* inadequate to explain several of the seminal results of modern neuroscience.

2. Definitions of Materialist, Idealist, and Dualist Views of Mind

To MAKE such a case, clarity of terms is important. Three general metaphysical perspectives ground theories of mind—materialism, idealism, and dualism—and these three categories are a logical place to start if we are to test theories of the mind-brain relationship using neuroscience. In order to empirically test the adequacy of materialism, idealism, and dualism to account for the mind and its relation to the brain, we must state these metaphysical perspectives clearly. There are of course many varieties of each of these metaphysical positions, which are differentiated from one another by exquisite subtleties, but for clarity I'll choose definitions that have wide acceptance and that represent the core assertions of the position.

Materialism: I define materialism[12] as the view that all mental states supervene on material states of the brain; that is, mental states are determined by brain states, without remainder. Colloquially in this context, supervenience is the view that the brain causes the mind, and that there cannot be a mind change without a brain change. All variants of materialism—reductive, non-reductive, and eliminative—*hew to this view.*

Idealism: I will use the ontological (rather than epistemic) sense of idealism:[13] idealism is the view that the foundation of all reality is mental, not material. From the ontological idealist perspective, mind is the only real substance, and matter (e.g., the brain) is ultimately an idea in the mind.

Dualism: Dualism,[14] broadly defined, is the view that mind and matter are both real, and are radically different kinds of thing, neither one assimilable to the other.

3. Ways in Which Materialism, Dualism, and Idealism May Be Tested Empirically

IN A scientific study, careful selection of the hypothesis is necessary for meaningful interpretation of the data. What, exactly, is the question asked? How are the results of the research to be interpreted? How does the design of the investigation allow confirmation or refutation of hypotheses at issue? This framework is particularly important in neuroscience research, because the metaphysical issues are extraordinarily subtle, yet metaphysical clarity is crucial to understanding the results. To address the main question—"Which metaphysical framework—materialism, idealism, or dualism—is most consistent with the results of neuroscience?"—three fundamental metaphysical questions can be asked using the methods of science. (To the objection that metaphysical theories cannot be investigated by the methods of natural science, I point out that metaphysics is the study of existence *qua* existence, and natural science is the study of what exists in nature. The validity of a metaphysical theory of the mind-brain relationship depends on (among other things) the correspondence between the metaphysical theory and the results of neuroscience.) The three salient questions we can ask to clarify the metaphysical framework—materialist, idealist, or dualist—of neuroscience are:

1. *Is the mind metaphysically simple?* By this I mean, does the mind have parts, in the sense of parts or properties that are distinct and separable from itself? Can the mind be split or divided into parts that are no longer the original mind? Metaphysical simplicity means singleness of the mind itself, and not necessarily singleness of the effects of the mind on the world. *Materialism*

predicts that the mind, like the brain from which it arises, is metaphysically complex, in the sense that it is divisible into parts, just as any material thing has dimensions and physical relations and as such can be split in one sense or another. *Idealism* predicts that the mind is metaphysically simple, because immaterial spirit, lacking spatial extension, does not have internal physical relations or separable parts, which are characteristic of material, but not mental, things. *Dualism* predicts that some aspects of mind are divisible, but others are not. Dualism posits that the immaterial powers of the mind are metaphysically simple, while material powers of the mind are divisible in some sense, as matter is.

2. *Are there immaterial powers of the mind?* By this I mean, does the mind have powers that are not caused by the brain? *Materialism* predicts uniform material causation—the brain causes the mind. Neither the mind as a whole nor individual powers of the mind can exist without brain and brain function. *Idealism* predicts uniform immaterial causation—mind is the root cause of all reality, and brain and brain function are ideas in the mind without independent reality. *Dualism* predicts that some aspects of mind are caused by the brain, while other aspects are not. Dualism predicts that mind and/or several of its powers exist independently of the brain, in the sense that they are not caused by activity of the brain.
3. *Is free will real?* On this question, *materialism* is almost invariably determinist, and it follows from physical determinism that the will is wholly determined by neurochemistry and the evolutionary and individual history of the organism. While some materialists are "soft determinists" (who claim materialism is compatible with a weaker notion of free will), materialist determinism precludes the robust, commonsense idea of libertarian freedom, where choices are within the agent's power. *Idealism* is nondeterminist, or at least entails denial of physical determinism by material processes. Idealism posits a broad and radical freedom of will. *Dualism* posits genuine free will, associated with the immaterial powers of mind, but admits influence on the will by material processes. In the dualist perspective, we are "tempted" by material influences (e.g., alcohol, images, hormones), but we have libertarian freedom to choose to accept or reject temptation.

There is good reason to believe that the human capacity for reason and will,[15] as well as the human capacity for language,[16] imply that many cognitive traits of human beings cannot be studied in animals. The similarities and differences between the human mind and the animal mind are still the subject of much research and controversy.[17] Thus, we will restrict our inquiry to research on the human mind.

We will now consider several seminal experiments in modern neuroscience in light of these metaphysical questions. In these categories—metaphysical simplicity, immateriality, and free will—it should be noted that some of these seminal experiments in neuroscience point to one or more metaphysical implications—Sperry's split-brain research, for example, points to the immateriality of intellect and will as well as to metaphysical simplicity of these powers. Libet's research on free will also points to immateriality of will. I have chosen the metaphysical implications that seem to follow most directly from each area of neuroscience research.

4. Is the Mind Metaphysically Simple?

4.1 Sperry and Split-Brain Research

Epileptic seizures are caused by unpredictable electrical discharges in groups of neurons in the brain. The location, spread, and manifestation of the seizures vary enormously. Some seizures are virtually undetectable. Other seizures involve a small group of neurons in one location in the brain, and the seizure remains in that location and doesn't spread to other parts of the brain. Some seizures start in a specific location in the brain and then spread, like a fire, to both cerebral hemispheres, causing not just a minor localized twitching of a few muscles (from a localized seizure focus) but a severe generalized convulsion, with complete loss of consciousness.

Researchers in the early twentieth century realized that many seizures that propagate from a small focus in one hemisphere of the brain to both hemispheres of the brain (generalized seizures) travel across the corpus callosum, a large bundle of white matter fibers (axons) that connects the two cerebral hemispheres. This observation suggested a new approach to the treatment of some generalized seizures that were not treatable by usual methods such as anti-seizure medications: neurosurgeons could cut the corpus callosum (an operation called a commissurotomy)[18] and thereby prevent propagation of seizures and prevent generalization of a minor seizure to a major seizure. Remarkably, despite that fact that the corpus callosum is a large structure that contains nearly all of the direct connections between the hemispheres of the brain, it seemed that the corpus callosum could be cut without causing a major neurological disability, although subtle perceptual and motor abnormalities were noted in some laboratory studies on animals[19] and in studies of humans who had undergone commissurotomy for seizures.[20] In fact, it is not uncommon for people to be born without a corpus callosum and still have reasonably normal neurological functioning in everyday life.[21]

Although the subtle perceptual and motor disabilities of patients who had undergone disconnection of their cerebral hemispheres were well known beginning in the 1940s, detailed systematic study of the neurological disabilities caused by commissurotomy only began with neurophysiologist Roger Sperry, who in the 1960s studied epilepsy patients who had undergone commissurotomy. Sperry carried out a series of ingenious experiments which allowed detection and analysis of subtle abnormalities that were unapparent in everyday life and even on routine neurological exam by experienced neurologists. Sperry was awarded the Nobel Prize in Physiology or Medicine for this work.

Sperry and other investigators found that patients after commissurotomy had perceptual and motor disabilities of which they were generally unaware in everyday life.[22] The disabilities correlated with the isolation of each hemisphere from the other. For example, patients for whom speech was mediated only by the left hemisphere were unable to name objects shown to them in the visual field mediated by their right hemisphere, yet using their left hand (mediated by the right hemisphere) they could point out the object they saw from a group of objects. It was apparent that each hemisphere of the brain mediated specific functions, such as movement of the contralateral side of the body, vision in the contralateral half of the visual field, speech (usually the left hemisphere) and spatial judgment (usually the right hemisphere). In commissurotomy patients, perceptual information could be processed by the hemisphere directly connected to the relevant sensory organ, but not by the disconnected opposite hemisphere. On occasion, some patients would experience movements of one side of their body that seemed in conflict with that of the other. For example, a few patients with alien hand syndrome would simultaneously attempt to unbutton their shirt with one hand while buttoning it with the other.[23]

It is noteworthy that these disabilities were always *perceptual and motor* in nature. Despite the discernible but subtle disabilities experienced by these patients, Sperry and other investigators noted a simultaneous *unity* of mind—unity of self-awareness, unity of capacity for reason and purpose—unaffected by commissurotomy.[24] For example, Justine Sergent studied two patients with commissurotomy who had different information presented to disconnected hemispheres, so that correct responses to the total information could only be made if there was a unity of awareness despite the disconnection of perceptions. Sergent noted:

> Both patients performed significantly above chance in 7 of the tasks (spatial orientation, calculation, lexical decision), suggesting that the information divided between the hemispheres could be united, related, and acted upon in a unified manner. This performance was achieved in spite of each hemisphere being unaware of the information received by the other, as typically observed in "split-brain" patients. The results indicate the coexistence of perceptual disunity and behavioural unity, and they suggest that, even when the two disconnected hemispheres receive different information, the commissurotomized brain works as a single and unified organism.[25]

This unity of mind even with splitting of perception is the common daily experience of commissurotomy patients, including those in my own neurosurgical practice. Commissurotomy patients do not have "split" identity or split sense of self, and they are certainly one person in any meaningful sense. In fact, the occasionally disconcerting perceptual disabilities experienced by these patients (e.g., alien hand syndrome) would only be disconcerting to a unified person. There would be nothing unusual about split perceptions in two different people. Two minds would not find split perception strange. The very strangeness of the split perception characteristic of commissurotomy patients presupposes fundamental unity of mind.

Other investigators have noted this paradoxical unity of mind despite splitting of the brain. In 2017, Pinto et al. published a paper titled *Split Brain: Divided Perception but Undivided Consciousness*. In the two split-brain patients that they studied, they observed the following:

> … the canonical textbook findings that a split-brain patient can only respond to stimuli in the left visual half-field with the left hand, and to stimuli in the right visual half-field with the right hand and verbally, are not universally true. Across a wide variety of tasks, split-brain patients with a complete and radiologically confirmed transection of the corpus callosum showed full awareness of presence, and well above chance-level recognition of location, orientation and identity of stimuli throughout the entire visual field, irrespective of response type (left hand, right hand, or verbally). Crucially, we

used confidence ratings to assess conscious awareness. This revealed that also on high confidence trials, indicative of conscious perception, response type did not affect performance. These findings suggest that severing the cortical connections between hemispheres splits visual perception, but does not create two independent conscious perceivers within one brain.[26]

In a separate study, Pinto et al. propose a model of "conscious unity, split perception" in which the conscious patient is a *unified* agent who experiences two parallel unintegrated streams of information:

> We argue that the classical view [that commissurotomy causes split consciousness] may not hold for several reasons. First, some of the defining features also occur in healthy adults with unified consciousness (hemispheric specialization, inability to explain own actions, and split attention). Second, the most convincing argument against unified consciousness in split-brain patients (the response × visual field interaction) does not hold for all split-brain patients. Third, in the absence of any convincing proof of split consciousness, unified consciousness should be the default position. Both the patients and the people nearest to them claim that consciousness is still unified in the patient. Moreover, their everyday behavior confirms this. Thus, the claim of destroyed conscious unity is extraordinary, and requires extraordinary evidence.[27]

Pinto's stress on the everyday experience of split-brain patients is particularly salient, and is consistent with my experience as a neurosurgeon. After commissurotomy, patients retain a fully unified consciousness. The perceptual and motor disabilities that do occur are subtle and inconsistent, and to some extent overlap with perceptual and motor variations in normal people who have connected hemispheres.

One way of describing Pinto's model of conscious unity with split perception in commissurotomy patients is to say that the patient's higher cognitive power—the sense of self and the power of intellect, reason, judgment, and will—remain fully unified, while mental powers of sensation, perception, and movement may show disconnection.

Hemispherectomy,[28] which is a neurosurgical operation distinct from, but related to, commissurotomy, demonstrates retention of unity of mind despite complete removal of an entire hemisphere of the brain (to stop intractable seizures). The removed hemisphere is often quite abnormal to begin with, hence the seizures that arise from it, but it is generally in many ways quite functional, and complete removal of the hemisphere does *not* consistently impair mental function to a substantial degree, but sometimes even *improves* it, probably as a consequence of better seizure control.

A related operation, called functional hemispherectomy,[29] was developed to circumvent bleeding complications common with traditional hemispherectomy. Functional hemispherectomy entails the cutting of tracts that connect lobes of the brain—often, the temporal, frontal, parietal and occipital lobes are disconnected from each other, but not removed. In functional hemispherectomy, the brain may be cut into four or five disconnected pieces. Although the neurological and psychological outcomes of hemispherectomy and functional hemispherectomy are not as well studied as those of commissurotomy, it is clear that the same retention of unity of consciousness prevails.[30] There is no evidence of splitting of consciousness—meaning intellect, will, and unitary sense of self—despite complete removal of major portions of the brain or widespread disconnection of the tracts between many brain regions.

Although in some cases there are subtle perceptual disabilities, splitting the brain does *not* split the sense of self, the intellect, or the will.

4.2 Metaphysical Implications of Split-Brain Research

This understanding of the mind that emerges from studies of split-brain patients—that some powers of mind such as perception and movement can be split, while others, such as unitary sense of self, reason, intellect, and will cannot—is clearly inconsistent with *materialism*. If mind is caused by brain, without remainder, a materialist would naturally expect commissurotomy to split the mind as well as the brain, and indeed the claim for split consciousness as a consequence of commissurotomy has been a dominant theme in the neuroscience literature of commissurotomy.[31] Yet several more recent studies of split-brain patients make it clear that the most surprising outcome of commissurotomy is *how little the mind is split when the brain is split*. Dividing the brain has no impact on the sense of self or abstract reasoning.[32] Only *perception* splits, and even that is difficult to discern without meticulous testing. The patient's daily experience, sense of self, and exercise of powers of reason and will remain unitary, despite disconnection of the two halves or even several lobes of his brain. Materialism is inconsistent with unity of mind in split-brain patients.

The fact of "conscious unity despite split perception" is consistent with the *idealist* understanding of the mind. Idealism certainly predicts that conscious unity—spirit, devoid of parts, is not divisible, at least by material means. But idealism as a metaphysical framework for understanding the neuroscience of commissurotomy falls short in its failure to predict split *perception*.

The *dualist* perspective offers a more satisfactory framework for understanding the neuroscience of split-brain surgery. The neurological consequences of commissurotomy are inherently dualist—the sense of self and the capacity for reason and will remains unified, while perception splits.

5. Are There Immaterial Powers of the Mind (Powers Not Caused by the Brain)?

5.1 Phrenology and Cerebral Localization

The salient advance in neuroscience in the nineteenth century was the realization that some neurological functions, such movement, tactile sensation, vision, and speech, were mediated in very specific regions of the cerebral cortex.[33] Movement is mediated by a strip of cortex in the posterior part of each frontal lobe. Tactile sensation is mediated by an adjacent cortical strip just posterior to the motor area. Vision is mediated in small regions of the medial occipital cortex, and speech is mediated in a small region of the inferior-lateral frontal lobe and in the posterior part of the upper temporal lobe and lower parietal lobe (usually on the left). Even more remarkable was the discovery that motor and sensory abilities mapped linearly—as a homunculus—on the cortex. The leg motor area is at the top of the motor strip near the fissure between the brain hemispheres, the trunk muscles on the cortex below it, the arm muscles on the cortex below that, and the facial muscles at the bottom, on the cortex near the fissure that separates the frontal lobe from the temporal lobe. The visual field is mapped in two dimensions on the visual cortex; the upper left portion of the visual field is mediated by the lower right bank of the visual cortex—like an inverting mirror.

This remarkable and precise mapping of many perceptual and motor abilities on the cortex of the brain led some nineteenth-century neuroscientists to propose that *all* neurological functions could be mapped, including powers of abstract thought, such as contemplation of justice, logic, courage, etc.

Lacking any technology such as CT or MRI to see the living brain, many of these scientists—phrenologists[34] they called themselves—made an inference that seems to us bizarre but in context was not entirely unreasonable. It is known that the shape of the skull hews quite closely to the shape of the underlying brain—in some circumstances, the gyri of the brain correspond to grooves on the inner surface of the skull. Given the inference that a highly developed cortical representation for, say, mathematical ability or preference for mercy, would probably be associated with a large lump of cortex in the "mathematics area" or the "mercy area," phrenologists developed a detailed science of inferring higher mental proclivities from skull shape. Phrenologists palpated bumps and dips in patients' skulls to diagnose personality traits.

Phrenology flourished in the nineteenth and early twentieth century, but ultimately came to naught. It is commonly thought of as a pseudoscience, but that's not quite right. It had some hallmarks of genuine science—a plausible (for the nineteenth century) inference that all mental powers map to the cortex (just as speech, perception, and locomotion were known to map), a not-unreasonable inference that skull shape follows more or less on brain shape, an extensive collection and interpretation of data, and the creation of professional societies and the publication of research. It is vitally important to understand that the primary reason phrenology failed was not that phrenologists were evil fools (although some were). Phrenology failed *because the assumption that higher intellectual powers of the mind are mapped to specific cortical regions was incorrect.* There is no "justice area" or "logic area" or "courage area" on the cortex or anywhere in the brain. The distinction between the (nonexistent) brain localization of intellectual powers of mind and the genuine precise localization of motor and sensory powers of mind is remarkable. Neuroscientists can pinpoint regions of the cerebral cortex that mediate movement of the hand, but they cannot pinpoint any region of the brain that mediates the intellectual content of what is written by the hand. There is no "opinion area" of the cortex by which opinions are conceived that corresponds to the hand area by which opinions are written.

The failure of phrenology revealed a striking fact about the relation between the brain and the mind: motor and sensory powers of the mind are mapped precisely in the brain, but rational powers—the capacity for reason, abstract thought, and free will—have no specific brain localization.

In light of this dichotomy, it has been customary to ascribe higher intellectual functions to "association areas" of the brain, which are large tracts of the frontal, temporal, parietal, and occipital lobes that lack specific perceptual or motor function, but this ascription is really *ad hoc*. There is no cogent evidence that higher intellectual function arises *from* these association areas in any way akin to the way that perception and muscle motion arise from mapped cortical regions. In neurosurgical parlance, cortical regions that subserve perception and muscle movement are called "eloquent cortex" and need to be protected from surgical injury at all costs. Cortical regions (association areas) that subserve no known function are called "non-eloquent cortex," and often can be sacrificed for surgical reasons (e.g., to excise a seizure focus or a tumor) without neurological disability.

5.2 New Phrenology

The salience of this remarkable observation—that rational powers of mind do not map to the brain in any way akin to mapping of locomotion and perception—is not, in my view, sufficiently appreciated by the neuroscience community. New MRI technology, called functional MR Imaging (fMRI), permits scientists and clinicians to see regional changes in blood flow in the brain in real time, and it is believed that these blood flow changes correspond to neuronal activity and thus permit mapping of neuronal activity in living subjects. Many neuroscientists have attempted to map abstract intellectual thought to "non-eloquent" brain regions using fMRI, but the results have been inconsistent and are open to various and often conflicting interpretations. A cogent critique of these efforts to map intellect to brain has been made by several commentators,[35] and quite prominently by psychologist and cognitive scientist William Uttal in his book *The New Phrenology: The Limits of Localizing Cognitive Processes in the Brain*.[36] Uttal decries the paucity of analysis of the assumptions that underlie localization research. He critiques the difficulty of specifying and even defining cognitive processes, and points out the conceptual fallacies and methodological conundrums entailed in research on cerebral localization of cognitive processes. What does an increase or decrease in local cerebral blood flow and the associated neuronal changes mean with respect to cognition and higher levels of thought? What is the basis for inferring *causation* of mental states by correlation with neuronal activation? Given the unbridged explanatory gap between brain activity and subjective mental experience—i.e., the explanatory gap between third-person brain events and first-person experience—what can be meaningfully inferred from patterns of neuronal activation and suppression?

Neuroscientist and philosopher Raymond Tallis highlights the nonsensical logical assumptions that underlie much of the new phrenology. He asks:

> Let us suppose there were no limit to the precision of imaging. Let us suppose also the kinds of localizations seen on fMRI scans... were robust. And let us suppose that the separate psychological states or functions to which the brain activity is supposed to correspond are real entities rather than ad hoc constructions. And let us, finally, suppose that we have explained how that which has been teased apart comes together in the conscious moment... what then would fMRI tell us? If we could obtain a complete record of all neural activity, and we were able to see the firing state of every individual neuron, would this advance our understanding in the slightest? Would the record of neural activity be as useless at telling us what it is like to be conscious as a complete printout of his genome at telling you what it is like to be with your friend? Would (human) consciousness be... "explained"?[37]

Tallis points out that fMRI would explain consciousness only if patterns of brain activation were *identical* with consciousness, which is philosophically dubious. Correlations loosely demonstrated by fMRI between brain activation and the mind do not demonstrate causation. At most, correlations between brain activity and the mind demonstrate a *necessary* relationship between mind states and brain states, without demonstrating *sufficiency*. Tallis argues that the "necessary but not sufficient" paradigm for the relation between brain and mind states is buttressed by the observation that a person's behavior becomes more explicable in material terms the more neurologically damaged the person is. He points out that a seizure (or a stroke) fits the materialist model better than *living* with epilepsy or *living* with a stroke. There is an irreducible complexity and subjectivity in lived human experience that materialist "sufficiency" arguments fail to capture, and seem hopelessly unable to capture. The new phrenology seems as failed a project as the old phrenology.

Even the methodological challenges facing the new phrenology are daunting. Bennett et al. recently used fMRI imaging to map the correlation between cognition and brain activity of a *dead salmon*—and (using statistical anomalies inherent to the imaging program) were able to generate a brain map: "The salmon was shown a series of photographs depicting human individuals in social situations with a specified emotional valence. The salmon was asked to determine what emotion the individual in the photo must have been experiencing."[38]

Bennett and his co-authors pointed out that even with the use of statistical methods routinely used in fMRI data analysis, random noise in the fMRI data can suggest neuronal activity that does not in fact exist.

Even though fMRI is certainly much farther advanced as a methodology than phrenology, the conceptual pitfalls inherent to correlating brain states with mind states remains the same. Modern technology provides a window into the brain, but like phrenology, the modern approach is implicitly predicated on materialist assumptions about causation of mental states. Indeed, inferences drawn from fMRI imaging of brain states may be even more misleading than phrenology in a sense, in that modern technological sophistication conceals dubious materialist assumptions that underlie the prevailing metaphysical paradigm in modern neuroscience. Metaphysical misconceptions are less evident when covered with a patina of technological sophistication. There is little evidence that efforts to map higher mental function with fMRI imaging are based on more credible metaphysical inferences than phrenologists' efforts to map character traits by bumps on skulls a century ago.

What emerges from the debacle of old and new phrenology is that higher cognitive powers such as intellect do not map to the cerebral cortex in any way similar to the well-established high-fidelity mapping of locomotion and perception. While the inference that a coherent causal chain of brain events gives rise to locomotion and perception is lavishly supported by two centuries of research, such an inference for material causation based on mapping cannot be made for higher intellectual powers of the mind. Intellect seems to have a radically different relation to brain function than locomotion and sensation.

A plausible inference that can be drawn from this dichotomy is that intellectual powers of the mind are not related to the brain in the same way perception and locomotion are related. An inference that intellectual powers of mind are in some sense immaterial, as compared with the materially caused powers of perception and locomotion, is not unreasonable here.

5.3 Wilder Penfield: Can We Evoke the Mind by Cortical Stimulation?

Wilder Penfield was a neurosurgeon and neuroscientist who worked in the mid-twentieth century. He founded the Montreal Neurological Institute and Hospital and pioneered the surgical treatment of epilepsy, and he made seminal contributions to the science of brain physiology, epilepsy, and brain mapping.[39] His work remains a cornerstone of the modern neuroscience.

Penfield's clinical and research focus was the mapping and excision of seizure foci in the brains of locally anesthetized (awake) patients, and in his career he performed 1,132 brain operations on awake patients.[40] The brain itself has no capacity for sensation of pain, and local anesthetics can be injected into the scalp to make the surgery painless. With the patient awake while Penfield was operating, he used the patient's responses to his meticulous mapping by electrical stimulation of the cortex and of deeper brain structures to locate and safely excise areas of the brain that were causing seizures. In order to excise the epileptic foci safely, Penfield mapped the epileptic focus and adjacent brain in an effort to locate functional brain regions that might be damaged by resection of the epileptic focus. Penfield carefully recorded hundreds of thousands of responses to stimulation in all regions of the cerebral hemispheres over four decades of surgery, and his techniques are used routinely in neurosurgery today. Every day in the operating room I use techniques he developed and instruments he designed, and he revolutionized our understanding of brain function and its relation to the mind.

As his research progressed over decades, Penfield gained deep insight into the functional organization of the brain based on the correlation between brain stimulation during surgery and its effect on mental states. He was particularly interested in the question raised by his mentor Sir Charles Sherrington, an Oxford neuroscientist and Nobel laureate widely considered the greatest neuroscientist of the early twentieth century. Penfield pondered the "Sherringtonian alternatives": is the mind composed of two elements, or one? Sherrington, who concluded that dualism is the best framework in which to understand the mind, had noted, "That our being should consist of two fundamental elements offers, I suppose, no greater inherent improbability than that it should rest on one only."[41]

Penfield asked the same question: "Can the brain explain the man? Can the brain achieve by neuronal action all that the mind accomplishes?... Either brain action explains the mind, or we must deal with two elements."[42]

In thousands of electrical stimulations of the cerebral cortex of awake patients during surgery, Penfield often found negative responses—for example, transient inability to speak when the speech areas of the left hemisphere were stimulated. There were four (and only four) kinds of *positive* responses—muscular movement, sensation, interpretative perception, and recall of conscious experience (memory).[43] (Interpretative perception refers to a patient's ability to ascribe emotional content to accompanying perceptions.)

Penfield noted a consistent dichotomy between what he called "brain action," which meant movements, sensations, perceptions, and memories that he could evoke by stimulating the brain with an electrode, and "mind action," including mental states like self-awareness, volition, and reasoning, which he could not evoke by brain stimulation and thus seemed independent to some degree from brain states. Given that large tracts of the cerebral cortex were thought to be responsible for abstract reasoning, Penfield was amazed that he was never able to evoke any "mind-action":

> During brain action, a neurophysiologist can surmise where the conduction of potentials is being carried out and its pattern. It is not so in the case of what we have come to call mind-action.

> And yet the mind seems to act independently of the brain in the same sense that a programmer acts independently of his computer, however much he may depend upon the action of that computer for certain purposes.[44]

Over decades of research and clinical experience, Penfield concluded:

> For my own part, after years of striving to explain the mind on the basis of brain-action alone, I have come to the conclusion that it is simpler (and far easier to be logical) if one adopts the hypothesis that our being does consist of two fundamental elements... the mind must be viewed as a basic element in itself... [t]hat is to say, [the mind] has a continuing existence.[45]

In addition to interpreting the results of his own research, Penfield studied the life and writing of Hippocrates, and he was impressed by the fact that his own research seemed to validate Hippocrates's dictum that "To consciousness the brain is the messenger."[46] Penfield came to see that ascription of mind-action to the brain is a scientific error: "To suppose that consciousness or mind has localization is a failure to understand neurophysiology"[47] and "it is the mind (not the brain) that watches and at the same time directs."[48]

Penfield did identify what he called lower and higher brain-mechanisms, which correspond to automatic sensory-motor mechanisms (lower) in the diencephalon, to which he drew an analogy to a "brain computer,"[49] and to arousal and basic interpretative abilities (higher) in the cortex.

The following example should help to clarify what Penfield meant by "mind." He described the intraoperative responses of a seizure patient, who clearly seemed to be using his upper and lower brain mechanisms much as a person would use a computer. Penfield would suppress the patient's speech area with electrical current, and the patient repeatedly tried to speak about a concept he had in his mind despite being unable to elaborate speech:

> He remained silent, then expressed his exasperation by snapping the fingers and thumb of his right hand. That he could do without making use of the special speech mechanism [which Penfield was suppressing with his electrode]. Finally, when I removed my interfering electrode from the cortex, he explained the whole experience with a feeling of relief, using words that were appropriate to his thought. He got the words from the speech mechanism when he presented concepts to it. For the word "he," in this introspection, one may substitute the word mind. Its action is not automatic.[50]

Penfield understood that there were two different processes at work—those of the mind and of the brain:

> I can only say that the decision came from his mind. Neuronal action began in the highest brain-mechanism. Here is the meeting of the mind and the brain. The psychico-physical frontier is here. The frontier is being crossed from mind to brain. The frontier is also being crossed from brain to mind since the mind is conscious of the meaning of the stream of consciousness. The neuronal action is automatic as it is in any computer.[51]

There are "two elements" involved in the production of mental states, the mind and the brain, and Penfield struggled for years to understand what the "mind"—that which he could not evoke by stimulation—really is. He started with the dictionary definition—"[the mind is] the element... in an individual that feels, perceives, thinks, wills and especially reasons."[52]As his research progressed, he understood the mind more deeply:[53]

> It is what we have learned to call the mind that seems to focus attention. The mind is aware of what is going on. The mind reasons and makes new decisions. It understands. It acts as though endowed with an energy of its own. It can make decisions and put them into effect by calling upon various brain mechanisms. It does this by activating neuron-mechanisms... Mind decides what is to be learned and recorded... The mind conditions the brain. It programs the computer so that it can carry out an increasing number of routine performances. And so, as years pass, the mind has more and more free time to explore the world of the intellect, its own and that of others.[54]

In time he came to define "mind" more succinctly—the mind is the *person*, and in this sense the person is distinct from his brain. While he found this evidence for the dualist hypothesis surprising—he had started his career with a materialist bias—Penfield concluded that the materialist hypothesis was less consistent with the evidence than the dualist hypothesis. He observed:

> Does [dualism] seem to be an improbable explanation? It is not so improbable, to my mind, as is the alternative expectation—[the materialist hypothesis] that the highest brain-mechanism should itself understand, and reason and direct voluntary action, and decide where attention should be turned and that the computer [i.e., the brain] must learn, and record, and reveal on demand.[55]

Penfield concluded, on the basis of his research and clinical experience, that the mind is separate from the brain and the brain functions more or less like a biological computer. The brain is, in Hippocrates's words, the "messenger" between the mind, which is immaterial, and the material world.

It is noteworthy that Penfield distinguished the interpretation of perceptions from the mind per se, and localized perceptual interpretation in the association cortex: "The gray matter of the interpretive-cortex is part of a mechanism that presents interpretations of present experience to consciousness."[56]

The correspondence between Penfield's discovery (of the material basis for perception and perceptual interpretation, in contrast to the lack of a material basis for mind understood as the personal or rational aspect of man) and the Thomistic model of material powers of perception and perceptual interpretation (i.e., the Aristotelian *sensus communis*)[57] and the immaterial powers of intellect and will, is striking. More will be said about the Thomistic model later in this essay.

5.4 Wilder Penfield: Why Are There No 'Mind' Seizures?

In addition to the enormous body of surgical research discussed above, Penfield studied epilepsy for many decades, and collaborated with pioneering epilepsy neurologists such as Herbert Jasper and Theodore Rasmussen at the Montreal Neurological Institute.[58] In his evaluation and treatment of many thousands of patients with seizures, Penfield noted a remarkable fact about epilepsy: *there are no 'mind' seizures*. Penfield noted:

> There is no area of gray matter, as far as my experience goes, in which local epileptic discharge brings to pass what could be called "mind action"... there is no valid evidence that either epileptic discharge or electrical stimulation can activate the mind... If one stops to consider it, this is an arresting fact. The record of consciousness can be set in motion, complicated though it is, by the electrode or by epileptic discharge. An illusion of interpretation can be produced in the same way. But none of the actions we attribute to the mind has been initiated by electrode stimulation or epileptic discharge. If there were a mechanism in the brain that could do what the mind does, one might expect that the mechanism would betray its presence in a convincing manner by some better evidence of epileptic or electrode activations.[59]

By "mind," Penfield meant *the person*, the fundamental unitary sense of self and the capacity for reflection, reason, and abstract thought, which seems to correspond more or less to what Aristotle called the intellect and will. Penfield contrasted mind in this sense with emotions, memories, and sensory and motor activities, which were clearly mediated by specific localizable brain activity, and which could be readily evoked by seizures or by electrical stimulation of the brain during surgery. A term that I think corresponds more or less to Penfield's concept of "mind" is "intellect" or "abstract thought." Penfield's observation that there are no seizures that evoke intellect or abstract thought remains valid today.

To see this, consider that there is a broad spectrum of recognized seizure types. I will use here the I.L.A.E. classification scheme that is widely used in neurology:

Partial (focal) seizures—without impairment of consciousness:

1. Simple: motor, sensory, psychical, autonomic
2. Complex—with impairment of consciousness
3. Partial seizures that subsequently generalize

Generalized seizures:

1. Absence seizures
2. Myoclonic seizures
3. Tonic, tonic-clonic and atonic seizures

Unclassified (rare seizures that do not fit easily in established categories)[60]

Simple partial (focal) seizures arise from discrete locations in the brain and do not cause loss of consciousness. Focal seizures can entail twitching of a limb or perceptions such as tingling on the skin or flashes of light. *Complex partial (focal) seizures* resemble simple seizures, but they entail complex manifestations with loss of awareness, and can involve periods of blank staring, repetitive purposeful motions such as picking at clothing or repeating words. Rarely, complex seizures can manifest as prolonged complex activities, such as walking or doing complex repetitive tasks. *Generalized seizures* may or may not arise from specific brain regions, but progress to involve the entire brain and cause loss of consciousness. The classic generalized clonic-tonic seizure in which a patient fall unconscious with twitching of muscles is a kind of generalized seizure.

Among the simple focal seizures (see the first group above), there are some rare types of seizure—*psychical seizures*—that at first appear to entail abstract thought, but do not in fact do so. They were described by Penfield, who recognized that although they entailed specific thoughts as the core of the ictal content, they were not "activation of the mind." Psychical seizures[61] are traditionally classified in four categories: hallucinatory, perceptual, emotional, and forced thinking.

Hallucinatory seizures, which are the most common of this (rare) seizure type, arise from the temporal lobe and usually entail ictal memories of personal experiences, which the patient re-experiences. The memories are always perceptual, in the sense that patients see, smell, feel, or hear things from the past. Patients remain awake during the experience and are aware of the hallucinatory nature of the seizure.

Perceptual seizures are transient impairments of real concurrent perceptions, frequently with emotional content and a sense of déjà vu.

Emotional seizures entail sudden ictal emotions such as fear, terror, or well-being, without significant cognitive content.

Note that all three of these rare psychical seizure types *lack abstract intellectual content.* Emotional and perceptual content is the rule.

The fourth type of psychical seizure, called *forced thinking seizure,* is the closest to a putative intellectual or "mind" seizure, but on close examination is not, because forced thinking seizures lack *abstract* content. Forced thinking seizures are seizures in which the patient experiences intrusive thoughts or words that are not memories per se. There is often a strong emotional content to the experience. The semiology of forced thinking seizures resembles that of obsessive-compulsive disorder, which is not a manifestation of epilepsy. In forced thinking seizures, the mental content is perceptual or emotional, and notably is concrete, in that thought is directed to objects, situations, and emotions, without abstraction.

Forced thinking seizures frequently entail compulsive thoughts to do a task, and a systematic classification of cognitive seizures, which includes forced thinking, includes[62] inability to speak, inability to name objects, inability to understand language, inability to name or interpret sounds, inability to repeat speech, inability to read or write, inability to remember, déjà vu, sensory hallucinations, perceptual illusions, sense of dissociation or disconnection from environment, repeatedly intrusive thoughts (with concrete content such as a task or a memory), inability to do mathematics or write, left-right confusion, and neglect.

What does not occur in forced thinking seizures or in any kind of cognitive manifestation of epilepsy is *evocation of abstract thought*—there are no forced calculus seizures or forced logic seizures or forced morality seizures.

5.5 Penfield's Observations on the Evocation of Mental States by Brain Stimulation

We can summarize Penfield's observations about evocation of mental states by brain stimulation as follows:

A. *Mental states that can be evoked by seizures and/or brain stimulation:*
   1. Motor movements (e.g., localized twitching of muscle groups, generalized rhythmic contractions of all muscles, purposeful simple movements such as walking)
   2. Perceptions (e.g., smells, flashes of light)
   3. Emotional states (which can include "psychical" seizures that entail obsessive concrete thoughts with strong emotional content about objects, activities, hallucinations, or a sense of déjà vu)
   4. Memories (of specific events, objects, people, etc.)

B. *Mental states that cannot be evoked by seizures and/or brain stimulation:*
   1. States of the intellect (mental states such as abstract reasoning, mathematical calculation, and logic, which are directed to universals rather than to particular objects)
   2. States of the will (goal-directed decisions to act based on abstract ethical considerations)

Practically every imaginable category of evoked neurological activity has been observed by doctors and researchers, as reflected in the broad scope of recognized seizure types, but there is no category for *intellectual* seizure or for *seizure of the will*. Seizures *always* involve either complete unconsciousness, or specific activation of a non-abstract neurological function—flashes of light, smells, sensations on the skin, jerking of muscles, specific memories, or strong emotions and forced *concrete* thoughts. As Penfield noted, this is, from the materialist framework, unexpected, given the fact that in the materialist predicate much of the cortex—the association cortex—is devoted to abstract thought. Association areas, which are the regions of the cortex imputed to be the origin of abstract thought in the materialist paradigm, comprise most of the cerebral cortex.

There is a clear parallel between the phenomenology of epilepsy and the results of intraoperative stimulation of the cerebral cortex. The brain causes muscle movements, sensations, perceptions, perceptual interpretation, memories, and emotions, but the brain does not cause the mind, understood as reason, will (I will discuss Penfield's research on will below), and self-awareness.

Penfield, who was a materialist prior to his ground-breaking study of the mind-brain relationship, finished his career as an emphatic dualist. He recognized that the complete absence of "mind-action" evoked by seizures or by direct brain stimulation is one of the strongest lines of evidence in neuroscience that the mind—the personal and rational aspect of man—is immaterial. This independence of mind from brain activity, and the dependence of sensation, perception, imagination, memory, locomotion, and emotion on brain activity is consistent with idealism and dualism generally, but it is not consistent with materialism. In particular, Penfield's results correlate with *Thomistic* dualism (about which more will be said below) with stunning accuracy.

5.6 Craniopagus Twins

An anecdotal but illuminating example of the application of neuroscience to the question of the immateriality of some aspects of mind is the case of Krista and Tatiana Hogan, who are craniopagus twins.[63] The sisters were born joined at the head, and they share portions of their brains. They cannot be safely separated surgically because they share vital blood vessels. Their brains are connected by a thalamic bridge, which allows them to control sensation and movement in several limbs. Krista has motor control and sensation in both of her own legs *and control of Tatiana's left leg*. Tatiana has motor control and sensation in both of her own legs and arms and has motor control and receives sensation from Krista's right arm. Notably, their common control of several limbs seems to only entail unconscious movements, not deliberate conscious movements. They share some visual sensations and taste, and even some thoughts—both girls will laugh together without speaking, as if they share funny thoughts or mental images.

The issues regarding which powers of thought the girls share are subtle,[64] but it is clear that the two girls *are separate human beings in every meaningful sense*. They share some perceptual and motor abilities, but they remain two distinct people. In fact, it is their distinctness that is the basis for our amazement at the perceptions and motor control they share—if they were one person, and not two people, it would be unexceptional for "them" to share in common sensations or control of limbs.

Notably, it does not seem that they share abstract thought—there is no report that they share concepts such as arithmetic or logic that they learn in school. A study in which one child was taught an abstract concept—multiplication for example—when the other was not would be of value in teasing out the neurological basis for reason and abstract thought. As our knowledge of the neurological abilities of these girls stands now, they share some aspects of perception, motor control, imagination, and emotions. Despite the anatomical connection of their brains, there is no evidence that they share intellect, which argues for the immaterial nature of abstract thought.

5.7 Owen and fMRI Studies of Patients in Persistent Vegetative State

In the 2006 issue of the journal *Science*, Adrian Owen, a physiologist at Cambridge University, published with his colleagues a study titled "Detecting Awareness in the Vegetative State." Owen performed fMRI studies on a woman with massive brain damage from an automobile accident. She had no evidence by clinical exam of any mental function at all. She was in persistent vegetative state (PVS), which means that she lacked any evidence for any mental function. PVS is in this sense the deepest and most persistent form of coma. It implies a human body without a mind—the meaning of "vegetative."

Owen placed this woman in an MRI machine and did fMRI study of her brain while he asked her questions through a set of headphones. fMRI images changes in blood flow in regions of the brain that are known to correlate with brain activity. To his surprise, he found that she had quite a bit of brain activity—despite the near-destruction of her brain—when he asked her to think about certain things, such "think about playing tennis" or "think about walking through your house."

He compared this activity in her nearly destroyed brain to brain activity in normal volunteers who were asked the same questions. The brain activity in the woman in PVS and in the normal volunteers was *indistinguishable*. Her brain activity, as measured by fMRI, was normal, despite her apparent deep coma and her massive brain damage.

Owen reasoned that the brain activity in the woman in PVS might be merely the result of reflex in response to the sounds in the headphones, rather than from any genuine understanding of the requests or any actual thoughts. So he repeated the experiments with the words scrambled, so the sounds were the same, but the meaning of the questions was missing. The brain activity of the woman and of the volunteers stopped when meaning was removed from the requests. *The woman understood what she was being asked to think about*, despite massive loss of brain tissue and the deepest level of coma.[65]

Owen and others have repeated this work using a variety of strategies for detecting active mental states in "vegetative" patients.[66] For example, investigators asked a patient in PVS to focus his attention on one of two different pictures held continuously in front of him. fMRI revealed shift patterns of activation of his brain that indicated he was indeed concentrating on different pictures when asked, despite the fact that the pictures in front of him were not moved. This response was also found in normal volunteers. This implied that the patient was able to understand and concentrate on a specific object in his visual field despite having such severe brain damage that he was diagnosed as having no mental activity at all.

Several investigators have been able to use fMRI responses to communicate with patients in PVS. Using practiced "yes" and "no" responses on fMRI, patients in PVS have been able to give detailed personal and family information subsequently confirmed to be true and patients have correctly answered location questions (e.g., "Are you in a hospital or a supermarket?"). Researchers have been able to establish communication that entails verifiable answers to detailed questions.[67]

While persistence of abstract thought in many patients in PVS is not *direct* evidence for the immateriality of the intellect—after all, fMRI directly detects blood flow changes associated with brain activity—it is strong evidence for the weakness of the correlation between brain damage and the ability to think abstractly.

The argument that this fMRI research does not imply the immateriality of preserved abstract thought because only *material* brain blood flow changes are measured (rather than direct confirmation of abstract thought) is not tenable. All inference to the thoughts of others is behavioral—we have no direct access to thoughts not our own. This fMRI technique is also behavioral, in the sense that it measures a behavioral variable (changes in brain blood flow), which is not qualitatively different from interpreting spoken or written (behavioral) answers to questions.

Ordinary motor and perceptual function in these PVS patients is profoundly impaired—they cannot perform any meaningful motor activities and usually (to the limits of clinical testing) appear to be blind (the aforementioned case of the patient concentrating on pictures is exceptional), and large regions of association cortex (the putative sites of abstract thought in the materialist paradigm) are destroyed. Yet they retain remarkable abilities to think abstractly, and this retention of the capacity for abstract thought despite massive brain damage suggests a fundamental dichotomy between causal prerequisites of perception and movement on the one hand and abstract thought on the other.

While material activity is indeed the measure of abstract comprehension in the patient with PVS, the fact that this is the case in the context of such dramatic brain damage undermines the materialistic assumption that the complex physical structures in the brain are responsible for the capacity for abstract thought. Since comprehension can continue even with severe diffuse damage to the brain, we must conclude either (a) that the remaining profoundly damaged brain is all the material complexity needed to cause abstract thought (which in itself would be quite surprising), or (b) that there is another immaterial component in operation which is needed to explain abstract thought.

The salient implication of research on fMRI studies in PVS patients is that thought is not impaired by brain damage in the same way that motor function or perception is impaired. This is a specific prediction of both idealism and dualism, and is difficult to reconcile with a completely materialist understanding of the mind.

5.8 Near-Death Experiences

A detailed discussion of near-death experiences (NDEs) is beyond the scope of this chapter (see Gary Habermas's chapter in this volume for a more detailed discussion of the evidence). It suffices to say that the NDE literature is extensive and clearly points to the veracity of a subset of NDEs that involve verifiable knowledge.[68] The argument for immateriality of at least some aspects of mind is greatly strengthened by this research and can support either idealism or dualism. Materialism is clearly inconsistent with this research. Although NDE experiences tend to align to some extent with individual personal and cultural beliefs prior to the event, there are elements of the experience that are common across history and cultures. These include an awareness of death, a sense of peace, a sense of leaving the body and traveling through a tunnel, a sense of "going to a light" accompanied by a feeling of love and acceptance, an encounter with deceased loved ones, a life review, and a reluctance to return to the body. People who experience an NDE report a vividness and a sense of intense reality that one does not generally encounter in dreams or hallucinations.[69]

From the perspective of neuroscience, NDEs are very difficult to study systematically, given the unpredictability and chaotic nature of the circumstances in which they occur. There is, however, a large body of scientific literature on the topic, and there are prospective studies under way that aim to characterize the experiences rigorously.

The ubiquity of NDEs is undeniable. Materialists have offered a number of explanations for the phenomenon. Materialist explanations have ranged from psychological (birth memory, wishful thinking, depersonalization, psychological expectations, hallucinations, dissociation, sensory isolation, REM sleep intrusion, and even adaptation via evolutionary psychology) to physiological (anesthetics, temporal lobe seizure, hypoxia, hypercarbia, endorphins, "dying brain," neurotransmitter imbalances, ketamine effects, and cortical inhibition).[70] None of the materialist explanations is credible, as each explanation, at best, accounts for only a small part of the full experience of many individuals, and is quite inconsistent with many other characteristics of the experience. Indeed, the plethora of materialist conjectures suggests the inadequacy of materialist explanations.

The most parsimonious explanation, and the explanation most consistent with the data available on NDEs, is that some of these experiences are manifestations of persistence of mind despite cessation of brain activity. Much remains to be learned about this phenomenon, and it is likely that some (perhaps most) NDEs have prosaic explanations, but their ubiquity and frequent corroborated verity suggest that it is only materialistic bias, rather than scientific evidence, that precludes the acceptance in the neuroscientific community of NDEs as clear evidence for dualism.

In contrast to the inadequacy of the materialist framework to account for NDEs, both idealism and dualism account for them naturally, as they posit, to a greater or lesser extent, existence of the mind independent of the brain.

6. Is Free Will Real?

A central claim of modern materialism is that determinism is true and, therefore, that free will is an illusion.[71] For the materialist, all mental states are caused by brain states, which in turn are caused by physical processes determined by the laws of physics and by natural history. There is no room for free will in the materialist framework, at least in the incompatibilist view of free will held by many materialists.

Compatibilists (who are usually materialists of one variety or another) believe that determinism is true but that free will is genuine, at least in a sense.

The opposing view is that of libertarian free will, which is the view that human will is not determined—that is, not mandated—in any way by physical processes. All participants in this discussion agree that the will is *influenced* by matter—alcohol can bend the will with surprising ease—but believers in libertarian free will insist that man in some circumstances retains the genuine ability to choose.

There have been several neuroscientific studies that have examined the question of free will. We will discuss two here: the work of Wilder Penfield and of Benjamin Libet.

6.1 Wilder Penfield and Electrophysiological Evocation of Will

As noted above, Wilder Penfield performed hundreds of brain operations for epilepsy on awake patients using local anesthesia. In many of his operations, he would stimulate the brain with an electrical probe to look for seizure foci and to map critical areas of the cortex.

Penfield found, to his amazement, that while he could elicit a spectrum of movements, perceptions, memories, and emotions by stimulating the brain, he was never able to stimulate the *will* of the patient. He noted:

> When I have caused a conscious patient to move his hand by applying an electrode to the motor cortex of one hemisphere, I have often asked him about it. Invariably his response was: "I didn't do that. You did." When I caused him to vocalize, he said: "I didn't make that sound. You pulled it out of me." When I caused the record of the stream of consciousness to run again and so presented to him the record of his past experience, he marveled that he should be conscious of the past as well as of the present. He was astonished that it should come back to him so completely, with more detail than he could possibly recall voluntarily. He assumed at once that, somehow the surgeon was responsible for the phenomenon... There is no place in the cerebral cortex where electrical stimulation will cause the patient to believe or decide.[72]

That is, whenever he stimulated a mental state, the patient *always* knew that the mental state was imposed from without, and always knew that he (the patient) did not will it. Penfield could not stimulate the will of the patient, in countless thousands of brain stimulations over decades of surgical experience. Penfield concluded that the will, the sense of agency, is an immaterial power of the mind, not a material power of the brain. In this sense, it is reasonable to infer that Penfield concluded that the will is free of material determinism.

6.2 Libet and Free Won't

Perhaps the most well-known and important research on the neuroscience of free will was done by neurophysiologist Benjamin Libet[73] in the mid-twentieth century. Libet's research is described in more detail elsewhere (see Cristi Cooper's chapter in this volume), but I will summarize it here.

Libet was fascinated by the *timing* of mind states vis-à-vis brain states. He asked the question: "What is happening in the brain at the moment of a thought?" To study free will, Libet devised an ingenious series of experiments. He attached electrodes to the scalps of normal volunteers to record their brain waves, and asked them to decide (whenever they wished) to push a button and then to push the button. He placed a clock in front of them so they could note the exact moment they made the decision to push the button (with a resolution of 10 milliseconds or so).

He found consistently that the first indication of a decision was an unconscious brain wave—called the "readiness potential"—that occurred 400 milliseconds *before* the subject was aware of an intention to act. At 400 milliseconds *after* the readiness potential, the subject would decide consciously to push the button, and 200 milliseconds later he would push the button.

It seemed at first glance that the "free" decision to push the button wasn't really free. It seemed that the brain generated the decision unconsciously, *prior* to the subject's awareness.

The choice was really dictated by the brain. Free will, it seemed, was an illusion.

However, Libet carried the experiment a step further. He told the subjects to veto the decision to press the button immediately after the decision is made—to say, that is, "I'm going to push the button… no, I'm not!"

Libet found that *the veto was electrically silent*. That is, there was no new brain activity associated with the veto—the only brain activity that correlated with the decision was the pre-conscious readiness potential.

The veto appeared to be free will, and there was no brain wave corresponding to it. It appeared to be an immaterial act of will. Libet said, of the electrically silent veto, that he didn't exactly prove free will, but he proved "free won't." Libet concluded that we are bombarded continuously with preconscious motives generated by the brain—represented by the readiness potentials—but that we have the libertarian freedom to accept or reject the preconscious motives. The motives are generated by the brain, but the acceptance or rejection of the motives is independent of brain activity, at least as he measured it. Libet wrote:

> The role of conscious free will would be, then, not to initiate a voluntary act, but rather to control whether the act takes place. We may view the unconscious initiatives for voluntary actions as "bubbling up" in the brain. The conscious will then selects which of these initiatives may go forward to an action or which ones to veto and abort, with no act appearing. This kind of role for free will is actually in accord with religious and ethical strictures. These commonly advocate that you "control yourself." Most of the Ten Commandments are "do not" orders.[74]

Libet noted that this neuroscientific view of free will corresponds nicely to the classical Christian concept of temptation and original sin. We are bombarded by motives generated by our physical nature, over which we have no direct control. But we retain the libertarian (immaterial) freedom to accept or reject these motives. We are tempted, often against our will, but we can genuinely choose our course of action. He summarized his view of the metaphysical implications of his research:

> My conclusion about free will, one genuinely free in the non-determined sense, is then that its existence is at least as good, if not a better, scientific option than is its denial by determinist theory. Given the non-determinist theories, why not adopt the view that we do have free will (until some real contradictory evidence may appear, if it ever does). Such a view would at least allow us to proceed in a way that accepts and accommodates our own deep feeling that we do have free will. We would not need to view ourselves as machines that act in a manner completely controlled by the known physical laws. Such a permissive option has also been advocated by the neurobiologist Roger Sperry.[75]

Libet's work has been replicated and generally confirmed over the past few decades, using fMRI imaging of the brain. Notably, his results are routinely misrepresented by neuroscientists working from a materialist perspective. It is frequently asserted that Libet demonstrated that free will is an illusion. But his decisive study of the immaterial veto of the motive is not widely acknowledged, and it was this veto that Libet thought was the salient result of his research. Libet provided compelling neuroscientific evidence for libertarian free will—the immaterial acceptance or rejection of a material motive.

Libet's research is consistent with the idealist and dualist understanding of the material passions (temptations) and the immaterial (spiritual) free will. Clearly, a mental state (in this case, a veto) without a corresponding brain state is incompatible with materialism.

7. Summary Thoughts on the Three Metaphysical Questions

Philosopher Roger Scruton, commenting on the incoherence of much of contemporary research in cognitive neuroscience, quipped that neuroscience is "a vast collection of answers, with no memory of the questions."[76] The questions forgotten in neuroscience are metaphysical, which hinge on the question: What is the best metaphysical framework with which to understand the relationship of the mind to the brain? That metaphysical question is testable—it *must* be testable—by neuroscience. It must be testable because all science is done within a metaphysical framework, and the more consistent the framework is with reality and the more clearly the framework is articulated, the closer the scientific understanding will hew to reality. An unsatisfactory metaphysical framework can skew research design and send scientists off on pursuits that lead away from, not toward, a deeper understanding of nature. Metaphysical predicates can obscure as well as enlighten.

The need for a cogent philosophical basis for neuroscience has been a cornerstone of the work of M. R. Bennett (a neuroscientist) and P. M. S. Hacker (a philosopher). In several works over the last few decades,[77] they have attacked the philosophical underpinnings of much of modern neuroscience. Their analysis is sharp and devastating. Detailed discussion of it is beyond the scope of this chapter, but their essential critique is that scientific inquiry must be based on metaphysical frameworks *that make sense*.[78] Bennett and Hacker call this rectification of the philosophical basis of neuroscience a work of "conceptual hygiene,"[79] and indeed conceptual hygiene is a central theme of this volume.

How does the neuroscientific evidence weigh in on metaphysics? The difficult thing is to frame the scientific questions in a way that will shed metaphysical light. It can be done, I think, by singling out metaphysical issues that are amenable to empirical tests by well-known and validated experiments. The three metaphysical questions I raised earlier in this chapter will fit the bill, and I will summarize the results of several seminal experiments in modern neuroscience that help to answer these metaphysical questions.

7.1 Is the Mind Metaphysically Simple?

The work of Roger Sperry and other investigators[80] on split-brain surgery and on hemispherectomy for epilepsy support the inference that the higher powers of the human soul—especially what Aristotle called the intellect and will[81]—are metaphysically simple, in the sense that they cannot be divided by material means. The most remarkable insight provided by split-brain research is how *little* effect such a radical operation has on the higher powers of the mind. It is very difficult to reconcile the persistent unity of self despite disconnection of major parts of the brain, and unity of intellect and will, with the materialist framework.

On the other hand, there are fascinating perceptual consequences—perceptual disconnection—that result from this surgery. The most parsimonious way of understanding the conscious unity with perceptual disunity is that there is a fundamental unity to the higher intellectual powers of mind, which implies metaphysical simplicity. Yet for lower powers, such as perception, there is a pronounced dependence on brain states.

7.2 Are There Immaterial Powers of the Mind (Powers Not Caused by the Brain)?

Several findings of modern neuroscience point to immateriality of the intellect and will. There is a clear and rather striking distinction between highly localized cortical representation of perception and motor movement, and the non-localization of intellect and will. Wilder Penfield was unable to evoke "mind" by electrical stimulation of the cortex, despite ease of evocation of perceptions, memories, motor movement, and emotion. Consistent with this observation, Penfield's conclusion that there are no "mind" seizures—no seizures that evoke abstract (intellectual) thought remains true, and remarkable. This observation is consistent with our modern understanding of epilepsy—there are no seizures that evoke abstract intellectual thought. Studies of craniopagus twins suggest that separate individuality of consciousness persists in twins who share brain tracts. Adrian Owen observed that in persistent vegetative state, despite massive diffuse brain damage, patients are sometimes able to think at high levels, which supports the inference that some aspects of the mind are not linked to brain states in the same way that perceptual or motor powers are linked. Finally, near-death experiences, if veridical, demonstrate independence of mind from matter, and provide strong evidence for immateriality of the mind.

How are these findings to be reconciled? Perhaps the most apt formulation is that for sensation, perception, motor movement, imagination, memory, and emotion, specific brain states are necessary and sufficient, while for intellect and will, brain states are (ordinarily) necessary but not sufficient. Intellect and will—the characteristic powers of the rational soul in Aristotelian psychology, are not powers caused by matter.

The most parsimonious explanation for the perceptual/intellectual dichotomy is that intellect, in contrast to perception, does not arise from brain states, but requires a source other than matter.

7.3 Is Free Will Real?

Penfield noted that he was unable to evoke a sense of personal agency (will) by electrical stimulation of the cortex during awake brain operations. Libet showed that the decision to veto a preconscious voluntary act was not itself associated with brain activity. Both results are consistent with the view that the will is not generated by brain states but is free in the sense that is it not determined by matter.

There is considerable evidence from neuroscience that the mind is metaphysically simple in the sense that it is not divisible in the same way that matter is divisible, that the intellect (understood as the capacity for abstract thought) is immaterial, and that the will is free in the sense that it is not completely determined by material processes. These attributes are not consistent with the materialist view of the mind-brain relationship. Idealism and dualism provide logical frameworks for neuroscience that hews most closely to the scientific evidence, and the most promising framework is, in my view, Thomistic dualism.

8. Thomistic Dualism

NEUROSCIENCE SHOWS us that there is an inherent *dualism* in the relation of the mind to the brain, and for that reason I believe that dualism provides a more satisfactory framework for neuroscience than idealism. Furthermore, the dualism revealed by neuroscience has a very specific structure. Sensory, perceptual, locomotive, mnemonic, and emotional powers of the mind hew tightly to brain states, but intellectual and volitional powers do not. This empirical dualist structure of the mind-brain relationship hews closely to Aristotle's psychology, and for that reason I favor Thomistic dualism as the best framework for neuroscience. I will review the Thomistic understanding of the soul here, and for a more thorough explanation of Thomistic dualism, see James Madden's chapter in this volume.

8.1 The Three Powers of the Human Soul

For the Thomistic dualist, the soul is to the body as the form of a chair is to the chair. Form is the principle that makes a thing real and makes it what it is. Thus, the soul is the principle that makes a body alive. The mind, as understood by moderns, is several powers of the soul.

In the Thomistic view, the powers of the human soul are in three categories: vegetative, sensitive, and rational.

Vegetative powers mediate growth, nourishment, reproduction, and all basic unconscious physiological functions necessary for life. Vegetative powers are material powers of the soul, which means that they are generated by the matter of the brain and body.

Sensitive powers mediate sensation, perception, imagination, memory, locomotion, and emotion. For example, vision is caused by action of the eyes, the optic nerves, and the visual cortex of the brain. Emotional states are caused by the hypothalamus and the limbic system of the brain.

Rational powers are the contemplation of abstract concepts by the intellect and the actions of the will based on abstract reasoning. Rational powers of the human soul are *immaterial* in the Thomistic view, which means that intellect and will are not caused by matter. Intellect and will are certainly influenced by matter—in the Aristotelian phrase, all that is in the intellect was once in the senses—but the exercise of abstract thought, and the will that follows on abstraction, is not caused by the brain or by matter of any sort.

Thomistic dualism accounts naturally for the material properties of sensation, perception, imagination, memory, locomotion, and emotion, and accounts as well for the immateriality of the intellect and will, and accounts for metaphysical simplicity of the soul and for libertarian free will. Thomism, while avoiding the fatal pitfalls of materialism, provides a framework of material and efficient causation that corresponds to the intimate relation of some mental states to brain matter, while preserving a coherent account of the immateriality of intellect and will. The relations between matter and soul—how they work in perception, imagination, comprehension, and volition—were questions that Aristotle and St. Thomas addressed at length, and these questions of matter and form provide a fertile landscape for neuroscience research.

8.2 The Four Causes Applied to Neuroscience

The cornerstone of the Thomistic (hylomorphic) understanding of the mind is the Aristotelian Four Causes, which can be applied to neuroscience with relative ease. In this view, every mind-body state can be explained by four causes: a formal cause, a final cause, a material cause, and an efficient cause. In fact, consideration of all four causes is necessary to comprehensively understand a mind-body state. The *formal cause* is the intellect—the belief or reason that motivates the act. The *final cause* is will—the decision to act. The *material cause* is the organic apparatus—the brain, nerves, and muscles. The *efficient cause* is the physiological process that accompanies the belief, the decision, and the act.

Note that the Thomistic dualistic framework incorporates *everything* currently included in materialistic neuroscience—all of the neuroanatomy (material cause) and all of the neurophysiology (efficient cause)—while at the same time offering a coherent explanation for the higher mental functions—intellect (formal cause) and will (final cause)—that materialistic explanations are inadequate to explain. Thomistic dualism integrates the material and immaterial causation inherent to mind-brain states and does so in a way that is logically coherent, philosophically rigorous, and entirely consistent with modern neuroscience.

9. Conclusion

In the Thomistic paradigm, traditional neuroscience can be done with explicit focus on material and efficient causes, and neuroscientists can begin to explore the relationship of brain states to formal and final causes—to the immaterial intellect and will—with genuine insight and with scientific rigor. With coherent metaphysics and meaningful questions, the prodigious insight neuroscientists provide can lead us away from materialist fallacies and to a deeper understanding—a genuine understanding—of the brain and of the mind.

Notes

1. For Sherrington, see Wilder Penfield, *The Mystery of the Mind: A Critical Study of Consciousness and the Human Brain* (Princeton, NJ: Princeton University Press, 2015), 4.
2. Penfield, *Mystery of the Mind*, 73–100.
3. Benjamin Libet, *Mind Time: The Temporal Factor in Consciousness* (Cambridge, MA: Harvard University Press, 2004), 182.
4. R. W. Sperry, "Mind-Brain Interaction: Mentalism, Yes; Dualism, No," *Neuroscience* 5, no. 2 (Feb. 1980), 195–206.
5. Karl Popper and John C. Eccles, *The Self and Its Brain: An Argument for Interactionism* [1977] (London: Routledge, 1983).
6. Jerry A. Fodor, *The Language of Thought* (Cambridge, MA: Harvard University Press, 1975).
7. Patricia S. Churchland, *Touching a Nerve: Our Brains, Our Selves* (New York: W. W. Norton & Company, 2013).
8. William Ramsey, "Eliminative Materialism, *Stanford Encyclopedia of Philosophy*, 2022, https://plato.stanford.edu/entries/materialism-eliminative/.
9. Francis Crick, *The Astonishing Hypothesis: The Scientific Search for Soul* (New York: Touchstone, 1994), 3.
10. Julien Musolino, *The Soul Fallacy: What Science Shows We Gain from Letting Go of Our Soul Beliefs* (Amherst, NY: Prometheus Books, 2015), 73.
11. Daniel C. Dennett, *From Bacteria to Bach and Back: The Evolution of Minds* (New York: W. W. Norton & Company, 2017), 22–23.
12. Daniel Stoljar, "Physicalism," *Stanford Encyclopedia of Philosophy*, Summer 2023, https://plato.stanford.edu/entries/physicalism/.
13. Paul Guyer and Rolf-Peter Horstmann, "Idealism," *Stanford Encyclopedia of Philosophy*, Spring 2023, https://plato.stanford.edu/entries/idealism/.
14. Howard Robinson, "Dualism," *Stanford Encyclopedia of Philosophy*, Spring 2023, https://plato.stanford.edu/entries/dualism/.
15. Edward Feser, *Aquinas: A Beginner's Guide* (Oxford, UK: Oneworld Publications, 2009), ch. 4.
16. Robert C. Berwick and Noam Chomsky, *Why Only Us: Language and Evolution* (Cambridge, MA: MIT Press, 2015).
17. Mortimer Adler, *Ten Philosophical Mistakes* (New York: Touchstone, 1985), 45–49.
18. Julian R. Youmans, *Neurological Surgery* (Philadelphia: W. B. Saunders, 1990), 4314–4322.
19. R. W. Sperry, "Some Developments in Brain Lesion Studies of Learning," *Federation Proceedings* 20 (1961).
20. A. J. Akelaitis, "Studies on the Corpus Callosum II: The Higher Visual Functions in Each Homonymous Visual Field Following Complete Section of the Corpus Callosum," *Archives of Neurology and Psychiatry* 45, no. 5 (May 1941): 788–796, doi:10.1001/archneurpsyc.1941.02280170066005; A. J. Akelaitis et al., "Studies on the Corpus Callosum III: A Contribution to the Study of Dyspraxia and Apraxia Following Partial and Complete Section of the Corpus Callosum," *Archives of Neurology and Psychiatry* 47, no. 6 (June

1942): 971–1008, doi:10.1001/archneurpsyc.1942.02290060109008.

21. M. F. Bedeschi et al., "Agenesis of the Corpus Callosum: Clinical and Genetic Study in 63 Young Patients," *Pediatric Neurology* 34, no. 3 (2006): 186–193.
22. Mitchell Glickstein and Giovanni Berlucchi, "Classical Disconnection Studies of the Corpus Callosum," *Cortex* 44, no. 8 (2008): 914–927.
23. Xiaoyu Gao et al., "Alien Hand Syndrome Following Corpus Callosum Infarction: A Case Report and Review of the Literature," *Experimental and Therapeutic Medicine* 12, no. 4 (2016): 2129–2135.
24. Roger W. Sperry, "Forebrain Commissurotomy and Conscious Awareness," *Journal of Medicine and Philosophy* 2, no. 2 (1977): 101–126.
25. Justine Sergent, "A New Look at the Human Split Brain," *Brain* 110, no. 5 (1987): 1375–1392.
26. Yair Pinto et al., "Split Brain: Divided Perception but Undivided Consciousness," *Brain* 140, no. 5 (2017): 1231–1237.
27. Yair Pinto, Edward H. F. de Haan, and Victor A. F. Lamme, "The Split-Brain Phenomenon Revisited: A Single Conscious Agent with Split Perception," *Trends in Cognitive Sciences* 21, no. 11 (2017): 835–851.
28. Youmans, *Neurological Surgery*, 4284.
29. J. Schramm et al., "Pediatric Functional Hemispherectomy: Outcome in 92 Patients," *Acta Neurochirurgica* 154, no. 11 (2012): 2017–2028.
30. Margaret B. Pulsifer et al., "The Cognitive Outcome of Hemispherectomy in 71 Children," *Epilepsia* 45, no. 3 (2004): 243–254.
31. Glicksein and Berlucchi, "Classical Disconnection Studies"; Gao et al., "Alien Hand Syndrome."
32. Sergent, "A New Look at the Human Split Brain"; Pinto et al., "Split Brain: Divided Perception but Undivided Consciousness"; Pinto, de Haan, and Lamme, "The Split-Brain Phenomenon Revisited."
33. Walther Riese and Ebbe C. Hoff, "A History of the Doctrine of Cerebral Localization: Sources, Anticipations, and Basic Reasoning," *Journal of the History of Medicine and Allied Sciences* 5, no. 1 (1950): 50–71.
34. John Van Wyhe, *Phrenology and the Origins of Victorian Scientific Naturalism* (London: Routledge, 2017).
35. Raymond Tallis, *Aping Mankind* (London: Routledge, 2016), 33–50.
36. William R. Uttal, *The New Phrenology: The Limits of Localizing Cognitive Processes in the Brain* (Cambridge, MA: MIT Press, 2001).
37. Tallis, *Aping Mankind*, ch. 3.
38. Craig M. Bennett, Michael B. Miller, and George L. Wolford, "Neural Correlates of Interspecies Perspective Taking in the Post-mortem Atlantic Salmon: An Argument for Multiple Comparisons Correction," *Neuroimage* 47, Suppl. 1 (2009): S125.
39. Lady Diana Ladino, Syed Rizvi, and José Francisco Téllez-Zenteno, "The Montreal Procedure: The Legacy of the Great Wilder Penfield," *Epilepsy & Behavior* 83 (2018): 151–161.
40. Penfield, *Mystery of the Mind*, 28–31.
41. Charles Sherrington, *The Integrative Action of the Nervous System*, 2nd ed. (1947), quoted in Penfield, *Mystery of the Mind*, 4.
42. Penfield, *Mystery of the Mind*, 3–4.
43. Penfield, *Mystery of the Mind*, 28–31.
44. Penfield, *Mystery of the Mind*, 79–80.
45. Penfield, *Mystery of the Mind*, 80–81.
46. Penfield, *Mystery of the Mind*, 7.
47. Penfield, *Mystery of the Mind*, 7.
48. Penfield, *Mystery of the Mind*, 109.
49. Penfield, *Mystery of the Mind*, 49.
50. Penfield, *Mystery of the Mind*, 53.
51. Penfield, *Mystery of the Mind*, 53.
52. Penfield, *Mystery of the Mind*, 11.
53. Penfield, *Mystery of the Mind*, 6–7.
54. Penfield, *Mystery of the Mind*, 86.
55. Penfield, *Mystery of the Mind*, 5.
56. Penfield, *Mystery of the Mind*, 35.
57. Bahia Guellaï et al., "Sensus Communis: Some Perspectives on the Origins of Non-synchronous Cross-Sensory Associations," *Frontiers in Psychology* 10 (2019): 523.
58. Herbert H. Jasper, "History of the Early Development of Electroencephalography and Clinical Neurophysiology at the Montreal Neurological Institute: The First 25 Years, 1939–1964," *Canadian Journal of Neurological Sciences* 18, no. S4 (1991): 533–548.
59. Penfield, *Mystery of the Mind*, 77–78.
60. Lewis P. Rowland and Timothy A. Pedley, eds., *Merritt's Neurology* (Philadelphia: Lippincott Williams & Wilkins, 2005), 814.
61. J. R. Whitten, "Psychical Seizures," *American Journal of Psychiatry* 126, no. 4 (1969): 560–565.
62. International League Against Epilepsy, "Cognitive Seizure," *Diagnosic Manual*, 2022, https://www.epilepsydiagnosis.org/seizure/cognitive-overview.html.
63. Jordan Squair, "Craniopagus: Overview and the Implications of Sharing a Brain," *University of British Columbia's Undergraduate Journal of Psychology* 1 (2012);

Annalisa Coliva, "Comments on Peter Langland-Hassen's 'Craniopagus Twins and the Possibility of Introspective Misidentification,'" presented in online conference, Feb. 2013, available at https://consciousnessonline.files.wordpress.com/2013/02/coliva-commentary1.pdf. See also Michael Egnor, "What the Craniopagus Twins Teach Us About the Mind and the Brain," *Evolution News & Science Today*, November 24, 2017, https://evolutionnews.org/2017/11/what-the-craniopagus-twins-teach-us-about-the-mind-and-the-brain/ and "BC's Hogan Twins Share a Brain and See out of Each Other's Eyes," CBC, accessed 2023, https://www.cbc.ca/cbcdocspov/features/the-hogan-twins-share-a-brain-and-see-out-of-each-others-eyes.

64. Peter Langland-Hassan, "Introspective Misidentification," *Philosophical Studies* 172, no. 7 (2015): 1737–1758.
65. Adrian M. Owen et al., "Detecting Awareness in the Vegetative State," *Science* 313, no. 5792 (September 8, 2006): 1402.
66. Adrian M. Owen, Nicholas D. Schiff, and Steven Laureys, "The Assessment of Conscious Awareness in the Vegetative State," in *The Neurology of Consciousness*, eds. Steven Laureys, Olivia Gosseries, and Giulio Tononi (San Diego, CA: Academic Press, 2016), 155–166.
67. M. M. Monti et al., "Willful Modulation of Brain Activity in Disorders of Consciousness," *New England Journal of Medicine* 362, no. 7 (2010): 579–589.
68. Kenneth Ring and Madelaine Lawrence, "Further Evidence for Veridical Perception during Near-Death Experiences." *Journal of Near-Death Studies* 11, no. 4 (1993): 223–229.
69. Bruce Greyson and Ian Stevenson, "The Phenomenology of Near-Death Experiences," *American Journal of Psychiatry* 137, no. 10 (Oct. 1980): 1193–6, doi:10.1176/ajp.137.10.1193.
70. Bruce Greyson, Emily Williams Kelly, and Edward F. Kelly, "Explanatory Models for Near-Death Experiences," in *The Handbook of Near-Death Experiences: Thirty Years of Investigation*, eds. J. M. Holden, B. Greyson, and D. James (Santa Barbara, CA: Praeger, 2009), 213–214.
71. Gregg D. Caruso, *Free Will and Consciousness: A Determinist Account of the Illusion of Free Will* (Lanham, MD: Lexington Books, 2012); Susan Blackmore et al., *Exploring the Illusion of Free Will and Moral Responsibility* (Lanham, MD: Lexington Books, 2013); Daniel M. Wegner, *The Illusion of Conscious Will* (Cambridge, MA: MIT Press, 2002).
72. Penfield, *Mystery of the Mind*, 76–77.
73. Libet, *Mind Time*; Benjamin Libet, "Do We Have Free Will?," *Journal of Consciousness Studies* 6, nos. 8–9 (1999): 47–57.
74. Libet, *Mind Time*, 149.
75. Libet, *Mind Time*, 156.
76. Roger Scruton, "Brain Drain," *The Spectator*, March 17, 2012, https://www.spectator.co.uk/article/brain-drain/.
77. Maxwell R. Bennett and Peter Michael Stephan Hacker, *Philosophical Foundations of Neuroscience* (Oxford: Blackwell, 2003); Maxwell R. Bennett et al., *Neuroscience and Philosophy: Brain, Mind, and Language* (New York: Columbia University Press, 2007); Maxwell R. Bennett and Peter Michael Stephan Hacker, *History of Cognitive Neuroscience* (New York: Wiley-Blackwell, 2008).
78. Bennett and Hacker, *Philosophical Foundations of Neuroscience*, 12.
79. Bennett et al., *Neuroscience and Philosophy*, 74.
80. Sperry, "Mind-Brain Interaction: Mentalism, Yes; Dualism, No"; Akelaitis, "Studies on the Corpus Callosum II"; Akelaitis et al., "Studies on the Corpus Callosum III"; Bedeschi et al., "Agenesis of the Corpus Callosum"; Glickstein and Berlucchi, "Classical Disconnection Studies"; Gao et al., "Alien Hand Syndrome"; Sperry, "Forebrain Commissurotomy and Conscious Awareness"; Schramm et al., "Pediatric Functional Hemispherectomy"; Pulsifer et al., "The Cognitive Outcome of Hemispherectomy."
81. Feser, *Aquinas: A Beginner's Guide*, ch. 4.

14. Free Will, Free Won't, and What the Libet Experiments Don't Tell Us

Cristi L. S. Cooper

1. Introduction

When considering the mind-brain problem, an important aspect of the human mind is an individual's agency—the capacity of the person to make independent, free choices. Most people come to what they believe about free will by way of philosophical or theological convictions. In fact, studies of what people think about free will indicate that a majority of the "folk" (the non-philosophers and non-experts) believe that they are in control of their decisions: their "self" tells their brain what to do and not vice versa. In contrast, a majority of neuroscientists believe that the brain is in charge, that it is not merely the organ that is used to carry out the decisions, but is in fact the decider itself.

The views of philosophers on free will span a spectrum from non-deterministic (or "libertarian") to deterministic. A non-deterministic view, as the name implies, posits that decisions are not determined by the physical world and that humans are truly free to choose between one outcome or another. A deterministic view, on the other hand, holds that all of our decisions are causally pre-determined by brain states, entirely through physical processes, which excludes the possibility that the mind has any independent power of its own. In other words, non-deterministic views of free will leave open the door to mind-brain dualism while deterministic views of free will typically do not.

While some would argue that the existence of free will is incompatible with determinism (i.e., we can't possibly have free will if our decisions are pre-determined by or dictated by our matter), some believe that free will and determinism are mutually compatible, that even under determinism it is possible for people to be free, i.e., able to make decisions according to their desires, as long as their decisions are not impeded by external or internal coercion. This *compatibilist* view constitutes a middle ground.

Interestingly, there are Christians that would favor the compatibilist view on free will as well, depending on their view on and convictions about God's sovereignty. There are also many other nuanced stances on free will; see Robyn Repko Waller's white paper for more details.[1]

One's belief in free will and the manner in which it affects one's view of moral responsibility is important to consider. Quite often, people reason like this: "If I believe that I possess free will, then I have the responsibility to make the right moral decisions. If I don't have free will, then my brain is in control of me and I can do whatever I want and blame my brain." The sociological ramifications if the vast majority of people actually thought they weren't responsible for their actions are important to consider, and worthy of further research, but they are beyond the scope of this chapter. I raise this issue to acknowledge the importance of the topic of free will, without necessarily endorsing the conclusion that moral responsibility requires subscribing to a libertarian form of free will.

When one is specifically looking to science (e.g., biology, neuroscience) for what it can offer on a topic (as we are doing in this chapter), one must let the science speak for itself. In fact, a cursory review of the literature on the neuroscience of free will might cause one to think that the data supports a deterministic view of the will—one in which the brain is entirely in control of making a person's decisions. The predominance of this view is largely due to about forty years of neuroscientific study based on a publication in 1983 by Benjamin Libet. The purpose of this review is to summarize the data surrounding the Libet study and to indicate why many modern researchers believe that the conclusions of Libet and his followers deserve serious skepticism.

2. The Neuroscience of Free Will

Neuroscientists have been interested in finding the neural correlates of the will for a very long time, probably since the very beginning of brain research. The idea that the will could entirely be localized in brain tissue is something that philosophers, psychologists, and neuroscientists have debated for years, and I will not attempt to summarize the state of the debate in this chapter as it is well done elsewhere.[2] However, the findings of a pioneering study in the field in 1983 on the readiness potential (RP) by Benjamin Libet were interpreted at the time to support free-will skepticism (the idea that free will doesn't exist) and gave rise to an entire field of experimental investigations that continue to the present day.

Some go so far as to say that his study and subsequent studies prove that free will does not exist at all—that our decisions are entirely initiated and controlled by the matter of our brain. Others believe that too much has been made of this study and that further experimentation has left the question of the nature of free will wide open. Either way, Libet's experiments are so foundational to the "neuroscience of free will" that nearly every review of the subject begins with a description of his work. A 2021 review by Aaron Schurger et al. drove this point home: "It is difficult to overstate the degree to which the conclusions of Libet's papers on the RP have permeated the intellectual *zeitgeist*."[3] In the following pages, we will take a close look at the 1983 study; then we will look at studies that have been done in the past forty years that examine various components of the initial experimental paradigm; and, finally, I discuss the current state of the Libet-style of free will studies.

3. The 1983 Study

Before the time of Libet et al.'s 1983 study, German scientists had recorded an interesting and reproducible change in brain potential in the time immediately preceding a voluntary muscle movement.[4] Specifically, through electroencephalographic (EEG) recordings, they observed a slowly increasing surface-negative cortical potential leading up to voluntary muscle movement that begins up to 1–1.5 seconds before the movement.[5] They called this the *Bereitschaftspotential* or the "readiness-potential"

(RP). Given the relatively long time of the RP, Libet et al. were interested in the question of whether *conscious awareness* of the urge or intention to act also appears with similar timings.[6]

To address this question, Libet et al. devised a novel strategy for studying voluntary movements and the intention or will to make those movements. Subjects were asked to voluntarily and spontaneously move their fingers or wrists of their right hands whenever they freely wanted to do so. At the same time, they also were asked to note the position of a spot of light traveling around in a circle on a screen—its "clock-position"—when they were first aware of an urge to move. By doing this, the experience of the timing of the awareness (= time "W" for the "wanting" or "will" or intending to act) was turned into a visually related spatial image, similar to reading and recalling the clock-time for any experience. Subjects were also asked to report their awareness of "actually moving," which they called "M." During the experiment, an electroencephalogram (EEG) measured electric field potentials and an electromyogram (EMG) recorded activated muscle in the arm. Using all of this information, Libet et al. then related the time of the onset of the readiness potential (RP) to the time of the awareness of the urge to move (W), the awareness of moving (M) and the actual movement itself (as recorded by the EMG).

If conscious (or free) will were to be in control of producing a movement, then one would expect time W to exist before, or at the onset, of the RP in order to command the brain to perform the desired act. However, Libet et al. found that the negative shift in RP began *before* the corresponding W value by an average of 350 milliseconds (ms). They interpreted this to mean that the brain initiates a voluntary, spontaneous movement *unconsciously*, before there is any (at least recallable) subjective awareness that an intention to act has already been initiated. This finding kicked off the next forty years, up to the present day, of scientists referring to the Libet experiment as being the seminal experiment in the field that showed that there is no free will.

4. Free Won't

Interestingly, this was not necessarily Libet's own interpretation after more years of study, as summarized by himself in 1999.[7] This was due, in part, to a series of experiments that began with a finding made during the 1983 study. The scientists noted that after time W, there was another 150–200 ms before the beginning of the movement as signaled by the EMG. The authors suggested that this might be sufficient time for a conscious veto. While the study's major finding led researchers to believe that voluntary movements are unconsciously dictated by the brain before a subject has any awareness of a "wanting" or will to move, subjects indicated that periodically throughout the experiment there were times when they decided not to move at the last moment. Consciously changing one's mind had not been evaluated due to the experimental set-up—the lack of a movement in this case resulted in the absence of a trigger to record any RP that may have preceded the subject's veto. Therefore, there was no way to ask whether the RP preceded a movement that would end up being vetoed later.

However, in follow-up experiments, the final decision to act could still be consciously controlled during the approximately 150 ms after awareness of conscious intention (W) appeared. Subjects could "veto" motor acts during a 100–200 ms time period before an intended time to move.[8]

Reflecting on these results in 1999, Libet wrote, "The conscious veto may *not* require or be the direct result of preceding unconscious processes. The conscious veto is a *control* function, different from simply becoming aware of the wish to act."[9] What did he mean by this? His explanation rests on a distinction between the awareness of the decision to veto and the content of that awareness:

> The possibility is not excluded that factors, on which the decision to veto (control) is *based*, do develop by unconscious processes that precede the veto. However, the *conscious decision to veto* could still be made without direct specification for that decision by the preceding unconscious processes. That is, one could consciously accept or reject the programme offered up by the whole array of preceding brain processes. The *awareness* of the decision to veto could be thought to require preceding unconscious processes, but the *content* of that awareness (the actual decision to veto) is a separate feature that need not have the same requirement.[10]

5. The Readiness Potential

In the time since Libet's 1983 study, scientists have interrogated the readiness potential to better understand exactly what it is. In 1999, Patrick Haggard and Martin Eimer tested Libet's hypothesis that there would be a causal relation between the W judgment and initiation of the brain potentials (during the RP) by investigating whether random variation in the time of the former was accompanied by covariation in the latter. As Haggard and Eimer pointed out, covariation of causes and effects is a characteristic feature of causal relations. However, experimentation showed that the timing of the onset of the RP did not co-vary with the timing of a W judgment. They interpreted the lack of covariance to mean that the readiness potential can be ruled out as the unconscious cause of the conscious state upon which W judgment depends. Furthermore, they believed that this showed that W judgments reflect events related to the carrying out of a specific movement and not more abstract processes occurring upstream of *selection* of a specific movement.[11]

In 2010, Judy Trevena and Jeff Miller published a study directly testing the functional significance of the "movement-preceding negativity" (as they called the readiness potential, referring to the surface-negative cortical potential).[12] Using the experimental paradigm of Libet, Trevena and Miller recorded the EEG activity of subjects who were asked to generate occasional spontaneous movements or choose not to move at all at specific points in time. They hypothesized that the negativity preceding decisions to move should be larger when movements are actually made than when they are not made, assuming that this would be the case if the negativity is specifically associated with a brain preparing to cause a movement. However, in experiments, they found no evidence of the predicted difference: there was no evidence of stronger electrophysiological signs before a decision to move than before a decision not to move, although they did observe clear negativity (showing that their experimental set-up did indeed work).

They concluded that the negativity must not necessarily reflect preparation for movement and might occur merely due to ongoing attention to or involvement with a task producing occasional spontaneous movements. They believed that this conclusion was clearly at odds with the view proposed by Libet et al., i.e., that this negativity always reflected subconscious preparation for movement by the brain. In their view, this finding weakened the argument, based on Libet's 1983 results, that voluntary actions are initiated unconsciously. Despite this disagreement, Trevena and Miller did not suggest that neural activity didn't underlie conscious decision-making; in fact, most, if not all, of the scientists studying free will in the experiments mentioned above and later in this chapter do believe in a deterministic model of free will.

Since the publication of Libet's paper in 1983, psychologists have also used the lateralized readiness potential (LRP) in studies such as these because it has seemed to be more specific to motor preparation than the RP.[13] The LRP is a negative potential that can be observed over the motor cortex contralateral to the hand making the movement. In Haggard and Eimer's 1999 study, they also measured the LRP in their experiments to detect covariance, this time comparing LRP and W. This time, they did find that the W judgment covaried with the onset of the LRP, suggesting that the LRP may have a causal relation to W. Because the LRP is a relatively late event in the sequence of physiological changes that lead to action, they suggested that the LRP onset represents the stage when representation of abstract action is translated into representation of specific movement. In other words, it may indicate the starting point of conscious awareness of motor action.[14] However, this finding has since failed to be replicated. Alexander Schlegel et al. found no within-subject covariation between LRP onset and W with a larger group of participants and several variations in analytical method.[15]

So if neither the readiness potential (RP) nor the lateralized readiness potential (LRP) are *causally* linked to the moment of conscious will, then what do they represent? Schlegel et al. (2015) performed an interesting experiment on post-hypnotic patients to test whether the RP precedes the feeling of wanting to act or whether it could be present if this feeling was absent.[16] They reasoned that if endogenously initiated volitional movement can occur in the absence of the *feeling* of wanting to move (W), then conscious intention is not a cause of volitional action in all cases. In addition, they believed that if the presence of a readiness potential precedes such a movement, then the readiness potential must reflect neural processes that are independent of conscious will.

In a set of experiments, hypnotic patients were asked to squeeze a stress ball. Upon being awakened from the hypnotic state, patients were told a cover story as to why their arms moved—that the experimenter was calibrating the electromyography (EMG) electrodes on their forearms. They were told this to remove any suspicion that they had been asked to squeeze a ball when they watched video recordings of themselves. They were then asked to write down whatever they recollected about their hypnosis session. Only subjects who believed the cover story and said that they did not experience feelings of conscious will were included in the analysis.

In these subjects, the researchers were able to elicit volitional movements in the absence of conscious feelings of will. Interestingly, the researchers found that the RP still occurs when subjects make self-timed, endogenously initiated movements due to a post-hypnotic suggestion. These movements were made without a conscious feeling of having willed those movements themselves. They conclude:

> Since both an RP and LRP occur even when subjects perform a motor act without being conscious of having commanded it (due to the post-hypnotic suggestion, they perceive that the ball squeezes happened due to external forces), the RP and the LRP may be unrelated to the subjective experience of intentional movement. While the

> exact nature of the neural processes reflected by the RP remains unclear, we can conclude that those processes are not specific to conscious willing.[17]

However, even though the RP and the LRP do not appear to be causally linked to the moment a person is aware of a will to move and the RP appears even when one never has a conscious will to move, the negative potential preceding movements remains nonetheless. So what is it? Jeff Miller et al. rightly reasoned that if the negative potential arises from some brain process other than unconscious movement preparation, then the fact that it appears before the conscious decision to move would not necessarily support the claim that movement preparation begins unconsciously.[18] Correlation does not necessarily imply causation. Miller et al. compared the EEGs observed when subjects participating in Libet-paradigm experiments *did* versus *did not* monitor a clock display while waiting to make spontaneous key-press movements. Their results show that when subjects monitored a clock, there was a reliable negative potential on EEG, but in the no-clock condition, there was no such effect. They concluded that much or all of the premovement negative EEG shift can be attributed to the subjects' observation of the clock used to report event times.[19]

The previously mentioned review on the readiness potential by Aaron Schurger et al. argues that recent advances in our understanding of the RP call for a reassessment of its relevance for understanding volition and the philosophical problem of free will.[20] In the "Concluding Remarks" section, the authors write, "If recent models of the RP are on the right track, we cannot infer from the existence of the phenomenon that it reflects an actual signal in the brain that, in individual trials, has the characteristics of the RP, or that has causal efficacy. Because of this, one cannot infer that we lack conscious free will based on the temporal profile of the RP."[21]

6. The Question of "W"

IN 1999, Haggard and Eimer questioned the nature of the unconscious and conscious events described by Libet.[22] They conjectured that, because the movements made by the subjects in Libet's study were always fixed (i.e., there was only one kind of movement to be made, no decision about which hand to move or anything more complicated than *when* to move), the reports of W or M might have reflected something as general as arousal. Alternatively, they suggested that the subjects' reports might have reflected information specific to the actual movement to be made.

To better understand the nature of the W and M judgments, this study compared reports made by subjects carrying out fixed movements with those made by subjects carrying out free choice movements, in which subjects chose freely on each trial between two voluntary actions. They hypothesized that, in a serial model of action generation, earlier judgments would be expected in free movement conditions because a decision about which movement to make (or which hand to move in this case) would be more time-consuming than a decision about when to move an already chosen (fixed movement) hand.

In experiments, however, they observed no difference in either W or M judgments, or in movement-related brain potentials, between fixed and free movement conditions. They did not think the finding was surprising in the case of M judgments, suggesting that awareness of when an action is performed must be tied to when the specific movement involved is executed. They did, however, find the lack of difference in W judgments surprising, and concluded that the data suggest "that people's awareness of initiating action relates to preparing a specific movement, rather than a general abstracted state of intending to perform an action of some kind." In other words, W must not represent a wanting or desire to move one hand versus another—not an exertion of the will *per se*—but rather the awareness of preparing a specific movement of whatever kind.

William P. Banks and Eve A. Isham further call into question the meaning of W in a 2009 study.[23] They proposed that the reported W would not be determined by neural events related to the generation of the readiness potential, but that instead, W is the time participants select based on available cues, especially the apparent time of response. They tested the hypothesis that the critical cue for judgment of intention is perception of the response, thus reversing the assumed causal relation between intention and action. To do this, they used delayed-response feedback to create an illusion that the response was later than it actually was. They reasoned that if the perceived time of response was an important factor in judging one's intention, then a delay in the perceived time of the action would result in a delay in the reported time of W. However, if the contrary were true and W measures an event that occurred prior to the response, it would be constant no matter what false information was presented to influence the apparent time of response.

After running subjects through a protocol adapted from the Libet paradigm, with the addition of a beep delayed by a computer-generated random sequence at 5, 20, 40, or 60 ms, participants were asked to report the position of a cursor on a clock face at the moment they made the decision to move. Interestingly, they found a reliable effect of delay of feedback in the hypothesized direction. In other words, when the beep was delayed (i.e., when their auditory cue that they had pressed the button was delayed), they reported a delay in when they first felt the wanting to move W.

They concluded that a large component of or possibly the entire judgment of W is retrospectively inferred after the fact from the response, or "postdicted." Furthermore, they interpreted their findings to mean that the intuitive model of volition is overly simplistic in that it assumes a causal model by which an intention is consciously generated and that it causes an action. However, because W can be manipulated after it occurs, it cannot possibly be causal for the action in a straightforward manner.

7. Where the Libet Paradigm Stands

After scientific interrogation of the Libet experimental paradigm over the last forty years, scientists know much more about the readiness potential and the moment of conscious will but don't seem any closer to agreeing as to the significance of many aspects of the original findings. To summarize the experiments described here, the readiness potential that was first described by Libet as being an upstream, unconscious causal signal for initiation and invocation of conscious will has been cited for years as being the key evidence that unconscious brain states cause conscious will. This study has been taken to be the key neuroscientific evidence that we truly don't have free will. Libet himself later came to believe that the readiness potential may start a specific voluntary action, but that the conscious will controls volitional outcome (what some have called "Free Won't").

Experiments show that the readiness potential (RP) does not co-vary with the will (W) to act, ruling out a causal role of the RP on W,[24] and that a decision to move does not cause a larger EEG negativity compared with a decision to not move, indicating that it must not be associated with preparation for movement at all.[25] In hypnotic patients, the readiness potential is associated with preparing to move even when patients have no desire or will to move, indicating that the RP is independent of processes associated with a conscious will.[26] Interestingly, pre-movement EEG shifts disappear when subjects don't watch a clock, calling into question the role that watching a clock plays in the Libet results.[27] W, or the moment someone feels a "will" or "wanting" to make a movement seems to be unaffected by having a choice, which seems counterintuitive.[28] Lastly, W can also be manipulated after it occurs, calling its significance into question.

The studies described here constitute just a handful of the studies that call into question the significance of the Libet findings on the free will question. This is significant and important considering that, at the popular level, non-neuroscientists use Libet's studies to support a deterministic view of the mind.[29]

Indeed, although the vast majority of neuroscientists believe in a deterministic view of free will, many of them do not believe that Libet's experiment can be shown to do away with free will. In a useful review in the journal *Neuroethics*, Andrea Lavazza and Mario De Caro provide examples of conceptual confusion surrounding neuroscientific studies on human agency. Their conclusion is spot-on: "Our aim here has only been to suggest that, when one comes to the issue of human agency, great caution should be used before drawing bold philosophical, political and social conclusions from neurological findings, whose correct interpretation and value are still extremely controversial."[30] This is the aim of this chapter, too.

They add that there are "plenty of excellent reasons for believing that neurobiology will continue to enrich our understanding of many features of the human mind, but there are no good reasons for thinking that it is going to fully explain them all."[31] Neuroscience has told and will continue to tell us much about the brain as the organ that carries out the actions of the mind, and we have to continue to have conversations about what the science means. The scientists would do well to remain open to philosophical considerations that span the spectrum of thought on free will and, likewise, the philosophers would be wise to ground their ideas in what has been demonstrated regarding the human brain and body.

Among the goals of this book are to bring together scientists, philosophers, psychologists, sociologists, computer scientists, and others to discuss the mind-body (or mind-brain) problem. Neuroscientists and philosophers alike agree that such a collegial approach is precisely what is needed. At the end of Waller's white paper, she writes:

> Of course there are limits to what psychology and neuroscience can tell us about metaphysical matters such as determinism and dualism. To this end, scientists and philosophers would do well to continue to foster interdisciplinary dialogue on what, exactly, we mean when we speak of determinism (and varieties thereof) as well as how nonexperts conceive of free will and moral responsibility and the impact of that conception on their social behavior.... Continued collaboration of the humanities and sciences promises new insights regarding our lives as free and responsible agents.[32]

An interdisciplinary approach to such complicated questions indeed is conducive to progress, as scientists, philosophers, and non-experts bring different perspectives and insights to the table. Different perspectives breed more creative ideas, better hypotheses, and more thorough analyses of data. In fact, inclusion of dualist views of the mind may inspire questions and/or interpretations of data that may ultimately lead to a better understanding of the mind.

Whether we truly possess "free" will, or rather are entirely pre-determined by brain states, is important to consider for many reasons, among which is to help answer the question of whether we are of two natures (physical and spiritual) or just one. As shown here, that question is not answered by the Libet experiments, and the exciting news is that the field is wide open for scientific investigation.

Notes

1. Robyn Repko Waller, *Free Will: Recent Work on Agency, Freedom, and Responsibility* (John Templeton Foundation, 2019), https://www.templeton.org/wp-content/uploads/2020/09/FreeWill-JTF-Final.pdf.
2. Waller, *Free Will.*
3. Aaron Schurger et al., "What Is the Readiness Potential?," *Trends in Cognitive Sciences* 25, no. 7 (July 2021): 559, https://www.sciencedirect.com/science/article/pii/S1364661321000930.
4. Hans H. Kornhuber and Lüder Deecke, "*Hirnpotentialänderungen bei Willkürbewegungen und passiven Bewegungen des Menschen: Bereitschaftspotential und reafferente Potentiale*" ["Brain Potential Changes in Voluntary and Passive Movements in Humans: Readiness Potential and Reafferent Potentials"], *Pflüger's Archiv für die gesamte Physiologie des Menschen und der Tiere* 284 (1965): 1–17.
5. Kornhuber and Deecke, "Brain Potential Changes."
6. Benjamin Libet et al., "Time of Conscious Intention to Act in Relation to Onset of Cerebral Activities (Readiness-Potential). The Unconscious Initiation of a Freely Voluntary Act," *Brain* 106, no. 3 (September 1983): 623–642, https://doi.org/10.1093/brain/106.3.623.
7. Benjamin Libet, "Do We Have Free Will?," *Journal of Consciousness Studies* 6, nos. 8–9 (1999): 47–57.
8. Benjamin Libet, "Unconscious Cerebral Initiative and the Role of Conscious Will in Voluntary Action," *The Behavioral and Brain Sciences* 8 (1985): 529–566.
9. Libet, "Do We Have Free Will?," 53. Emphasis in original.
10. Libet, "Do We Have Free Will?," 53.
11. Patrick Haggard and Martin Eimer, "On the Relation between Brain Potentials and the Awareness of Voluntary Movements," *Experimental Brain Research* 126 (1999): 128–133.
12. J. Trevena and J. Miller, "Brain Preparation before a Voluntary Action: Evidence against Unconscious Movement Initiation," *Consciousness and Cognition* 19 (2010): 447–456.
13. M. G. H. Coles, "Modern Mind-Brain Reading: Psychophysiology, Physiology, and Cognition," *Psychophysiology* 26 (1989): 251–269; M. Eimer, "The Lateralized Readiness Potential as an On-Line Measure of Selective Response Activation," *Behavior Research Methods, Instruments, & Computers* 30 (1998): 146–156.
14. Haggard and Eimer, "Brain Potentials and the Awareness of Voluntary Movements."
15. Alexander Schlegel et al., "Hypnotizing Libet: Readiness Potentials with Non-conscious Volition," *Consciousness and Cognition* 33 (2015): 196–203.
16. Schlegel, "Hypnotizing Libet."
17. Schlegel, "Hypnotizing Libet," 201.
18. Jeff Miller, Peter Shepherdson, and Judy Trevena, "Effects of Clock Monitoring on Electroencephalographic Activity: Is Unconscious Movement Initiation an Artifact of the Clock?," *Psychological Science* 22, no. 1 (Jan 2011): 103–109.
19. Miller et al., "Effects of Clock Monitoring."
20. Schurger et al., "What Is the Readiness Potential?"
21. Schurger et al., "What Is the Readiness Potential?," 567.
22. Haggard and Eimer, "Brain Potentials and the Awareness of Voluntary Movements."
23. W. P. Banks and E. A. Isham, "We Infer Rather Than Perceive the Moment We Decide to Act," *Psychological Science* 20, no. 1 (2009): 17–21.
24. Haggard and Eimer, "Brain Potentials and the Awareness of Voluntary Movements."
25. Trevena and Miller, "Brain Preparation before a Voluntary Action."
26. Schlegel, "Hypnotizing Libet."

27. Miller et al., "Effects of Clock Monitoring."
28. Haggard and Eimer, "Brain Potentials and the Awareness of Voluntary Movements."
29. Annaka Harris, *Conscious: A Brief Guide to the Fundamental Mystery of the Mind* (New York: HarperCollins, 2019).
30. Andrea Lavazza and Mario De Caro, "Not So Fast. On Some Bold Neuroscientific Claims concerning Human Agency," *Neuroethics* 3 (2010): 39.
31. Lavazza and De Caro, "Not So Fast," 39.
32. Waller, *Free Will*, 49–50.

15. On the Limitations of Cutting-Edge Neuroscience

Joseph Green

1. Introduction

Neuroscience is one of the fastest growing scientific fields. Increasing our understanding of how the brain works is often regarded as one of the most significant challenges of the twenty-first century. Recent neuroscientific discoveries have been celebrated step by step in the media as a result of their significance. Yet, to this day, no major technology company has been able to turn scientific knowledge of the brain into profits. Engineering the brain has proven extremely difficult and no mind-reading devices or mind-controlled tools have yet been invented. The day when the brain is an engineered system still seems a long way off.

The gap between popular expectations of neuroscientific knowledge and our ability to manipulate the brain as an engineered system is an important one rarely addressed. Here we explore the latest neuroscience and neurotechnology discoveries in an attempt to chart the current scientific frontier of this fast-evolving field of science. The present paper demonstrates how the popular expectation of high-performance neurotechnology, often based on the technocratic propaganda of eminent entrepreneurs such as Elon Musk,[1] is probably misplaced. Though heady claims have recently been made, mind-reading devices and artificial memories incorporated into brain networks are still more a part of science fiction films than science.

Our next step will be to determine when this frontier is likely to impact philosophical debates within philosophy of mind and metaphysics. Philosophical problems related to the mind-body problem, which for centuries have been out of reach for scientists, may now seem to fall within the sphere of neuroscience. But is that really the case? In reality, even in the most optimistic case, neuroscience may be decades away from being able to inform critical ideas in philosophy of mind. In other words, philosophical theories addressing the mind-body problem are, for the most part, unlikely to be validated or invalidated by neuroscientific findings in the upcoming years.

Due to this distance between the frontier of neuroscience and most philosophical theories, it follows that no specific philosophical theory should be preferred over another on the basis of scientific findings. This is not to say that no matter what scientific findings are made in the future, the science will still underdetermine the philosophy (a position termed the "autonomy thesis"[2]), but rather that at present scientific theories do not constrain most philosophical ones.

This implies that the current dogma that pervades neuroscience—materialist monism that can be simply stated as "we are nothing but our brains"—is established by intellectual pressure rather than solid scientific evidence. In order to quantify this, we briefly explore recent developments in promising neuroscientific directions, which in the years to come could produce important results for limiting philosophical theories of mind and body. Some of these are further developed in other chapters towards which we point the interested reader.

2. The Knowledge Gap in Neuroscience

According to a recent research paper titled "The Seductive Allure of Neuroscience Explanations," when people are presented with explanations of psychological phenomena or behavior, they are more inclined to trust them if they include neuroscientific material.[3] Surprisingly, this is true even when the neuroscientific knowledge supplied is clearly unrelated to the topic under discussion. This anomaly regarding neuroscience explanations is part of a broader category of cognitive biases that affect the vast majority of the population.[4] A cognitive bias is a consistent flaw in our perception that causes us to misunderstand something. In other words, we are led to believe something is not what it is. The example above is an example of what is known as a framing bias. This happens when we make a decision or come to a conclusion based on how information is presented rather than on evidence.

In an effort to foster scientific advancement, neuroscientific breakthroughs are frequently celebrated in the news. It is not surprising, then, that our perception, which has been conditioned to resonate with such a message, clings to the framing bias just described. Some recent scientific discoveries are astounding and deserve to be promoted as major achievements; however, the increased promotion of scientific discoveries has the unintended consequence of biasing people toward believing scientific explanations—neuroscientific explanations in particular—even when they are false, as demonstrated by the study mentioned above.

This invites the question of whether neuroscience, like other scientific fields, is being depicted in a way that exaggerates its explanatory potential. While the way science is presented by journalists is constantly being corrected,[5] it still seems that people are led to believe that the current state of neuroscience is significantly more advanced than it is. This preconception is a contributing component to the framing cognitive bias, and it highlights a divergence between the frontier of the neuroscience field and people's perceptions of it.

This divergence in people's perceptions is the result of a variety of reasons. The way the media depict neuroscientific discovery is part of this, as is the pressure on scientists to acquire financing for their studies. However, we shall argue that the most significant factor contributing to it is the gap between neural technologies' ability to act on the brain and our understanding of neural mechanisms. Our technological ability to record from the brain, perturb it, and genetically manipulate it has outpaced our ability to understand the neural mechanisms it exploits to function. In other words, while new technologies have enabled us to operate on the brain and offer new areas of inquiry, they have yet to produce a mechanistic understanding of how the brain functions.

Certainly, a number of spectacular outcomes have been obtained, and we will discuss some of these below, but these technologies have not yet been followed by breakthroughs of comparable size in our fundamental understanding of how the brain functions. We refer to this gap between our technological ability to act on the brain and our understanding of its mechanisms as the *knowledge gap*, and show how it fuels the popular misconception that our understanding of the brain is more advanced than it is.

Consider the technological ability to systematize distinct cell types in specific brain areas. Many followers of popular science know that brain neural networks are made up of excitatory and inhibitory neurons. However, these neurons come in hundreds of different kinds that are grouped into cell types depending on their shape and other attributes. Recent technological developments known as optogenetics, which we shall discuss in more detail below, enable the activation of all neurons of a certain cell type. This extraordinary skill, one would imagine, is accompanied by a deep understanding of why and how the brain uses specific cell types for computations. However, this is mistaken; our ability to affect the brain does not directly translate into comprehension of its principles, and this frequently neglected knowledge gap leads people to believe that our current understanding of brain circuits and functions is deeper than it is. Media then contribute to people's misperceptions by celebrating technological achievements without depicting our limited understanding, but the true cause of such misperception is to be attributed to this gap in our understanding, which is more prominent in neuroscience than in other fields.

As a result, we will first examine how such mechanistic understanding has evolved in fields other than neuroscience. This will allow us to build a baseline against which we may compare other scientific domains. Then, we will highlight this gap by reviewing recent neurotechnological advancements and relating them to examples that demonstrate our poor understanding of brain circuits and dynamics. This will allow us to trace the frontier of neuroscience as a field, assisting us in evaluating its relationship to important philosophical positions.

3. Neuroscience: A Mature Science?

Ramon y Cajal, a Nobel laureate in 1906, is often credited with establishing the "neuron doctrine," the concept that the functional unit of the brain is the neuron.[6] Since then, neuroscience has evolved into the fast-paced field that it is today. How does this progression compare to that of other scientific fields? Several researchers have sought to quantify and rank scientific fields based on their qualities over the previous century.

According to Dean Keith Simonton, one of the fundamental characteristics, across scientific fields, is the importance of theories: "Strong theories have a crucial function" in that they "define a discipline's central nomenclature and concepts, and they determine what facts are important and what not. Furthermore, theories provide the foundation for generating the predictions and hypotheses that guide empirical research in a highly specific direction."[7] According to Simonton's analysis, fields such as physics and chemistry are at the top of the hierarchy, driven by strong theories, while fields such as sociology are at the bottom. Because of the inherent complexity of biological systems and the lack of theories that guide empirical investigation, life science fields such as neuroscience occupy midway positions.

Thomas Kuhn proposed another influential way of defining scientific fields. He demonstrated how mature scientific fields are led by paradigms—a set of hypotheses and facts that form the field's backbone—to the point where some disciplines can be classified as pre-paradigmatic and others as paradigmatic. According to his definition, a scientific field is paradigmatic when "some accepted examples of actual scientific practice—examples which include law, theory, application, and instrumentation together—provide models from which spring distinct coherent traditions of scientific research."[8] In other words, a paradigmatic field is distinguished by coherence and tuning of research directions.

According to these assessments, neuroscience falls somewhere between paradigmatic and pre-paradigmatic sciences, or, to use Simonton's terminology, somewhere in the middle of his science ranking. One of the reasons is that neuroscientific theories are not yet developed enough to have broad community acceptance and, more significantly, to create testable predictions. Neuroscientific theories do not yet have the ability to construct frameworks (or paradigms) that can drive the field as a whole. In other words, neuroscience is still an experimentally driven field in which experimental findings, rather than ideas, drive the pursuit of its boundaries. This is not to suggest that theoretical efforts have not been relevant and spreading in recent decades; nonetheless, these efforts are relatively new when contrasted to the theoretical traditions of disciplines such as chemistry or physics. Neuroscientific theories rarely predict the outcome of an experiment; rather, theories and models are most often used backwards to explain the outcome of experiments and conceptually organize experimental findings *a posteriori*. This differs from other sciences in certain aspects.

A popular parallel is with some fields of physics. For example, in the recent decade, two significant physics predictions were confirmed: gravitational waves and the Higgs boson. The former was predicted by Einstein in 1916 and discovered in 2015. The latter was predicted by Higgs in 1964 and discovered in 2012 at the CERN large hadron collider. These two discoveries were made feasible by two paradigms that allowed physics to become a paradigmatic science throughout the last century. The first is the theory of general relativity, while the second is the standard model of high-energy physics.

It is evident how the theoretical forefront of these fields can drive experimental findings with a lag of decades. This is not true in neuroscience or biological fields in general. Theoretical frameworks are not as powerful in such fields, and it is therefore a mistake to think of our understanding of the brain in similar terms. When we think of scientific paradigms such as quantum physics or organic chemistry, we think of theories and their technological applications; nevertheless, the theoretical aspect of neuroscience is not as established. It's simply different in that our theoretical understanding of the brain isn't much further along than the experiments that have been conducted on it.

The aspects just discussed propose a method for assessing a science's maturity: track its empirical frontiers and see if theoretical frameworks are ahead or behind. This would imply that hypotheses would be available to explain the various possible outcomes of the same experiment. In that spirit, we will now investigate the state of neuroscience as a field and the degree of what we termed the knowledge gap—the gap between our technological ability to act on the brain and our understanding of it.

4. Recent Successes in Neuroscience

RECENT BREAKTHROUGHS in neuroscience have been defined by significant technological advances. The combination of novel materials, electronic devices, and genetic alterations has increased our ability to record and perturb brain activity. Furthermore, modern machine learning technologies provide promising ways for decoding and predicting many variables from brain activity, such as limb movements (the motion of an arm reaching out to grab an object), binary choices (the decision to move right or left), and so on. Three examples that best portray these achievements and encapsulate these advancements are: prosthetic limbs, optogenetics, and calcium recordings. While additional notable examples might be offered, these three exemplify the achievements of modern neurotechnologies; thus, we will explore them in greater depth.

4.1 Brain Machine Interfaces and Prosthetic Limbs

Recent days have clarified our ability to decode limb movements (arms and legs) from neural activity. Brain-machine interface (BMI) studies in monkeys have shown that animals can control cursors on a screen with the simple activity of their brain and similarly, whenever monkeys move their arms, it is possible to tell the position of the arm from the brain's neural activity.[9] These findings show not only that neural population recordings—probing the neural activity of many neurons at a time—enable us to read out behavioral movements of the animal,[10] but also that in principle we can move physical objects with the simple activity of our brain.

Brain-machine interfaces, now mainly developed through implants in monkeys, seem to be very promising. Indeed, similar technologies, coupled to surgical innovations and enhanced rehabilitation, are pushing forward the field of prosthetics.[11] Recently, humans have been shown to be able to control prosthetic robotic arms directly through their brains. Although many challenges are yet to be faced to make this an everyday reality, it is possible to have surgical implants in the brain that control arm movements.

The problem is that these technologies require medical and surgical expertise to correctly install and operate, are very costly, and involve potential risks. Furthermore, they may not be permanent solutions and may work only temporarily. Research is advancing steadily and, despite the fact that most of these technologies require brain-invasive implants (chips implanted in the brain), it is still possible that future discoveries will enable us to partially control limb movements with noninvasive signals such as EEG, a technology that reads electric activity outside the skull.[12]

4.2 Calcium Recording Technologies

The last couple of decades have seen incredible progress in genetic techniques. Our ability to process, study, and manipulate the genetic code has increased incredibly. Some of these techniques have had important applications in neuroscience. In the last decade, specific technologies have been developed where it is possible to record the activity of large neural populations simply as a movie.

Indeed, there exist genetic technologies nowadays for which it is possible to have neurons light up (literally emit luminous signals) whenever they are active (i.e., they emit neural spikes). Neurons in the brain communicate through synapses by emitting electric discharges—spikes—which travel through the synapses to reach other neurons. The chemistry of neurons is such that in order to emit spikes neural cells exploit their calcium channels to exchange calcium with the external environment. Recent genetic techniques have allowed scientists to genetically encode light-emitting calcium indicators in neurons so that, when the cell intakes calcium to generate spikes, it also emits light.[13] This jaw-dropping genetic manipulation requires animals to be genetically modified but can target specific brain areas and specific cell types (such as excitatory or inhibitory neurons). As a result, it is possible to record with a camera (or other photon-detecting technologies) these luminous signals and image neural activity. Formally the signal recorded is the rate of calcium intake of neurons which strongly correlates, because of the process just described, with the neurons' spike generation.

This technique, often deployed on rodents (mice and rats), has also been deployed on other animals. One example is the zebrafish. The body of the zebrafish is transparent, and therefore it has been possible to record, from outside of the animal without invasive techniques, all the neural activity throughout its body. This technique allows researchers to record the simultaneous activity of thousands and thousands of neurons while the animal is freely moving in the water.[14]

4.3 Optogenetic Manipulations

Another technology based on principles similar to the one just described is one of the main foci of future neuroscience directions: optogenetics. In this technology the same calcium channels mentioned above (that mediate the intake of calcium in a neuron to allow spike generations) are bound to a light-sensitive molecule. In this way calcium channels become light-sensitive ion channels that can be activated by shedding light on them.[15] This technology can be applied to specific neural areas and neural cell types, and when these ion channels are activated by means of light beams the neurons in which they are present are activated. In other words, this technology allows one to activate neurons on demand by reaching them with a light beam. Recent and future neuroscience studies are therefore focusing on recording neural activity with and without optogenetic perturbations. This allows one to understand what happens when specific neurons in the brain are turned on, participating in the activity even when they are supposed to be silent.

It has already been shown that this procedure can impair learning and performance of animals in behavioral tasks, and the specific areas and circuits that impair learning are currently under study.[16] Similarly it has been possible to optogenetically manipulate neural circuits controlling emotions such as fear. In a recent study, researchers have shown that the activation of a specific area critically reduced fear induced by specific signals in rodents.[17] Another study focused on social interactions[18] and registered a clear increase in aggressive behaviors in rodents (attacks, offensive upright posture, and chasing) when optogenetic manipulations were targeting specific neurons, i.e., oxytocin neurons. These results pave the way to future studies that may enable us to uncover the neural circuit pathways and hierarchy contributing to the generation of behaviors, emotions, and social interactions.

5. Limitations within Neuroscience

The three technologies just discussed demonstrate our extraordinary ability to record and act on neural activity from the brain—they truly are incredible. Some of these have obvious philosophical implications. For example, the fact that certain information can be decoded from the brain implies that such information is encoded in some way in neural circuits. It does not, however, mean that the brain circuits where activity is recorded are also the origin of this information. Furthermore, the fact that perturbations of neural activity result in changed mood and disrupt learning means that, at least in rodents, some aspects of emotions and learning ability can be affected by perturbing material brain tissue.

In the future, such perturbation may even be able to influence an animal's decisions in decision-making tasks. In these activities, the animal must usually select between two images or sounds that have a specific property (e.g., has a stronger contrast in the case of two images), and by selecting the correct one, the animal receives a reward (e.g., some drops of juice). It is possible that some optogenetic perturbations will influence the animal to choose one option over the other. Even in this instance, however, it would not be possible to say where the choice originated in the animal brain, but only that such information can be acted upon as it is communicated via neural impulses. We shall return to this topic later in our discussion.

To come back to our main point, there is a scientific gap between the technological developments that allow us to do incredible experiments on the brain and our current understanding of the underlying neuronal dynamics and mechanisms. In the same vein as that of the last three examples, we now present three more cases that demonstrate our poor understanding of brain systems.

5.1 C. elegans

One of the neural systems that has attracted a lot of attention in neuroscience is that of *C. elegans*. This is a species of worm that has been extensively studied. The interest in this tiny worm's system arises mainly from the fact that it has only 302 neurons and that the animal can perform very few behaviors. One of its primary behaviors is chemotaxis, which means motion based on sensing specific chemicals when seeking out food sources, avoiding noxious substances, and finding mates. Furthermore, all neurons and connections among them have been completely mapped in *C. elegans* so that we fully know its connectome. Additionally, most of its genetic properties have been investigated.[19]

Despite knowing so much about such a simple neural system, it is fair to say that we yet do not understand how the animal performs the behaviors we observe. We do not understand how its decisions are generated and how neural dynamics control what it does. Of course, there have been several breakthroughs in *C. elegans* research, but the expectation that we would by now understand "everything" of this small system has largely gone unmet. Making sense of neural dynamics in *C. elegans* is an open problem to which researchers keep contributing year after year.[20]

The fact that our understanding of a system such as this, which we can modify and study in detail, is still so limited may or may not be surprising, but it is an indicator of how far we might yet be from making sense of the dynamics in the human neural cortex. One cubic millimeter of cortical tissue usually accommodates between 10^5 and 10^6 neurons. At present, we do not know how these neurons are connected in detail, nor even their genetic properties at large. Therefore, the lesson that we may learn from *C. elegans* research is that the neural dynamics in a system as complex as the human brain may still be out of reach of our understanding.

5.2 Computer Chips

Someone might be inclined to think that the case of *C. elegans* is not a good indicator of our ability to make sense of neural dynamics, simply because a worm, after all, is not a very intelligent animal. Its behavior is limited to performing very few activities and one could hope that by analyzing systems—brains—which perform more intelligent operations, it would be possible to get further. In a recent study researchers have shown that if we were to analyze, with the current theoretical frameworks and analysis tools, a microprocessor (the "logic chip" in a computer) as if it were part of a brain, then we would understand barely anything of it.[21]

In this study Eric Jonas and Konrad Paul Kording showed that even after analyzing the "activity" and "connectivity" of the chip, even by knowing almost everything about it, it would yet be practically impossible to deduce what operation the chip is performing and even more so how the chip is performing such operations. Opening such a "black box" as a microprocessor is hopeless unless we know its design principles and functioning.

5.3 Distributed versus Localized Coding

Lastly, we bring an example that uncovers a fundamental unknown in neuroscience. This is often termed as a dichotomy: "segregation" versus "integration," or "modularity" versus "integrated processing." The question is whether, at least in first approximation, we should think of the brain hierarchically, such that different areas process different types of information independently (modularity or segregation), and then this information is combined, or non-hierarchically, such that processing is instead more distributed and integrated across the brain.[22] This question, towards which many research lines converge, has left scientists divided. No consensus has been achieved. While the idea that different cortices (e.g., visual, olfactory, motor) are dedicated to specific functions has been popularized and to some extent confirmed, the neuroscience community keeps being puzzled by findings which are left unexplained.

Recently, for example, researchers have found that in the visual cortex of rodents most of the activity could not be explained by (i.e., correlated with) the visual scenes that the animal was observing, but rather was traced to the different movements that the animal was performing (e.g., running, whisking, rearing).[23] It seems likely that information relative to animal's movements is distributed across the brain, and theoretical models are emerging on why it might be so. But this nonetheless stands as an example of why we cannot easily pin down specific organizational principles of the brain when it comes to information flow. While much is being discovered about the activity and connectivity of specific brain areas and circuits, not much is yet understood of how information is actually processed and integrated. Someone may wonder whether and how these studies performed in animal systems inform our understanding of the human brain. The answer is certainly "Yes," although warnings should certainly be issued.

Rodents and monkeys are mammals and share much of their brain structure and genetic code with humans. The advantage of performing research on animals is the ability to use invasive methods to record neural activity. In humans, fMRI imaging research is one of the leading tools to image the brain. This technique records the blood flow in the brain. As areas which are active tend to require an elevated blood supply, it is possible to infer which areas respond to specific stimuli or tasks. Significantly, it is possible to record the entire brain at once with this technique.

On the other hand, one of the limitations to fMRI research is that subjects are required to be still inside the fMRI machine for the imaging to be successful; therefore, their movements cannot easily be studied. For the same reason, it is quite difficult to study fMRI in monkeys because they cannot be easily instructed to work with fMRI equipment; also, they have a smaller brain.[24] These limitations, and separation of techniques between human and animal research, are real obstacles in our understanding of how the brain processes and integrates information, and the result is that we do not yet know in general whether to think of neural circuits from a modular or integrated perspective across the brain.

The three examples above demonstrate how limited is our understanding of how brain circuits function and generate neural dynamics. The goal here is not to demonstrate how neural theories fall short of explaining specific features of brain dynamics, but rather to sketch our existing understanding of neural circuits in broad strokes. Our capacity to predict neural dynamics, or how the brain would respond to stimuli or manipulations (such as the optogenetic perturbations described above), is seriously limited. It is easy to argue that there is currently no brain theory with prediction powers equal to theories in other physical fields, such as organic chemistry, general relativity, or material science.

Still, the reader may wonder: Isn't it "knowledge" of brain dynamics if we can read out—decode—information such as arm movements from neural activity? Most of the time, the answer is "No." It is true that we can read information from neural recordings about motions, judgments, and sensory encoding; this demonstrates that specific information is represented in the brain, which is very important. However, this only establishes a correlation between neural activity in a certain brain area at a specific time and movements, decisions, or sensory information: it demonstrates that neural activity carries such information in part. It does not explain much about how such information was generated, reached that specific brain area, and was used by brain circuits.

In some ways, it's like a child enjoying an artist's painting and realizing that it depicts a starry night, without any idea of how the artist brilliantly employed painting techniques to create such a masterpiece. Even less so does the child comprehend how the texture of the canvas absorbs color pigments and how light striking such pigments reveals the color that his eyes are subsequently able to detect. Similarly, a true understanding of the brain would entail knowing how neuronal parts function in order to generate the neural activity we record. Our ability to classify a specific neural activity pattern in the visual cortex as a sensory response to Van Gogh's *Starry Night* rather than to a Picasso painting does not explain how our eyes, retina, and dedicated neural circuits process visual sensory information to generate such painting-specific neural activity. It simply states that the brain reacts differently to a Van Gogh work than to any other painting.

Another example that highlights our limited understanding of neural information processing comes from recent developments in machine learning. In the last decade, many machine learning applications have outpaced our imagination of what an algorithm can do. Recognizing faces or objects, driving cars, and playing complicated games like Go are just a few examples of the machine learning revolution. Yet the neural networks used to solve these tasks are deemed "black boxes" because we yet don't understand how these networks solve these tasks.

Despite knowing every single detail of these engineered neural networks (their architecture, input, cost function, learning rule, and so on), still we don't understand *how* they generate the representations they exploit to solve these tasks. A mechanistic mathematical characterization is still eluding our understanding. At this moment (though this may change in five to ten years from now), it is fair to say that these are treated as emergent phenomena waiting for a theoretical description. These machine learning algorithms represent, in a way, a humbling point for neuroscientists: if in the case of artificial neural networks we can know everything of the system yet remain unable to fully understand their mechanisms, how much more is this true for brain neural circuits, where we still don't know the details of their connectivity, the way they learn, and much of the biological machinery they leverage to produce neural activity?

All these considerations and examples are only briefly touched upon here, though they deserve a lengthier discussion. We point the interested reader, who may wish to further pursue some of the ideas here mentioned on the shortcomings of neuroscience, to a recent volume published by neuroscientist Matthew Cobb where further arguments are presented.[25] The gap between theoretical understanding and contemporary achievements of neural technologies (see our first set of examples, in the previous section) and the weakness of our current understanding of neural mechanisms (our second set of examples, in this section) underscores the shortcomings. The scientific frontier of neuroscience is a jagged line which quickly proceeds towards futuristic achievements, yet the understanding of neural circuits in the way they generate neural information is far behind.

Some have explained this by comparing the brain to a building and saying that neuroscience nowadays is much about describing the building and its architecture but not yet much about the people that live in the building and what happens within its rooms. The mistake, in these terms, is to confuse structure with function. Saying whether a room is a bedroom or a kitchen doesn't tell us much about how a person sleeps or cooks.

6. Neuroscience and the Mind

So far, we've laid the groundwork for assessing the state of neuroscience as a field. We can now examine whether it informs the contemporary philosophical debate about the mind-body problem.

The first thing to note is that the use of the language of neuroscience is a fundamental barrier to shifting gears and including philosophical content in our conversation. Neuroscience, like most natural science, discusses things from a "physicalist" standpoint: it proceeds as if there is nothing beyond the material world. This position has the justification that the physical sciences' focus of investigation is the physical world and their technique is developed for this purpose. Nonetheless, in recent history, the scope of physical science appears to have become thought of as the boundary of our understanding. Questions that have puzzled the greatest minds throughout history (from Aristotle through Pascal and beyond) appear to have recently become nonsense for the greatest part of the scientific community. The limits of our scientific method have become the limits of our thinking. But, if we allow our minds to stray beyond such boundaries, the first question that arises is: Is the physical world all there is? In neuroscientific words, this means: "Are we nothing but our brains?"

This is the gateway to the mind-body problem and a long philosophical tradition. If we could prove that the physical brain is all there is, then a position of monistic materialism would have to be upheld.[26] However, if such proof is not accessible, a wide range of philosophical possibilities must be considered and analyzed. Indeed, some philosophical frameworks claim that there are non-physical properties of the brain, and describe the mind as the recipient of such non-physical properties. This distinction between brain and mind is not only lexically useful but also allows for the formulation of two of the most important questions in the philosophical tradition of the mind-body problem. These are frequently expressed in terms of states of the brain (physical states) and states of mind (mental states):

- The ontological question: What are mental states and what are physical states? Is one a subclass of the other, or vice versa?
- Is there a link between physical and mental states? Do mental and physical states interact with or influence one another?

These questions are beyond the scope and approach of neuroscience as a modern scientific field, and it is not our intention to address them in depth. Rather, we will now consider whether, if at all, neuroscience can contribute to the philosophical debate surrounding such issues.

One method to answering these problems is to think of neuroscientific information as a description of physical states with no influence upon or access to mental states. That is, physical and mental experiences are completely distinct, and neuroscience focuses solely on the former. Following this approach, neuroscience would inform the philosophical debate based on whether the specific philosophical framework under examination allows physical states (and their properties within such framework) to have the characteristics highlighted by current neuroscientific discoveries. A philosophical theory suggesting that knowledge about each conceivable choice prompted to an animal or human should not have a corresponding physical condition, for example, would be in sharp contrast to current neuroscientific investigations capable of reading the animal's decision from its brain.[27] Similarly, theories claiming that physical states cannot embody any type of abstract thinking, such as a strategy for executing a task, would clash with a different line of neuroscientific inquiry.[28]

While an examination of multiple philosophical positions and their connections to neuroscientific claims is beyond the scope of this chapter, it is important to note that the vast majority of philosophical camps in the mind-body debate can accommodate all neuroscientific findings to date with little difficulty. This thesis can be articulated in terms of George Bealer's *autonomy thesis*, which claims that philosophical study can answer most of the important questions in philosophy without relying on the sciences. This independence of philosophical frameworks from the sciences has been claimed by scientists as well, albeit in different ways. For example, Francis Crick and Christof Koch admitted in a neuroscientific publication that the most widespread opinion among neuroscientists is that the nature of consciousness is "a philosophical problem, and so best left to philosophers." This reflects the widely held belief among neuroscientists that metaphysical ideas should be avoided and left for "late-night conversations over beer."[29] While these remarks reveal only scientists' and philosophers' attitudes on questions in the mind-body field, these opinions do seem to capture a truth about the broad independence of most philosophical frameworks from neuroscientific findings. A more detailed treatment of this can be found in the 2018 *Philosophia Christi* essay by J. P. Moreland.[30]

As a result, it is reasonable to conclude that in this debate—the mind-body problem—neuroscience as a field, and neuroscientists as researchers, would be better off taking an agnostic stance rather than a monistic materialist one. Indeed, the lack of neuroscientific theories explaining the functioning and features of neural circuits represents a fundamental deficiency in our understanding of the brain (and mind) that is easily identified. Because of this limitation, neuroscience should take a metaphysically agnostic stance, accepting its limitations. Individual researchers' postures may lean either way and partially reflect their viewpoint and vision regarding science and beyond, but their ideas should not be mistaken for scientific findings or, more importantly, scientific proofs.

What appears to have occurred is that, because the majority of scientists subscribe to physicalism, the field as a whole has come to embrace it without examining the evidence for this conclusion. Given this widespread consensus among scientists, monistic physicalism—which implies that we are nothing more than our brain—appears to be presented as a dogma or doctrine rather than a personal belief. In other words, monistic physicalism is often implicitly deemed something you subscribe to as a member of the scientific community.

7. Building a Bridge between the Brain and the Mind

While it may be true that most neuroscientific research is not yet capable of constraining or informing mind-body philosophical theories, a few areas are rapidly becoming instrumental in the construction of a bridge (and a common language) between neuroscience and philosophy of mind. These neuroscientific topics are critical because they are likely to deliver results relevant to the mind-body debate sooner rather than later. Three such examples are related to important philosophical issues: the problem of self-consciousness, the problem of intentionality, and the binding problem.

The first is concerned with the unique human ability to reflect on oneself, known as introspection. We are not only, like many animals, conscious, but we are also aware of our own consciousness—we are *self*-conscious. It is not merely that we have qualia (conscious experiences of things), that there is something it is like for us to see a sunset, smell a rose, or feel pain. Through introspection, we are also aware *that* we have such experiences, and thus we can study and talk about our mental life. This ability to reflect on ourselves is sometimes called the first-person perspective: I am not only aware of being happy or feeling pain, but I am aware that *I* am happy or that *I* am in pain. So I have a point of view not only on the world, but also on my own point of view. This is expressed by my use of first-person terms, such as "I," "me," "my," etc.

In recent years, neuroscientific theories of consciousness have arisen along separate but related lines. Two important competing theories, in particular, have been actively advanced: the global cognitive workspace[31] and integrated information theory.[32] While these theories do not directly address the problem of self-consciousness but, rather, the more general notion of phenomenal consciousness, they yet appear to build an interesting framework for future discussions.

The second issue, intentionality, has been conceptually related to the concepts of agency and "aboutness," the idea that mental states can be about things other than themselves, forming a relationship that is unique and has no counterpart in the physical world. For example, one may think that Budapest is in Hungary (one has a thought about Budapest), or one may desire to see this historic city and, as a result, decide to have a vacation there (leading to an action). No purely physical description of a state, including a state of the brain, implies that the state is about something beyond itself, or that the state could function as a reason for a further thought or action. Yet neuroscience is interested both in what brain states "represent" (in some sense of that term) and in the explanation of our actions.

The idea of agency has been related, in neuroscience, to decision-making frameworks in which humans or animals are tasked to make decisions and their neural representations are linked to their intentions and behaviors. In this case, too, there is a significant gap between the neuroscientific method and the issues addressed by philosophy. When we speak of "representation," this assumes but does not explain intentionality: Do the brain states themselves have it, or is it a distinct but correlated mental state? Likewise, when we speak of agency in neuroscience, this assumes but does not explain the idea of having reasons for an action: Can brain states be reasons, or again, is it only mental correlates of those brain states? Yet, as both philosophy and neuroscience appear to investigate closely related phenomena, an important bridge appears to be emerging along these lines.

In both cases, significant steps have been made toward developing a unified vocabulary that can inform both neuroscience and philosophy. These fields, however, also demonstrate how neuroscience and philosophy of mind have not yet entirely merged into a mature neurophilosophical form. Overall, this might be interpreted as a good reason to think that existing neuroscientific understanding does not constrain philosophical views. As a result, it can be considered a point in favor of neuroscience needing to take a step back from adopting a distinct philosophical stance, as described above.

The binding problem is a third important topic in both neuroscience and philosophy of mind. When processed by specific brain circuits, multiple perceptions (for example, color and shape in vision) are "bound" to the same object. Similarly, cross-modal binding refers to the difficulty of integrating percepts from distinct senses (also known as modalities, such as visual and auditory) via neural information processing.

According to neuroscientific studies, brain activity patterns during multimodal sensing hint at a reciprocal and "competitive" interaction between multimodal and unimodal regions that underpins the perceptual processing of simultaneous inputs from various sensory modalities.[33] Yet results on how cross-modal information is integrated to create a unified perception are limited and mainly confined to study different elements of visual perception (e.g., texture, shapes, contrast).[34] For centuries, the unity of percepts has been a central issue in philosophy of mind. Recently, the binding problem, which includes issues about agency and consciousness, has been explored (primarily in relation to vision) in light of neuroscientific results.[35] Yet a clear mechanistic explanation of how multiple percepts are bound into a coherent experience through sensory processing is lacking. This critique raises crucial issues about the limitations of our present neuroscientific understanding, which is consistent with our discussion above.

Nonetheless, current investigations advance intriguing links between neuroscientific results and philosophical ideas, creating a potentially fruitful route for future neuroscience-philosophy interaction. Areas of neuroscience that are philosophically significant and may advance rapidly in future years include the aforementioned binding problem, the previously discussed problem of segregation versus integration, the interaction between the two hemispheres of the human brain, the study of mental diseases, the study of individuals in various medical conditions (comatose state), and several more. We are excited to see how advancements in these areas can contribute to future discussions and help to construct a more robust bridge between philosophy and neuroscience that extends beyond physicalist perspectives.

8. Conclusion

We began this chapter by demonstrating how people have a cognitive bias toward believing neuroscientific explanations, even when they are irrelevant to the topic at hand. While media efforts have exacerbated this framing bias, it has deeper roots, and at least two factors influence how people perceive neuroscience. The first is that we tend to overestimate our knowledge of how the brain works. Between our technical capabilities and our understanding of the brain, there is a considerable knowledge gap. Given recent technical breakthroughs, it is reasonable to believe that the scientific community now understands the mental activity captured by cutting-edge technology, yet this is incorrect. Our theoretical understanding of the brain, unlike that of other physical systems, remains extremely limited.

This mismatch between popular expectations of our predictive theoretical power and the actual state of the neuroscience field is one of the reasons why people have faith in neuroscientific explanations. The second factor is the overreaching posture of neuroscientists, who, like their counterparts in several other scientific fields, tend to adopt materialistic monism and push for a unified view of the world that reduces yet unexplained problems to simple statements (e.g., "we are nothing but our brains") in which neuroscience can explain everything. This posture, which, as we show, lacks a solid foundation of scientific proofs, is far too comprehensive for neuroscience. We conclude that these two elements must be corrected in order to accurately evaluate neuroscientific findings and explanations.

We argued that neuroscience is not yet as mature as other disciplines, i.e., is not yet a developed paradigmatic field, because it lacks a strong theory that allows predictions and unifies findings in a common framework. We traced an approximate frontier of neuroscience as a scientific field and exposed the limited theoretical understanding of brain mechanisms in the process. Then we attempted to show how far this frontier is from informing philosophical issues concerning the mind-body problem. While the mind-body debate was only touched on briefly, a few promising future directions that may further connect neuroscience to philosophy were suggested. Overall, we concluded that neuroscience is far from informing metaphysical theories, and that, despite being an extremely fast-paced field, it does not yet limit the philosophical mind-body debate to a specific position. As a result, neuroscience as a field should be agnostic (not taking a position) toward any metaphysical perspective.

While some neuroscientists will undoubtedly share their opinions regarding the philosophical significance of their discoveries, their endorsement of a physicalist viewpoint ("we are nothing but our brains") should be interpreted as a personal statement rather than one based on scientific findings. A major philosopher of the last half-century, Roger Scruton, commented on a similar scientific posture with the following words:

> There is a widespread habit of declaring emergent realities to be "nothing but" the things in which we perceive them. The human person is "nothing but" the human animal; the law is "nothing but" relations of social power; sexual love is "nothing but" the urge to procreation; altruism is "nothing but" the dominant genetic strategy described by Maynard Smith; the Mona Lisa is "nothing but" a spread of pigments on a canvas, the Ninth Symphony is "nothing but" a sequence of pitched sounds of varying timbre. And so on. Getting rid of this habit is, to my mind, the true goal of philosophy. And if we get rid of it when dealing with the small things—symphonies, pictures, people—we might get rid of it when dealing with the large things too: notably, when dealing with the world as a whole.[36]

We look forward to future discoveries and the day on which neuroscience and philosophy can shed their "nothing buttery"[37] attitude, and attempt to thoroughly grasp the uniqueness of the human brain.

Notes

1. Abhinav Kulshreshth, Abhineet Anand, and Anupam Lakanpal, "Neuralink — An Elon Musk Start-up Achieve[s] Symbiosis with Artificial Intelligence," *2019 International Conference on Computing, Communication, and Intelligent Systems* (2019): 105-109.
2. George Bealer, "On the Possibility of Philosophical Knowledge," in *Philosophical Perspectives 10: Metaphysics*, ed. James E. Tomberlin (Cambridge, MA: Blackwell, 1996).
3. D. S. Weisberg et al., "The Seductive Allure of Neuroscience Explanations," *Journal of Cognitive Neuroscience* 20, no. 3 (2008): 470–477.
4. Thomas Gilovich, Dale Griffin, and Daniel Kahneman, eds., *Heuristics and Biases* (Cambridge, UK: Cambridge University Press, 2002); Kathleen Hall Jamieson, "Crisis or Self-Correction: Rethinking Media Narratives about the Well-Being of Science," *Proceedings of the National Academy of Sciences* 115, no. 11 (2018): 2620–2627.
5. Jamieson, "Crisis or Self-Correction."
6. To this day the vast majority of neuroscience focuses on studying neurons despite the proven importance of other brain's components such as glia cells.
7. Dean Keith Simonton, "Scientific Status of Disciplines, Individuals, and Ideas: Empirical Analyses of the Potential Impact of Theory," *Review of General Psychology* 10, no. 2 (2006): 104.
8. Thomas S. Kuhn, *The Structure of Scientific Revolutions*, 2nd ed. (Chicago: University of Chicago Press, 1970), 10.
9. E. R. Oby et al., "Intracortical Brain-Machine Interfaces," in *Neural Engineering*, 3rd ed., edited by Bin He (Cham, Switzerland: Springer Nature, 2020), 185–221; Xiyuan Jiang et al., "Structure in Neural Activity during Observed and Executed Movements Is Shared at the Neural Population Level, Not in Single Neurons," *Cell Reports* 32, no. 6 (August 2020), https://doi.org/10.1016/j.celrep.2020.108006.
10. Simon Musall et al., "Single-Trial Neural Dynamics Are Dominated by Richly Varied Movements," *Nature Neuroscience* 22, no. 10 (Oct. 2019): 1677–1686.
11. Taylor J. Bates, John R. Fergason, and Sarah N. Pierrie, "Technological Advances in Prosthesis Design and Rehabilitation Following Upper Extremity Limb Loss," *Current Reviews in Musculoskeletal Medicine* 13, no. 4 (Aug. 2020): 485–493.
12. Han Yuan and Bin He, "Brain-Computer Interfaces," *IEEE Transactions on Biomedical Engineering* 61, no. 5 (May 2014): 1425–1435.
13. Lin Tian et al., "Imaging Neural Activity in Worms, Flies and Mice with Improved GCaMP Calcium Indicators," *Nature Methods* 6, no. 12 (Dec. 2009): 875–881; Jasper Akerboom et al., "Optimization of a GCaMP Calcium Indicator for Neural Activity Imaging," *Journal of Neuroscience* 32, no. 40 (Oct. 2012): 13819–13840; Carsen Stringer and Marius Pachitariu, "Computational Processing of Neural Recordings from Calcium Imaging Data," *Current Opinion in Neurobiology* 55 (2019): 22–31.
14. D. Kim et al., "Pan-Neuronal Calcium Imaging with Cellular Resolution in Freely Swimming Zebrafish," *Nature Methods* 14, no. 11 (Nov. 2017): 1107–1114.
15. Erika Pastrana, "Optogenetics," *Nature Methods* 8, no. 1 (Jan. 2011): 24–25; News Staff, "Stepping Away from the Trees for a Look at the Forest," *Science* 330, no. 6011 (Dec. 17, 2010): 1612–1613.
16. S. Ramirez, S. Tonegawa, and X. Liu, "Identification and Optogenetic Manipulation of Memory Engrams in the Hippocampus," *Frontiers in Behavioral Neuroscience* 7, article 226 (17 January 2014), https://doi.org/10.3389/fnbeh.2013.00226; Claire E. Geddes, Hao Li, and Xin Jin, "Optogenetic Editing Reveals the Hierarchical Organization of Learned Action Sequences," *Cell* 174, no. 1 (June 28, 2018): 32–43.E15.
17. Hyung-Su Kim et al., "Selective Control of Fear Expression by Optogenetic Manipulation of Infralimbic Cortex after Extinction," *Neuropsychopharmacology* 41 (April 2016): 1261–1273.
18. Sergey Anpilov et al., "Wireless Optogenetic Stimulation of Oxytocin Neurons in a Semi-Natural Setup Dynamically Elevates Both Pro-social and Agonistic Behaviors," *Neuron* 107, no. 4 (Aug. 19, 2020): 591–593.
19. Jun Yoshimura et al., "Recompleting the *Caenorhabditis elegans* Genome," *Genome Research* 29, no. 6 (June 2019): 1009–1022.
20. Jimin Kim et al., "Whole Integration of Neural Connectomics, Dynamics and Bio-Mechanics for Identification of Behavioral Sensorimotor Pathways in *Caenorhabditis elegans*," *bioRχiv*, Aug. 3, 2019, https://doi.org/10.1101/724328; Monika Scholz et al., "Predicting Natural Behavior from Whole-Brain Neural Dynamics," *bioRχiv*, Jan. 15, 2021, https://doi.org/10.1101/445643.

21. Eric Jonas and Konrad Paul Kording, "Could a Neuroscientist Understand a Microprocessor?," *bioRχiv*, Nov. 14, 2016, https://doi.org/10.1101/055624.
22. Gustavo Deco et al., "Rethinking Segregation and Integration: Contributions of Whole-Brain Modelling," *Nature Reviews Neuroscience* 16 (2015): 430–439; Jessica R. Cohen and Mark D'Esposito, "The Segregation and Integration of Distinct Brain Networks and Their Relationship to Cognition," *Journal of Neuroscience* 36, no. 48 (November 30, 2016): 12083–12094; "Understanding Principles of Integration and Segregation Using Whole-Brain Computational Connectomics: Implications for Neuropsychiatric Disorders," *Philosophical Transactions of the Royal Society A: Mathematical, Physical and Engineering Sciences* 375, no. 2096 (May 15, 2017), royalsocietypublishing.org/doi/10.1098/rsta.2016.0283.
23. Musall et al., "Single-Trial Neural Dynamics Are Dominated by Richly Varied Movements," *Nature Neuroscience* 22, no. 10 (Oct. 2019): 1677–1686; Carsen Stringer et al., "Spontaneous Behaviors Drive Multidimensional, Brain-Wide Population Activity," *Science* 364, no. 6437 (Apr. 19, 2019): 1–11, science.org/doi/pdf/10.1126/science.aav7893.
24. Krishna Srihasam et al., "Non-Invasive Functional MRI in Alert Monkeys," *Neuroimage* 51, no. 1 (May 15, 2010): 267–273.
25. Matthew Cobb, *The Idea of the Brain: The Past and Future of Neuroscience* (New York: Basic Books, 2020).
26. Unless a non-material substance were found elsewhere (e.g., in a different universe, if such exists, or in forms not tied to the physical structure of brains).
27. Nicholas A. Steinmetz et al., "Distributed Coding of Choice, Action and Engagement across the Mouse Brain," *Nature* 576, no. 7786 (Dec. 2019): 266–273, https://www.nature.com/articles/s41586-019-1787-x.
28. Silvia Bernardi et al., "The Geometry of Abstraction in Hippocampus and Pre-Frontal Cortex," *Cell* 183, no. 4 (Nov. 12, 2020): 954-967.E21, https://doi.org/10.1016/j.cell.2020.09.031.
29. John Horgan, "Can Science Explain Consciousness?," *Scientific American* 271, no. 1 (July 1994): 91.
30. J. P. Moreland, "The Fundamental Limitations of Cognitive Neuroscience for Stating and Solving the Ubiquitous Metaphysical Issues in Philosophy of Mind," *Philosophia Christi* 20, no. 1 (2018): 43–51.
31. Stanislaus Dehaene and Lionel Naccache, "Towards a Cognitive Neuroscience of Consciousness: Basic Evidence and a Workspace Framework," *Cognition* 79, nos. 1–2 (Apr. 2001): 1–37, https://doi.org/10.1016/S0010-0277(00)00123-2.
32. Giulio Tononi et al., "Integrated Information Theory: From Consciousness to Its Physical Substrate," *Nature Reviews Neuroscience* 17 (July 2016): 450–461, https://doi.org/10.1038/nrn.2016.44.
33. Khalafalla O. Bushara et al., "Neural Correlates of Cross-Modal Binding," *Nature Neuroscience* 6, no. 2 (Feb. 2003): 190–195, https://doi.org/10.1038/nn993.
34. Sebastian Schneegans and Paul M. Bays, "New Perspectives on Binding in Visual Working Memory," *British Journal of Psychology* 110, no. 2 (May 2019): 207–244, https://doi.org/10.1111/bjop.12345; Emma Wu Dowd and Julie D. Golomb, "The Binding Problem after an Eye Movement," *Attention, Perception, & Psychophysics* 82, no. 1 (Jan. 2020): 168–180, https://link.springer.com/article/10.3758/s13414-019-01739-y.
35. See two articles by Eric LaRock: "Disambiguation, Binding, and the Unity of Visual Consciousness," *Theory & Psychology* 17, no. 6 (Dec. 1, 2007): 747–777, https://doi.org/10.1177%2F0959354307083492; "Hard Problems of Unified Experience from the Perspective of Neuroscience," in Mihretu Guta, ed., *Consciousness and the Ontology of Properties* (New York: Routledge, 2019), 223–240.
36. Roger Scruton, *The Soul of the World* (Princeton, NJ: Princeton University Press, 2014), 39–40.
37. Mary Midgley uses the phrase "nothing buttery" in her criticism of reductionism, in works such as *The Ethical Primate: Humans, Freedom and Morality* (London: Routledge, 1994). However, while Midgley uses the phrase, she did not coin it. It was used before her by Donald MacKay, another critic of reductionism, whose most famous work is perhaps *The Clockwork Image* (Downers Grove, IL: Intervarsity, 1974), and it may have been used by others before MacKay.

16. Revising Our Pictures of Emotions

Natalia Dashan and David Gelernter

1. An Unusual Health Problem

ALICE IS not a happy 23-year-old. She sees herself as a kind and social person who goes out of her way for others. But she has been flustered at work in her teaching position, and all her friends seem to bother her, all the time. She feels they take more than they give. When Alice goes out to make new friends, they quickly bother her too. Her friends tell her they like her, but she notices they sometimes find her overwhelming to be around and do not respond to her messages. Not only does Alice not know what to do, she cannot identify if anything is even wrong.

Brian is not a happy 41-year-old. He has established himself in a lucrative computer science career and has a pleasant disposition towards life, but he is lonely. He has not had a longstanding relationship with a woman in over twelve years. His unhappiness is not at the forefront of his consciousness, but he knows something is wrong, thought he cannot identify what. He does not know what to do.

Charlie is not a happy 29-year-old. He is competent as a mid-level manager at his finance firm and makes a steady income that he shares with his family, but instead of spending time with them, he spends each evening watching the news for hours. He does not enjoy this—in fact, the news stresses him—but he feels an obligation to stay informed. When visiting friends, instead of enjoying intimacy, he interrogates them on their voting patterns. He stays up late debating politics online to the detriment of his health and relationships. He is aware that his family is concerned, but he does not know what to do.

At first glance, these may seem like three completely unrelated situations—three unhappy people in different occupations at different stages of their lives. At second glance, we can see a pattern: each person's unhappiness stems from unsatisfying relationships. Perhaps they can be taught skills that would help them with their interactions.

At third glance, there is something major in common. Each of these people, in his own way, is facing a major distortion in his mental model of what it means to have an emotion. These are not minor interpersonal hiccups. Each of these cases may take months or years to unravel, as the core of the problem—the conceptual distortion—is so distant from how it manifests in day-to-day life. Interpersonal training may help smooth out a subset of interactions, and cognitive behavioral therapy may teach a person to introspect and regulate his emotions. But unless examined head-on, the perceptual distortion of emotions will persist.

What are perceptual distortions of emotions? And how are perceptual distortions different from merely different points of view? Let us examine Alice, Brian, and Charlie, and notice how their different perceptual distortions (or "false-filters") are not merely different points of view, but significantly different meaning-making metaphors that shape how they interact with the world.

Alice sees emotions as responses triggered in her by other people. She does not conceive herself as forging her own emotions, but rather they are a reaction in her body created by other people's words and actions. Thus, her emotions are not something she has control over. She operates under the metaphor of being a puppet, and the strings are held by those around her. It is not surprising that she gets angry at every wrong move made by her loved ones and feels strung around by her students. They have the power to make her happy or unhappy, at their will, and if they misjudge or blunder, she has no power to set things right.

Brian sees emotions as dangerous temptations that derail him from his carefully constructed, comfortable existence. He has set up a routine that works well for him and allows him to succeed in his programming career. Emotions are scary and make him weak. He does not open up or examine himself because to do so is to open a Pandora's box of everything he has been avoiding for years. He has trouble letting anybody in because he does not even let himself in. Although he is generally considered a nice guy, he avoids attending to other people's feelings because for him, assuming care means entanglement, and entanglement means a loss of control that could derail his carefully constructed life. He lives as if emotions are something to be reined in, like snakes that need to be grabbed and caged.

Charlie sees emotions as rewards for good behavior, or punishments for bad behavior. In his eyes, he was born sinful and has to prove his right to exist. Because Charlie never feels like he has done enough good, he pushes away any positive feelings because he does not feel he deserves them. In fact, he seldom feels any emotions other than shame, guilt, anxiety, and inadequacy, as he takes on more responsibilities to prove his worth and then fails to meet his own standards. Watching the news is a side effect of this false-filter. Charlie has trouble distinguishing between what is his responsibility and what is not, and takes on the entire world's pain as his own.

With these three examples, we can see that each person does not merely have a different point of view. Rather, each person has a distinct way of constructing his emotional reality, and therefore his reality. For Alice, emotions are pleasure or torment bestowed onto her by other people. For Brian, emotions are inconveniences that must be constrained. For Charlie, emotions are rewards or punishment for virtue or sin. Each of these three people has a different blueprint for how his emotions should govern his own behavior, which of his emotions are trustworthy signals about how the natural world works, which of his emotions are trustworthy signals about how the social world works, how other people should be judged on their emotions, and how his emotions signal how well he is doing in life.

Not only do Alice, Brian, and Charlie feel different feelings at different moments, but they each have a distinct interpretive architecture for creating meaning from those feelings.

The relevant aspect that makes each of Alice's, Brian's, and Charlie's emotion metaphors a distortion is not that it is completely incorrect; rather, each is a distortion because it is *incomplete*. Each of these metaphors neglects important realities about how emotions are created, both in the interpersonal space and in the body. Each of these metaphors neglects the ways that emotions can be shaped, controlled, and converted into other emotions. Each of these metaphors, in one way or another, renders the person a slave to his emotions, rather than freed by his emotions to experience the full complexity and exaltation of the universe. And each of these metaphors limits each person's appreciation for his community, his loved ones, and himself.

Think about your own mind for a moment. How do you construct emotions? What metaphors do you intentionally or unintentionally use to interpret your reality?

If you find yourself stuck on the exact definition of an emotion, know that you are not alone. Continue with the exercise using your intuition, even without an airtight definition. "Emotion" is a word that everybody uses, and yet nobody seems to know precisely what it means. Even emotions researchers have trouble answering this question, as seen by the internal debates within the field. *The Atlantic* quotes Doctor Joseph LeDoux, the director of the Emotional Brain Institute at New York University: "It's been said that there are as many theories of emotions as there are emotion theorists."[1]

There have long been schemas trying to identify, categorize, and explain emotions. The Hellenistic philosophers had the Stoic passions. Hippocrates popularized the four humors—though at different times, the number was contested. When Paul Ekman proposed universal facial expressions in the 1970s, he was working in an intellectual environment where it was debated whether this was possible, or whether facial expressions were culturally derived.[2]

One thing, however, is not debated: The stakes are high. How a person conceptualizes his emotions may affect all areas of his life. It may affect how a person sees himself, how he sees other people, how he manages his energy, and how he reacts to failure and conflict. It may affect his interpersonal relationships and his relationship with himself. It affects his motivation and work output. It could mean the difference between joy, stability, and flourishing versus despair, insanity, and deprivation. Countless brilliant souls have failed to reach their potential and contribute to society not because of any material or monetary deficiencies, but because they were not able to control their emotional lives with the granularity that was needed.

An individual's emotional world is complex and comprises many biological and social factors. Genetics, epigenetics, nutrition, and socioeconomic factors are all relevant,[3] but these will not be the focus of this essay. Another factor, however, will be: the norms and memes that individuals have absorbed from their social spheres. These memes sometimes accurately reflect existing research findings, but often, they have no grounding in reality.

Alice, Brian, and Charlie did not create their false-filters on their own. False narratives about emotions do not arise out of nowhere. They are propagated through films, books, and the discourses of academics and public intellectuals. It is clear that despite the growing body of psychology papers in journals, many productive and illuminative insights have not left the academy. When an idea does leave, it is often reduced to a caricature, and often the idea that finds its way into the mainstream is not the most profound idea or the most true, but the most marketable by business-savvy promoters.

Philip Zimbardo's famous 1971 Stanford Prison Experiment is one such example, garnering decades of notoriety and ultimately being publicly contested in 2018.[4] This experiment concluded that human beings naturally and spontaneously turn cruel when granted even negligible authority. This idea propagated through the public consciousness for half a century, before it was publicly unveiled that the participants were not acting in a neutral state of nature, but were instructed to be cruel. Situations like this enable people to create false-filters based on false information.

At the other end, there exist insightful findings that could improve people's lives, but people have never heard of them. Cultural memes and norms have not been updated to reflect the new research. For all its effect on updating people's ideas of how they should interact with each other, the new research may as well not exist.

What is clear is that decades of careless academic and media output have led both researchers and the general population astray in terms of what emotions are and how you can modify them to get along with people, be productive, and generally lead a good life. These errors have led people to become more anxious, more unstable, and less able to collaborate. Instead of a culture of peace, love, steadiness, charity, and growth, we have been creating a culture of neuroticism, anxiety, blame, and avoidance.

This has been done in two ways. The first way is by dismissing as unimportant major elements of the mind that are instead crucial to human functioning and flourishing. Major psychological movements in different time-periods have each decided that indispensable parts of the mind can be ignored completely. The behaviorists ignored cognition. The computationalists ignored biology and phenomenology almost completely. And after Jungianism waned in popularity, the idea that a moral framework or ideals can significantly affect cognitive structures from the top level has largely been avoided in clinical therapies.

The second way is by repeating ideas that are untrue until they become memes in the culture, even though they have nothing to do with modern developments in psychology and neuroscience.

Together, these two ways have generated an unusual health problem. We have many pictures of emotions, and yet none that encompass all the known research, none that lead to productive results, and none that allow people to experience the full richness of their environment with the full breadth of their emotional or sensory sensitivity.

Yet this does not have to be the case. We have enough research in closed books that we can spread on the table to improve people's lives. We can develop, and people can live by, metaphors that make them flourish. Kurt Vonnegut once said, "Live by the harmless untruths that make you brave and kind and healthy and happy." But the development of healthy and harmless untruths is not automatic; not all untruths are equal. For healthy untruths to germinate, we must be vigilant about destructive and limiting lies we tell ourselves, and be open to a world bigger than the world we see.

2. Blurred Images

LET US start with the faulty metaphors—the blurred images that people use to guide their lives. What common metaphors have been propagated in society? What are the metaphors people currently live by? What are the engineering diagrams that they use to imagine themselves in relation to their own bodies, and the bodies around them?

One familiar conception is the mind as software and the body as hardware. The mind can then be decoupled from the body and put into a mechanical machine with no difference between the original and the artificial version, the same way that a software program can be copied from one machine to another. This idea of mind as software goes back at least to Alan Turing's seminal 1950 paper, "Computing Machinery and Intelligence,"[5] and has propagated through academic communities since. This paper described the "Imitation Game," which over the years became known as the "Turing Test." The imitation game proposes a test for artificial mind; should a human be unable to distinguish a machine's replies to posed questions from a human's replies to those questions, then the machine can be said to have a kind of behavioral human intelligence.

Can any problems emerge from an overdependence on this idea? An overdependence can lead people to neglect the role their bodies play in regulating their thinking and forget the physio-psychological feedback loops that govern their existence. Am I angry or am I hungry? Am I sad because somebody hurt me, or because I hurt them? How do I tell? Accurately identifying the cause of an emotion has consequences for the solution (a hungry person needs not a therapy session, but a sandwich) and yet the mind-as-software model erases these distinctions. In that model, the body is merely an input stream of data to the mind, where the data is analyzed and action taken. All feelings and thoughts are reduced to computations. Such a person may consider himself to be very logical, but then find himself in unpleasant situations he does not have the tools to diagnose. He may snap at his colleagues after a long day of work and then mandate a demoralizing company meeting to correct their errors. He may never learn that the cause of his irritation was not their mistakes, but the empty water cooler.

Here is another common notion: the mind as a social construction. René Girard in his theory of social memetics proposes that a person's desire is constructed from witnessing another person's desire.[6] Thus, when we see a beautiful flower, we do not want it for its own sake, but rather because somebody else wants it; and when we see a man swimming in the ocean, we do not admire his physique because we have a natural proclivity to admire it, but because advertising agencies convince us to believe that this is admirable. An overdependence on this idea may lead a person to experience heightened anxiety, as he loses his sense of agency over his actions and environment. If he can be so easily controlled by others, how can he control himself? If everything can be so easily programmed and reprogrammed in his mind, how can he trust his own thoughts? A person who subscribes to this theory is likely to be emotionally fragile and anxious without knowing why. This metaphor cannot be correct, because it excludes his ability to shape his own world.

Another popular conception is: the human as animal. Although technically true, an oversubscription to this theory can also lead to pathology. The frame is that there are behaviors that are "natural" that we ought to lean into, and behaviors that are "unnatural" that we ought to lean out of. Though this theory often purports to be based on the natural sciences, purveyors almost never cite recent literature, and so "human as animal" often becomes not an homage to cutting-edge animal studies, but rather a weapon to excuse aberrant behavior as natural, or to punish people who do not comply with consensus. Drinking to excess with strangers may become "natural" once somebody decides this, even though lowering inhibitions around strangers may not seem instinctive at all. Meanwhile, spending the day alone reading may be considered "unnatural," even though animals generally have an instinct for curiosity. "Human as animal" can thus easily become a tool to control conduct rather than articulate the mechanics of our reality.

We know that the depictions of the mind discussed above—mind as software, mind as social construction, human as animal—are distortions, because we can point to entire areas of the mind that are not accounted for. This means that many cause-effect mechanisms cannot be accurately modeled. We also know they are distortions because they lead to negative health outcomes when followed. Emotions, when properly calibrated, expand a people's understanding of themselves and the universe, rather than producing corruption or limitation. But while we can see that such distortions are wrong, how do we know what is right? How do we know which depictions of emotions are merely poetic descriptions, or tools to control behavior, and which have real correspondence with the social and biological realities of emotion? That is, when it comes to emotion, how do we know what is true?

3. What Does It Mean to Be a Fish?

WE CAN start at the beginning. "I think, therefore I am." This seminal observation from Descartes, in his *Discourse on Method*,[7] resonates with people because it *feels* true. If you think, you can feel yourself thinking; therefore, you exist.

But is the converse true? If you exist, must you think? Or should we take this a step further? Perhaps it is not, "I think, therefore I am," but rather "I feel, therefore I am."

Would you still be alive if you could not think, in the conventional meaning of the term? Would you still be alive if you could not speak or argue or do mathematical computations? You would still feel. People who have recovered from comas describe their internal emotional states. People who experience basilar migraines can tell you about the sunspots that appear in their eyes, their hands growing numb, and finally their loss of speech and ability to solve basic arithmetic—on an average weekday. But they still feel, often intensely.

This conscious feeling of what it means to be alive is often ignored, and yet this feeling, and the different textures of these feelings and the transitions between them, form your existence. Without this feeling of consciousness, you would be a robot or a zombie.

What goes on in your mother's head when she tells you to check the attic for bats, for the twelfth time that day? What goes on in your father's head when he picks you up an hour late from school because he "saw the most electrifying bird"? And what does this have to do with their health?

It has everything to do with it. The neural and biological events in your mother's mind and body that make her bat-anxious do not start when she notices she is anxious, and they do not end once she has made a decision—"ask my son to check the attic"—like a program that has finished parsing its inputs to create one final output.

The thoughts in her mind have no start or end. Rather, they *are* her. They are the texture in her mind. They are her experience of living. If you compared her internal texture to your hypothetical father's, you would see two very different phenomenological landscapes.

Your mother's landscape is a rugged, winter terrain filled not just with bats, but snakes, ants, and fire. She is perched on a jagged stone, jumping from rock to rock so as to not fall into a ravine. At the end of her journey, miles away, she sees a thirty-foot scorpion ready to eat her unless she is fully prepared. Your father's landscape is a lush plain enveloping him on a hilltop. He is lying on his back, looking at the sky and the creatures that fly towards him. He does not move except to scratch an itch, as he has his handy bow and arrow to shoot down anything that may harm himself or his avian friends.

The two parents view external events completely differently and perceive threat very differently. The activities in your father's mind and body—the hormones, the electrical charges, and ions, etc.—combine to create for him a pleasant, calm life. Meanwhile, your mother experiences a chaotic, anxious life. Both experience a conscious *feeling*. Your mother *feels* the texture of her thoughts, before her decision to annoy you, during her decision, and after her decision. This difficult-to-pin down internal feeling that she has when a biological or emotional state manifests in her brain has a name: *affect*. The term is used by researchers in multiple disciplines, and a branch of psychology, *affective psychology*, is devoted to pinning down what this *feeling* of consciousness may be.[8]

The changes between every mental state, the "tides of mind"[9] that occur every second of every day of a person's existence, determine the quality of that person's life. Whether or not a person is happy has less to do with their success than with their possession of pleasant affective states for a proper proportion of the time.

And yet, if you use "affect" in front of somebody as a noun, they are more likely to correct your grammar ("Did you mean *effect*?") than to converse with you about how a bird made them feel glorious inside. If you ask them to tell you how they live their lives, moment-by-moment, they would not be able to tell you. If you ask which tasks make them feel particularly unhappy, they would not know. If you ask which affective states start a chain reaction into more and more unpleasant states, they would not be able to tell you what they are, let alone how to stop doing it. But why not? You would expect human beings—who are notorious for examining themselves, as evidenced by our obsession with personality tests—to want to talk about this. You would expect them to take inventory of what goes on in their own minds.

One problem is that affective psychology, like consciousness more generally, is an inherently difficult topic to study. To use an analogy, affective psychology is not a fish asking what is water; it is a fish talking about what it feels like to be a fish. What is it supposed to say? What frame of reference is it supposed to use? Our only means to communicate with others—words, graphs, images—are already a few layers removed from the level of consciousness. And thus, in order to discuss matters of consciousness, we cannot reach a satisfying directness.

We have to use language, metaphors, and pictures, and hope that other people have enough shared experiences with the subjects of the words and metaphors to understand what we are talking about. It is imperfect, but people desperately try to express their feeling anyway—with whatever words they can.

4. Language as Proxy

When looking at metaphors, we can examine the actual words and expressions used.

People say "keep yourself under control" or "keep yourself together" or "keep it in," but when we talk about showing emotions, we say "Come on, spill it!" The impression is that people are containers, with a valve that controls how much of their emotions they allow to leave at any time. A person with high emotional control would have a finely calibrated valve, only letting the finest stream out at any instance, like a teapot, whereas a less sophisticated individual would be a constantly overboiling pot of water, never able to control his emotions. Somebody may tell him, "Get a grip," i.e., squeeze the valve to keep the liquid from pouring out.

Here is another expression: "Try to be less emotional." We have all heard this. But what does it mean to be emotional? "The crowd reacted emotionally and started shouting." Under this frame, there is a dichotomy between "logic" and "emotion." Emotion means any showing of passionate expression, and "logic" means restraint. This is a crude, binary model, with no nuance about what "emotion" and "logic" actually are, and one defined to be bad and the other good.

Another category of idioms uses spatial imagery. "On the ball," "on the fence," "off one's rocker," and "get carried away," are all ways of identifying an emotion through one's place in space. The emotion is described not as something inside the person; instead, the person's emotional state is described as the person's relationship to spatial reality.

Sometimes emotions are externalized into gods, spirits, or forces of nature. Emotion is sometimes an arrow someone shoots into you. Cupid has a bow and arrow even in ancient sculpture. If "something just struck me" and it is bad, it is like "an arrow in the chest." A sudden bad emotion is associated with sudden death—like an arrow-shot. A leading character in the Hebrew Bible gets such bad news, so unexpectedly, that he drops dead on the spot. But good emotions are also "a bolt from the blue"—and a thunderbolt is like an arrow. Everyone knows what "lovestruck" means. Everyone knows what happens when Cupid shoots you: you fall in love and die.

We are moved by emotion. Touched by emotion. The emotion seems to reach inside and grab us. It is a force that can kick us to the curb. It can stab us. It can blind us.

This difficulty in characterizing emotion—what it is and where it comes from—is not purely linguistic. Jaak Panksepp in his landmark *Affective Neuroscience: The Foundations of Human and Animal Emotions*, traces the pathways of emotions in the body.[10] There are important and complicated questions. Are emotions caused inside us, or are they caused by our environment? If they are caused by the environment, is there a processing delay? How long is the delay?

An extremely common idea is that emotions are "triggered." An external event happens, and your emotions are your response to the event. You see a snake and your fear circuit kicks in. Your palms start to sweat, your heart races, and you feel "fear." Or you see a bunny get hit by a car. Now you feel "despair."

But emotions do not really happen this way. Much of your emotional life is context-dependent. You would not be afraid of the snake in your room if this was your pet snake. In this way, emotions are not just responses to stimuli, but are complex aggregations of your memories, current physical state, and expectations.

When you feel an emotion, often you do not have a choice when it comes up, but then you can decide for how long you want to feel the emotion, whether you want to change it, and whether the emotional response is the right one for the situation. You can over time learn to build your emotions, rather than reacting to your circumstances, and based on which emotions you want to feel, you can build entire frameworks to help you feel them more often.

5. Happiness is a Well-Timed Drink

ON TOP of the affective framework exists the cognitive framework. This acts as a traffic controller between your affective states. What do you feel when? And how do you transition between your feelings?

Let us look at an illustration around just one emotion—happiness, to see how this is done.

We can start with the complexity and contention of what happiness even *is*. The metaphors of the *zeitgeist* determine which emotional categories we consider to be important, and which we neglect entirely. What we consider to be "happiness" is often less a proper consideration of what is a desirable affect, and closer to what we have seen advertised in Hallmark cards and Hollywood movies.

But there does exist a conception of happiness we can pin down. If you were to approach every individual around the world and ask him what he wants more than anything, perhaps he will say that he wants success, or a family, or a Maserati. But very likely, at the end of the day, he will say that he wants to be happy. He may give it a different name—contentment, or fulfillment, or meaning, or duty. He may say that he does not want these things for himself, but rather for his loved ones. He may argue that these are different from happiness—that they are a deeper, richer, more long-lasting joy than the mere word "happiness" implies.

But people everywhere discuss "happiness" and its variations—which implies that, behind the rhetoric, there is something important that people are trying to express. There is an idea at the center of the words.

How to be happy? In our attempt to find happiness, often it slips right through our fingers, as if we were grasping water. And the harder we grip, the less we have. Maybe we obtain happiness the way we obtain anything else of value: we endure pain over many years until we *earn* it. We cannot have it until we *deserve* it. Or perhaps happiness is not something we earn, or even pursue. Perhaps happiness is something that finds *us* in due time, the way a drink eventually finds its way to our table after we order it at a restaurant. Or maybe happiness is something we are entitled to.

Notice what we have just done. In the last paragraph alone, we used three completely different metaphors for happiness that may or may not have anything to do with neurological, physical, or metaphysical reality. And yet, the metaphors construct mental pictures of emotion that are not static images, to be viewed and then replaced in a week by other, more interesting ones. Rather, the image in our mind is a blueprint that affects our behavior.

What domains does an image-blueprint affect? The image affects how you perceive your own emotions and your ability to manage and change those emotions.

If somebody believes that happiness is something we "earn" from the external world if we are sufficiently deserving, he is likely to never find happiness. He is likely to live a life of anxiety and self-flagellation. Sometimes a life of astonishing achievement—but not a life of joy. Every time a positive affect would start to manifest, he would push it away, as the feeling would come part and parcel with affective feelings of shame, inadequacy, and dishonor. Such a person may be like the pre-Reformation Luther, his life an effort to absolve himself of original sin. Or he may be a chemist, working tirelessly in his lab at the expense of his family on his quest to prove himself worthy through his next big discovery. Or it may be a child with an unusually strict caretaker. In any case, the picture of happiness is the same: it is external to you, a reward to be "earned."

If somebody believes that happiness is something that arrives if we want it badly enough, he may constantly make wishes without taking any action towards them. And even once he gets what he says he wants, he will still not be happy. Because in his image, happiness is always something "out there" versus something inside himself. Even if he gets the object of his desire, he does not allow the feeling of joy to find its way inside him for any measurable duration of time. He may not know that this is possible to do, let alone that this is essential for feeling joy.

If somebody believes that happiness is something they have a right to, they may also seriously squash their opportunities for having it. They may act in ways that are inconsiderate or even selfish, as they see any shift in what they believe to be their internal contentment as a personal infringement. They may push people away who might bring them much deeper joy than they have previously experienced, in the name of maintaining their inner status quo at any given moment. They do not see the relationship between conflict, sacrifice, and long-term fulfillment.

These are three completely different ways of framing the concept of happiness in language. None are based in modern neuroscience,[11] and yet they are propagated over and over, to the point where each of these common frameworks causes distress to people every day.

Happiness is one example of an emotion people conceptualize very differently from each other. When you add up different conceptions for different emotions, you end up with people who see the world quite differently from each other. They have entirely different phenomenological landscapes inside their minds.

One person may be content, whole, connected. Another person may feel himself to be in a completely different universe from everybody else: alone, despairing, falling. And then in order for them to advance as people, their path forward involves gaining different skills from each other. One person may need to learn to be responsible for other people. Another person may need to learn to be more responsible for himself.

How you see your own emotions also determines how you see other people's emotions. If you believe that your emotions are something manifested inside yourself, and somebody around you is crying, you may want them to buck up and get over it. They may stimulate your sadistic tendencies. Your affective response to the tears is disgust. However, if you believe that most emotions are caused by an external event, if you see a person crying, your response to the tears may be anger at whatever created the tears.

Describing this as a mere difference in values is not the correct framing. Both people may believe that it is bad to "make someone cry." Both people may believe that it is good to have self-control, and good to not let other people affect you too much. Both people may get angry when they perceive injustice. But regarding their phenomenological response to the same stimuli, not only is their reaction completely different, but even their interpretation. It is almost as if they are living in different realities.

If you believe that your emotions are changeable, you are likely to be much more optimistic in your life. You are likely to find creative ways to make yourself happy. If you believe your emotions are not changeable, you are more likely to feel stuck, with little hope of escaping.

6. A Better Way

How do we use what we know about emotion to improve people's lives? One major way is to directly observe emotion, as if observing any other natural phenomenon, and then examine your guiding metaphors—which you imagine to be sound ones but may not be—and play with ideas, if need be, to come up with different ones.

We can apply this process to six simple assertions about emotions that have been derived from—and can be verified by—direct observation.

One: The body experiences continuous feelings as a result of our touching or directly sensing the environment. We might feel a light, cool breeze on our hands, or slipperiness when we step from a rug to a polished wood floor, or pain and a sharp movement backward when someone punches us in the face, or a change of wheel-grip when we drive onto gravel. We may feel acute and expanding joy when we pick up a puppy, or relief when we have walked many miles on a hot day and are finally able to sit down, or a deep and well-rounded happiness when we hit a baseball or tennis ball powerfully, in just the way we had wanted to.

These physical feelings are, for the most part, continuous. Most are minor and we barely notice them, or miss them completely. We only notice relatively intense feelings. Such a feeling might quickly disappear or might linger. Ordinarily, we experience many such feelings at the same time, and are able to keep them separate in our mind. But several simultaneous feelings might each have a distinct phenomenal profile of its own (for example, the spring of a diving board combined with the approach of a pleasant turquoise sparkle.)

These feelings occur as a matter of course when the body moves through the world that encompasses it. We might imagine the encompassing world as a medium, and physical feelings as the "friction" of the body moving through it.

Two: The mind, likewise, experiences continuous feelings as it moves through the physical *and mental* worlds. Someone shouts something obnoxious at you, and you feel angry—a physical event (someone shouting) mediated by a mental event. We see the abrupt rise of a military plane in a jet-assisted takeoff and feel impressed and happy. We feel the final piece of a proof fall into place as we walk across the campus, and our satisfaction feels, in many ways, like physical satisfaction. We show a woman a painting we made, and she claps at the sight of it, and our head explodes with pride and pleasure. We suddenly remember what we are supposed to be doing tomorrow at this time, and fear gathers in niches throughout the mind. Someone slaps us with a flounder, and we throw him back hard against a wall without exactly meaning to, because we are surprised and angry. (We do not like being slapped with flounders.)

This continuum of mental responses to the physical and mental worlds, to the external and internal environments, consists of *emotions*. Again, they vary in intensity and duration. You only notice the sharp ones. You can experience many at once. You can feel a new one while an old one remains active.

Three: We have *physical feelings* continuously. They are the body's response to the experience of living. We have *mental feelings* continuously also. They are the *mind's* response to the experience of living. A physical feeling is a "feeling" in the simplest sense, and a mental feeling is an emotion.

Four: Simultaneous thoughts do not mix. Or, if they feel like they are mixing, often emotions are mixing. Meanwhile, simultaneous emotions mix all the time. *Proud anger* might be a different feeling from *confused anger*, and both might be different from plain *anger*. Metaphorically, we might think of each distinguishable emotion as a lamp (or an LED element) of a different color. A distinguishable emotion with its own LED might be a compound emotion, like proud anger.

Five: An emotion might be faint or intense, or anything in between—like lamplight. An emotion might be too brief to notice or might go on for months or years. Two separate emotions might be recognizable as themselves but might also mix, producing a distinctive new emotion—as two lamps of different colors are clearly recognizable, but their projected light might mix. Separate colors form a continuum, so there are always enough colors for each emotion to have its own.

Six: Thoughts form a sequence (which might include blank spaces). Once a thought has had its moment in the sun, at the center of our attention, it backs off into memory. A continuous train of "spent thoughts" enters memory—but in memory the train might come apart, and emotions might be rearranged or superimposed.

Where do emotions fit in this model? Emotions modify and give texture to thoughts, *nonsequentially*. We might envision a (*long*) sequence of different-colored lamps, one lamp for each emotion. At the exact point when a mind feels the start of an emotion, the lamp corresponding to that emotion turns on. (Even if we have never felt this emotion before, a corresponding lamp exists.) If the emotion grows stronger and weaker, the intensity of the corresponding lamplight changes. Each lamp projects its own corresponding beam of light.

Now, we take our lamps (mounted on, say, a long metal rod) and bend the rod into a circle. The mind's train of thoughts runs through the center of the circle. One thought, say, "That tree is a maple," might be illuminated by the emotions *angry* and *irritable* simultaneously. We therefore think our thought in an *angry, irritable* way. Thus the thought is *modified* by these emotions, as a noun is modified by adjectives. We might think, accordingly: "That is one ugly, annoying maple!" Or, "Naturally Schwartz has to plant his maple right *there*, where its enormous, rapacious root system will suck all the water out of my soil and kill my delphiniums and steal my girlfriend!" These might not be *reasonable* fears, but our thinker is angry and irritated.

The *angry* and *irritable* lamps might remain lit for a long time, and might modify many thoughts. Or they might blink on for a moment, and modify one thought or none.

Why is this a good metaphor? Visualizing it brings a person closer to what is happening in his body, rather than farther away. Imagining emotions in this way may lead to more emotional granularity and control, rather than less.

7. Revisiting Our Friends

When we look at Alice, Brian, and Charlie, what can we say about each of their blurred pictures of emotions, and thus of reality? Do we have more tools with which to analyze them?

Alice is a friendly, open, social person, and generally likes her job teaching, but is frustrated with her social relationships. She partially subscribes to the social construction theory. Alice does not see the people around her as independent biological entities, but rather as symbols of her own worth and goodness. When people compliment her and perform favors for her, Alice sees herself as good, but when this stops, she sees herself as a bad person. Although she is kind and goes out of her way for people, she often has trouble seeing them as independent entities with their own unspoken needs. In fact, she barely sees her own needs or her own biological self, and often loses sleep and forgets to eat. She keeps in her head an implicit ledger of tit-for-tat interactions and gets upset whenever she feels she is on the wrong end of a deal. Her friends get overwhelmed when they have to explain that they did not mean to hurt her feelings, but were instead taking actions that felt reasonable to them. They feel bad about this because Alice is a nice person, but get tired of justifying themselves, and pull away, thus leading Alice to feel even more insecure.

What should Alice do? Perhaps she could find a practitioner who will explain that her feelings are legitimate and that she is not a bad person despite how others treat her. From the practitioner she might learn that she is in control of what her standards are for the people around her, and that boundaries can be healthy, not selfish. In a bad case, she might further enshrine her concept that she has been wronged by everybody who cares about her, and that they deserve her anger.

Alice may also do something very different. She may learn about her own biological needs, and her own material biological reality, to round out her current social view of herself. In learning about her own biological needs, Alice will become more attentive to the needs of others. She will grow more understanding when another person is too tired, hungry, or distracted to properly tend to her, and thus she will be less angry at other people. She will also be more caring and forgiving towards herself, with less of a need for other people's validation, help, or attention. Thus, her social relationships and her relationship to herself will improve.

What about Brian the programmer? What false truths does Brian subscribe to? Brian seems, on the surface, to be the opposite of Alice. Brian is emotionally distant. He is aloof. He is introverted, whereas Alice is extroverted. But inside, Brian and Alice have something important in common. They are both filled with emotions that they do not know how to control. Brian seems aloof because when he feels too many emotions, he short-circuits and turns away from them. Thus, he is filled with emotions that he does not acknowledge. In Brian's model of the world, emotions are scary things that, if looked at too closely, get out of control. Although Brian keeps a polished persona in front of his friends and co-workers, he has trouble with intimate relationships, and often sabotages them before they get too deep.

What should Brian do? Perhaps opening up is a way for Brian to let more opportunity into his life. Perhaps Brian can learn that strong emotions are not something to be afraid of. But if told that he is repressed and needs to open up more, Brian may likely say that he feels rather open already, or as open as he can be.

It is likely that Brian needs something entirely different. Brian can learn that emotions are not something to be controlled or released at all. Energy is something that can be controlled or released, and though emotions can affect your energy, they are not your energy. Emotions are something that you cannot help but feel, but then you have choices in what you do with the feeling. Separating these two concepts of energy and emotion would allow Brian to maintain the control that he cares so much about, while allowing him to have a rich and pleasant emotional life.

What about Charlie, the political-news addict? Charlie is laden with so much guilt that he cannot feel worthy of love and care unless he has proven himself. He dives into national politics not because he has a passion for the subject but because he wants to prove himself good. As a result, he alienates those around him, thus proving himself more virtuous and them more selfish, while also furthering a spiral of self-loathing.

What should Charlie do? It would be a mistake to miscategorize Charlie's compulsions as genuine interests and thus encourage him to dive further into them. It would be better to teach Charlie that he does not own all the world's pain, and should focus on caring for himself. Then he can learn to enjoy life again.

This last sentence about Charlie captures the entire point of this essay. The ideas about our emotions that are worth spreading are the ideas that help people live more enjoyable lives, form productive relationships, and build thriving communities. We know this is true because the alternative—self-conceptions that create boredom and unnecessary suffering, self-conceptions that tear apart intimate relationships, self-conceptions that create social conflict—is unthinkable.

Notes

1. Julie Beck, "Hard Feelings: Science's Struggle to Define Emotions," *The Atlantic*, 24 Feb. 2015, https://www.theatlantic.com/health/archive/2015/02/hard-feelings-sciences-struggle-to-define-emotions/385711/.
2. Paul Ekman and Dacher Keltner, "Universal Facial Expressions of Emotion: An Old Controversy and New Findings," in *Nonverbal Communication: Where Nature Meets Culture*, eds. Ullica Segerstråle and Peter Molnár (Mahwah, NJ: Lawrence Erlbaum Associates, 1997), 46.
3. Jaak Panksepp, "At the Interface of the Affective, Behavioral, and Cognitive Neurosciences: Decoding the Emotional Feelings of the Brain," *Brain and Cognition*, 52, no. 1 (2003): 4–14.
4. Benedict Carey, "Psychology Itself Is under Scrutiny," *The New York Times*, July 16, 2018, https://www.nytimes.com/2018/07/16/health/psychology-studies-stanford-prison.html.
5. A. M. Turing, "Computing Machinery and Intelligence," *Mind* 59, no. 236 (October 1950): 433–460, https://doi.org/10.1093/mind/LIX.236.433.
6. René Girard, Jean-Michel Oughourlian, and Guy Lefort, *Things Hidden since the Foundation of the World*, translated by Stephen Bann and Michael Metteer (Stanford, CA: Stanford University Press, 1987).
7. René Descartes, *Discourse on Method* [1637] (New York: Collier Macmillan, 1986).
8. M. A. Hogg, D. Abrams, and G. N. Martin, "Social Cognition and Attitudes," in *Psychology*, eds. G. N. Martin, N. R. Carlson, and W. Buskist (Harlow, UK: Pearson Education Limited, 2010), 646–677.
9. David Hillel Gelernter, *The Tides of Mind: Uncovering the Spectrum of Consciousness* (New York: Liveright Publishing Corporation, 2016).
10. Jaak Panksepp, *Affective Neuroscience: The Foundations of Human and Animal Emotions* (Oxford: Oxford University Press, 1998).
11. Daniel Gilbert, *Stumbling on Happiness* (Toronto: Vintage Canada, 2007).

17. A Case for the Relational Person

C. Eric Jones

1. Introduction

When social psychologists conduct research, they often fail to see the role that philosophical assumptions play in the conclusions they draw. All too often, it is assumed that social psychological theories and empirical data alone establish an important result, but a closer look reveals that the researcher has not articulated the specific philosophical context being used as an interpretive framework.

This can be a real limiting factor on the progress of good science, as scholars have observed how particular interpretive frameworks can severely hinder the social psychologist's ability to make rich and robust scientific insights. For example, according to Richard Ryan and Kirk Brown, "The failure of most modern 'cognitive' psychology to posit any needs at all, and rather to treat all motives as 'equal'..., has led to an impoverished depth to our models of human behavior."[1] This "impoverished depth" may be remedied by first acknowledging the philosophical assumptions undergirding the conclusions, and then critically examining the explanatory power of these assumptions.

To correct for the lack of explicitly stated assumptions about human nature, we will begin by stating clearly that this chapter reviews certain psychological findings using a relational ontology. A relational ontology claims that there are things which cannot be adequately understood without considering their connections to other things. Just as understanding sonship requires a reference to parenthood, so personhood cannot be understood apart from its wider network of relationships.

This interpretive framework and the resulting conception of the person is in sharp contrast to the atomistic and egoistic narrative, which has been the dominant, though rarely explicitly considered, conception of human nature assumed in social psychology research for more than a half century. On the atomistic view, just as we can distinguish different atoms in the atomic table, we can adequately understand individuals in isolation from one another; on the egoistic view, human motivations ultimately reduce to human (genetic) survival interests.

With these two contrasting conceptions of the human person in full view, we see that the atomistic/egoistic ontology sits in an uncomfortable tension with the data, whereas a relational ontology has the resources to support a rich and coherent interpretation of the data. Moreover, it resonates with our commonsense lived experience.

Connecting with the broader themes of this anthology, we may further consider the metaphysical foundations undergirding these two views of the person. The atomistic/egoistic conception of the person is rooted in a materialist metaphysic, wherein the survival of our selfish genes is primary and the non-material relations humans establish with each other are considered derivative and even "illusory," at best in indirect service of our genes. This is precisely the conclusion of Darwinist thinkers Michael Ruse and E. O. Wilson: "Human beings function better if they are deceived by their genes into thinking that there is a disinterested objective morality binding upon them, which all should obey."[2]

Because of their materialist commitment, Ruse and Wilson find it necessary to claim, in other words, that humans deceive themselves into believing there is an objective morality. Presumably the two men believe that their skeptical take on morality makes the best sense of their data, but if they were to critically examine their philosophical assumptions, they might reach a different conclusion. A more natural and elegant interpretation accounts for moral behavior as an outworking of a genuine relational morality.

This chapter will consider this alternative view, that a person is in pursuit of a lived experience characterized by meaningful and purposeful community engagement. The pursuit of the person is ultimately one of immaterial, mental, abstract connection and dynamic interaction with others. In the relational view of the person, the immaterial relations in view are ontologically primary and thus most naturally considered "real." The relational ontology therefore sits best within a non-materialist metaphysic. Thus, if in our analysis of the social psychology research we come to prefer the relational conception over the atomistic/egoist conception of the person, we have in addition provided good reason to prefer a non-materialist over a materialist metaphysic.

The rest of this chapter reviews atomistic/egoistic and relational views of the person so that the reader will be able to contrast the ability of the two views to account for the person as described by various research efforts. Research is then presented to help the reader decide the central question of the chapter: Which is the more tenable view of the person—the relational one or the atomic/egoistic one?[3] To begin, let us more precisely define the two contrasting views of the person used throughout the chapter.

2. The Person: Two Views

Materialistic views of the person are pervasive within the empirically based research of psychology. The most philosophically explicit of those views commonly define people as determined, egoistic, atomistic, with their meaning or purpose in life reduced to genetic transmission, i.e., biological reproduction. One of the most interesting aspects of the materialistic views of the person for our discussion is what I refer to as *the illusion of personhood*, well represented by Lonnie W. Aarssen's hierarchy of needs.[4] His hierarchy labels as illusory the parts of life people find most important and significant and labels as real and ultimate the material, physical resources, defense needs, and *instrumental* social needs that lead to genetic transmission.

Note that this perspective does not deny the impact of social and intellectual pursuits, transcendent activities, after-life beliefs, and corresponding behaviors, or meaning- and purpose-related thoughts and behaviors; the theory merely labels them as distractions and delusions. Aarssen acknowledges the fact that human minds engage in these activities, which are too obvious to deny, but he relegates them to the service of gene transmission and denies them any inherent meaning. So, in addition to affirming a model of egoism and atomism, materialistic models of the individual typically include no inherent *telos* (aim, purpose) or meaning in life. As Richard Dawkins bluntly puts it, "We are survival machines—robot vehicles programmed to preserve the selfish molecules known as genes."[5] Here, the self is seen as a utilitarian shell on a singular quest to achieve gene replication. According to these more explicitly stated materialistic explanations, various activities of life to which people attribute meaning and purpose fall within the categories of distraction or delusion.

Aarssen's hierarchy is also a good example of inserting a new philosophical framework when none is required. For instance, if one accepts that people can have meaningful, non-instrumental relationships with others, and if the data are consistent with this understanding, no new philosophical framework is needed that redefines relationships as *always* instrumental. So why does Aarssen insert this framework? In this case, it seems that materialistic assumptions about human nature have been brought in from outside the field of psychology.

The same assumptions are often imported into biological literature, e.g., in the following passage by Richard Dawkins: "We are machines built by DNA whose purpose is to make more copies of the same DNA...This is exactly what we are for. We are machines for propagating DNA, and the propagation of DNA is a self-sustaining process. It is every living object's sole reason for living."[6]

Joel Green says that Dawkins's statement represents a "scientific view of humanity."[7] It is better said that the statement represents a philosophical view of human purpose for which many scientists have a personal preference. Though scientists are certainly entitled to their personal philosophical assumptions about human nature, good scientists will recognize when the data in their field are inconsistent with what one would expect from those assumptions. As the research is presented, it is important to keep in mind that materialistic assumptions about the person and mind are not scientific, but are personally preferred philosophical views. And in the case of psychology, which is our subject here, the materialistic preferences appear to be largely unsupported by psychological data, as the rest of this essay will show.

The *strong relational view of the person* presented here is almost completely unknown within psychology and therefore is one of the most infrequently used models defining the person within psychology. The existing research discusses to greater or lesser degrees the implications of others for the self as the self thinks of the self, thinks of others, or engages with others. Some theorists even use the term relational self, though I am proposing something beyond their idea that others are an important factor for the person. I am proposing an *ontologically relational* view of the person, saying that development into personhood and the ultimate experiences of life essentially depend on connections with others. That is, relationships with others are not merely a helpful option in life, but an essential factor for which there is no substitute for reaching personhood and experiencing fulfillment in life.

In contrast to the dominant atomistic (or "entity") view of the person and the egoistic view of the functioning of the person, Brent Slife, Greg Martin, and Sondra Sasser have suggested that persons are best characterized by strong relationality.[8] The aspect of strong relationality emphasized in this discussion is functional interdependence as seen in systems. Many systems are structurally such that the removal or ineffective functioning of one part can cause the entire system to become less effective, or in extreme cases not to function at all. For example, to the extent that a fuel pump works poorly or not at all, the performance of the entire car is compromised, regardless of how well all other parts of the car are functioning. For a more detailed example, we turn to the human body.

The list of functions a human hand can perform is seemingly endless. A hand can use a pencil to write, hold a knife to cut, hold a fork to feed a person, grasp a doorknob and turn it to open a door, pet a dog, shake another hand, or give a "high five." Yet as much as the hand may be thought of as a distinct part of the body, the hand is functionally nothing on its own.

The structural and functional interconnectivity is as apparent as it is comprehensive. If you grasp with your left hand your right forearm while you flex your right hand, you notice the muscles of your forearm are required for movement within your hand. So, if you desire to write something you will need your forearm muscles, not just your hand. Of course, you may also need to stretch and reach your pen and require muscles in the upper arm and back. Or you may want to go to your front door and open it for someone, in which case you likely will use muscles throughout the body. Further, muscles are not the only other part or subsystem of the body necessary to conduct relatively simple functions. The hand needs the entire circulatory system to carry nutrients and oxygen to the muscles. The hand needs the digestive system to supply the nutrients. The hand needs the respiratory system to supply the oxygen. Frequently, the hand finds eyes helpful to direct its movement. And it needs the nervous system and a brain to process and coordinate all of this. Absent structural connectivity and proper functioning of any one of numerous parts, the functionality of the hand is diminished (such as with the absence of eyes) or ceases (as with the absence of the brain). Visually the hand may seem to be a separate entity; however, the structure and functionality of the hand is inextricably intertwined with and dependent upon many other parts of the body.

Recognizing the reliance on other body parts for its functioning brings to light several other notable characteristics related to the functioning of the hand. Two principles are the focus here. First, the hand is defined by and is essentially dependent upon the rest of the body. In the extreme sense of a severed hand, the hand lacks definition of identity and purpose. It is only really a hand as we understand it in relation to the rest of the body. Second, as much as the hand relies on other parts to function, the hand primarily functions for the good of other parts. That is, the hand is intended to and primarily carries out functions for other parts of the body. The hand rarely functions for itself. And as much as the hand serves the needs of the rest of the body, the rest of the body serves the needs of the hand.

This sort of interdependence also means that the functioning of the whole system is the ultimate goal for all parts of the system. Given the primacy of system functioning, the "self-actualization" of any one part is not important if the other parts are not equally developed. And, even if the parts are all well developed, the connections among them must develop and operate well. Therefore, development and maturity are based upon not only the development of each part, but the progressive development of the interconnections of the body. Even after reaching full maturity, each part of the body must maintain connections and continue in the "serving" of the other parts for the body to operate effectively.

The analogous argument is that people are strongly and intangibly interconnected with each other by way of thoughts, behaviors, emotions, and memories that result in the psychological co-constitution of persons by other persons and that psychological research provides evidence of this interconnectedness. The relational person has certain parallels to the hand as part of the system of the body. One notable exception to the body metaphor is that parts of the body do not possess any degree of free will, whereas (I argue) the person does. This means that as much as people are ontologically essential for each other, people can choose to nurture their psychological connections and move toward optimal functioning, sometimes embrace and sometimes ignore those connections and live an underwhelming life, or attempt to disconnect as much as possible and experience significant dysfunction in life. Regardless of the choices made, a person's psychological interconnections are essential for personhood and are a foundational part of our nature. Free will may significantly affect outcomes of our lived experience but has no bearing on the relational aspect of human nature.

As people are part of a social system, they will exhibit broad categories of psychological phenomena. Evidence will show that counter to atomism, people need each other to form a self-concept, to experience satisfaction, to experience well-being, and to develop over time toward maturity. Research will also show that counter to egoism, we not only receive from others, but give back or reciprocate toward others. This sort of giving and receiving is documented in research, but that research is usually restricted to close relationships when giving and receiving is reciprocal between the same two people. From the relational model of the person, giving and receiving can be reciprocal but it can also be indirect, extending beyond two-person interactions. For example, I may help someone I know, and another person may provide me similar help rather than the person I helped helping me directly.

The relational view expects that the more one gives to others while still having one's own needs met, the more well-being one will experience. In contrast, the atomistic and egoistic views of the person would suggest that getting from others more than we receive from them will result in more well-being, since it results in higher levels of survival advantage—which is how "well-being" is understood in the atomistic and egotistic views. As this essay will show, psychological research supports the relational view's highly interdependent nature of the person by showing that we cannot become who we need to be in any significant way without others and that others similarly need us to become who they need to be. Therefore, the atomistic view is not only not the right view of an individual, but that the whole idea of the *individual* is mistaken. The person is not an individual or discrete entity at all, much as the hand is not an entity unto itself. The person, and the hand, are parts of a larger system and each is best understood within that conceptual framework.

3. The Life of the Relational Person

First, before reviewing the relevant research, we repeat the point made above that conducting and reporting research in psychology requires a researcher to answer fundamental questions concerning human nature. Whether or not these questions are answered deliberately or outside of conscious awareness, and whether they are answered in a philosophically consistent manner or not, they are answered. Second, assumptions about human nature not only inform the thinking of researchers at the various steps of conducting and reporting a single study; they orient one's interpretations of entire programs of research. At times, the social and cognitive psychological research being reviewed here is interpreted at both levels from a materialistic evolutionary perspective without allowing for competing interpretations. In contrast, the developmental and positive psychological research is typically interpreted from perspectives that assume that the person is less atomistic and egoistic, though again the connection to any underlying philosophy is rarely explicitly stated.

Here I am presenting selected but representative research that, I contend, can be better interpreted from the viewpoint of the ontologically relational, agentic person who experiences real meaning in life rather than the materialistically oriented life of atomism, egoism, and determinism that is devoid of inherent meaning. Though only a few studies are being presented and reinterpreted here, hundreds could be presented similarly, given the time and space to do so.[9] It is important to acknowledge, however, that the data presented in these studies is not intended by the original authors to support relationality, and that the inferences I am drawing from the data are my own, not those of the authors of the research.

3.1 The Relational Person in Community

This review of the research is based on the view that people live as (1) interdependently constituting/constituted, (2) interdependently satisfied/satisfying persons, (3) in community with one another. The review of data will establish the kind of person and the purpose of the person for which the mind is functioning. Please note that though the three categories are intertwined and overlapping, they are mentioned separately for convenience's sake, not because each is a philosophically or practically orthogonal concept. That is, the lines could be drawn differently among the evidence for the constitution, the development, and the satisfaction of the person, but those lines of distinction are somewhat arbitrary and the absolute distinctions among them are false.

Further, the 1) interdependent constitution section of research and the 2) interdependent lived experience section of research together essentially represent the person in community. That is, one can see the 3) communal lived experience of the person portrayed in the two sections of the reinterpreted research supporting 1) interdependent constitution and 2) interdependent lived experience of the person.

The next two sections on interdependent constitution and interdependent lived experience will highlight the empirical evidence supporting the co-constitution and life satisfaction interdependence of the person as experienced within a meaningful and purposeful community. Further, this research will show that the person is not intended to think egoistically. The person is intended to work neither egoistically nor altruistically, but in a way that transcends the self-other distinction upon which those concepts rest.

3.2 The Interdependently Constituted/ Constituting Person in Development

In addition to not being atomistic, the relational person transcends the self/other distinctions that are foundational in most research and theory within the field of psychology. The relational person avoids the two extremes of being either fully constructed by others or being self-made. The relational person has a self to begin life, so is not fully socially constructed, but is also not self-made as the self must have ongoing connections with others to become a full person and to experience life as intended. By transcending the self/other distinction, the relational person can account both for data that suggests people are socially constructed and for data that suggests social construction does not account for everything about a person. The dual picture, of a core self and a self co-constituted in connection with others, provides the most accurate view of the person and makes possible a recasting of the voluminous social thought and behavior research into a more cohesive narrative that describes a person's lived experience.

Since the relational person is neither frequently nor explicitly presented in research, we need to ask: What kinds of things are expected from an ontologically relational view of the structure of the person? Research should present evidence of ontological connectivity with others and the necessary use of those connections with others as related to a human *telos* of community. The ontological connectivity evidence should be noticeable in research that includes developmental progressions toward maturity. In other words, research should define development and maturity by presenting the person as non-egoistic, non-atomistic, and developing toward becoming a contributing member of community. Research should also show that in addition to being ontologically connected with others, those connections must be used well for one's lived experience to go well. This evidence should show how people with healthy relationships develop the necessary capacities for life in community, and should show how poor-quality relationships inhibit the development of essential capacities for life in community. The research interpretations that follow show how existing research fits within a relational person paradigm.

In developmental psychology, Lev Vygotsky, Daniel Levinson, Richard Lerner, and Urie Bronfenbrenner have centrally included the necessity of others for the developmental process of the person. Vygotsky's use of others in development processes is exemplified in his core concept, the "Zone of Proximal Development." This concept outlines various interdependent factors that describe specifically how the self relies on others to progressively develop.[10] Lerner[11] and Bronfenbrenner[12] both have described in great detail how the person is embedded in and reciprocally interacts with a community of others at levels of the family, community, society, and culture, and how these great many interactions result in the growth of the person physically, socially, and cognitively in every stage of life. A general overview of the developmental literature indicates that human development depends on others to attain language ability and skills, form a personal identity, and gain and exhibit bonding abilities with others.[13] These major advances for the person must be achieved for the person to become a contributing member of his or her community.

Attachment theory is one of the most empirically investigated lines of research, not just in developmental psychology, but in psychology generally. According to this theory, attachment begins with the dependence of an infant with a primary caregiver. The infant develops an "attachment style" based on the availability, attentiveness, and the responsiveness of the caregiver.[14] If the caregiver's responsiveness is reliable, then the infant will seek the proximity of the caregiver when distressed and develop a stable sense of security and safety, a secure attachment style. When caregivers are not reliably responsive, then the infant will turn to the strategies of anxiety and avoidance to relieve distress.[15] The degree of avoidance corresponds to the extent of emotionally distancing from others due to distrust of relationship partners. The degree of anxiety corresponds to the extent to which one worries that a relationship partner will not be available and supportive when needed. These attachment styles can persist throughout life[16] and pervasively affect a wide range of subsequent relationships.[17] A single caregiver can set one's attachment style (secure, anxious, or avoidant) as a relationship framework for the person. From those differences alone the person is significantly transformed from who they might have been.

The effects of the primary caregiver do not stop there, however. A multitude of other aspects of the person, ranging from self-concept to aggression levels, also depend upon how one thinks about and interacts with others. Therefore, attachment style directly or indirectly affects a myriad of characteristics of the person that are highly influenced and only possible through interaction with others. From our earliest development, people are co-constituted by others in obvious areas such as relationships, relational outcomes, and attachment style, and those outcomes affect not-so-obvious aspects of life, such as how we appraise, organize, and choose goals.[18]

One specific example of others participating in the construction of the person is the *Michelangelo Effect* research of Caryl Rusbult and colleagues. As much as Western culture focuses on aspects of the self-made person, many powerful interpersonal processes contribute to the becoming of the person. The Michelangelo Effect details how desired skills and traits of one's ideal self are sharpened in close dyadic relationships through the interpersonal processes of affirmation and verification.[19] By becoming aware of a partner's ideal attributes, one can help "sculpt" the person into progressive approximations of that person's ideal self. The idea is that people need close others to properly "sculpt" them into the person they want to be, because a person cannot achieve the same results alone. The outcomes of the sculpting become more powerful as the interdependence between the dyad becomes more pronounced, so that what begin as temporary adaptations emerge as stable components of the partner's ideal self.[20] This line of research speaks with some precision to the processes and outcomes related to the co-constitution of the person.

Research also exists detailing that people think of self and others in ways reflecting co-constitution. Arthur Aron and colleagues produced data showing that people confuse or blur the boundaries with close others because they overlap cognitive representations of self and close others.[21] In effect, the self cannot tell what aspects of the person are characteristic of the self and what aspects are characteristic of the close other. This "inclusion of other in self," or merging self with another, directly opposes the atomistic view of the individual and points to a more relational view of the person. In addition, Christian Waugh and Barbara Fredrickson proposed and found support for the idea that positive emotions experienced with another resulted in increased self-other overlap and a more complex understanding of the other.[22] Data also show effects at the group level. John Dovidio, Samuel Gaertner, Alice Isen, and Robert Lowrance report that people who perceive self-other overlap form more inclusive social categories,[23] and they are significantly more likely to see higher similarity between ingroup and outgroup members.[24] Building on the self-expansion model of Aron at the individual level, support exists for the concept of inclusion of community in self. Debra Mashek, Lisa Cannaday, and June Tangney were able to measure a psychological sense of community or a connectedness of self and community, showing people connect and become part of something bigger than themselves with not only a single close other but with a surrounding community.[25]

The pattern of results from this "self-expansion" literature repeatedly shows that the boundaries blur between self and close others, and even between self and larger groups. However, the researchers' perspective is that the direction of the effect is that the other or the community is being included in the self. Perhaps the effect is better described as the self becoming part of a more communal entity (whether just a pair of persons or a larger group) which provides a proper relational context for the self. "Self-expansion" theory should be called the "self-subsumed-into-a-larger-social-entity" theory because "self-expansion" ends in the person seeing themselves as part of a larger whole. Dovidio et al. report that members of groups "see both groups as one superordinate group," and that couples "emphasized more communal themes of shared relationship attributes" and "used more 'we' language and less 'I' language when describing their relationships."[26] This is not about "self-expansion" *per se*, but about the connections of the self with a larger social context being properly recognized. It is the orienting of the self in line with a dyad or group, not the other or the group being funneled into one person. Choosing and using the term "self-expansion" itself as an interpretive framework or label for this research shows just how implicitly individualistic the culture of social psychology is.

3.3 The Interdependently Satisfied/Satisfying Person

Though psychologists infrequently state a teleology for the person, Lerner has done so. He proposes that "adaptive developmental regulation results in the emergence among young people of an orientation to transcend self-interest and place value on and commitments to actions supportive of their social system."[27] Lerner is suggesting that optimal development involves persons who not only receive from but also are able and willing to give back to their families, communities, and societies. Again, this is not a happy coincidence. Acts of relational reciprocity result in optimal development because the actions are aligned with a relational ontology of the person.

Since the relational person is not frequently and explicitly presented in research, we need to ask: "What kinds of things are expected from an ontologically relational view of the person's lived experience?" Research evidence should be consistent with a lived experience related to a human *telos* of community. This evidence should be noticeable in research that includes meaning, purpose, and well-being. In other words, if the research can explain well-being, maturity, and meaning by presenting the person as non-egoistic, non-atomistic, and in need of community, then the relational person is not an illusion.

Just as the nature of the relational person is not atomistic, so the motivation and functioning of the relational person is not completely egoistic and does not always use others instrumentally. The structural interdependence of the person means that the motivation and functioning of the person also transcends the self-other distinction. Therefore, the clear distinction between the self-oriented motivation of egoism and the other-oriented motivation of altruism is moot. The relational person necessarily gives to others and receives from others as part of the design of the larger system, though not necessarily in directly reciprocal two-person exchanges. To the extent that the person has healthy connections and gives to others as well as receives from others, the person will experience high levels of well-being, meaning, and purpose. To the extent that people think and act inappropriately egoistically, they will experience lower levels of well-being, meaning, and purpose. The following review presents research showing that a life well lived is what is expected from a relational view of the person.

One of the earliest and most persistent phenomena in psychology is the *pursuit of self-esteem*. Egoism clearly predicts that people's desires and actions will be directed to obtaining higher levels of self-esteem, and so many find egoism to have an intuitive appeal. Yet the data say otherwise. The pursuit of self-esteem undermines learning, relatedness, autonomy, and self-regulation.[28] Any short-term emotional gains are dwarfed by the long-term results of decreased mental and physical health, along with the aforementioned negative effects. In short, pursuing what has always been considered an egoistic goal results in a deterioration of one's psychological foundation.

In a sweeping review of research related to selfish and "otherish" motivations, Jennifer Crocker, Amy Canevello, and Ashley Brown, describing the effects of giving (whether the giving is material or psychological), strike against egoism when they state, "People are psychologically constructed in such a way that giving to others can be rewarding despite its obvious material costs, and selfishness can be costly despite its immediate material benefits."[29] Along with their review of the effects of giving and taking, they include a review of selfish and otherish motivations that accompany giving and taking and find strong evidence for the benefits of otherish motivation and the costs of selfish motivation. The authors conclude that section of their review by stating "we located no recent studies demonstrating that otherish motivation has costs or selfish motivation has benefits for psychological well-being, physical health, or relationships,"[30] driving home the point that egoistic motivation will not help the person function well, and that egoism is not the primary motivation of the healthy person.

Moving from generalities to a specific example, we see that the results of research on the subject of *gratitude* over the last twenty years are consistent with the relational person. Gratitude has been shown to increase the psychological well-being factors of meaning in life,[31] optimism,[32] humility,[33] post-traumatic growth,[34] resilience,[35] and emotional well-being factors such as life satisfaction,[36] subjective well-being,[37] and positive affect.[38] Compared to the psychological and emotional well-being factors, social well-being factors have shown even stronger and more consistent effects. The social factors of responsiveness of relationship partners,[39] relationship commitment,[40] improved sense of belonging,[41] and prosocial behavior directed at beneficiaries and third parties[42] are all positively associated with gratitude, whereas gratitude is negatively associated with psychopathology.[43] The general pattern, i.e., that gratitude yields increased positives and lower negatives, supports a relational view of the person. The fact that, of the well-being factors, the social factors are the strongest, provides another example that even though the benefits to the individual are the most investigated, the social aspects frequently are the most important.

Another facet of egoistic interpretation is the idea that people tend to associate with those who can benefit them in terms of resources or power. However, research on the *Friend Number Paradox* shows that the assumption that people want to be associated with others who have large social connections is false. When looking for a long-term relationship, people want those with smaller social circles.[44] The importance of a relationship is not about the ability of the other to provide an instrumental advantage. It is about being connected to someone with whom there is time to relate and with whom a deep interpersonal connection can develop. Studies also show that sharing experiences increases well-being and positive affect, that when close others respond actively and constructively the effects are additionally enhanced, and that when active and constructive responding is the norm in relationships, relationship well-being increases.[45] These simple yet profound findings show how the atomistic, egoistic view of the person runs counter to well-being and how other-oriented, ongoing relationships facilitate well-being.

Another blow to the assumed egoistic motivations of the individual is the work on the effects of purpose and meaning on well-being. William Damon, Jenni Menon, and Kendall Bronk found that "beyond-the-self" purpose was related to greater well-being and life satisfaction whereas self-oriented purpose was not.[46] In a study exploring well-being and purpose among a Chinese population, data show that the most social purpose of the six purposes tested, social dedication, resulted in the highest level of well-being.[47] Data further revealed that the individual-related purposes of social recognition and pleasure quest were unrelated and negatively related to well-being, respectively. This study extended previous research showing that social dedication is related to high levels of well-being[48] by also showing that social dedication promotes continuous prosocial behavior and thus may play a role in maintaining positive community functioning.

Just as social dedication has positive effects beyond a single person, so too do the other-praising emotions of elevation, gratitude, and admiration. The relationship-building functions of gratitude have already been discussed, so we can move to the other two concepts. Elevation is defined as one's response to moral excellence and has been shown to motivate prosocial and affiliative behavior toward others, typically not toward the person modeling the virtuous behavior.

Admiration of another's virtue motivates self-improvement, but the improvement is not limited to the virtues displayed by the person who is the model. Several studies across the three concepts show that interacting with or observing excellence in others inspires people to improve their relationships (due to gratitude), exhibit more moral behavior (i.e., display elevation), and work to improve themselves (due to admiration).[49] These other-praising emotions provide additional evidence that people engage in behavior highly beneficial to others and indirectly to themselves and, in contrast to atomism, show how people are parts of an interdependent system, not discrete entities. Separate but related research shows that communally oriented people with tendencies to give care rather than receive it experience greater self-esteem, greater satisfaction and love in their relationships, and greater love for humanity in daily life.[50]

Reaching back in psychology's history to William James, Karen Horney, Abraham Maslow, and John Bowlby, we see theorists who acknowledge that people seek to develop and maintain social bonds. Research has also shown that attachment with caregivers and relationship with strangers are common across cultures. However, Roy Baumeister and Mark Leary were the first to elevate the belongingness motivation from an important factor to a psychological need.[51] Proposing that belongingness is a need means that it is at the same level as other needs, such as the physiological need for food; people must have it. According to Baumeister and Leary, the following are the criteria for a psychological need or fundamental motivation:[52]

a. it produces effects readily under all but adverse conditions;
b. it has affective consequences;
c. it involves direct cognitive processing;
d. it leads to ill effects (such as on health or adjustment) when thwarted;
e. it elicits goal-oriented behavior designed to satisfy it (subject to motivational patterns such as object substitutability and satiation);
f. it is universal in the sense of applying to all people;
g. it is not derivative of other motives;
h. it affects a broad variety of behaviors;
i. it has implications that go beyond immediate psychological functioning.

The Baumeister and Leary article presents evidence that people form social bonds quickly and easily, that people in every culture belong to groups, and that people develop group identities. Second, evidence shows that people are reluctant to dissolve relationships, even if the relationships have no instrumental value or if the relationships are temporary. Further, people also prefer to maintain contact with those with whom they once had relationships and prefer to stay in poor-quality relationships rather than disconnect or dissolve the relationships.

A third category of evidence shows that belongingness is the source of people's thoughts and conversations, such that people's relationships are a major source of emotions. Further, daily life connections and changes in social connections frequently become a primary source of emotions that people then act on and talk about. Disrupted relationships result in hurt feelings, loneliness, anger, sadness, and jealousy, and it is rare that the disruption of any relationship does not result in negative emotions, thought, and actions. Conversely, strengthening of relationships or social bonds usually results in happiness, joy, and contentment.

A final section of the Baumeister and Leary review discusses the effects of relationship deprivation on individuals. Evidence clearly shows people experience many adverse effects, including stress, depression, poor psychological adjustment, lowered ability to self-regulate, and compromised physical health, when experiencing inadequate social connections. Once in a state of lowered social connections, people typically obsess on their weak social connections and attempt to find ways to strengthen them. Baumeister and Leary present a convincing case, based on their review of hundreds of studies, that belongingness does indeed rise to the height of a psychological need.[53]

The establishment of the need to belong served as a catalyst for numerous studies exploring ideas related to belongingness. The need to belong breathed new life into research on social inclusion and social exclusion, and new lines of research in loneliness emerged. In particular, the investigation of ostracism (or more generally, social exclusion) became a thriving pursuit. For instance, the literature expanded to include findings that show social exclusion impairs the development of self-regulation,[54] decreases exertion of self-regulation,[55] increases cortisol levels,[56] increases activation of brain regions associated with pain,[57] reduces selective attention,[58] decreases memory or recall,[59] decreases prosocial behavior,[60] and increases aggression.[61]

Numerous other lines of research have shown that social rejection has negative effects on well-being by increasing self-destructive behaviors such as making risky choices, engaging in unhealthy behaviors, procrastinating, and being unable to delay gratification. Rejected individuals also become less prosocial, specifically less likely to donate money, volunteer, help after a minor mishap, and more likely to act aggressively. Taken together, the research on social exclusion, loneliness, and the need to belong clearly present the devastating effects of social disconnection for the person. The relational premise that others are essential for the person is solidly supported.

In perhaps the most impressive study ever done on the subject of flourishing, we see the positive side of community and the power of social connections. Beginning in 1938, the Harvard Grant Study followed 268 Harvard men who were sophomores in the classes of 1939, 1940, and 1941. This study is unique not only because the men have been followed throughout their lives, but because the longitudinal data are qualitative and quantitative, include detailed interviews at multiple points in the men's lives, cover physical, cognitive, emotional, and developmental data, and include data from various peers and family members for each man. After decades of research, thousands of data points, and over twenty million dollars spent, George Vaillant summarizes the Harvard Grant Study by saying, "In short, it was the capacity for intimate relationships that predicted flourishing in all aspects of these men's lives."[62]

Said even more succinctly, Vaillant states that the conclusion of the study is, "Happiness is love."[63] For a researcher of Vaillant's reputation to so briefly declare one lone meaning of a study of this quality, depth, and duration tells us that this singular finding alone is more than noteworthy. Love through relationships with others is life, as expected for the relational person.

4. The Relational Mind and the Brain

A CENTRAL theme of this anthology is the question: "What is the relationship between the mind and the brain?" By "the brain," what is meant is the organ inside our skulls, composed of an immense number of various sorts of neurons and the like, interacting with each other and other bodily systems electrically and chemically. What is meant by "the mind" is less clear, and various conceptions of it are rooted in competing metaphysical frameworks. For some materialists, "the mind" is nothing but the operation of the brain. For some dualists, "the mind" refers to a spiritual substance that interacts somehow with our physical brain and possibly our whole body. For some idealists, "mind" is the primary reality and "the brain" (along with the rest of the external world) is a collection of ideas in a transcendent mind (e.g., God) with which our individual minds interact. We will leave the careful metaphysical arguments to the metaphysicians, and focus chiefly on the data, but at the end will briefly reflect on which metaphysical conception of the mind fits most naturally with the data.

Apart from the different views of the mind's metaphysical nature, when the term "the mind" is used, we have in view the subjectively experienced cognitive aspects of a person (e.g., perceiving, thinking, understanding, learning, remembering). Clearly, the mind of a person serves and supports the functioning and purpose of the relational person so far reviewed; to keep this insight at the forefront of our discussion, we will coin and use the term "the relational mind." There should be no difficulty in imagining the association between the proper functioning of the relational person's non-material connections to others and the relational mind at the center of those connections. The previously reviewed research has already outlined such an association, but the following brief overview of the literature on the relational powers of the mind may more obviously present the mind's role in the function and purpose of the relational person.

Floyd Allport once proclaimed that socialized behavior is the supreme achievement of the brain. A more accurate contemporary statement might be that the crowning achievement of the brain's neural mechanisms is the support of the mind's engagement with social thought and behavior.[64] As we are putting aside for the moment the question of whether this behavior is "only" the operation of the physical brain, we can here reframe this statement more abstractly and consider instead how the relational mind (supported to some degree by the brain) enables social behaviors. Early in life, persons establish foundational abilities to know themselves, to know how others respond to them, and to regulate their actions to coexist with others. These abilities have been noted within the first forty-eight hours of life, when infants attend to human faces and voices more than any other visual or auditory stimuli.[65] These socially oriented abilities can generate massive amounts of social information for the relational mind to consider.

Baumeister has proposed that the social brain (which, again, we will read as "the relational mind") engages in thought primarily or ultimately to facilitate interpersonal connections.[66] In a theoretical perspective consistent with Baumeister's social brain proposal, Leary proposed that not only the self-esteem literature but all interpretations of research on the self should be recast, moving away from primarily egoistic motivations and results and acknowledging motivations and results oriented toward interpersonal connections.[67] These theoretical proposals and empirical findings suggest that a person not only requires a community to act and live as a fully human being, but that it is the very living in community that is the *telos* of humanity. Without the relational mind involved in the countless number of cognitive processes connecting a person to a community, the person does not properly develop and is not able to experience a fully human life.

For instance, according to Garth Fletcher and Nickola Overall, a general model within the close relationship literature that incorporates much of what is known about the functioning of the relational mind begins with goal evaluation, goal prediction, goal regulation, and goal satisfaction potential related to a potential partner.[68] Goal information is then passed through stored knowledge structures, experience summaries, including beliefs, expectations, interpersonal goals, and behavioral strategies. All of this is processed in a variety of automatic and controlled ways, generating other cognitions and emotions that may encounter varying levels of self-regulation before a behavioral outcome is selected and enacted. This general framework points out several broad areas of cognitive processing within relationships, each possessing many specific functions of the relational mind that are ongoing, essential operations for the person to develop into and experience personhood. This means the relational person requires a relational mind able to continuously carry out complex, interpersonally oriented activities in order to experience life in community.

Following the data means having to account for the complexity involved in the continual stream of social activity including, but not limited to, planning future social events, comparing social options, prioritizing social options, considering immediate versus long-term effects of social thought and action, assessing costs and benefits across various relationships, weighing costs and benefits of socially engaging or not, deciding to act or not, tracking progress of social goals, redirecting social goals, abandoning ineffective social goals, and recognizing social successes worthy of repeating. As the bulk of this paper has argued, these activities of the relational person can be best understood by reference to immaterial, abstract relations. Furthermore, as we have noted above, these data sit in tension with a purely materialistic metaphysic, where such immaterial entities are at best derivative or illusory, and are much more comfortable within a metaphysical framework that has a primary place for the mental and relational in its ontology.

In this section, we have explored how the relational mind is central to these activities of the relational person. While we cannot conclude much from this vantage point about precisely how the material brain supports the mind's involvement in these operations, the relational mind must in some way be engaging with these immaterial relations. This being the case, considering the broader possibilities afforded by a non-materialist metaphysic, we suggest that there is certainly good reason to consider that a portion of the relational mind may itself be immaterial in nature.

5. Conclusion

Is THE relational view of the person a sound one? The demonstration of this has been the central aim of this chapter. We have seen that a dogmatic commitment to materialist models of the human person (and by extension, the mind) fail to do justice to the findings of social psychology. If the relational view of the person is wrong, it seems these findings must be declared illusory. A more natural approach, surely, is to accept the data at face value and remain open-minded about the best philosophical conception of the person. If we do this, a relational view of the person clearly fares better than typical, atomistic-egoistic models. In turn, a non-materialist metaphysic is to be preferred, as it provides the more capable ontology to support the relational view of the person. Psychologists therefore have good reason to question whether materialism should serve as a non-negotiable assumption of their inquiries.

It is arguable that materialists in psychology face a dilemma analogous to Angus Menuge's Philosophical Dilemma for Physicalism.[69] If the materialists use clearly materialistic frameworks to explain social thought and behavior of the person, their explanations are significantly insufficient. On the other hand, if they more accurately and comprehensively explain social phenomena, they must leave behind the materialistically associated foundations of atomism and egoism for which they have such a strong personal preference. In an effort to be philosophically consistent, many materialistic theorists jettison meaning, purpose, and non-instrumental relationships in pursuit of the concrete *telos* of genetic transmission, but by doing so become inconsistent with psychological data. Based on the research reviewed here, philosophical consistency and consistency with psychological data can only be achieved by using a version of the person incompatible with materialism, i.e., the relational person.

In a remark very suited to our discussion here, Peter Ossorio said that we should understand "things as they seem, unless we have reason enough to think otherwise."[70] If we start with the view of the person most people experience, then we should start with something that resembles the relational person; and based on the available research, we see no clear reason to shift from it. Quite to the contrary, much evidence supports exactly what a relational view expects. The psychological data broadly support the relational view of the person and the corresponding operations of the mind. For an empirically grounded scientist, considering which conceptual model of the person best explains psychological data, it makes little sense to stay attached to the atomistic-egoistic model, with its anemic explanatory power, when research findings resonate so well with the relational model of the person.[71]

In sum, the empirical evidence presents a picture of the life of the person as one of essential interdependence with others based on multiple, frequent, and ongoing higher-order processes with which atomism and egoism are inconsistent. If the data and our lived experience are telling us that we must treat meaning, purpose, and relationality as if they are real, because when we do not live accordingly, we psychologically fall apart, then perhaps we should consider the non-materialistic, relational version of the person to be correct.

NOTES

1. Richard M. Ryan and Kirk W. Brown, "Why We Don't Need Self-Esteem: On Fundamental Needs, Contingent Love, and Mindfulness," *Psychological Inquiry* 14, no. 1 (Jan. 2003): 73. The elided internal reference is to R. M. Ryan, K. M. Sheldon, T. Kasser, and E. L. Deci, "All Goals Are Not Created Equal: An Organismic Perspective on the Nature of Goals and Their Regulation," in *The Psychology of Action: Linking Cognition and Motivation to Behavior*, eds. P. M. Gollwitzer and J. A. Bargh (New York: The Guilford Press, 1996), 7–26.
2. Michael Ruse and E. O. Wilson, "Moral Philosophy as Applied Science," *Philosophy* 61 (1986): 179.
3. This chapter presents evidence arguably consistent with a relational view of the person and perhaps a non-materialistic view of the mind. The extent to which the behavior and thought processes presented here are unable to be resourced by a materialistic foundation is discussed in greater detail in other chapters in this volume.
4. L. W. Aarssen, "Darwinism and Meaning," *Biological Theory* 5, (2010): 296–311.
5. Richard Dawkins, *The Selfish Gene* (Oxford: Oxford University Press, 1976), v.
6. Richard Dawkins, "Growing up in the Universe: Ultraviolet Garden," lecture 4, Royal Institution Christmas Lectures for Children (Dec. 1991), https://www.rigb.org/explore-science/explore/video/growing-universe-ultraviolet-garden-1991.
7. Joel B. Green, "Body and Soul, Mind and Brain," in *In Search of the Soul: Four Views of the Mind-Body Problem*, eds. Joel B. Green and Stuart L. Palmer (Downers Grove, IL: InterVarsity Press, 2005), 7.
8. Brent Slife, Greg Martin, and Sondra Sasser, "A Prominent Worldview of Professional Psychology," in *The Hidden Worldviews of Psychology's Theory, Research, and Practice*, eds. Brent Slife, Kari O'Grady, and Russell Kosits (New York, NY: Taylor and Francis, 2017).
9. In a broader sense, and consistent with the message of this chapter, much of the narrative of psychological research is vulnerable to the criticism voiced in Menuge's "Dilemma for Physicalism," which states that "the stricter accounts that remain faithful to the core doctrines of physicalism seem obviously inadequate, but more promising, relaxed accounts are no longer obviously physicalist at all" (see Angus J. L. Menuge, "Declining Physicalism and Resurgent Alternatives," in this volume). Psychology in particular exemplifies the problem of purely physicalist accounts of nature: stricter accounts that remain faithful to the core doctrines of materialism seem obviously inadequate to explain the central experiences of human life and the mental processes supporting those experiences, but more promising accounts to explain those same experiences and processes are no longer materialistic at all.
10. Lev Vygotsky, *Thought and Language* (Cambridge, MA: MIT Press, 1986).
11. Richard M. Lerner, *Concepts and Theories of Human Development*, 3rd ed. (Mahwah, NJ: Erlbaum, 2002).
12. Urie Bronfenbrenner, *The Ecology of Human Development* (Cambridge, MA: Harvard University Press, 1979).
13. Jack O. Balswick, Pamela Ebstyne King, and Kevin S. Reimer, *The Reciprocating Self* (Downers Grove, IL: InterVarsity Press, 2005).
14. John Bowlby, *A Secure Base: Clinical Applications of Attachment Theory* (London: Routledge, 1988).
15. Jeffry A. Simpson and W. Steven Rholes, *Attachment Theory and Close Relationships* (New York: Guilford Publications, 1997).
16. Chris R. Fraley and Phillip R. Shaver, "Adult Romantic Attachment: Theoretical Developments, Emerging Controversies, and Unanswered Questions," *Review of General Psychology* 4, no. 2 (2000): 132–154.
17. Mario Mikulincer and Phillip R. Shaver, "The Attachment Behavioral System in Adulthood: Activation, Psychodynamics, and Interpersonal Processes," *Advances in Experimental Social Psychology* 35 (2003): 53–152.
18. Robert A. Emmons, "Motives and Goals," in *Handbook of Personality Psychology*, eds. Robert Hogan, J. Johnson and S. Briggs (San Diego, CA: Academic Press, 1997), 485–512.
19. Caryl E. Rusbult, M. Kumashiro, K. E. Kubacka, and Eli J. Finkel, "'The Part of Me that You Bring Out': Ideal Similarity and the Michelangelo Phenomenon," *Journal of Personality and Social Psychology* 96, no. 1 (2009): 61–82.
20. Stephen Drigotas et al., "Close Partner as Sculptor of the Ideal Self: Behavioral Affirmation and the Michelangelo Phenomenon," *Journal of Personality and Social Psychology* 77, no. 2 (1999): 293–323.
21. Debra Mashek, Arthur Aron, and Maria Boncimino, "Confusions of Self with Close Others," *Personality & Social Psychology Bulletin* 29, no. 3 (March 2003): 382–392.
22. Christian Waugh and Barbara L. Fredrickson, "Nice to Know You: Positive Emotions, Self-Other Overlap, and Complex Understanding in the Formation of a New Relationship," *The Journal of Positive Psychology* 1, no. 2 (April 1, 2006): 93–106.

23. John Dovidio et al., "Group Representations and Intergroup Bias: Positive Affect, Similarity, and Group Size," *Personality & Social Psychology Bulletin* 21, no. 8 (August 1995): 856–865.
24. Kareem Johnson and Barbara L. Fredrickson, "'We All Look the Same to Me': Positive Emotions Eliminate the Own-Race Bias in Face Recognition," *Psychological Science* 16, no. 11 (Nov 1, 2005): 875–881.
25. Debra Mashek, Lisa W. Cannaday, and June P. Tangney, "Inclusion of Community in Self Scale: A Single-Item Pictorial Measure of Community Connectedness," *Journal of Community Psychology* 35, no. 2 (March 2007): 257–275.
26. Dovidio et al., "Group Representations and Intergroup Bias," 859.
27. Richard Lerner, Elizabeth M. Dowling, and Pamela M. Anderson, "Positive Youth Development: Thriving as the Basis of Personhood and Civil Society," *Applied Developmental Science* 7, no. 3 (July 1, 2003): 176.
28. Jennifer Crocker and Lora E. Park, "The Costly Pursuit of Self-Esteem," *Psychological Bulletin* 130, no. 3 (May 2004): 392–414.
29. Jennifer Crocker, Amy Canevello and Ashley A. Brown, "Social Motivation: Costs and Benefits of Selfishness and Otherishness," *Annual Review of Psychology* 68, no. 1 (January 3, 2017): 318.
30. Crocker et al., "Social Motivation," 318.
31. Y. Joel Wong et al., "Giving Thanks Together: A Preliminary Evaluation of the Gratitude Group Program," *Practice Innovations* 2, no. 4 (December 2017): 243–257.
32. Shelley Kerr, Analise O'Donovan, and Christopher Pepping, "Can Gratitude and Kindness Interventions Enhance Well-being in a Clinical Sample?," *Journal of Happiness Studies* 16, no. 1 (February 2015): 17–36.
33. Elliott Kruse et al., "An Upward Spiral between Gratitude and Humility," *Social Psychological & Personality Science* 5, no. 7 (September 2014): 805–814.
34. Xiao Zhou and Xinchun Wu, "Longitudinal Relationships between Gratitude, Deliberate Rumination, and Posttraumatic Growth in Adolescents Following the Wenchuan Earthquake in China," *Scandinavian Journal of Psychology* 56, no. 5 (October 2015): 567–572.
35. Isabel Maria Salces-Cubero, Encarnación Ramírez-Fernández, and Ana Raquel Ortega-Martínez, "Strengths in Older Adults: Differential Effect of Savoring, Gratitude and Optimism on Well-Being," *Aging & Mental Health* 23, no. 8 (August 3, 2019): 1017–1024.
36. Rene Proyer, Willibald Ruch, and Claudia Buschor, "Testing Strengths-Based Interventions: A Preliminary Study on the Effectiveness of a Program Targeting Curiosity, Gratitude, Hope, Humor, and Zest for Enhancing Life Satisfaction," *Journal of Happiness Studies* 14, no. 1 (March 2013): 275–292.
37. Sharon Southwell and Emma Gould, "A Randomised Wait List-Controlled Pre-Post-Follow-Up Trial of a Gratitude Diary with a Distressed Sample," *The Journal of Positive Psychology* 12, no. 6 (November 2, 2017): 579–593.
38. Steven Toepfer, Kelly Cichy, and Patti Peters, "Letters of Gratitude: Further Evidence for Author Benefits," *Journal of Happiness Studies* 13, no. 1 (March 2012): 187–201.
39. K. E. Kubacka et al., "Maintaining Close Relationships: Gratitude as a Motivator and a Detector of Maintenance Behavior," *Personality & Social Psychology Bulletin* 37, no. 10 (2011): 1362–1375.
40. Samantha Joel et al., "The Things You do for Me," *Personality & Social Psychology Bulletin* 39, no. 10 (October 2013): 1333–1345.
41. T. Diebel, C. Cooper, and Catherine Brignell, "Establishing the Effectiveness of a Gratitude Diary Intervention on Children's Sense of School Belonging," *Educational and Child Psychology* 33, no. 2 (March 8, 2016): 117–129.
42. Martin A. Nowak and Sébastien Roch, "Upstream Reciprocity and the Evolution of Gratitude," *Proceedings of the Royal Society, B: Biological Sciences* 274, no. 1610 (March 7, 2007): 605–610.
43. Alex M. Wood, Jeffrey J. Froh, and Adam W. A. Geraghty, "Gratitude and Well-being: A Review and Theoretical Integration," *Clinical Psychology Review* 30, no. 7 (2010): 890–905.
44. Kao Si, Xianchi Dai, and Robert S. Wyer, "The Friend Number Paradox," *Journal of Personality and Social Psychology* 120, no. 1 (January 2021): 84–98.
45. Shelly L. Gable et al., "What Do You Do When Things Go Right?: The Intrapersonal and Interpersonal Benefits of Sharing Positive Events," *Journal of Personality and Social Psychology* 87, no. 2 (August 2004): 228–245.
46. William Damon, Jenni Menon, and Kendall Cotton Bronk, "The Development of Purpose during Adolescence," *Applied Developmental Science* 7, no. 3 (July 1, 2003): 119–128.
47. Tong Wang et al., "Exploring Well-being among Individuals with Different Life Purposes in a Chinese Context," *The Journal of Positive Psychology* 16, no. 1 (January 2, 2021): 60–72.
48. Kendall Cotton Bronk and W. Holmes Finch, "Adolescent Characteristics by Type of Long-Term

Aim in Life," *Applied Developmental Science* 14, no. 1 (January 29, 2010): 35–44. Sara K. Johnson et al., "Configurations of Young People's Important Life Goals and Their Associations with Thriving," *Research in Human Development* 15, no. 2 (April 3, 2018): 139–166.

49. Sara B. Algoe and Jonathan Haidt, "Witnessing Excellence in Action: The 'Other-Praising' Emotions of Elevation, Gratitude, and Admiration," *The Journal of Positive Psychology* 4, no. 2 (March 1, 2009): 105–127.
50. Bonnie M. Le et al., *The Personal and Interpersonal Rewards of Communal Orientation* 30 (London, England: SAGE Publications, 2013).
51. Roy F. Baumeister and Mark R. Leary, "The Need to Belong," *Psychological Bulletin* 117, no. 3 (May 1995): 497–529.
52. Baumeister and Leary, "The Need to Belong."
53. Baumeister and Leary, "The Need to Belong."
54. Frode Stenseng et al., "Social Exclusion Predicts Impaired Self-Regulation: A 2-Year Longitudinal Panel Study Including the Transition from Preschool to School," *Journal of Personality* 83, no. 2 (April 2015): 212–220.
55. Roy F. Baumeister et al., "Social Exclusion Impairs Self-Regulation," *Journal of Personality and Social Psychology* 88, no. 4 (April 2005): 589–604.
56. Sally S. Dickerson, Tara L. Gruenewald, and Margaret E. Kemeny, "When the Social Self is Threatened: Shame, Physiology, and Health," *Journal of Personality* 72, no. 6 (December 2004): 1191–1216.
57. Naomi I. Eisenberger, "The Neural Bases of Social Pain: Evidence for Shared Representations with Physical Pain," *Psychosomatic Medicine* 74, no. 2 (February 2012): 126–135.
58. Wendi L. Gardner et al., "On the Outside Looking In: Loneliness and Social Monitoring," *Personality and Social Psychology Bulletin* 31, no. 11 (November 2005): 1549–1560.
59. Wendi L. Gardner, Cynthia L. Pickett, and Marilynn B. Brewer, "Social Exclusion and Selective Memory: How the Need to Belong Influences Memory for Social Events," *Personality and Social Psychology Bulletin* 26, no. 4 (April 2000): 486–496.
60. I. van Beest and K. D. Williams, "'Why Hast Thou Forsaken Me?' the Effect of Thinking about being Ostracized by God on Well-being and Prosocial Behavior," *Social Psychological & Personality Science* 2, no. 4 (2011): 379–386.
61. Mark R. Leary, Jean M. Twenge, and Erin Quinlivan, "Interpersonal Rejection as a Determinant of Anger and Aggression," *Personality and Social Psychology Review* 10, no. 2 (2006): 111–132.
62. George E. Vaillant, *Triumphs of Experience* (Cambridge, MA: Harvard University Press, 2012), 40.
63. George E. Vaillant, *Triumphs of Experience* (Cambridge, MA: Harvard University Press, 2012), 52.
64. Floyd Allport, *Social Psychology* (Cambridge, MA: Riverside Press, 1924).
65. Mark H. Johnson et al., "Newborns' Preferential Tracking of Face-Like Stimuli and Its Subsequent Decline," *Cognition* 40, no. 1 (1991): 1–19. Francesca Simion, Lucia Regolin, and Hermann Bulf, "A Predisposition for Biological Motion in the Newborn Baby," *Proceedings of the National Academy of Sciences—PNAS* 105, no. 2 (January 15, 2008): 809–813.
66. Roy F. Baumeister and Brad J. Bushman, *Social Psychology and Human Nature* (Belmont, CA: Thomson Higher Education, 2008).
67. Mark Leary, "A Functional, Evolutionary Analysis of the Impact of Interpersonal Events on Intrapersonal Self-Processes," in *Self and Relationships: Connecting Intrapersonal and Interpersonal Processes*, eds. Kathleen D. Vohs and Eli J. Finkel (New York: Guilford Press, 2006), 219–236.
68. Garth J. O. Fletcher and Nickola C. Overall, "Intimate Relationships," in *Advanced Social Psychology*, eds. Roy F. Baumeister and Eli J. Finke (Cary: Oxford University Press, 2010).
69. See Angus J. L. Menuge, "Declining Physicalism and Resurgent Alternatives," in this volume.
70. Peter G. Ossorio, *Place: The Collected Works of Peter G. Ossorio, Vol. III* (Ann Arbor, MI: Descriptive Psychology Press, 1998), 38.
71. Though beyond the scope of this chapter, a related, important question for psychological research is what is to be done when a framework so fundamentally fails to deliver? For a discussion of and answer to this question, see Angus J. L. Menuge, *Agents Under Fire: Materialism and the Rationality of Science* (Lanham, MD: Rowman & Littlefield, 2004), 28. Menuge states: "When a theory is that far off the mark, it is not merely that it should be superseded by a better theory, which invokes the same ontology. The ontology itself is the problem." Perhaps the only proper conclusion to draw about the materialistically founded person is that, ontologically speaking, no such thing exists, and we should move on to another ontological view of the person.

18. Evidential Near-Death Experiences

Gary R. Habermas

1. Introduction

Many of the essays included in this volume are occupied with rigorous metaphysical and epistemic considerations regarding the nature of persons. A different angle is considered here, namely, the cognate subject of whether there might also exist at least some initial, credible indications of afterlife consciousness. The data presented in this essay are more empirical in nature. While several different research categories could be entertained in such studies,[1] the chief focus at present is that of near-death experiences (NDEs).

Though research in this area is sometimes considered to be less than scholarly, one recent volume on NDEs is a collection of peer-reviewed articles and editorials on the subject, originally published in *Missouri Medicine: The Journal of the Missouri State Medical Association* and then published in an edited volume by the University of Missouri Press. The editor begins by commenting that in the United States alone, various estimates suggest that between nine and twenty million persons have had NDEs! While such estimates indicate very little of evidential value by themselves, they have generated serious interest from scholarly communities, especially when some of these NDE accounts produce "corroborated veridical recollections."[2]

The discussion of NDEs in this chapter offers a potential bridge to the arguments about personhood given elsewhere in this book. Evidence for human consciousness during near-death situations may constitute evidence for the thesis that human consciousness is not entirely dependent on brain or bodily states. For instance, corroborated data may indicate the presence of human consciousness during times where neither the heart nor brain registered any observable measurements. Further, such experiences apparently took place during NDEs that involved substantiated observations that almost certainly could not have been made from the person's bodily location, even if they had been fully conscious, healthy, and observing their surroundings at that time. The bulk of this essay seeks to present these situations and the potential evidence that may proceed from them, and then to address some major objections of both natural and non-natural sorts.

2. Evidential NDEs[3]

2.1 Debating NDEs

ARGUABLY THE most insightful dialogue on the possibility of veridical NDEs cases occurred from 2007 to 2008 over four full issues of the peer-reviewed *Journal of Near-Death Studies*.[4] The series featured a vigorous debate between the skeptical philosopher Keith Augustine (who specialized in NDEs) and several distinguished NDE scholars. In these spirited discussions, Augustine proposed many specific challenges, aimed especially at the evidential NDE reports. Augustine's respondents usually addressed his challenges in areas where they were the lead investigators. In each of the first three issues, Augustine provided both the chief essay and the final rejoinder.[5]

In his lead installment of three major articles, Augustine questioned several well-known NDE accounts, including Kimberly Clark Sharp's report on "Maria" (who while undergoing resuscitation after heart failure reportedly observed a tennis shoe on a third-floor window ledge of a Seattle-area hospital), Michael Sabom's report on "Pam Reynolds" (who reportedly had out-of-body perceptions during brain surgery), and NDE reports among blind persons. He also provided a brief overview of prospective experiments which attempted to ascertain whether NDErs were able to identify random visual targets placed high overhead.[6]

Augustine's second main article examined numerous cases of *false* perception in NDE accounts, including discrepancies with the physical world, encounters with living persons that never occurred, encounters with mythological creatures or fictional characters, and so on. He argued that perceptions of these sorts are much more compatible with hallucinations.[7]

In the last of his main essays, Augustine argued that other ND features, such as psychophysiological and cultural correlates, suggest that NDE imagery is solely the product of an individual's mind rather than of any supernatural reality. Throughout, Augustine argued against NDEs operating either beyond or independently of the human brain.[8]

In his rejoinder to the replies to his first article,[9] Augustine allowed that the most impressive species of NDE data would probably be more indicative of an afterlife thesis than of his hallucinatory hypotheses, if they could stand up to scrutiny. For instance, citing positively the work of Michael Potts, Augustine postulated that the following would be among the most helpful in establishing an otherworldly interpretation of NDEs: specific clothing details of those health care personnel who resuscitated the patient, the precise order of events during the resuscitation, emergency room details that could have been learned only by being present, and so on.[10] Although he holds that such details have yet to be provided, he granted that such evidence could, in principle, be provided and, if so, would be dispositive. "*If* there were evidence of the sort Potts outlined, *then* the data would contradict my critique of near-death veridicality studies; but, as Potts also noted, anything of the sort has yet to happen."[11]

One general comment directed to Augustine was that arguing for an afterlife was seldom the goal of NDE research.[12] But Augustine, again following Potts, replied that while technically these studies may not have aimed at establishing an afterlife, it would be very difficult to deny that such a conclusion would follow rather naturally if human consciousness were capable of functioning after death, especially if natural explanations failed to suffice, as a large number of studies seemed to claim.[13]

It is noteworthy that Augustine is not the only skeptical scholar who agreed repeatedly that evidential NDEs could potentially establish a case for an afterlife or refute naturalism.

Naturalistic psychologist Susan Blackmore similarly asserted that NDE evidence could show her view to be mistaken.[14] Accepting the evidential challenges especially by Augustine and earlier by Blackmore, the respondents in the *Journal of Near-Death Studies* responded by providing many particulars pertaining to their own areas of study.

2.2 Evidential NDE Corroboration within the Room

Before moving to an evidential case for NDEs, a brief response needs to be made regarding the issue raised by Augustine and other researchers concerning the prospective experiments that have attempted to determine whether any NDErs were able to detect random visual targets such as numbers placed high overhead. The critics enjoy pointing out that while some partial information has been collected here and there, no full identifications of the random numbers by NDErs have yet occurred.[15] Perhaps these scholars hope that these issues could serve to cast doubt on all NDEs.[16]

In response, as will be noted more specifically below, some half-dozen or more evidential NDE cases *have* included either the successful identification of specific numbers, or the recognition of other casually placed or thrown objects that happened to be found overhead, that could only be seen from a position nearer to the ceiling. These include the successful identification and repetition of a random twelve-digit numeral on the top of an overhead medical device where the incident was confirmed by a nurse who was present, a four-figure number, and two of the present author's own cases where smaller numbers were visible but could only be seen if the NDErs had been looking down from above their bodies.[17] In a somewhat related case, a five-figure number was also correctly identified.[18]

This list may be increased if the correct identification of unlikely or strange objects not seen from the ground but identified correctly by NDErs are also included. As Michael Sabom points out after many NDE interviews, the "inability to recall verifiable details was attributed by the person, time and time again, to the fact that his attention had been directed toward the unique and pleasant qualities of the experience, in overall amazement at what was occurring."[19] Given the corroboration that has been discovered, these examples indicate that perhaps skeptics should be a bit more reserved in their criticisms of the prospective cases, especially when comparable results have already been correctly observed by NDErs.

We noted above that Augustine had suggested (along with Potts) that evidential reports be produced regarding the specific clothing worn by resuscitation team members, plus the precise order of events that transpired during the process, and other emergency room details that could have been known only by one who was present at the time. Although Augustine and Potts declared that such details were nonexistent, their comments were clearly mistaken, and on several fronts, as will be shown here. On many occasions, it appeared to be more a matter of disbelieving or disregarding the details when they were actually produced. Certainly this would apply in the cases of correctly cited numbers and other odd objects that could only be observed from overhead!

Pertaining to accounts where emergency room clothing was identified by NDErs, Ken Ring and Madelaine Lawrence reported the intriguing account of Joyce Harmon, an ICU hospital nurse. On her first day back at work after vacation, she was a member of the medical team that successfully resuscitated a female patient whom she did not know. The very next day she saw the patient, who responded, "Oh, you're the one with the plaid shoelaces!" and explained that she observed them while watching the resuscitation from overhead.

Intriguingly, Harmon had just purchased the plaid shoelaces while on her vacation and had worn them to the hospital that day for the very first time. Though casual or mundane conversations by staff often occur in hospital settings, even during stressful times, the color of shoelaces does not appear to be a topic that would be likely be discussed or even noticed during a frantic resuscitation attempt.[20]

In another case,[21] a nurse practitioner of my acquaintance in a Midwestern hospital rushed to the scene involving a patient experiencing an emergency situation due to cardiac arrest, where she assisted in a successful life-saving procedure. A couple of days later an unknown patient introduced herself as the resuscitated patient. The latter explained that, during the resuscitation, she witnessed the rescue process from above her body and had observed a unique object worn on the nurse practitioner's clothing, which the patient described in minute detail. The nurse practitioner had borrowed the object the previous day and had already returned it to its owner.

The nurse practitioner was stunned most of all by the patient's intricate description of the object even in the middle of the hectic assistance provided during her cardiac arrest. Even if a conversation were overheard due to the resuscitation and recounted *so precisely* (quite a stretch in itself, if not impossible for someone in her condition), it is quite unlikely that all the details would have been described precisely during a cardiac arrest.

What about Augustine's and Potts's request for the knowledge of a precise *sequence* of events during a resuscitation? While there are *many* examples from which to draw, one of the most detailed was reported by emergency room pediatrician Melvin Morse, who recounted the case of a girl ("Katie"—actually Kristle Merzlock) who had nearly drowned. A physician was present at poolside, and it was documented that Kristle was without a pulse for at least seventeen minutes, as well as having no gag reflex, with fixed and dilated pupils, and was "profoundly comatose." Morse noted that this condition most probably indicated that at least her upper brain was not functioning at that time, and that irreversible brain damage had probably occurred.

Three days later, Kristle inexplicably revived. About three weeks later, in a follow-up exam, she took almost an hour to tell her entire story. Morse was incredibly impressed with her precise description of the emergency room and the sequence of events. She knew that a tall physician without a beard was the first one to enter the emergency room. Then she recounted that Morse, shorter and sporting a beard, had come in next and was chiefly responsible for resuscitating her. She also recounted that she had first been brought into a larger room, and then was moved into a smaller one, for X-rays. She knew that she had been intubated through her nose, although this procedure is more commonly done orally.[22]

Given that Kristle was unconscious the entire time, with her eyes closed, requiring mechanical ventilation in order to breathe over the next three days, this is an incredible report. Even if it were thought that Kristle somehow could have heard certain snippets of emergency room conversation, details like the physicians' physical characteristics and the sizes of the hospital rooms seem to require sight, although that is certainly not the best explanation here. The clear and confirmed sequence is also beyond typical jumbled memories. Morse declared that this experience changed his life, including his religious agnosticism.[23]

Morse also relates other evidential NDE cases that also included a sequence of events, including an eight-year-old girl who nearly drowned in a swimming pool after her hair was caught in the drain. Initially her parents, then an emergency medical team that arrived, and lastly, physicians in the emergency room, all administered CPR for more than forty-five minutes before her heart began beating once again. A short time later, she exhibited "full recall of the event" and was capable of recounting the entire extended process of resuscitation.[24]

Another sequential case involved a woman who experienced a cardiac arrest that lasted about four minutes. During that time, she reported being up above her body and looking down below. She provided a description of the proceedings of the resuscitation attempts performed in the ambulance on the way to the hospital. Perhaps most crucially, the paramedic crew members attested that she recalled several corroborated observations *that occurred precisely during her cardiac arrest and were reported afterwards.*[25] As documented below regarding cardiac arrest with ventricular fibrillation, both measurable heart and at least upper brain activity would be eliminated in just seconds, with lower brain activity ceasing just slightly afterwards.

Lastly, Augustine and Potts also requested emergency room descriptions that could only have been known by someone actually present. Again, many examples on record could be noted. In one case, a young patient had almost died after suffering a cardiac arrest. There were also problems with a machine that was being used during the resuscitation process. After the crisis had passed, the patient pointed out that the machine with the problems was actually unplugged. The nurse went and checked, and soon discovered that the patient was correct.[26] Another patient under general anesthesia claimed that she had observed the situation from the ceiling. She said that she was also able to watch another surgery in the next room over, where she saw a patient undergoing the amputation of their leg, which was then placed inside a yellow bag! The attending surgeon related that she "described it as soon as she woke up." The surgeon checked and found that the story of the amputation was true, and he attested that this was not possible naturally.[27]

Cardiologist Sabom documented ten specific examples where patients, sometimes without any measurable heartbeat, detailed many items regarding the layout of furniture and other objects in the hospital emergency room, specific observations regarding the instruments that were located there, and even detailed readings on the dials. Their observations also included some uncommon medical practices that took place, which would not normally be guessed. Sabom concluded that several of the verified observations were such that they could only have been observed visually, beyond any potential hearing.[28]

Greyson recalled another account that he investigated personally regarding events in the middle of an open heart surgery. The patient later described watching the scene from above his body, noting that, while his chest cavity was open, his surgeon began "flapping his arms as if trying to fly." When Greyson interviewed the surgeon, the latter explained his "peculiar habit." After washing and at other times during the surgery in order to keep his hands from any contamination, he often placed them against his chest and pointed out various things to his assistants, by using his elbows. The appearance was that of "flapping" his elbows up and down.[29] The cardiologist who was also present confirmed that the patient had accurately reported these details just "shortly after he regained consciousness following the surgery."[30]

The sorts of cases requested by Augustine and Potts, exhibiting accurate descriptions drawn from specifics in the NDEr's immediate vicinity, are hardly rare. Janice Holden provides a list of 107 specific, evidenced NDE cases reported by many authors, with the greatest number describing the person's immediate surroundings. Further,

Holden made three distinct improvements to these data: (1) She arranged the publications by copyright date, and omitted reports after 1975 that were either popular or autobiographical; (2) She ordered the NDE accounts from the weakest to the strongest cases according to the kind of evidence presented, with the best rating being given to those that were confirmed by outside sources; (3) She specifically singled out and recorded NDE accounts that contained any errors, though there were very few of these. She still ended up with these 107 reports.[31]

It should be noted that the majority of these detailed NDE accounts listed by Holden were recorded by major scholars, such as eleven cases from Kenneth Ring and Evelyn Valarino,[32] ten cases from Emily Cook, Bruce Greyson, and Ian Stevenson,[33] the ten mentioned from Sabom,[34] seven from Peter Fenwick and Elizabeth Fenwick,[35] and seven from Morse.[36] To these we might add many additional cases both before and after Holden's research,[37] along with a lengthy list taken from my own research and sometimes published elsewhere, collected over a period of several decades.

Throughout this discussion thus far, specific cases have been mentioned where it was stated specifically or implied strongly that cardiac arrest with ventricular fibrillation can indicate exceptionally important evidential situations. This is due to the medically well-established and recognized research which specifies that such heart stoppage initiates the immediate and measurable elimination of upper brain activity just seconds later during the persistence of this state. The cessation of lower brain activity occurs just very slightly afterwards.[38] Therefore, verified NDE data that occur within this time frame are exceptionally crucial in indicating the potential presence of consciousness beyond the quantifiable existence of the central nervous system. Literally *dozens* of cases have been recorded where corroborations inside a room[39] have been documented during just such a state.[40]

A startling number of these detailed and later verified observations of reported incidents exist, of which most apparently occurred precisely within the moments when measurable heart and brain activity were measurably absent. This is a major argument for continued consciousness during these moments. Other options have been argued and will be treated below, but the sheer unlikelihood of mistakenness or deceit of one sort or another in every one of the dozens of different situations makes appeal to mistake or deceit seem incredible, particularly when the described occurrence happened precisely during those minutes rather than before or after.

These examples appear to address clearly the specific evidential requirements of Potts and Augustine regarding events within the room where the NDEr was located, with details being reported during surgery or otherwise, where the best explanation is that the patient truly recalled events that were observed during that time, with each incident being confirmed by one or more persons who were actually present during that time. Due to the vantage point being claimed, the repeated testimony of patients that they were positioned above their bodies counts for something, too. The "arm-flapping" episode illustrates the many statements where the patient reported the insightful information very quickly after regaining consciousness, which contributes to the overall veridicality of this conclusion. Altogether, just from the narrated accounts or sources mentioned above, the total stands at well over *150 evidenced NDE cases in the immediate vicinity* of the patient.

2.3 Evidential NDE Corroboration at a Distance

Other skeptical authors prefer different sorts of corroboration. Despite the requests from

Augustine and Potts to produce very specific types of visual data *within* the emergency or operating room, or from the immediate vicinity of the near-death-experiencer, other skeptics, such as psychologist Susan Blackmore, prefer visual information reported by the NDEr from a distance away, outside the room, that was corroborated later. Such data would be helpful in order to rule out the patient's having learned the information from normal sense data drawn from the immediate proximity, even given the unlikelihood that such precise knowledge should be gained in that manner during cardiac arrest or general surgery.

For Blackmore, two specific varieties of cases would be the most evidential, and have the most potential to disprove her naturalistic hypothesis: "distant vision" where the NDEr could not have obtained the reported data from their presence in the resuscitation room, and accurate NDE testimony from sightless persons. However, she attested that, search as she could, she was never able to locate good examples of either sort.[41]

It is agreed that verifiable cases of the sort that Blackmore requested, especially those from well beyond the NDEr's immediate line of physical sight or hearing, could definitely provide yet another body of evidence that could help to establish a non-naturalistic cause for NDEs. Several such cases will be mentioned here.

Some skeptics, e.g., Augustine, have criticized (see above) Kimberly Sharp Clark's tennis shoe example, although subsequent research has buttressed her account.[42] Blackmore is actually positive here, pointing out that verified cases such as Sharp's could possibly constitute strong evidence for the disembodied NDE thesis, though she attests that she was "unable to get any further information" here.[43]

Another case involving a shoe found on a hospital roof was reported from all the way across the country (in Hartford, Connecticut) by Kenneth Ring and Madelaine Lawrence. The resuscitated patient claimed to have had an NDE in which she floated above her body and then watched the resuscitation attempt going on beneath her. Then she experienced being "pulled" through several floors of the hospital until she emerged near the building's roof, where she viewed the Hartford skyline. Looking down, she then observed a red shoe. When nurse Kathy Milne heard the story, she reported it to a resident physician, who mocked the account as a ridiculous tale. However, in order to ascertain the accuracy of the report, he enlisted a janitor's assistance, and was led onto the roof, where he found the red shoe! This occurred in 1985, and Milne was unfamiliar with the other tennis shoe account, which was published just shortly before.[44]

In the case of Kristle Merzlock mentioned earlier, the young girl who nearly drowned and was resuscitated by Morse: she reported more than the specifics of the resuscitation attempt and the sequential details from the emergency room. Upon regaining consciousness three days later, her intensive care nurses initially heard her recollection of having visited heaven, guided by an angel. Though there was no way to verify the angel, Kristle also testified that, although she was unconscious and hooked up in the hospital, she was "allowed" to observe her parents and siblings some distance away, at home for the evening. She provided exact details regarding where each person was located in the house, identifying the specific things they were doing, as well as the type of clothes that they were wearing. For instance, she identified that her mother was cooking roast chicken and rice for dinner. All of these particulars were subsequently confirmed very soon afterwards.[45]

One of the most detailed distance cases involved a patient (Tony) and his wife Pat ("a nursing supervisor"), who traveled to Milwaukee for a "complex" open heart surgery. Due to an upper body bleed, an additional open-chest surgery was planned for the very next day, but during that second prep, Tony's heart "was arrested for 30 minutes" and he also suffered a respiratory arrest at the same time. During this trauma, Tony reported correct observations regarding his wife in the hospital waiting room, followed by observations inside his home in Florida, located over 1,200 miles away! There he watched a friend who was house-sitting for the couple while they were in Wisconsin. He reported to his wife a sequence of several amazing, very specific, and out-of-the-ordinary details that he had observed both in the hospital and then in their home. These reports were confirmed later, both by discussions with the house sitter and by their own observations upon returning to Florida concerning the somewhat strange events that Tony had narrated earlier to his wife and friend. This case was also investigated later by another physician.[46] This incident involved a double heart and respiratory arrest and multiple distance sightings accompanied by more than one form of verification.

In two other instances, the physical distances between the patients and their seemingly incredible reported observations were approximately twenty and thirty miles away, respectively, and the details in both cases were confirmed by interviewers.[47]

Many other evidential NDE accounts have also included verification at a distance, with a number of these also being substantiated by subsequent interviews. Bruce Greyson, Emily Kelly, and Edward Kelly note from their NDE research that "60 people reported being aware of events occurring outside the range of their physical senses."[48] Greyson[49] reported that a number of distance cases were researched and confirmed, including examples by Pim Van Lommel et al.,[50] and by Penny Sartori, Paul Badham, and Peter Fenwick.[51]

Two journal articles, by Cook, Greyson, and Stevenson[52] and Kelly, Greyson, and Stevenson,[53] recorded a total of fifteen more NDEs with observations viewed at a distance. Jeffrey Long reported that the cases involved in these two NDE studies were concerned with "observing earthly events far from their physical bodies and beyond any possible physical sensory awareness." These accounts of "corroboration of the NDErs' remote observations" were made by others, resulting in the subsequent confirmation of the testimonies.[54] To these, we could also add still more confirmed distance testimonies that were reported by Morse,[55] Sabom,[56] and an anecdotal case by Marvin Ford.[57]

In addition to the above cases, there is a large number of reported and documented distance NDEs said to have occurred in the absence of any measurable heart or brain activity.[58] A number of the cardiac arrest cases include some of the strongest evidential scenarios.[59] Once again, as in the previous category of corroborated observation inside the room, it is exceedingly unlikely that *every last one* of another dozen cases exhibiting neither apparent heart nor brain activity can be meaningfully accounted for by data learned through other means, misperception or deception, coincidences, or mistakes, especially when the events described occurred precisely within the time interval of the medical crisis rather than subsequently.

Just these accounts alone add up to *more than 110 evidenced NDEs reported a distance away from the experiencer*! This testimony must be viewed as among the most convincing evidence of all. The attempts to explain away this wide variety of observed scenarios, especially the compound cases here, may well reveal the pervasive influence of certain critics' world views and an unwillingness to explain the data at hand.

2.4 Evidential NDE Corroboration concerning Previously Deceased Persons

Another body of evidence comes from cases where particular persons, often friends or loved ones but sometimes others, had died recently, but the death was unknown to the NDEr until the deceased persons appeared in the NDE or deathbed vision. Greyson notes three distinct species of these experiences: (1) where the death of the deceased individual occurred well before the NDE, but the death was not known by the NDEr before the occurrence of the NDE; (2) where the deceased person had died at the same time as the NDE experience, or "immediately beforehand," thus precluding previous knowledge of the death on the part of the NDEr (or others in the room); (3) where the deceased was a person whom the NDEr did not even know.[60]

In all three species of these NDEs (especially the last), information was purportedly imparted from the previously deceased individuals to the person who had the near-death experience, with subsequent confirmation of the reports. Over two dozen such examples were collected by Greyson,[61] and other researchers have also presented similar accounts.[62]

Regarding the three subtypes here, Greyson argues that since these accounts cannot be attributed to the NDEr's expectations or knowledge, subjective hallucinations should be ruled out. It should be noted additionally that the wide variety of evidential reports in this category that were later confirmed due to additional investigation argue even more strongly against subjective states explaining these NDE observations, since hallucinations and similar phenomena are subjective in nature, and thus do not properly account for reports of actual aspects in the world that could not have been observed even in healthy, fully functioning bodies in the vicinities of the objects observed.

Three examples will be mentioned here briefly. Cardiologist Van Lommel described the case of a man who had an NDE after a cardiac arrest, who related to him that during his NDE, he had met his biological father who had died many years earlier. At the time of the NDE, the patient did not recognize the man in his NDE; he had never met his biological father previously, and had never even known that his biological father was anyone other than the man he had called his father all his life and who had raised him. Later, on her deathbed, his mother revealed to him the identity of his biological father, and showed him a photograph, which he recognized as the image of the man he had seen in his NDE.[63]

Among the many cases recounted by Greyson, two will be mentioned briefly. A young nine-year-old boy named Eddie was seriously ill in a hospital. Recovering from a thirty-six-hour fever, Eddie immediately told those in the hospital room that he had been to heaven, recounting seeing his grandfather, an aunt, and an uncle there. But then his startled and agitated father heard Eddie report that his nineteen-year-old sister Teresa, away at college, was in heaven too, and she told Eddie that he had to return. But the father had just spoken to Teresa two days prior. Checking with the college, the father found out that his daughter had been killed in a car accident the previous day, but that the college could not reach the family at their home, presumably because of Eddie's hospital stay![64]

Greyson also recounted another case where a man named "Jack" was exceptionally ill in a hospital with double pneumonia. His nurse Anita had left that weekend to go home for her twenty-first birthday party. Soon after, Jack suffered a pulmonary arrest and experienced an NDE, where he saw Anita, who asked him to tell her parents that she was sorry that she had wrecked the new red sports car that they had just given her, and to tell them that she loved them. She likewise told Jack that he had to return but that she was remaining there. The accident had apparently occurred the same day that Jack went into arrest, or just hours beforehand. When Jack told another nurse about his NDE, she promptly ran out of the hospital room crying since she and Anita had been good friends.[65] Apparently, news of the story had not yet reached her.

Moreover, as in our previous two categories, some of the reports here concern cardiac arrest cases that most likely also extend beyond measurable heart and brain activity.[66] Further, some of these cases involved the communication of information not just from people who had died, but from people who had been dead for years or even decades. These two points address the potential objections that perhaps some remaining residual activity in the NDErs' nervous system could account for these corroborative reports, or that life after death is perhaps only of relatively brief duration. Once again, post-cardiac arrest cases plus the possibility of testimonies from irreversibly dead persons strengthen this category considerably, closing the door further and making natural explanations even more unlikely![67]

These compound evidential situations discussed here include about thirty cases of persons who reported being with previously deceased individuals (with confirmation from several types of corroborative details). They raise considerably the evidential bar in favor of the combined NDE data. No wonder Greyson asserts that "cases such as these provide some of the most persuasive evidence for the ontological reality of deceased spirits. Recent medical and societal advances in end-of-life care offer favorable opportunities for the further investigation of these cases."[68]

2.5 Evidential Corroboration from Shared NDEs

Still another type of substantiation is supplied by *simultaneous* or *shared* near-death experiences reported by at least two different people, and supported by accounts from physicians, nurses, or others. In most such cases, a healthy person shares part or all of the NDE experience of the ill person.[69] In one of Morse's cases, a seventeen-year-old boy named Shane died in a traffic accident. His fifteen-year-old deaf sister Cheryl "observed" the entire accident process, though she was a distance away at home when the event occurred. She observed Shane "flying through the air" and knew that he was dead. Cheryl sensed that her brother was contacting her without words, relating that he wanted to show her "something really cool." She then reported that the two of them "rose in the air, high above the scene of the accident." She accompanied her deceased brother to heaven, where they met previously dead relatives.

Her brother did not return, since he was deceased. But during the process, her parents could hear her talking to her brother while the other three family members were all at home, where Cheryl was present and conscious. Upon her "return," she brought back information that no one else knew, but which was verified subsequently. Shane had said to Cheryl repeatedly, "I know something that you don't know." Then he told his sister that their aunt was pregnant with a boy, though no one in the family knew that at the time.[70] Cheryl reported observing the event while her two parents witnessed her side of the conversation.

Raymond Moody described another shared NDE where five family members in Atlanta were present during the last moments of their mother's life. They were stunned as each of them simultaneously witnessed a bright light appear inside the room where they were all gathered.

Several lights morphed into an entranceway through which they watched as their mother's image or spirit apparently left her body and appeared to enter through the entrance! The joyous family members agreed that the entryway reminded them of Natural Bridge in Virginia's Shenandoah Valley. The family members told the hospice nurse, who replied that she had heard similar things before and "that it was not uncommon for the dying process to encompass people nearby."[71]

In yet another example, a woman who had experienced cardiac arrest for several minutes reported a string of corroborated items that occurred precisely during this state, which is highly evidential, given her lack of heart and brain function during that time. But to compound the matter further, her husband testified that he had shared the beginning of her NDE, witnessing her immaterial self actually rising above her body. One paramedic named Carl stated: "She was able to tell us word for word what we said, everything that we did physically to her, and was able to say it in such detail."[72]

Another out-of-the-ordinary example, from the present author's NDE collection, concerns two close family members who were both in hospitals in different locations halfway across the country from each other. They both reported a simultaneous NDE where they were with each other, even though neither knew the other was even in the hospital. When resuscitated, each reported their experiences to those in their respective rooms, with the similar descriptions from the two separate viewpoints being made known later to family members. Subsequent interviews confirmed these basic details.[73]

Other shared NDE cases are reported by various researchers.[74] There are not as many joint NDE cases as of other NDE cases, but we have mentioned about another dozen evidence-backed cases here.

2.6 Evidential Corroboration from Sightless NDErs

As if in answer to the challenge of Susan Blackmore (mentioned above) to produce NDE accounts from blind persons, Ken Ring and Sharon Cooper provided detailed reports of thirty-one blind NDErs. These cases produced several accurate testimonies, both from inside the room occupied by the patient and outside of it. There are fewer cases in this category, and the evidence is less than what has been provided elsewhere, yet the cases are significant, as it would be difficult for many of the specific items to have been known previously by the patients, through any of their physical senses.

One of the described episodes concerned a woman named Vicki, who had been blind from birth. During her NDE, Vicki reported color images, including a rendezvous with two close friends from her youth. Both of them also had been blind, and both had died previously. She reported that two other deceased friends and a deceased relative were also present. She provided accurate physical descriptions of each one, even though she had never seen any of them before.[75] She also provided details, such as a glimpse of the roof of the hospital and a description of some jewelry.[76]

Another NDE case involved a woman named Nancy, who became blind during an operation and remained completely blind afterwards. She had an Ambu bag placed over her nose and mouth to make her breathe. In her NDE she reported watching the process away from her body. Afterwards, she properly described the identities of two men standing down the hallway away from her, and gave the correct number of staff people around her. Her medical records, plus a testimony from one of the men, were "in substantial agreement" with Nancy's comments and agreed "in virtually every significant respect." Ring and Cooper characterized Nancy's case as very strong, "since it is backed up by independent witnesses and various forms of documentation."[77]

Brad had been blind since birth. He had stopped breathing for a few minutes and then noticed himself looking down on his body in the bed. He described details of another person in the room with him who went to get help, then experienced himself going up through the ceiling and out of the building, where he observed the rooftop. He related that he could see clearly and described the scene outside of his home by providing many details, including very specific information regarding the snow on the ground and how it had been plowed into heaps. A streetcar also drove by. In another case, Frank was a blind man who described the pattern of colors and designs on a necktie that he had received. Some of these details were confirmed in additional interviews.[78]

Ring and Cooper conducted extensive discussions with these blind individuals, and attempted to confirm their stories with others who were present. They concluded that "the blind persons in our study saw what they certainly could not possibly have seen physically. Our findings in this section only establish a putative case that these visions were factually accurate, and not just some kind of fabrication, reconstruction, lucky guess, or fantasy."[79] Others have reported further NDE cases in the blind,[80] but even restricting the number to the four cases discussed just above, we still have accumulated a total count, from the five categories discussed in the past five sections, of *more than three hundred evidential NDE cases.*

3. Additional Evidential NDE Corroboration

One volume of NDE reports especially—*The Self Does Not Die* by Rivas et al.— makes it exceptionally difficult for NDE naysayers, for this text contains more than one hundred confirmed NDE reports, adding many noteworthy cases to the overall numbers. According to the text's researchers, the major criterion for the NDE accounts included in this volume is that each episode *must* have been "directly confirmed by at least one other person."[81] Such strict requirements make this text all the more valuable.

This recent research contains fourteen corroborated cases from the vicinity of the NDEr, eighteen more accounts from a distance beyond the NDEr's senses (sometimes *very* far away), plus more than thirty-six additional reports from cardiac arrest patients. These last heart stoppage cases, of which there are more examples below, are certainly among the most impressive, because the patients were in states in which, "according to current materialist models, such perceptions should have been impossible."[82] Add to these testimonies a few dozen other cases that include shared NDEs, meeting deceased loved ones and friends while in the ND state, along with learning previously unknown information during NDEs. Additional "compound" cases are also included in this text, where NDEs were accompanied by physical healings, the acquisition of messages unaccompanied by spoken words, or other evidential occurrences. A total of sixty-six accounts in this work were not included in our earlier tallies.[83] Along with several additional cases not previously mentioned so far, and the more than three hundred cases already referenced in this essay, this brings the total count of corroborated NDE cases to nearly four hundred, across a wide variety of types and subtypes.[84]

There is much variety among the hundred-plus corroborated cases in *The Self Does Not Die*, too, including some of the strongest cases on record. For example, considering once again confirmed cases from inside the room, the NDE reports include an NDEr who suffered a cardiac arrest and did not respond to resuscitation attempts until he received a shot of epinephrine in his heart. He reported that he was "out of my body and floating above the trauma room." Peering down from above, he observed a quarter perched on top of an eight-foot-tall medical machine underneath him. He told the physician about the quarter and that it was dated 1985. The physician took a ladder to the ICU and while the nurses were watching, he retrieved the coin and verified that the date was exactly what the patient had recorded! The doctor later published the account.[85]

Another NDE patient who had suffered a cardiac arrest also related that she "had observed the room from above." In the process she noted a long, twelve-digit number listed on top of a high medical machine beneath her and, suffering from obsessive-compulsive disorder, memorized the number and repeated it to the nurse and others there, who wrote down the figure. When the patient no longer required the machine, a custodian set up a ladder in order to dust the top and then moved it out. The twelve-digit number was read, and it was the same figure that the witnesses had originally been told by the NDEr.[86] Later a nurse verified the story once again, stating that this incident was one of the most incredible occurrences that she had ever witnessed.[87]

Selecting additional examples from this volume consists largely of choosing from among dozens of potentially impressive reports. In another instance, a patient without heartbeat or brain activity for 20 to 25 minutes during hospital surgery correctly reported many details in the operating room, such as an anesthesiologist who came rushing back into the room after having left, as well as a collection of post-it notes attached to the physician's monitor that had accumulated during the surgery, but were not present prior to the beginning of the operation. The patient was declared dead and the body was prepared for an autopsy with the chest basically remaining open. The patient's wife was even informed of his death prior to his reviving spontaneously and making these accurate observations! One of the attending cardiologists, Dr. Cattaneo, wrote that there were "close to 20 minutes or more of no life, no physiologic life, no heartbeat, no blood pressure, no respiratory function whatsoever and then he came back to life and told us what you heard on the video.... 25 minutes or more with no cardiac pulmonary movement or brain function.... but it happened and I am a living witness of this case, I was there." Dr. Cattaneo added that while he had seen patients who still had some life in them and recovered, that "in this case there was no life." Dr. Cattaneo also commended the chief attending cardiac surgeon, who had given an exceptionally detailed medical account, and then attested that, "Dr. Ruby's description of this event at the time of this patient's surgery is absolutely correct.... these are the facts."[88]

In another case, a nurse who was present at a resuscitation attempt described a woman who had suffered a severe heart attack that had left her in a state of clinical death. The patient later reported occupying a location up above her body. While looking down from a corner of the room near the ceiling, she correctly reported a number of items, including an IV bottle that had accidentally smashed on the operating room floor and a nurse's especially prized hair clip that fell onto the floor, got stepped on, and was broken. The patient even identified and provided a description of the particular physician who had accidentally broken the clip![89]

Still another incident, this time reported from outside the room, involved a seventeen-year-old patient who was in a coma after a car accident left her with a severe brain injury, causing her to be flown to the ER. She reported having an NDE where she was up near the ceiling and yet was still able to watch her family members who were in the hospital cafeteria. Her father, a smoker, announced that he was going to light up a cigarette. Then the young woman watched her two grandmothers assert that they were also going to have a smoke with her father. This seemed incredible, given that one of her grandmothers (identified as her mother's mother) had never smoked in her entire life and had always proclaimed loudly over the years that no one would *ever* see her with a cigarette! Yet the patient witnessed the incident and told her mother that she had seen both of her grandmothers smoking, and her mother confirmed that this incident had indeed occurred.[90]

Another aspect of the overall state of the NDE evidence needs to be mentioned here, too. While we have described many of the different directions and angles from which the NDE corroboration emerges, the fact ought not be missed that dozens of individual NDE cases mentioned or referred to in this essay were confirmed by the presence of *more than one species or kind of evidence*. The examples can be multiplied almost at will. In several NDE cases without heartbeat, the individuals later provided complete *sequential* descriptions over a half hour or more. In several other cases, the individuals reported observations from many miles away – and at least three of these were also cardiac arrest cases, with one of these cases additionally being testified to by a healthy person as having been a shared experience. The dozens of accounts during at least the probable absence of both heart and brain activity, accompanied by evidence sometimes of more than one variety, such as involving triple confirmations, are all quite impressive! Verification was provided in each of these cases.

Moreover, a few cases of blind NDErs (some from birth) have emerged in the NDE literature, where the blind NDErs apparently "saw" items that were corroborated, before reverting back to their previous blind states.

In the some of the many cardiac arrest patients that presented plentiful examples of accurately observed data, at least one of these NDErs was reportedly detected by another healthy person; this corroboration in other cardiac arrest cases came *before* resuscitation ever began; while still other NDE testimonies came from events that clearly occurred *during* the state of cardiac arrest. Then there are the accounts where numbers of specific objects located above the heads of those in the room below were accurately reported by cardiac arrest patients. Many other testimonies involve meeting deceased loved ones or friends whose deaths were unknown, but where sometimes rather startlingly accurate knowledge was also communicated. The prospect of explaining all these *combinational* accounts seems especially troublesome for NDE deniers.

4. Do NDEs Provide Actual Details Regarding "Heaven" or the Nature of the Afterlife?

MANY NDERS claim and even believe firmly that they have visited some sort of exceptionally peaceful heavenly realm. Could that actually have been the case? Can it be concluded that these NDErs were *literally* on the "other side" or at least got a chance to see that realm, and ascertain true conclusions about it? Could it also be the case that since many unbelievers, avowed atheists included, have also claimed to have witnessed otherworldly and even wonderful environments as well, that these reports similarly constitute a warranted conclusion, with or without a conversion?[91]

While the testimonies on this subject are often quite moving as well as potentially convincing, a crucial distinction must be made.

Almost every one of the 300 to 400 evidential NDEs mentioned or referred to above concern testimony *pertaining to this world*. And even in those rare times where testimony may concern another world or form of existence, it rarely reaches the detailed level of corroboration that is frequently heard in the stronger cases.

Potential testimonies from a "heavenly realm" could still possibly warrant consideration of some sort or another. However, by the criteria employed in this essay, such reports would need to provide evidence of some kind. The things typically appearing in the heavenly reports, such as the incredible beauty of the music or the surroundings, ineffable feelings, heightened senses of awareness, or the correct identifications of heavenly figures such as angels, Jesus, or other religious figures are almost totally without corroborating evidence by which the reports might be evaluated.

Exceptions to this criticism would be cases where the NDEr reported comments from previously deceased individuals where verifiable information was imparted. But in these cases, the focus ought to be on the claims that may present confirmable details rather than other comments like the nature of the heavenly environment. Even if such long-deceased individuals may have added more generic comments (e.g., concerning the otherworldly environment or even that they were doing well), these latter remarks could simply have been matters of the NDEr's interpretation that were appended to the potentially verifiable statements within the testimonies. To state the matter more succinctly, it is the potentially verifiable statements that must be checked. Other sorts of comments in whole or part may simply be the NDErs' views or honest reflections. But except in the specific instances where the potential data can be checked, how else can it be ascertained as to which of the remarks or portions and snippets of otherwise reliable information can be known probabilistically to be correct and true?

Caution of a different sort needs to be inserted here, as well. When reports or comments cannot be verified, this does not establish that they are therefore false. Some or even all of the NDErs' report still could be true. But the question is: How could it be determined if this is the case, or which portions might be true?

Thus, to go back to the original question, it would appear that very few NDE accounts can provide verifiable information concerning heaven. Again, there could be any number of true reports there, but none that could be confirmed. Personal experience cannot be completely ignored, but neither can an individual's unverified testimony be taken as evidence for the details of afterlife existence. It may seem a harsh judgment, but unsubstantiated statements cannot be propagated as established data.

The same evaluation would likewise seem to apply to testimonies regarding negative NDEs, including even grotesque hellish accounts.[92] As just stated, these experiences actually could be true in whole or in part, but like the other testimonies, there do not seem to be ways to *demonstrate* that this is the case.[93]

Thus, heavenly sights and sounds, common meetings with family members or other loved ones though without corroboration, life reviews, or having positive, negative, or in-between experiences and feelings are personal and are rarely supported by evidence—especially not like the degree that has been discussed throughout this essay. Hence, while the experiences could have occurred in whole or in part and no doubt influence one's personal views, they cannot be used with any probability in order to establish specific religious doctrines or falsehoods.

What follows from these ideas is that NDEs may point to an afterlife of some sort, yet lack the ability to distinguish between individual religious perspectives. Thus, members belonging to various religious persuasions can basically stand shoulder-to-shoulder and endorse the evidential NDEs as phenomena that, while they do not settle theological differences between them, still agree that these occurrences are among the best reasons that refute philosophical naturalism's rejection of an afterlife. But NDEs cannot tell us which religious world view is true, as in questions like Eastern versus Western views, universalism, syncretism, and so on. Many religious ideas are potentially compatible with the general afterlife truth of corroborative NDEs, which is the chief consideration here.

That said, the differences in NDEs across religions and cultures are worthy of attention. There is a vast amount of data which indicate that cultural influences shape especially how these experiences are described and interpreted, especially where they talk about heaven. Examples abound here. The identifications of heavenly beings (except, possibly, the generic concept of angels) vary from culture to culture. (E.g., a Hindu might more likely see Krishna in an NDE, whereas a Christian might more likely see Jesus or St. Peter.) The particular religious messages that are heard, and delivered later by NDErs, can also reflect different theological frameworks. Further, particularly common experiences in some cultures, such as the sensation of traveling through a tunnel, appear to be absent in many Eastern NDE accounts, and for this reason are usually thought to be culturally inspired. Again, whether the "human greeters" in NDE accounts are male or female shows large discrepancies based on culture.[94]

Some researchers distinguish between primary, evidenced NDE features, which are more reliable, versus secondary ones that differ from culture to culture and according to other considerations like interpretations and "spins."[95] The differences between veridical factual reports and the interpretations drawn from them is exceptionally significant. Researchers agree that, in contrast to this-worldly, corroborated aspects of NDE consciousness, there is virtually no way to evidence the uncorroborated heavenly visions themselves.[96]

After assessing his own research computations regarding the differences between the more factual reports from the physical world and the NDErs' cultural or religious interpretations, Sabom reported that "religious beliefs appear to affect the interpretation but not the content of the NDE." Thus, NDErs often develop a deepening faith after their experience, but whether that change takes them in the direction of "Eastern religion, New Age spirituality, Christianity and so on—appears to be influenced by factors other than the NDE itself."[97] Sabom adds elsewhere that NDEs are therefore not indicators of who is "headed for heaven… and those headed for hell." He continues: "Thus, I do not believe that there is compelling evidence that NDE content or type is an accurate diagnostic of a person's religious beliefs, nor are they road signs pointing to a person's ultimate destiny."[98] Other researchers have reached similar conclusions regarding these cultural and interpretive factors.[99]

NDEs do provide some robust evidence for consciousness in near-death states, especially during more advanced stages such as when neither the NDErs' hearts nor their brains appear to be functioning, or when exceptional evidence is derived from previously deceased individuals. Many other veridical details have been presented, as well. The number and variety of these cases are versatile enough to appeal to those who prefer different sorts of evidence, as well, in various situations accompanied by strong data. Cultural factors along with personal interpretations generally appear to give shape to these experiences.[100] But alternative theses have been offered to challenge even the more evidential ideas, to which we turn now.

5. Naturalistic Rejoinders

It is not surprising that there have been many attempts to argue for naturalistic explanations of events given in NDE testimonies, and to oppose the notion that they support some kind of afterlife. Scholars and scientists of naturalistic inclinations often see claims about an afterlife as intrinsically connected with religion (a connection noted by Bertrand Russell, who thought that belief in immortality, like belief in God, was central to at least the Christian religion[101]), and therefore as potential challenges to various naturalistic positions. The connections between these religious ideas are not being pursued here, only that some of the reasons for opposing NDE data would initially resonate with many who oppose religion and its influence. This may account in large part for some of the chief opposition to NDEs.

Many scholars and scientists have thoroughly examined the naturalistic explanations of NDEs and compiled a wide variety of responses directed at these explanations.[102] While naturalistic critics have attempted to explain NDEs as nothing more than subjective experiences, hundreds of pages of rejoinders have been produced. A number of the major arguments on both sides were introduced at the outset of this essay, drawn from the four-issue dialogue in the *Journal of Near-Death Studies* in 2007–2008. But which position makes the most compelling arguments based on the available data? Some questions will help to point the way through the major issues.

For example, were the events that were recounted in ND reports witnessed personally by the NDEr *during* the event itself, or were they learned in some other way? It is undisputed that many of these corroborated ND reports clearly *concern* incidents that had occurred *during* the distressed medical states. But could all of these testimonies have been derived from naturally acquired information that was learned by some means *other than* observation via the actual NDE—perhaps through natural senses, or just prior to the resuscitation, or sometime *after* the particular occurrences took place?

Quite clearly, many of the more than 150 evidential near-death observations (mentioned above) from inside the room appeared to come from where the patient was positioned, and contained many observations that were very specific and unlikely to have come from remembered remarks of others: readings on machine dials, an unplugged medical device that no one but the NDEr had noticed, a "bird-like" arm-flapping surgeon in the operating room, the location of misplaced dentures, a nurse's plaid shoelaces. In some cases, the nature and even the wording of jokes told to relieve tension inside the room were recounted; in others, embarrassing incidents that happened during the episode were reported. Finally, occasional ND experiences were reportedly shared by other healthy persons in attendance.

Critics who reject a "supernatural" explanation for these observations have to explain how *all 150* of these same-room cases can be explained in a naturalistic way; yet it seems to be highly improbable that such explanations can carry the day in each of the cases.

Further, many observations clearly were made from the vantage point of the ceiling (precisely as claimed by the NDErs, who had no doubt that they were looking down at themselves from above their bodies), rather than from the hospital bed. When reports of ceiling-level, elevated viewpoints include corroborated information *from that position*, then claim for the higher location has to be favored.

This is especially the case when some of these reported items could *only* be seen from the upper angle, as in the several cases of looking down on accurately reported numbers.[103]

Moreover, studies such as Sabom's included a control group consisting of twenty-five cardiac patients who had not experienced a ND episode plus other related data. Eighty percent of the cardiac control group members made at least one major error in their imagined descriptions, whereas the NDErs' descriptions were "accurate in all instances where corroborating evidence was available."[104] Penny Sartori likewise utilized a control group, with similar results to those of Sabom: while the NDE descriptions were highly accurate with some errors, those who had been resuscitated but without NDEs were exceptionally inaccurate.[105]

The indications drawn from evidenced reports precisely during the NDE itself, the elevated viewpoint and angle of many of the perceptions, and the mistakes by those who did not observe events from those heights all point in the direction of the testimony originating during the NDE episodes. Other suppositions are far less likely.

Could resuscitation attempts "wake" the brain just long enough for the patient to observe certain elements in the room by natural means? Medical experts have asserted that consciousness does not return in just seconds, and does not fully return until after the heart has actually been restarted again, and then needs a long enough time to get past the initial confusion. Furthermore, in several cases, *no resuscitation efforts at all had even been administered* to the patient prior to the NDE observation, or such efforts had been *discontinued* prior to the NDE, clearly eliminating this particular natural option.[106] In some cases, Sam Parnia and Peter Fenwick add that EEG activity may even take hours to return.[107] Moreover these natural retorts of dying or waking brains during CPR producing in-room data would not even apply to many or even most of the strongest NDE reports, such as all of those taking place over a distance, or when healthy persons share the NDEs, or to seeing deceased friends or loved ones who impart unknown information, to those occurring to blind persons, in other words, to an entire host of evidential scenarios.

Were any evidential reports observed precisely *during* the periods of "brain death"? This question must be addressed at more than one level. Increasingly, the most evidential NDE cases seem to occur during a state of cardiac arrest due to ventricular fibrillation. It is exceptionally well-documented that during such a condition, the heart stops and at least higher (cortical) brain activity usually ceases in a matter of just ten to fifteen seconds or so afterwards.[108] Even lower brain stem activity likewise ceases very shortly afterwards,[109] and any lower-level brain activity that might possibly be present for slightly longer than upper level cortical consciousness—generally just seconds—would be insufficient to explain high-level consciousness anyway, and thus could not account for the clearest, most realistic experiences that NDErs have ever reported in their entire lives, not to mention the frequently-reported confirmations.[110]

About three dozen cases were cited in this study in which verified data were reported *after* such cardiac arrests began, well beyond the very brief brain shutdown window that began just seconds later, and which was often prolonged for many minutes beyond that. Further, many of the reported and confirmed incidents themselves actually occurred *precisely within* these time periods, as mentioned.

These evidential confirmations included observations that were reported from a viewpoint and angle just above the body. These reports identified individual people in the operating room and their clothing, reported the hearing of repeated frantic conversations pertaining an arrested heart, recorded medical machine readings, and recorded the order of a sequence of digits that could only be seen from above. Often, sequences of these details were repeated end-to-end. Given that these accurate observations occurred precisely during this time when the heart and brain were apparently flatlining, what is the likelihood that the details in every single case were discovered some other way at another time, while almost magically fitting into just these exact intervals? And could any such alternate explanations be demonstrated?

Add to these observations, as attested further by these researchers, that *even fully operating, healthy brains still could not explain the more than one hundred corroborative cases* from a distance outside an individual's line of physical sight that were mentioned in this essay or many of the over *three hundred total evidenced NDE observations* cited here. Then there are the cases where the NDEr reported encountering a deceased individual where the meeting was also evidential in nature. Although the NDEr returned to normal life, the purportedly deceased person in the encounter had been dead for some time, often for many years.[111] If this is indeed the best explanation for this last set of events, then these NDEs definitely extend *far* beyond the irreversible, biological death of the deceased individual's brain!

These last evidential considerations clearly bypass the issue of cardiac arrest altogether. Such confirmed cases increase the likelihood that at least some NDE accounts report actual data from conscious states that *extend far beyond the death of the physical body including the brain*. So the question of whether any NDEs extend during and beyond brain death may be answered affirmatively.

Several naturalistic rejoinders to the conclusion of afterlife have been offered. Due to the very nature of the topic, NDEs have a powerful advantage over virtually all other supernatural-natural issues that are debated by scholars. Due to the nature of this subject, there is a theoretical line in the sand that is automatically built into this particular discussion. Many if not most alternative hypotheses, including the most prominent ones, postulate conditions that are *internal* to the individual NDEr, such as oxygen deprivation, temporal lobe seizure, drugs, exaggerations, false statements, hallucinations, or either a dying or a waking brain. That is, they rely chiefly upon the *interior* physiological conditions and/or psychological states of the NDEr's mind or body. Thus, a large portion of NDE critiques depend upon *subjective* conditions mostly *inside* individuals, while often attempting simply to deny outside corroboration. Herein lies the chief rub for naturalist objections.

The central problem for the overall *naturalistic* position is that research has identified (limiting ourselves to the cases mentioned in this essay alone) well over three hundred cases of *external* circumstances which argue for the veracity of these NDEs. Many reports are drawn from the immediate or surrounding vicinity of the NDEr, while others are derived from a distance away, beyond the eyes and ears of everyone present, and most of all, beyond the senses of the NDEr herself. Some describe individual events and others involve a sequence of occurrences. Some of the NDEs are reportedly shared by healthy onlookers, and still others occur simultaneously. A few of the NDErs are blind, and others claim to have met deceased individuals who imparted evidential information that they and/or others did not know. Lastly, many NDEs are confirmed by multiple species of evidence.

Therefore, if even just a handful of the cases accurately report data from the "real world," it would seem that the *subjective* theses attempting to explain away these experiences would fail to account for all of the reports. This is precisely what the NDErs themselves have always claimed all along, often based on no more than their personal experience alone!

Here's the key: no matter what subjective, internal states the critics wish to discuss, by their very nature they cannot explain the existence of objective, externally corroborated NDEs. Something that exists objectively—*out there*—cannot be refuted or denied by internal human issues.

In short, here is the single, major problem with most if not virtually all natural theses against NDEs: they all utterly fail to explain the evidential NDE cases. In order for the natural suppositions to work, almost *every one* of *these three-hundred-plus corroborative accounts cited here alone must be mistaken*! But how likely is that, particularly given the careful scientific efforts to determine the accuracy of many dozens of them? But there must be no evidential remainder from the three-hundred-plus cases if the natural challenge is to succeed, for by their nature, *the internal suppositions are absolutely trumped by the external, evidential ones*!

Since a plethora of well-evidenced NDEs cover so many different angles and circumstances, "shotgun" natural explanations are sometimes offered to alleviate this difficulty. Perhaps some of these events are recorded in interviews that are too old, or are drawn from false memories, or the evidence has been exaggerated, or contains lies. These suggestions could potentially explain *some* of the NDE data. But it is highly dubious that the more than three hundred evidenced cases referred to in this chapter can be accounted for by such means. Many ND testimonies were shared or collected immediately upon the spot or at least very soon afterwards, and many of the specific details were confirmed by a variety of witnesses who participated in the occurrences, in addition to other checks and balances. Many truly exceptional NDE instances were singled out and checked even more meticulously by researchers. These evidential cases need to be explained.

It would seem, then, that Augustine's personal concession might come into play here, namely, that the presence of these sorts of confirmed NDE data, if they obtained, would both "contradict" his naturalistic thesis, as well as indicate the likelihood of an afterlife.[112] It was noted above that Blackmore[113] and Potts[114] also made similar concessions. Of course, these scholars do not think that claims have yet been confirmed,[115] but their hypothetical openness should sound a cautionary note to those who repeatedly close the naturalistic door that opposes NDE data no matter how great the quality or quantity of the corroborative reports that continue to accumulate.

The tide seems clearly to have turned in recent years, and much further than many thought possible. This is especially the case when such a wide variety and quality of different types of NDE reports emerge, all coming from different angles, particularly when multiple sorts of evidence affirm many of the accounts.

As with other somewhat similar debates, the naturalistic position certainly appears to occupy the far weaker position and seemingly by a long shot in this case. Yet it simply seems that the lesser stance held by naturalists here is all that remains! The internal position is generally chosen while the claim is made that there are no true external evidences among the three-hundred-plus NDE testimonies here, which appears to be a highly unlikely view. But what else can be said? This is precisely the line in the sand over which the natural positions appear to have tripped![116]

6. Non-Naturalistic Objections

There are those who doubt whether NDEs indicate the existence of an afterlife who do not rely on naturalistic arguments. Such critics often recognize the reality of extra-sensory perception (ESP) and other psychical phenomena (collectively referred to as "psi"). Michael Sudduth, for example, suggests that confirmed information gained during NDEs may come from the minds of living persons via telepathy, instead of from an afterlife.[117] Others concede the validity of psi data and even NDE reports while opting for impersonal versions of survival.[118] I will consider each of these positions in turn, indicating how they fare in relation to the data presented in this essay.

The first view, involving a hypothetical power called "super-psi" by some but labelled by Sudduth as "living-agent psi" (LAP),[119] postulates telepathic communication as the source of the NDEr's knowledge. For example, an NDEr who reported accurately what a nurse told her family members in a hospital waiting room could potentially have received true information by telepathy from the nurse or from any of the other persons who knew it earlier, making it unnecessary to conclude that the source was someone deceased.

Both earlier as well as more recent versions of the super-psi/living-agent psi hypothesis have been criticized through the years as a sort of "catch-all" position whereby even legitimate psi phenomena could be extended *virtually indefinitely* to cover almost any postulated phenomena with a wave of the hand. Former Brown University philosopher and former atheist C. J. Ducasse, who accepted the afterlife thesis, remarked in a much-reprinted essay that "this hypothesis [super-psi] has to be stretched very far."[120] More forcefully, former Oxford University philosopher H. H. Price asserted that the "Super-ESP" view requires that some persons have "ESP powers of almost unlimited scope," powers "much greater than our *other* evidence about those capacities would suggest." He added, "I do not think we have much other evidence to suggest that this sort of thing can be done even by very gifted ESP subjects," noting that similar problems also arise for the cosmic memory (more impersonal) view.[121]

More recently, similar critiques of these ideas have been raised, even regarding the more carefully constructed arguments made by Sudduth. Especially in more popular and less cautious publications, ESP suppositions from living persons sometimes seem to be expressed as if almost any information may be projected and obtained at will, with the "justification" being the mere suggestion that this or that person *could* be the source. When there is a relevant example, it is often stretched too far, so that psi by living persons too often lacks its own empirical evidence, *especially as related to specifically NDE evidential situations*, as just noted by Ducasse and Price.

Further, these theses have been judged to engage in special pleading, preferring exceptionally contorted and often roundabout explanations without passing rigorous evidentiary standards themselves, and may be unfalsifiable as well. Even Keith Augustine rather stunningly commented that "faced with compelling evidence for veridical paranormal perception during out-of-body NDEs, I think that the burden of proof would fall on proponents of a rather *ad hoc*, unfalsifiable, and blanket super-ESP hypothesis to demonstrate otherwise."[122] It also has been charged that ESP by living persons seems to be treated by much easier standards, while survival cases are challenged with minute scrutiny. Accordingly, philosopher Stewart Kelly states that living psi views are too often "challenged epistemically."[123]

Moreover, even given the reality of LAP, many more commentators have remarked that attributing corroborated NDEs to living ESP is a much less likely option. When evaluating whether LAP or survival explanations can account better for *specific* NDE scenarios, particularly the best of the three hundred NDE cases mentioned in this chapter, we must, it seems, concede that the afterlife thesis is strongly favored.

For example, the NDE phenomena most generally occur in conjunction with apparent death, are often reported during the time without measurable heart or brain activity, include observations being reported from vantage points above the body, and note data from perspectives that seemingly require such angles. Information gathered from the minds of the living is not the best explanation for seeing a twelve-digit figure or other numbers located out of sight except from above, or dozens of surgical procedures reported from the top-down angle.[124] Almost meaningless information, such as designs on shoelaces worn during life-and-death situations and specific though irrelevant words that were spoken, seem to be poor candidates for remembrance and retrieval. And what about unknown information such as an important machine in an operating room that no one noticed was unplugged until *after* the NDEr told the medical staff later to check for themselves? Further, evidential NDEs that reportedly involved previously deceased persons, including post-death visions, where the imparted information was unknown to apparently anyone until it was reported by these deceased individuals would appear to be especially difficult to explain on the LAP view.

Countless ESP scenarios can be supposed and more evidential NDE cases can be rehearsed as well. But it remains that in the *specific* ND arena, the afterlife scenario provides better and more specific explanations, especially when the LAP supposals do not present the same depth, ability to counter alternate views, and degree of evidence that are demanded of the NDErs.[125] While appreciating many details of Sudduth's thesis, Edward Kelly holds that the survival view is stronger, especially in some of the specific areas just mentioned, such as in the evidential cardiac arrest cases and with crisis apparitions. Kelly also thinks that Ducasse's argument for an afterlife even responded adequately to the auxiliary hypotheses proposed by Sudduth.[126] Finally, beyond these objections, additional critiques have been levelled at the LAP view.[127]

I turn now to the second non-naturalistic sort of objection mentioned above, in which psi data and even afterlife interpretations are conceded, but in which the conception of survival after death is impersonal. Sometimes this view is compared to Buddhist or Hindu ideas, such as ideas that involve rather distilled notions of existence after death, described by Rune Johansson as "perhaps a diluted, undifferentiated, 'resting' existence, more or less impersonal but still recognizable."[128]

In response to the possibility of impersonal survival, we note that there are multiple reasons for interpreting the NDE cases discussed in this chapter in terms of *personal* survival. The cases with the strongest evidence virtually always involve personal contact points in the physical world. The continuity of selfhood is maintained throughout, in that the very same person who has the NDE observes the evidential details, and then reports them later as a continuous whole. Deceased family members are observed and are recognized immediately without question. Key personal elements do not change in the relevant cardiac arrest scenarios with apparent loss of measurable brain input. This would extend as well to the so-called Peak in Darien cases[129] where this personal continuity is maintained with deceased loved ones who are also recognized though a number had died years before, with memories intact, and who then imparted crucial data that are confirmed later by the NDEr. Often the NDEr's personal memories are enhanced considerably.

There are other subjective reports, such as cases where the NDErs view their bodies from above and conclude that they are dead, or where mothers glimpse their new-born babies below, or know that they must return in order to raise their children at home. In cases where deceased loved ones who are said to have greeted them, the deep impressions of joyous recognitions show that the NDEr at least *believed* that these events were profoundly personal.

Additionally, there are major general problems for those who conceive of the afterlife in impersonal terms—what sort of confirmation could be possible here? In a truly impersonal state, how could losing oneself or passing beyond the personal realm ever be verified? And how could a subjective state even be reported later if "one" is truly unaware of "oneself"? At this point, an accusation of utilizing western logic often ensues! However, the problem remains—how to distinguish one's afterlife view so as to both make some sense, as well as establish reasons for one's position to be viable. This surely ought to be of concern in a chapter that from the outset evaluates potential afterlife *evidence*! The Catch-22 on totally impersonal afterlife positions is that there is a distinct shortage of reportable evidence available. But conversely, if evidence is *remembered* and *claimed* afterwards, then some distinctions are being made by persons and the state is not totally impersonal.[130] An additional point is that it is regularly denied that Eastern religions teach truly impersonal afterlife views in the first place![131]

Granted, NDEs by themselves cannot determine the longevity of consciousness after death—this topic proceeds beyond this essay in that not even evidential NDEs are that far-reaching. Peak in Darien cases reported from many years later perhaps provide a hint. But at best, this is not eternal life. Additional indicators have been pursued in this regard, such as by Christians who attribute these beliefs to Jesus' resurrection and its corollary arguments.[132] Regardless, impersonal afterlife beliefs are not evidenced in the sense treated in this essay, so they do not help in establishing an impersonal afterlife.

At every turn, then, NDE data are quite personal. Truly impersonal afterlife beliefs do not produce verifiable evidence concerning that actual state, let alone that which approaches the degree of NDE corroboration. Further, the claim of impersonal afterlife distinctions is betrayed seriously in philosophical and scientific terms at more than one point.

7. Conclusion

The available evidence for near-death experiences seems clearly enough to establish their evidential reality. Do these occurrences contribute any insights regarding mind-body issues, the major question in this volume? Is one view of human nature to be favored above others according to this NDE evidence? Options have been discussed in the NDE literature, for instance, by philosopher Mark Woodhouse.[133] Some have argued directly for a connection between NDEs and substance dualism.[134]

Aspects of NDEs, such as NDErs' perception of leaving their body and looking down at it from above, have caused many, experiencers and scholars alike, to posit a dualist view of mind and body. This would seem to result from the manifestation of material and immaterial aspects of the self.[135] That the NDEr identifies herself with the location of her consciousness up above, often without even initially recognizing the identity of the body below, furthers this notion.

While this seems to make the most sense of the present data, it has occasionally been pointed out that sometimes the fact that the physically stationary NDEr reports on events (which are later corroborated) in some remote location could favor a unified notion of the body, though this appears to be a minority position and it is seldom addressed.[136] The NDE notions do provide some helpful empirical hints, and dualistic concepts often take key roles in the discussions to the extent that these issues are discussed.

Another thing seems to emerge clearly from these discussions. It often appears that the major underlying conflict in these matters is *not* primarily about evidence but more about a momentous clash between worldviews. It quite often makes far more difference which metaphysical position the debater favors prior to the beginning of the discussion. If this is accurate, then it seems that even strong evidential considerations are less likely to change minds. In fact, once minds are made up, it is often simply amazing what sort of responses are often preferred just to keep from entertaining even the possibility of an afterlife.

Naturalistic worldviews have shown many signs in recent years of major foundational fissures. Yet, it appears that many naturalists would say or do anything to maintain their naturalism. But ignoring the quickly mounting data regarding corroborated NDE incidents, or simply responding with guffaws in order to avoid dealing with such information, fails to refute the NDE argument.

Non-naturalist options that question or deny a personal afterlife are held as well and have already been discussed above in some detail. While some of these views propose an impersonal afterlife, the personal view of afterlife appears to be more favored by the evidence.[137]

In the dozens of cardiac arrest examples mentioned or listed in this essay, a number included rather incredible evidential corroboration of observed details after the measurable cessation of heartbeat, cortical brain waves, and even lower brain stoppage just slightly afterwards. In other words, some NDE cases occurred as nearly as can be ascertained during the specific time during which each of these heart and brain processes were apparently non-functioning or had flatlined.[138]

These brain termination stages occur roughly in tandem, an exceptionally brief time apart. Thereafter, the presence of meticulous and sequential evidenced reports have followed from the environment where the patient was located. On other occasions confirmed details were observed from a distance away, clearly beyond the range of even a healthy patient's physical senses. Adding to the mix have been reported corroboration from meetings with long-deceased friends and loved ones that were accompanied by additional, unknown information. A few cases also have been gathered from those who were blind, as well as other NDEs that appear to have been shared, witnessed, or corroborated by healthy onlookers, and so on.

This discussion encompassed more than just a few intriguing cases here and there. Clearly, so many multiple alternative explanations and extenuating "what if" scenarios would have to obtain in order to explain these hundreds of relevant cases, that naysayers rapidly reach a difficult impasse.

Additionally, alternative rejoinders clearly have not made as many gains in the most recent conversations. In fact, it could be argued that the number of medical NDE studies centering on alternative theses has largely become less plentiful in recent years, with the majority of essays clearly allowing for the possibility, if not the likelihood, of the NDE data. Further, when the NDE thesis is critiqued, comparatively little attention is often devoted to the corroborative accounts anyway, and especially not to the best-established examples. This is very intriguing, especially when both sides in this debate, including some of the key opponents, have conceded the crucial importance of explaining the best-evidenced claims. Many cases across the spectrum are so exceptional that the ones that were left out of this study could almost as easily have replaced those that were included with minimal loss of good data!

There is simply a large plethora of evidential cases that are backed by strong corroboration.

After all, these corroborated cases are precisely the keys to refuting both the natural and non-natural alternatives to the NDE conclusions drawn here. The bottom line, then, appears to be this: the competing theses have not come close to successfully explaining the well over three hundred evidential cases narrated or listed here, especially the strongest ones. These latter examples pack far more punch than do their alternate counterparts.[139] This is precisely why, when speaking of the cardiac arrest cases accompanied by corroborated data, Rivas, Dirven, and Smit assert that, "It is for this reason that the reliability of the kind of case that appears in this chapter is so fiercely challenged by materialists."[140] These authors add that if this domino falls, the critics know that their cause at this point is lost.[141] Even while attempting not to overstate the strength of the overall case, major researchers have rated the NDE data as indicating the probable reality of consciousness at least beyond the initial cessation of heart and brain function.[142]

It certainly seems as though the naturalistic and non-naturalistic alternative explanations possess far less explanatory power and that their critiques of survival are far more *ad hoc*. In the light of this, it often appears that those who hold opposed positions chiefly desire to preserve their worldview commitments at all costs, driven by a strong dislike of any "spiritual" or "religious" options. But one thing seems clear: the alternative positions do not fare well when attempting to refute the strongest NDE evidential cases. Hence, the alternative views are by far the weaker explanations here; it does not even appear to be a close contest.[143]

Notes

1. These somewhat related options could concern, for examples, near-death experiences that often—though not exclusively—involve near-death situations where one's consciousness might be perceived as existing apart from one's bodily location. Deathbed visions might include seemingly otherworldly perceptions at the end of one's life. Post-death visions (often called after-death communications) often consist of very brief visions of a recently deceased loved one (most often a husband or wife) where messages of being alive and well appear to be imparted. Many experiences across these categories also include evidential considerations that, except in the cases of near-death experiences, are rarely mentioned in this essay. Cross-cultural deathbed visions are treated by Karlis Osis and Erlendur Haraldsson, *At the Hour of Death* (New York: Avon, 1977), besides many more popular works that often mix these categories, such as John Burke, *Imagine Heaven: Near-Death Experiences, God's Promises, and the Exhilarating Future the Awaits You* (Grand Rapids, MI: Baker, 2015); Robert L. Wise, *Crossing the Threshold of Eternity: What the Dying Can Teach the Living* (Ventura, CA: Regal, 2007); Trudy Harris, *Glimpses of Heaven: True Stories of Hope and Peace at the End of Life's Journey* (Grand Rapids, MI: Baker, 2008); Pete Deison, *Visits from Heaven: One Man's Eye-Opening Encounter with Death, Grief, and Comfort from the Other Side* (Nashville: Thomas Nelson, 2016); Bill Guggenheim and Judy Guggenheim, *Hello from Heaven* (New York: Bantam, 1997).
2. John C. Hagan III, *The Science of Near-Death Experiences* (Columbia, MO: University of Missouri Press, 2017), xi, xiii, 3–4.
3. A portion of this section is edited and adapted from the author's essay, "Evidential Near-Death Experiences,"

in *The Blackwell Companion to Substance Dualism*, eds. Jonathan Loose, Angus J. L. Menuge, and J. P. Moreland (Oxford: Wiley-Blackwell, 2018), 227–246.

4. *Journal of Near-Death Studies*, 25, no. 4 (Summer 2007); 26, no. 1 (Fall 2007); 26, no. 2 (Winter 2007); and 26, no. 3 (Spring 2008). As of this writing, all the issues are available as free pdf downloads from the UNT Digital Library, https://digital.library.unt.edu/.
5. Augustine's contribution to the fourth issue in the series is not an article but a letter to the editor, responding to a letter from Neal Grossman in the same section. For the entire exchange between Augustine and Grossman, see *Journal of Near-Death Studies*, 26, no. 3 (Spring 2008), 227–243. Besides these articles, Augustine develops similar ideas against dualism and the afterlife, along with Yonatan I. Fishman, in "The Dualist's Dilemma: The High Cost of Reconciling Neuroscience with a Soul," in Michael Martin and Augustine, eds., *The Myth of an Afterlife: The Case Against Life after Death* (Lanham, MD: Rowman and Littlefield, 2015), 203–292, such as 203–204, 211–226, 253–271, 274, 277. Likewise, John Martin Fischer and Benjamin Mitchell-Yellin, *Near-Death Experiences: Understanding Visions of the Afterlife* (New York: Oxford University Press, 2016). Less assertive than Augustine and Fishman is the essay by Benjamin Mitchell-Yellin and John Martin Fischer, "The Near-Death Experience Argument Against Physicalism: A Critique," *Journal of Consciousness Studies*, 21 (2014): 158–183, especially where possible concessions are made to the NDE argument (158–159, 163–169, 171–173, 180–182). Also, Fischer, "Near-Death Experiences: To the Edge of the Universe," *Journal of Consciousness Studies*, 27 (2020), 166–91. Responses in the same issue of this journal that also emphasize the crucial importance of the corroborative NDE cases were made by Janice Holden and M. Woollacott, "Reasonable People can Disagree: A Response to 'Near-Death Experiences: To the Edge of the Universe,'" *Journal of Consciousness Studies*, 27 (2020) and Joseph Komrosky, "Challenging Physicalism with a Logical Analysis of Evidence for Near-Death Experiences with Out-of-Body Experiences," *Journal of Consciousness Studies*, 27 (2020).
6. Keith Augustine, "Does Paranormal Perception Occur in Near-Death Experiences?" *Journal of Near-Death Studies*, 25 (2007): 213–234.
7. Keith Augustine, "Near-Death Experiences with Hallucinatory Features," *Journal of Near-Death Studies*, 26 (2007), especially 3–10, 17–30.
8. Keith Augustine, "Psychophysiological and Cultural Correlates Undermining a Survivalist Interpretation of Near-Death Experiences," *Journal of Near-Death Studies*, 26 (2007): 90–122.
9. Keith Augustine, "'Does Paranormal Perception Occur in Near-Death Experiences?' Defended," *Journal of Near-Death Studies*, 25, no. 4 (Summer 2007), 261–283.
10. Augustine, "'Does Paranormal Perception Occur in Near-Death Experiences?' Defended," 269. Michael Potts's study, cited by Augustine, is "The Evidential Value of Near-Death Experiences for Belief in Life after Death," *Journal of Near-Death Studies*, 20 (2002): 250–251.
11. Augustine, "'Does Paranormal Perception Occur in Near-Death Experiences?' Defended," 269 (Augustine's emphasis).
12. Such as Bruce Greyson, "Comments on 'Does Paranormal Perception Occur in Near-Death Experiences?,'" *Journal of Near-Death Studies*, 25 (2007): 237–238.
13. Augustine, "'Does Paranormal Perception Occur in Near-Death Experiences?' Defended," 264, 269, 280.
14. Susan Blackmore, *Dying to Live: Near-Death Experiences* (Buffalo: Prometheus, 1993), 113, 125, 128, 262–263.
15. Augustine, "Does Paranormal Perception Occur in Near-Death Experiences?," 230–234.
16. Incidentally, in what Augustine terms a "hardly surprising development" he ends this discussion by commenting that distant prayer "has absolutely no effect on the health of hospitalized patients" as if to suggest that these prayer examples are "similar" (his word) and supportive of the lack of NDE results ("Does Paranormal Perception Occur," 234). Strangely enough, Augustine fails to mention several major peer-reviewed double-blind prayer studies that oppose his comment here, published in medical journals prior to the appearance of his articles, but which produced excellent results, such as: Randolph C. Byrd, "Positive Therapeutic Effects of Intercessory Prayer in a Coronary Care Unit Population," *Southern Medical Journal*, 81 (July 1988): 826–829. Rather stunningly in the "Abstract" to this medical journal essay (826), Byrd declares that, "These data suggest that intercessory prayer to the Judeo-Christian God has a beneficial therapeutic effect in patients admitted to a CCU." Cf. also Randolph C. Byrd with John Sherrill, "On a Wing and a Prayer," *Physician*, 5 (May-June, 1993), where on page 16, Byrd likewise notes the statistically significant improvement

in 21 of 26 monitored categories for the patients who received prayer! Further, see also William S. Harris, Manohar Gowda, Jerry W. Kolb, Christopher P. Strychacz, James L. Vacek, Philip G. Jones, Alan Forker, James H. O'Keefe, and Ben D. McCallister, "A Randomized, Controlled Trial of the Effects of Remote, Intercessory Prayer on Outcomes in Patients Admitted to the Coronary Care Unit," *Archives of Internal Medicine*, 159 (October 25, 1999): 2273–2278. Moreover, Candy Gunther Brown approvingly notes these two studies along with other relevant research items in her volume, *Testing Prayer: Science and Healing* (Cambridge, MA: Harvard University Press, 2012), especially chaps. 2–6. (This volume is endorsed on the dust cover by one present and another former member of the Harvard Medical School. This last point is *not* made to somehow "demonstrate" this thesis by citing key claims on its behalf, because such a conclusion would *not* follow, but rather to forestall the objection that findings such as these are somehow religious propaganda of some sort.) Again, each of these medical articles was published prior to Augustine's comments.

17. Though unpublished, my two cases were related to me in personal conversations where notes were taken.
18. This case was also related to me in personal conversations where notes were taken.
19. Michael B. Sabom, *Recollections of Death: A Medical Investigation* (New York: Harper and Row, 1982), 86.
20. Kenneth Ring and Madelaine Lawrence, "Further Evidence for Veridical Perception During Near-Death Experiences," *Journal of Near-Death Studies*, 11 (1993): 223–229.
21. Some noncrucial details have been purposely changed so as not to share personal items.
22. Many precise details are reported in Titus Rivas, Anny Dirven, and Rudolf H. Smit, *The Self Does Not Die: Verified Paranormal Phenomena from Near Death Experiences*, ed. by Janice Miner Holden, trans. by Wanda J. Boeke (Durham, NC: International Association for Near-Death Studies, 2016), 105–109.
23. Melvin Morse with Paul Perry, *Closer to the Light: Learning from Children's Near-Death Experiences* (New York: Random House, 1990), particularly 3–9.
24. Morse with Perry, *Closer to the Light*, 32–33.
25. Rivas, Dirven, and Smit, *The Self Does Not Die*, 109–112, emphasis added.
26. Melvin Morse with Paul Perry, *Transformed by the Light: The Powerful Effect of Near-Death Experiences on People's Lives* (New York: Random House, 1994), 201.
27. Rivas, Dirven, and Smit, *The Self Does Not Die*, 58–59.
28. Sabom, *Recollections of Death*, 86–115; also 64–75.
29. Bruce Greyson, "Comments on `Does Paranormal Perception Occur in Near–Death Experiences?'" *Journal of Near–Death Studies*, 25 (2007): 237–39.
30. Emily Williams Cook, Bruce Greyson, and Ian Stevenson, "Do Any Near-Death Experiences Provide Evidence for the Survival of Human Personality after Death? Relevant Features and Illustrative Case Reports," *Journal of Scientific Exploration*, 12 (1998): 399–400.
31. Janice Miner Holden, "Veridical Perception in Near-Death Experiences," in *The Handbook of Near-Death Experiences*, eds. Janice Holden, Bruce Greyson, and Debbie James (Santa Barbara, CA: Praeger, 2009), 187–211. Holden's chart along with the list of these careful improvements are all found on pages 193–199.
32. Kenneth Ring and Evelyn Elsaesser Valarino, *Lessons from the Light: What We Can Learn from the Near-Death Experiences* (New York: Insight, 1998), Chaps. 2–3, plus 224–226 present some even stunning cases.
33. Cook, Greyson, and Stevenson, "Do Any Near-Death Experiences Provide Evidence for the Survival of Human Personality after Death?" 384–399.
34. Sabom, *Recollections of Death*, Chap. 7, plus 64–65.
35. Peter Fenwick and Elizabeth Fenwick, *The Truth in the Light: Investigations of Over 300 Near Death Experiences* (London: Headline Books, 1995), 31–35.
36. Morse with Perry, *Closer to the Light*, 4–9, 24–26, 152–154; Melvin Morse, "Near-Death Experiences and Death–Related Visions in Children: Implications for the Clinician," *Current Problems in Pediatrics*, 24 (February 1994): 55–83.
37. Some of the sources cited by Holden at least seem to include still more cases, such as Morse, "Near-Death Experiences and Death-Related Visions in Children," especially a few undescribed hints on 70–74; likewise in Morse and Perry, *Closer to the Light*, 32–36 (three examples), 51–55, 106–107 (two examples), 152–154. Additional cases are found in Morse with Perry, *Transformed by the Light*, 114–118 (two examples), 180–181, 201. After the publication of Holden's excellent list, many more researchers have published additional evidenced NDE reports, as mentioned throughout this chapter, such as Chris Carter, *Science and the Near Death Experience: How Consciousness Survives Death* (Rochester, VT: Inner Traditions, 2010); Jeffrey Long with Paul Perry, *Evidence of the Afterlife: The Science of Near-Death Experiences* (New York: Harper Collins, 2010), which

provides nine arguments for an afterlife based on NDE testimony (Chaps. 3–11); Jeffrey Long with Paul Perry, *God and the Afterlife: The Groundbreaking New Evidence for God and Near-Death Experience* (New York: Harper Collins, 2016). An outstanding overview of sources, along with many evidential examples, and responses to skeptical objections is provided by J. Steve Miller, *Near-Death Experiences as Evidence for the Existence of God and Heaven: A Brief Introduction in Plain English*, with a Foreword and commendation by Jeffrey Long (Acworth, GA: Wisdom Creek, 2012). Holden has also been responsible for additional studies in these areas, as well, one example of which is the excellent volume above that she edited, *The Self Does Not Die* (2016).

38. Further documentation is provided below. For now, the testimony of cardiologists and others is found in Rivas, Dirven, and Smit, *The Self Does Not Die*, especially 55–58.
39. As well as outside the room in many of these instances, as documented below.
40. Specific examples are provided in Rivas, Dirven, and Smit, *The Self Does Not Die*, especially Cases 3.6 (62), 3.7 (62–68), 3.8 (68), 3.9 (68–70), 3.10 (70–71), 3.11 (71–78), 3.12 (78–79), 3.13 (80), 3.14 (81), 3.15 (81–82), 3.16 (82–83), 3.17 (83–84), 3.18 (84–86), 3.19 (86–87), 3.20 (87–88), 3.21 (88–89), 3.22 (89–90), 3.24 (90–91), 3.25 (91–92), 3.26 (92), 3.27 (93), 3.28 (93–95), 3.29 (95–104), 3.30 (104–105), 3.31 (105–109), 3.32 (109–112), 3.33 (112–113), 3.34 (113–114), 3.35 (114–116), 3.36 (116–120).
41. Blackmore, *Dying to Live*, 125–133.
42. Kimberly Clark Sharp, "The Other Shoe Drops: Commentary on 'Does Paranormal Perception Occur in Near-Death Experiences?'" *Journal of Near-Death Studies*, 25 (2007): 245–250; Rivas, Dirven, and Smit, *The Self Does Not Die*, 32–34; Greyson 2016, personal communication; Sharp, 1994, personal conversation (twice).
43. Blackmore, *Dying to Live*, 127–128.
44. Ring and Lawrence, "Further Evidence for Veridical Perception During Near-Death Experiences," 226–227.
45. Morse, "Near-Death Experiences and Death-Related Visions in Children," 61; Morse with Perry, *Closer to the Light*, 6–7; Morse with Perry, *Transformed by the Light*, 22–23.
46. Rivas, Dirven, and Smit, *The Self Does Not Die*, 42–43. The physician who investigated this case further was Barbara R. Rommer (*Blessing in Disguise: Another Side of the Near-Death Experience* [St. Paul, MN: Llewellyn, 2000], 5–7). Rommer asserted that Tony "was clinically dead for thirty minutes" (5) and mentioned this case under the heading "Objective Evidence."
47. Rivas, Dirven, and Smit, *The Self Does Not Die*, 159–160, 165–167.
48. Bruce Greyson, Emily Williams Kelly, and Edward F Kelly, "Explanatory Models for Near-Death Experiences," in *The Handbook of Near-Death Experiences: Thirty Years of Investigation*, eds. Janice Miner Holden, Bruce Greyson, and Debbie James (Santa Barbara, CA: Praeger, 2009), 230.
49. Bruce Greyson, "Seeing Dead People Not Known to Have Died: 'Peak in Darien' Experiences," *Anthropology and Humanism*, 35 (2010): 159–171.
50. Pim Van Lommel, R. van Wees, V. Meyers, and I. Efferich, "Near-Death Experience in Survivors of Cardiac Arrest: A Prospective Study in the Netherlands." *The Lancet*, 358 (2001): 2039–2045.
51. Penny Sartori, Paul Badham, and Peter Fenwick, "A Prospectively Studied Near-Death Experience with Corroborated Out-of-Body Perceptions and Unexplained Healing," *Journal of Near-Death Studies*, 25 (2006): 69–84.
52. Cook, Greyson, and Stevenson, "Do Any Near-Death Experiences Provide Evidence for the Survival of Human Personality after Death? Relevant Features and Illustrative Case Reports," *Journal of Scientific Exploration*, 12 (1998): 377–406.
53. Emily Williams Kelly, Bruce Greyson, and Ian Stevenson, "Can Experiences Near Death Furnish Evidence of Life After Death?" *Omega*, 40 (2000): 513–519.
54. Long with Perry, *Evidence of the Afterlife*, 208, endnote 9.
55. Morse with Perry, *Transformed by the Light*, such as 116–118, 180–181.
56. Sabom, *Recollections of Death*, 111–113.
57. Marvin Ford with Dave Balsiger and Don Tanner, *On the Other Side* (Plainfield, N.J.: Logos, 1978), 178–180.
58. In Rivas, Dirven, and Smit, *The Self Does Not Die*, these examples include Cases 3.13 (80–81), 3.15 (81–82), 3.22 (89–90), 3.31 (105–109), and 3.33 (112–113). Note that in each of these five examples here, there were confirmed reports both in the room as well as outside of it, hence these cases were included in the earlier list (endnote 27), as well.
59. Some of this wide variety of additional cases include more accounts in Rivas, Dirven, and Smit, *The Self Does Not Die*, Cases 2.2 (31–32), 2.5 (35–36), 2.6 (36–37), 2.12 (42–43), 3.32 (109–112), 5.5 (140–141), 7.3 (159–160), cf. 3.1 (58–59).

60. Greyson, "Seeing Dead People Not Known to Have Died," 162. There are some similarities here to Emily Williams Kelly, "Near-Death Experiences with Reports of Meeting Deceased People" *Death Studies*, 25 (2001): 229–249, especially 239.
61. Greyson, "Seeing Dead People Not Known to Have Died," 162–169.
62. Kenneth Ring, *Life at Death: A Scientific Investigation of the Near-Death Experience* (New York: Coward, McCann, and Geoghegan, 1980), 207–208; Eben Alexander, *Proof of Heaven: A Neurosurgeon's Journey into the Afterlife* (New York: Simon and Schuster, 2012), 162–164, 168–169; Morse, *Transformed by the Light*, 114–115; Editor John Myers includes at least one older, anecdotal example that involved a total of three persons who died at the same time in various parts of the United States, with the details being confirmed in follow-up investigating (*Voices from the Edge of Eternity* (Old Tappan, NJ: Fleming H. Revell, 1971), 97–99. Gary Habermas and J. P. Moreland also reported earlier a number of these same cases (*Beyond Death: Exploring the Evidence for Immortality* [Wheaton, IL: Crossway, 1998], 162–164; earlier ed.: *Immortality: The Other Side of Death* [Nashville: Thomas Nelson, 1992], 78–80).
63. Pim Van Lommel, "About the Continuity of Our Consciousness," *Advances in Experimental Medicine and Biology*, 550 (2004): 122.
64. Greyson, "Seeing Dead People Not Known to Have Died," 167.
65. Bruce Greyson, *After: A Doctor Explores What Near-Death Experiences Reveal about Life and Beyond.* New York: Macmillan/St. Martin's Press, 2021), 132–133.
66. Such as Rivas, Dirven, and Smit, *The Self Does Not Die*, Case 5.5 (140–141); Van Lommel, "About the Continuity of Our Consciousness," 122; Maurice Rawlings, *Beyond Death's Door* . . . . (Nashville: Thomas Nelson, 1978), 17–23.
67. In Rivas, Dirven, and Smit, Case 5.5 (*The Self Does Not Die*, 140–141). During her NDE, the cardiac-arrested person reported being with a brother whom she did not know she had. She was told subsequently that he had died approximately thirty or more years earlier. In Rawlings (*Beyond Death's Door*, 17, 21), the cardiac-arrested individual reported meeting a person who had died about forty-seven years previously. The NDEr had never met or seen a photograph of the person, but identified her immediately when shown a group photo. Van Lommel's previous account concerned an NDEr who had met and correctly identified his biological father whom he had previously known nothing about and who had been deceased for many years ("About the Continuity of Our Consciousness," 122). Elisabeth Kübler-Ross relates one of her cases of a girl who recounted meeting a brother that she did not know she had, whom she found out later had died 12 years earlier (*On Children and Death* [New York: Macmillan, 1983], 208; also Kübler-Ross, *On Life After Death* [Berkeley, CA: Ten Speed, 2008], 28.) Eben Alexander explains that during his own NDE he met his biological sister, who had died as an adult some ten years earlier. Alexander had been adopted as an infant and had never met her or seen a photo of her. Later he was given a photo and recognized the woman as the person he had met in his NDE.
68. Greyson, "Seeing Dead People Not Known to Have Died," 169.
69. Morse, *Transformed by the Light*, 168–170, 173–180 where many different sorts of accounts are narrated; Morse, "Near–Death Experiences and Death-Related Visions in Children," 72; L. Houlberg, "Coming Out of the Dark," *Nursing*, 22, no. 2 (February 1992); Miller, *Near-Death Experiences as Evidence for the Existence of God and Heaven*, particularly 66–68.
70. Morse, *Transformed by the Light*, 170–174; Morse, "Near-Death Experiences and Death-Related Visions in Children," 72–73.
71. Raymond Moody with Paul Perry, *Glimpses of Eternity: Sharing a Loved One's Passage from this Life to the Next* (New York: Guideposts, 2010), 11–12, 77, 80, 81. The story is also retold in Miller, *Near-Death Experiences as Evidence for the Existence of God and Heaven*, particularly 66–67.
72. Rivas, Dirven, and Smit, *The Self Does Not Die*, Case 3.32, 109–112.
73. These interviews were conducted by me via personal conversations where notes were taken.
74. Rivas, Dirven, and Smit, *The Self Does Not Die*, containing two examples: Case 4.3 (129–132) and Case 7.6 (165–167); Morse, *Closer to the Light*, two more possible examples: 127–128, 130–132; Morse, *Transformed by the Light*, for a half-dozen more possible accounts, 174–180.
75. Kenneth Ring and Sharon Cooper, "Near-Death and Out-of-Body Experiences in the Blind: A Study of Apparent Eyeless Sight," *Journal of Near-Death Studies*, 16 (1997): 101–147. The account of Vicki's NDE is reported on pages 108–112, 114.
76. A more detailed and slightly more evidential study is presented in Kenneth Ring and Sharon Cooper, *Mindsight: Near-Death and Out-of-Body Experiences in*

the Blind (Palo Alto, CA: William James Center for Consciousness Studies, 1999), especially the treatment of the visual aspects (Chap. 4, 41–96) and corroborative evidence (Chap. 5, 97–120). Vicki's initial case above where she described her two deceased friends is found on pages 41–59 (especially 45–47), 50–51, 55, including a limited amount of confirmation.

77. Ring and Cooper, *Mindsight*, 109–120, with the three quotations located on pages 115, 117, 109, respectively.
78. Ring and Cooper, "Near-Death and Out-of-Body Experiences in the Blind," with these two cases on pages 112–114 and 120–122, respectively.
79. Ring and Cooper, "Near-Death and Out-of-Body Experiences in the Blind," 124. Similar conclusions are also found in Ring and Cooper, *Mindsight*, 97, 120 (the latter chiefly concerning Nancy's case).
80. Such as Miller, who provides some cases drawn especially from van Lommel plus Ring and Cooper (*Near-Death Experiences as Evidence for the Existence of God and Heaven*, 70–73).
81. Rivas, Dirven, and Smit, *The Self Does Not Die*, xxvii.
82. Rivas, Dirven, and Smit, *The Self Does Not Die*, 126.
83. Confirmed in personal correspondence from editor Janice Miner Holden, 2016.
84. As noted above, a number of these cases were also cited in Habermas, "Evidential Near-Death Experiences," 238–239, along with several cases drawn from the author's own research collection that have been added here. These totals reflect only the counts that are mentioned or cited in this chapter, rather than a total from all NDE literature.
85. Rivas, Dirven, and Smit, *The Self Does Not Die*, Case 2.6, 36–37.
86. Rivas, Dirven, and Smit, *The Self Does Not Die*, Case 2.5, 35–36.
87. As confirmed in an interview afterwards.
88. Rivas, Dirven, and Smit, *The Self Does Not Die*, Case 3.11, 71–78, with the quotations on pages 74–77, with long answers being written out in great detail.
89. Rivas, Dirven, and Smit, *The Self Does Not Die*, Case 3.9, 68–70.
90. Rivas, Dirven, and Smit, *The Self Does Not Die*, Case 2.14, 44.
91. As in Burris Jenkins with Jess E. Weiss, "I Was an Atheist—Until I Died," in *The Vestibule*, ed. Jess E. Weiss (New York: Pocket Books, 1972), 29–39. Burris's account involved some mild confirmation of the physical world (34, 36), along with a conversion from atheism to Christianity (35–37). Perhaps the best known of these testimonies from a former atheist to Christianity is the truly fascinating account by Howard Storm, *My Descent into Death: A Second Chance at Life* (New York: Doubleday, 2005).
92. Nancy Evans Bush, a recognized authority in this subarea of NDE study, helpfully delineates several types of negative or hellish NDEs. These might include a range from frightening or disturbing NDE elements, to experiences of despair and meaninglessness, all the way to rather classic hellfire reports, including demons and dark pits (71–72). Various studies reported by Bush report quite a wide range of the percentages of those who experience these and similar negative states. The actual, reported research of this type extends anywhere from several reports with no such negative cases at all, to percentage measurements of these distressing cases such as the following: 12%, 15% (2 studies), 18%, 20%, 23%, 29%, 33%, 60%, and two overviews where the numbers were apparently much higher but where percentages were not provided. These accounts included varying numbers of participants, with Bush noting that surveyors recognized that many who had these experiences often "notoriously" avoided listing or discussing them (70), that at least some early researchers themselves did not want to ask or know about these distressing examples (71), and that these experiences "are underreported" (81). Bush's conclusion is that the actual percentages of negative reports are possibly best estimated as being somewhere in the mid- to high-teens (81). (Bush, "Distressing Western Near-Death Experiences: Finding a Way through the Abyss," in Holden, Greyson, and James, eds., *The Handbook of Near-Death Experiences*, 63–86.) Another more recent treatment that reviews what is known about these phenomena is Nancy Evans Bush and Bruce Greyson, "Distressing Near-Death Experiences: The Basics," in Hagan, *The Science of Near-Death Experiences* (Columbia, MO: University of Missouri Press, 2017), 93–101.
93. Bush also seems to state at least a similar conclusion ("Distressing Western Near-Death Experiences," 81), as do Bush and Greyson ("Distressing Near-Death Experiences: The Basics," 100).
94. Comments such as these may be observed in the research of many observers, such as Sabom, *Recollections of Death*, 184–186; Michael Sabom, *Light and Death: One Doctor's Fascinating Account of Near-Death Experiences* (Grand Rapids, MI: Zondervan, 1998), 134–137, 140–146, 213–214; Bush, "Distressing Western Near-Death Experiences," 100; Morse, *Transformed by the Light*, 119–120; Long with Perry, *Evidence of the Afterlife*, 158; Miller, *Near-Death Experiences as Evidence for the Existence of God and Heaven*, 41, 97–98, 148, 195 (note 23); Habermas and Moreland, *Beyond Death*, 178–180.

Cf. Karlis Osis and Erlendur Haraldsson, *At the Hour of Death* (New York: Avon, 1977), 92, 193–194, 198. A number of authors in the Holden, Greyson, and James, eds. volume, *The Handbook of Near-Death Experiences*, also affirm these items, including the following: Bush, "Distressing Western Near-Death Experiences," 69, 81; Janice Miner Holden, Jeffrey Long, and B. Jason MacLurg, "Characteristics of Western Near-Death Experiences," 120; Allan Kellehear, "Census of Non-Western Near-Death Experiences to 2005: Observations and Critical Reflections," 135, 140–142, 150; cf. 130–131; Greyson, Kelly, and Kelly, "Explanatory Models for Near-Death Experiences," 148, 214–215. Two critics with opposite perspectives from one another also agree with many of the cultural conclusions mentioned here: H. Leon Greene, *If I Should Wake before I Die: The Medical and Biblical Truth about Near-Death Experiences* (Wheaton: Crossway, 1997), 109–110, 113–114, 204; Blackmore, *Dying to Live*, 115–120, 134. For detailed critiques of both Greene and Blackmore, see Habermas and Moreland, *Beyond Death*, Chap. Nine, particularly 200–218.

95. Such as Bush and Greyson, "Distressing Near-Death Experiences," 100; Long with Perry, *Evidence of the Afterlife*, 141, 169; Morse, *Transformed by the Light*, 119; cf. Osis and Haraldsson, *At the Hour of Death*, 190, 193.
96. Sabom, *Recollections of Death*, 238–239; Ring, *Life at Death*, 238–239; Bush and Greyson, "Distressing Near-Death Experiences," 100.
97. Sabom, "The Shadow of Death (Part Two)," in the *Christian Research Journal*, 26 (2003): 42–51, with the quotation taken from page 46; cf. the Synopsis on page 45. (Part One of this article appeared in 26 (2003): 12–21.
98. Sabom, *Light and Death*, 214.
99. Additional related research has been cited by Habermas and Moreland, *Beyond Death*, especially 178–180, 217–218.
100. So the presence of subjective aspects here is of no help to various postmodern or other efforts to overemphasize these interpretive elements in all religious or other respects, since the strong evidence for an afterlife of some sort would provide one of several exceptionally powerful building blocks in such religious discussions.
101. As seen in Russell's often-anthologized 1927 lecture, "Why I Am Not a Christian," in Douglas W. Shrader and Ashok K. Malhotra, *Pathways to Philosophy: A Multidisciplinary Approach* (Saddle River, NJ: Prentice Hall, 1996), 300–308, especially 300–301, 304. Similar ideas are mentioned in Russell, *Bertrand Russell Speaks his Mind* (New York: Avon, 1960), 19–21.
102. Discussions of various perspectives are contained in the following studies: Blackmore, *Dying to Live*, 4–8, 50–53, 106–112, 114, 202–225, 261–264; Michael N. Marsh, *Out-of-Body and Near Death Experiences* (Oxford: Oxford University Press, 2010), especially Chaps. 7–9; Michael Marsh, "The Near-Death Experience: A Reality Check?" *Humanities*, 5, (2016): 1–25; Edward F. Kelly, "Review of Michael N. Marsh's Out-of-Body and Near-Death Experiences: Brain-State Phenomena or Glimpses of Immortality? *Journal of Scientific Exploration*, 24 (2010): 729–737; Greyson, Kelly, and Kelly, "Explanatory Models for Near-Death Experiences," 214–234; Ring, *Life at Death*, 206–217; Sabom, *Recollections of Death*, 151–178; Morse with Perry, *Closer to the Light*, 183–193; Miller, *Near-Death Experiences as Evidence for the Existence of God and Heaven*, 31–48, 105–108; cf. Osis and Haraldsson, *At the Hour of Death*, 45–59. The critiques and responses by Augustine and many NDE researchers have already been addressed earlier in this chapter.
103. On Sabom's testing these NDErs' perceptions, see *Recollections of Death*, 113–115.
104. Sabom, *Recollections of Death*, 71, 113–115, 184.
105. Sartori, *The Near-Death Experiences of Hospitalized Intensive Care Patients: A Five Year Clinical Study* (Lewiston, New York: Edwin Mellen, 2008); also Long with Perry, *Evidence of the Afterlife*, 71.
106. Rivas, Dirven, and Smit, *The Self Does Not Die*, 120–123; Long with Perry, *Evidence of the Afterlife*, 80.
107. Sam Parnia and Peter Fenwick, "Near Death Experiences in Cardiac Arrest: Visions of a Dying Brain or Visions of a New Science of Consciousness" *Resuscitation*, 52 (2002): 5–11, particularly page 7.
108. See especially Jaap W. De Vries, P. F. Bakker, G. H. Visser, J. C. Diephuis, and A. C. van Huffelen, "Changes in Cerebral Oxygen Uptake and Cerebral Electrical Activity During Defibrillation Threshold Testing," *Anesthesia and Analgesia*, 87 (1998): 16–20; M. J. Aminoff , M. M. Scheinman, J. C. Griffin, J. M. Herre, "Electrocerebral Accompaniments of Syncope Associated with Malignant Ventricular Arrhythmias," *Annals of Internal Medicine*, 108 (June 1988), 791–796; R. Pana, L. Hornby, S. D. Shemie, S. Dhanani, J. Teitelbaum, "Time to Loss of Brain Function and Activity During Circulatory Arrest," *Journal of Critical Care*, 34 (2016): 77–83, particularly 77, 82; Pim Van Lommel, "Near-Death

Experience, Consciousness, and the Brain: A New Concept about the Continuity of our Consciousness Based on Recent Scientific Research on Near-Death Experience in Survivors of Cardiac Arrest," *World Futures*, 62 (2006): 134–151, especially 138–141, 148–149; Pim Van Lommel, R. van Wees, V. Meyers, and I. Efferich, "Near-Death Experience in Survivors of Cardiac Arrest: A Prospective Study in the Netherlands," *The Lancet*, 358 (2001): 2039–2045, particularly 2043–2044; Bruce Greyson, "Western Scientific Approaches to Near-Death Experiences," *Humanities*, 4 (2015): 775–796; Greyson, personal communication, 8/21/16; Parnia and Fenwick, "Near Death Experiences in Cardiac Arrest," especially 5–7, 9–10; Sabom, "The Shadow of Death (Part Two)," 15, 19, 21, endnote 39; Rivas, Dirven, and Smit, *The Self Does Not Die*, 55, 58, 124–125; Mario Beauregard and Denyse O'Leary, *The Spiritual Brain: A Neuroscientist's Case for the Existence of the Soul* (New York: Harper Collins, 2007), 154, 156–157; Long with Perry, *Evidence of the Afterlife*, 102, cf. 104.

109. Van Lommel, "Near-Death Experience, Consciousness, and the Brain," especially 138, 141; Parnia and Fenwick, "Near Death Experiences in Cardiac Arrest," 6, 9; Rivas, Dirven, and Smit, *The Self Does Not Die*, especially 55, 124, according to comments from cardiologists.
110. Greyson, personal communication, 8/21/16; Sabom, personal communication, 2006; van Lommel, "Near-Death Experience, Consciousness, and the Brain," 138–139; Van Lommel, R. van Wees, V. Meyers, and I. Efferich, "Near-Death Experience in Survivors of Cardiac Arrest," 2044; Parnia and Fenwick, "Near Death Experiences in Cardiac Arrest," 6, 9–10; Rivas, Dirven, and Smit, *The Self Does Not Die*, 55–56.
111. Greyson, "Seeing Dead People Not Known to Have Died"; Kelly, "Near-Death Experiences with Reports of Meeting Deceased People."
112. Augustine, "'Does Paranormal Perception Occur in Near-Death Experiences?' Defended," 69.
113. Blackmore, *Dying to Live*, 262.
114. Potts, "The Evidential Value of Near-Death Experiences for Belief in Life after Death," 250–251.
115. To their credit in the sources immediately above, Blackmore, Potts, and Augustine, to greater or lesser extents, each acknowledge that future evidence could continue to challenge their theses, as well.
116. The claim may be made that this evaluation is too positive, with NDE testimonies being taken too straightforwardly. Yet much care has been taken throughout to employ dozens of caveats. Further, it has been explained that all evaluations here are probabilistic in nature. The critical comebacks have also been entertained. In probabilistic terms, it is difficult to know what else can be concluded here.
117. Michael Sudduth, *A Philosophical Critique of Empirical Arguments for Postmortem Survival* (New York: Palgrave Macmillan, 2016).
118. As with some examples in Edward F. Kelly, Adam Crabtree, and Paul Marshall, eds., *Beyond Physicalism: Toward Reconciliation of Science and Spirituality* (Lanham, MD: Rowman and Littlefield, 2015).
119. Sudduth, *A Philosophical Critique of Empirical Arguments*, outlined on pages 18–20. It should be noted that Sudduth eschews the "super-psi" moniker as being a rather derogatory reference.
120. Ducasse, "Is Life After Death Possible?" (1948) in John R. Burr and Milton Goldinger, *Philosophy and Contemporary Issues*, 4th ed. (New York: Macmillan, 1984), 389.
121. Price, "The Problem of Life After Death," (1968) in William L. Rowe and William J. Wainwright, *Philosophy of Religion: Selected Readings*, 3rd ed. (Fort Worth, TX: Harcourt Brace, 1998), 534, Price's emphasis.
122. Augustine, "'Does Paranormal Perception Occur in Near-Death Experiences?' Defended," 280. Similar comments are made by Augustine and Fishman, "The Dualist's Dilemma," 282, note 52.
123. Besides Ducasse, Price, and Augustine, see also the careful nuance in detail provided by David Kelly's review of Sudduth's work above in the *Journal of Scientific Exploration*, 30 (2016), 591–592; the many details provided in the discussion by Rivas, Dirven, and Smit, *The Self Does Not Die*, 226–232, 236–237; correspondence with Stewart Kelly referring to Sudduth and similar views, dated 2/21/21.
124. It is relevant here that even though the members of Sabom's control group were veteran open heart survivors sometimes with vast experiences with the relevant procedures, their descriptions were significantly less accurate than were those who reported that they actually viewed their own situations from above (Sabom, *Recollections of Death*, 83–86).
125. Ducasse, "Is Life After Death Possible?" 382, 387–389; Price, "The Problem of Life After Death," 533–538; Beauregard and O'Leary, *The Spiritual Brain*, cf. 168–171 with 152–156; Rivas, Dirven, and Smit, *The Self Does not Die*, 221–222, 227–230, 234–236; Habermas and Moreland, *Beyond Death*, 191–197;

correspondence from philosophers who research in these areas, including William Hasker, 1/20/21; J. P. Moreland, 1/18/21; Jeffrey Koperski (1/22/21), and Stewart Kelly (1/22/21).

126. In David Kelley's review of Sudduth's work, *Journal of Scientific Exploration*, 591–592, 594.

127. Other evaluations can only be sketched here. Some agree that this view involving living psi could be exceptionally difficult for naturalists to explain. But researchers have also remarked that dualists need not have problems with the majority of the LAP assertions. Worldview perspectives must likewise be addressed because positions favoring theism on other grounds, for example, could alone provide additional pathways though these charges. Some philosophers have issues with the configuration of Bayes Theorem in that how prior probabilities are figured could skew Sudduth's thesis by themselves. (Various of these comments are found in Augustine, "'Does Paranormal Perception Occur in Near-Death Experiences?' Defended," 264, 269, 280; J. P. Moreland, "The Soul and Near-Death Experiences: A Case for Substance Dualism," in *Raised on the Third Day*, eds. David Beck and Michael Licona (Bellingham, WA: Lexham Press, 2020), Chapter 2; William Hasker's treatment in the *Notre Dame Philosophical Review: An Electronic Journal* (November 30, 2015), accessed at https://ndpr.nd.edu/news/the-myth–of-an-afterlife-the-case-against-life-after-death/ on February 4, 2021; David Kelly's review, *Journal of Scientific Exploration*, 588, 593; philosopher Neal Grossman as cited by Rivas, Dirven, and Smit, *The Self Does Not Die*, 228–232; also 227, 234; additional correspondence from Hasker, 1/20/21; J. P. Moreland, 1/18/21; Jeffrey Koperski (1/22/21), and Stewart Kelly (1/22/21).

128. Rune E. A. Johansson, *The Psychology of Nirvana* (Garden City, New York: Doubleday, 1970 [Anchor Books ed.]), 59; also 13, 55, 57–58, 112, 131–132.

129. Masayuki Ohkado, "On the Term 'Peak in Darien' Experience," *Journal of Near-Death Studies*, 31 (Summer 2013): 203–211.

130. Quite notably, even from a very sympathetic scientific position, physicist Fritjof Capra states that science cannot go as far as this impersonal realm. This critique is clear—we cannot have it both ways (*The Tao of Physics*, 2nd ed., rev. (New York: Bantam, 1983), especially 127–129. As J. P. Moreland states from a different angle: "the epistemological argument, if successful, shows that the denial of an enduring self is guilty of self-referential inconsistency." See Moreland, "An Enduring Self: The Achilles Heel of Process Philosophy," *Process Studies*, 17 (Fall 1988): 193–99; see also William Hasker, "Resurrection and Mind-Body Identity: Can there be Eternal Life without a Soul," *Christian Scholar's Review*, 4 (1975): 319–325; Beauregard and O'Leary, *The Spiritual Brain*, 153–157.

131. Rune E. A. Johansson, *The Psychology of Nirvana* (Garden City, New York: Doubleday, 1970), 13, 55–59,112, 130–132; Frederick H. Holck, ed., *Death and Eastern Thought* (Nashville: Abingdon, 1974), 14–16, 101–102; 121–130, 193–194; Farnáz Ma'súmián, *Life after Death: A Study of the Afterlife in World Religions* (Oxford: Oneworld, 1995), 50, 116–117; Harold Cowan, ed., *Life after Death in World Religions* (Maryknoll, NY: Orbis, 1997), 76, 89; Christopher Jay Johnson and Marsha G. McGee, eds., *How Different Religions View Death and Afterlife* (Philadelphia: Charles, 1991), 44–48, 52, 93, 170, 195–201, 209–216, 224; 324; cf. other religious views, 195–201, 209–216.

132. Such as Habermas, *The Risen Jesus and Future Hope* (Lanham, MD: Rowman & Littlefield, 2003), Chapters 4–6. As a side note, the argument for Jesus' resurrection and beyond that in this source is based on the "Minimal Facts" that are well evidenced and hence accepted as historical even by very skeptical biblical specialists. In its own way, the argument here for evidential NDEs has likewise been based on a minimal foundation of data that favor after-death consciousness.

133. Mark Woodhouse, "Near-Death Experiences and the Mind-Body Problem," *Anabiosis: The Journal for Near-Death Studies*, 1 (1981): 57–65.

134. Such as J. P. Moreland, "The Soul and Near-Death Experiences: A Case for Substance Dualism," in Beck and Licona, eds., *Raised on the Third Day*, 15–36; Rivas, Dirven, and Smit, *The Self Does Not Die*, 238.

135. Woodhouse, "Near-Death Experiences and the Mind-Body Problem," 61–63.

136. Cf. Mark Woodhouse, "Near-Death Experiences and the Mind-Body Problem," *Anabiosis: The Journal of Near-Death Studies*, 1 (July 1981): 57–65.

137. For an overview, see Habermas and Moreland, *Beyond Death*, Chaps. 8–9 for more details, including specific critiques of some "non-naturalistic" options.

138. Specific details are reported in van Lommel, R. van Wees, V. Meyers, and I. Efferich, "Near-Death

Experience in Survivors of Cardiac Arrest," 2043–2044; Parnia and Fenwick, "Near Death Experiences in Cardiac Arrest," 6–10; Van Lommel, "Near-Death Experience, Consciousness, and the Brain," 138–142, 148; Sabom, "The Shadow of Death (Part Two)," 15, 19, 21, endnote 39; Rivas, Dirven, and Smit, *The Self Does Not Die*, 125; Long with Perry, *Evidence of the Afterlife*, 102.

139. See Long's insightful comments at this point (*God and the Afterlife*, 194).
140. Rivas, Dirven, and Smit, *The Self Does Not Die*, 123.
141. Rivas, Dirven, and Smit, *The Self Does Not Die*, 240.
142. Sabom's conclusion is that this overall case is "highly probable" (Sabom, "The Shadow of Death," (Part Two), 50). Greyson has also made some strong evidential statements, as we have seen (such as Greyson, "Seeing Dead People Not Known to Have Died," 169.). Long makes some of the strongest comments of all (*Evidence of the Afterlife*, 124, 201–202), including that the overall evidence for God and an afterlife is "overwhelming" (*God and the Afterlife*, 196).
143. The author wishes to thank longtime friends and researchers Bruce Greyson, Mike Sabom, and Jan Holden for their exceptionally helpful discussions, information, and bibliographic support through the years as well as during this research, though without necessarily implying their agreement with me in the comments here.

Unit 4: Information, Computation, and Quantum Theory

19. Information and the Mind-Body Problem

Angus J. L. Menuge

1. Introduction

In a typical introduction to philosophy class, all that students learn about dualism is a thumbnail caricature of the views of René Descartes—apparently the only dualist that ever lived.[1] (See Goetz and Taliaferro's chapter in this volume for a more accurate description of the wide range of options available to dualists.[2]) Descartes, we learn, suggested that mind and body were substances of fundamentally different kinds, with the body, but not the mind, located and extended in space. Yet without the common medium of space, it seems impossible to understand how mind and body could causally interact. How can something that moves in space influence, or be influenced by, something that has no spatial dimension? This standard statement of the mind-body problem is used to dismiss not only Descartes's philosophy of mind, but dualism in general. Physicalism, which claims that human beings are purely physical substances, therefore seems to win by default, especially because it is widely held to present a more "scientific" view of the world.

Meanwhile, the actual situation in the philosophy of mind is far different. This chapter argues that, for a number of reasons, dualism is undergoing a renaissance. As we will show in our brief review, the best that could be claimed for physicalism is that it has reached a stalemate with dualism, because physicalist theories face problems at least as severe as the mind-body problem (section 2). Further, dualist philosophers have shown that the mind-body problem can be neutralized, although such blocking responses fall short of an illuminating account of mind-body interaction (section 3). To make progress, we first need to understand that the mind-body problem consists of a number of distinct elements, all of which must be addressed by a compelling theory (section 4). My positive proposal does not claim to provide such a theory in a fully-developed form. Instead, it focuses on (in my view) one vitally important parameter, which that theory should include. I make the case that *information* is a promising intermediary between mind and body, because it can exist in both abstract (mental) and concrete (physical) forms, and I argue that human beings are gifted with an innate capacity to translate between these two forms of information, converting volitions into motor commands and nerve signals into conscious experiences. This innate capacity initially operates at a very basic level, accounting for very simple actions and sensations, but it can be further refined by experience (section 5).

My proposal includes several models that could help to explain mind-body interaction. These models will be discussed in due course. However, I am far from claiming that my models constitute a full theory. They depend at crucial points on an analogy between the transmission of information in computer science and mind-body interaction, and that analogy certainly does not capture everything that we need to understand in order to solve the mind-body problem. And these models require further development in order to grasp their testable implications for neuroscience. So I close this chapter with a discussion of the limitations of the models here proposed and some indications of how they might be further developed to guide neuroscientific inquiry (section 6).

2. Physicalism and Dualism in Stalemate?

For the most part, physicalists continue to think that the mind-body problem eliminates any version of substance dualism as a serious option in the philosophy of mind. In particular, they raise what is known as the "pairing problem."[3] They note that while it is easy to pair physical causes with physical effects, substance dualists have a hard time explaining why mental causes are paired with particular physical effects or why physical causes are paired with particular mental effects. The pairing problem arises because of an asymmetry between typical cases of purely physical causation and mind-body interaction as described by dualists. In familiar cases of purely physical causation, it is easy to pair a particular cause with a particular effect by appealing to the properties of the spatial medium they share (e.g., the geometrical properties of space help explain why each of a pair of guns hit the particular targets they do when they are fired). But in mind-body interaction, if the mind is not spatially related to any particular body, it becomes mysterious why a mind directly interacts with one body rather than any other.

Consider an example. Suppose two marksmen, M1 and M2, shoot at two different targets, T1 and T2, and both M1 and M2 hit their targets. If we want a physical explanation of why the bullet from M1's gun hit T1 and the bullet from M2's gun hit T2, it is easy to provide: M1's gun was pointing at T1, and when M1 pulled the trigger a bullet was ejected from M1's gun in a continuous path through space in the direction of T1, while M2's gun was pointing at T2, and when M2 pulled the trigger a bullet was ejected from M2's gun in a continuous path through space in the direction of T2. In this physical explanation, the position and direction of the guns and targets in a common medium of space explain why M1's bullet hits T1 (and not T2), while M2's bullet hits T2 (and not T1). So we can pair each cause with its effect.

But now suppose we are interested in psychological explanation. M1 intended to hit T1 and was happy when he did; likewise M2 intended to hit T2 and was happy when she did. If M1's immaterial mind is not, like M1's body, located in space, why does it affect M1's body rather than M2's? Why is M1's intention to shoot T1 paired with M1's finger firing M1's gun and hitting T1, and not with M2's finger firing M2's gun and hitting T2? Likewise, why is M2's intention to shoot at T2 paired with M2's finger firing M2's gun and hitting T2, rather than with M1's finger firing M1's gun and hitting T1? And why does the physical event of a bullet hitting T1 make M1 happy, rather than M2, and the physical event of a bullet hitting T2 make M2 happy, rather than M1?

The pairing problem for substance dualism arises because, absent a shared medium between one mind and one body, it is not clear why a particular mind interacts with a particular body. Why does my intention to raise my arm make my arm go up, and not Joe's across the street? And why, when I stub my toe, do I feel pain "in my toe," rather than Joe feeling that pain? The physicalist appears to have an easy time solving this problem: if the mind in some sense is the brain, then it is the contiguity of the brain with a particular body that explains why intentions cause events in that body, and why events in that body can cause experiences like pain in a particular mind. Substance dualists need a solution to the pairing problem if their account is to be taken seriously.

Thus, substance dualists must address a "how" question—*how* do mind and body causally interact? But physicalists are in no better shape, because they face an arguably even more fundamental "what" question: *What* are the distinguishing marks of the mental, such as consciousness and intentionality? It is not widely appreciated outside of the philosophy of mind how complete has been the failure of physicalism in addressing this question.

Over a quarter of a century ago, John Searle lamented the failure of physicalism to account for consciousness: "I believe there is no other area of contemporary analytic philosophy where so much is said that is so implausible.... In the philosophy of mind, obvious facts about the mental, such as that we all really do have subjective conscious mental states and that they are not eliminable in favor of anything else, are routinely denied by many, perhaps most, of the advanced thinkers in the subject."[4]

The problem is that all physical sciences describe the world in impersonal terms (from a third-person perspective), and nothing in those descriptions implies or predicts the existence of subjective states of consciousness. No theory based purely in the physical sciences can account for why there is something it is like for a person to be in pain. And the brain is a complex aggregate of physical parts, arranged in purely external relations, which appears to imply that consciousness must be decomposable into separable parts, such as neurons or collections of neurons. Yet an individual's states of consciousness seem to be inseparably unified by an internal relation to a single subject. Thus, if I hear the bird's roosting in the trees, smell the aroma of a barbeque and see a beautiful sunset, all of these experiences belong to one and the same subject. And it makes no sense to speak of an ownerless experience or of transferring one experience from one person to another: even if two people X and Y have qualitatively identical experiences of pain, it is not possible for X to have Y's pain or *vice versa*.

Likewise, none of the physical sciences imply that any physical state of a system has intentionality: that the state is *about* something else. But a thought can be about something else, e.g., the Swiss Alps. And the physical powers of something are limited by time, space, and causality. Thus, distant physical events cannot instantaneously cause a physical effect here,[5] future events cannot cause a physical effect now, and physical relations between two items require both items to exist in the actual universe. But I can think about the Eiffel Tower right now without the Eiffel Tower physically influencing my brain in any way, I can think about future events (like a vacation), and I can think of actually non-existent entities, like hobbits.

To be sure, some physicalist neuroscientists may claim that states of the brain "represent" items in the environment, but all this means is that the physical states of the brain reliably co-vary with those items, in much the way that the height of the mercury in a thermometer reliably co-varies with the temperature. Such covariance is not sufficient for intentionality, because the states of the brain require an interpretation in order to point to anything beyond themselves. John Searle famously made this point in his critique of "Strong AI," the view that suitably programmed computers would have their own understanding and intelligence.[6] We could easily design a computer system which, using a vast database, maps input Chinese symbols (questions) to output Chinese symbols (answers) without understanding either the questions or the answers. The reason is that these symbols must be interpreted by a being that grasps what these symbols mean (their sense and reference), but this interpretation is not provided by the symbols themselves. Likewise, even if physical states of the brain are understood as symbolic of environmental variables, still there must be some entity capable of interpreting those symbols, and, in our experience, this ability is confined to conscious subjects. It is in part because we do not believe our computers to be conscious subjects that we doubt that their states are intrinsically about anything. When we say those states are about something, this is merely derived intentionality, intentionality that derives from human goals and interpretations: those states are about something for us, because we are supplying the goals and interpretations that the computer seems unable to provide.

Physicalists have produced many ingenious proposals designed to address the problems of consciousness and intentionality, and space prohibits a detailed discussion. But it is fair to say that they have not been successful. So while dualism faces the problem of how mind and body interact, physicalism cannot even account for what the mind is. This might suggest a stalemate, but in fact dualists have already made some progress in addressing the mind-body problem.

3. Breaking the Deadlock

Dualist responses to the mind-body problem are of two main kinds. There are *defensive* responses, designed to blunt the force of the objection but without offering any positive account of mind-body interaction, and there are *constructive* responses that try to show that dualism has the resources to develop such an account. In the next few paragraphs I present five of the most effective defensive responses, while noting that they still allow physicalists room to press two important objections that need to be addressed; the constructive responses will be set forth later in the essay.

The classic mind-body problem assumes that the inability of an implicitly Cartesian variety of dualism to explain how mind and body causally interact is a strike against interactionist substance dualism in general. But there are many problems with this assumption. First, not all causation requires a medium, e.g., electromagnetic radiation does not require the physical medium of ether. Second, dualists can maintain that the mind does occupy a space, without, like a physical body, excluding (other) physical bodies from that space, and, following Augustine and Kant, that the mind can plausibly be located wherever there are sensations. It is then at least less mysterious why the mind principally interacts with just this or that particular body.[7]

Third, they can note that we do not have to know how two things causally interact in order to have good reason to think that they do. For centuries, people observed that when temperatures dropped, water froze. It would be absurd to claim that they had no good reason to believe that a drop in temperature causes water to freeze simply because they did not have modern molecular theory to explain how it happens. The same is true for psychophysical causation, as Richard Swinburne argues. Swinburne points out that for centuries we observed that sticking a pin in someone (a physical cause) produces a sensation of pain (a mental effect), and if we are convinced that conscious experiences cannot be reduced to physical events in the brain, we have good reason to believe that irreducible physical-to-mental causation occurs, even if we have no theory about *how* it is possible. By parity of reasoning, there is no problem of mental-to-physical causation either: "[I]f we are justified in believing that brain events sometimes cause conscious events even though those events are of different kinds, that these events are events of different kinds cannot be a good a priori objection to the claim that conscious events sometimes cause brain events."[8] The same argument holds, if "substances of different kinds" is substituted for "events of different kinds." We don't have to understand *how* mental and physical substances interact to have excellent evidence *that they do*.

Fourth, all worldviews that accept causation at all must eventually recognize a base level of causation that is either epistemically or ontologically unanalyzable and must be accepted (for now or eternally) as a brute fact. This applies just as much to purely physical relations (e.g., those between elementary particles) as it does to psychophysical ones.

Fifth, as David Hume noticed, in cases of apparent causation, all we observe is the constant conjunction of events, which does not demonstrate a conceptual connection between them (one cannot deduce an effect from a cause).[9] So there is a sense in which we *never* see how it is that a cause should produce a given effect: that is just the way it is. Since this is true of purely physical cases, the physicalist cannot advance it as a special problem for psychophysical causation.

In light of the many rejoinders available to the dualist to the classic mind-body problem, William Hasker concludes that the causal intelligibility objection to dualism deserves the "all-time record for overrated objections to major philosophical positions."[10]

Yet physicalists do have two more serious objections. They can pose a potent explanatory dilemma for dualism, which challenges the dualist to provide an account of mental causation that is not simply mysterious. And they can plausibly argue that mental causes, such as intentions to act, are at the wrong level of abstraction to explain the bodily movements that constitute an action on any particular occasion.

Physicalist Objection 1: An Explanatory Dilemma for Dualism

Physicalists charge that the very idea of an explanation of action implies that we must produce a causal mechanism; they think that if the mechanism is credible, it must be some system of parts governed by event causation. Such a mechanism is clearly at home in a physicalist ontology, so the only plausible explanation of action is a physicalist one.

Of course, the dualist feels justified in rejecting this demand for a mechanism because it appears to be question-begging: for the dualist, the mind just is not a mechanism. But from this, many dualists have drawn the (in my view) erroneous conclusion that the best way to reject the physicalist bait is to say that an agent's power to produce at least basic acts is a primitive and unanalyzable one. This is the position of classical agent causation theory (CACT), according to which, when an agent *a* does a basic act like raising his arm, there is "*nothing at all* by doing which *a* causes his arm to rise."[11] That is, there is no simpler act (that is, no simpler *intentional* movement or activity) than the arm-raising itself by means of which *a* raises his arm. The defender of CACT is seeking to avoid the physicalist view that all causation reduces to passive event causation; his goal is to defend the idea of a *sui generis* form of agent causation, where the agent initiates a new causal chain via his own power, and not because some prior event caused him to do so. This also allows the proponent of CACT to defend libertarian free will, according to which agents have primary responsibility for their actions. CACT is also defended by pointing out that even in physics, the causal powers of elementary particles may be brute facts and so it cannot be objectionable in principle to appeal to primitive causal powers.

However, at this point, the physicalist will claim that appeal to agent causal power is inadequate because it does not explain why this particular action was done, or why it was done at this particular time. Yet one can concede that CACT is inadequate while still maintaining that our actions result from the powers of mental substances. As E. J. Lowe argues, it is not plausible that agents simply produce an action A without first generating a volition to do A.[12] The issuing at a time of a volition to do A (at that or some later time) explains why A rather than some other action was done, and why A was done then (or at that later time) and not at some other time.

But, in my view, volitions do not by themselves go far enough because they do not explain the connection between the intention to do an action of type A and the various sequences of movements which, on particular occasions, constitute A. This leads to a second objection to dualism, the abstraction problem.

Physicalist Objection 2: The Abstraction Problem

On a thousand occasions that I will to raise my arm, there are a thousand physically different sequences of movements. The cause (a volition to raise my arm) is much less specific than any of its alleged effects. The physicalist will charge that there is therefore still a problem of causal specificity because, on any particular occasion, the alleged cause (I will to raise my arm) is too abstract to account for the detailed movements which (on that occasion) constitute my raising my arm. He will go on to point out that the preceding events in the brain and nervous system are of just the right level of specificity to account for these movements. It may then be argued that not only are specific physical causes better suited to explain specific physical effects, but also that they make alleged mental causes redundant.

A fairly standard dualist response (and one that I have defended) is to say that while neurophysiological causes explain specific movements, the volition is needed to account for why not only these, but also a vast number of alternative sequences of movements, constitute the same type of action.[13] While I believe this is correct, I do not think it answers the fundamental physicalist challenge: How does an abstract mental cause (a volition to raise one's arm) make *any* causal contribution to the particular arm-raising? Indeed, Lowe concedes, "I would not pretend to have explained *how* a person can, by willing, effect changes in his or her body—and maybe this is something that is destined forever to remain a mystery for us."[14] I find this troubling, because the physicalist can argue that if, in every case, all the causal work in producing arm-raisings is done by preceding neural events, the volition is excluded from any causal role. Then the fact that appealing to our volitions helps us to group movements into types of action will be dismissed as a convenient fiction, eliminable in principle if not in practice by the neurophysiological account.

What this suggests is that appeal to levels of explanation will not suffice to disarm the physicalist unless we can say more about *how* it is that volitions produce movements. And there is a parallel concern about physical-to-mental causation. I may stub my toe in many different ways and the precise pattern of neurophysiological signals generated may be different in each case, but it is possible that these different patterns produce qualitatively indistinguishable feelings of pain. How is it that so many physically different causes can produce the *same* type of mental effect? The dualist must explain why an abstract mental event (a volition or a pain) is correlated with concrete physical events (a sequence of movements, a set of nerve signals) yet without taking the physicalist bait requiring a physical mechanism. We need an illuminating but non-physicalist account of how abstract mental volitions produce concrete physical effects, and how concrete nerve signals produce abstract feelings of pain.

4. Dimensions of the Mind-Body Problem

Our discussion thus far already suggests that the mind-body problem has many components. Here I will distinguish four: (A) the correspondence problem; (B) the abstraction problem; (C) the pairing problem; and (D) the causal "oomph" problem. An adequate theory of mind-body interaction needs to address all four of these sub-problems.

4.1 The Correspondence Problem

In cases of psychophysical causation, why does one *type* of effect correspond to another *type* of cause? Why does the intention to raise my arm (that type of mental state) cause an arm-raising (that type of action) and not a leg-kicking or a head-turning (other types of action)? Why does stubbing my toe (that type of physical cause) cause a pain "in my toe" (that type of mental state) and not a feeling of euphoria or a dream about studying in Budapest (other types of mental state)?

4.2 The Abstraction Problem

Given that the content of mental states is highly abstract (e.g., I intend to raise my arm), how can this be causally relevant to the highly specific, concrete series of movements constituting a particular arm-raising? How can the intention to do an action of type A have any relevance to producing the complex sequence of movements which constitutes an action of type A on an occasion, especially when the particular items in this sequence (individual muscle movements) are typically not in our voluntary control (we typically could not voluntarily produce them individually)? Going the other way, given that the toe-stubbing is a highly specific bundle of neurophysiological events, how does it produce abstract pain qualia that might have been just the same for many different sets of neurophysiological events (for example, I stub my toe on the wall, not the doorstep; or my toe has been amputated but I have phantom toe-stubbing pains, etc.)? As one intention to do A can be translated into an indefinite number of sequences of movements that fulfill that intention, so many neurophysiological causes can be translated into the same (qualitatively indistinguishable) type of pain experience.

4.3 The Pairing Problem

When I intend to raise my arm, why does *my* arm go up, rather than Joe's across the street? For that matter, why does any part of my body move rather than some other physical object like a chair, a car, or Jupiter? How is my mind integrated with my body in such a way that my intention to move directly moves only my body? The dualist must answer this question, because in the dialectical context, the physicalist has an easy time of it: since, on his view (crudely put), the mind is the brain, the physical contiguity of that brain with a particular body and its lack of contiguity with other bodies makes it easy to see how the brain moves just this body. And on the physicalist view, the local character of pain (its association with a particular body) can be explained in the same way.

4.4 The Causal "Oomph" Problem

Why does *anything* happen? Why does a mere intention have any causal "oomph" at all? Why does a physical event in my toe have the ability to cause any mental event (such as a pain quale with a specific intentional content)?

Reflection on the nature and transmission of information suggests it has the potential to address (at least in part) all four aspects of the mind-body problem. I will develop this intuition and then propose two models to account for mind-body interaction.

5. Information as Intermediary

The nature of information makes it a promising intermediary between mind and body, because information can exist in both abstract (mental) and concrete (physical) forms. For example, the information necessary to find the greatest common divisor of two integers can be held in an abstract form, as a thought in someone's mind, or in a concrete physical form, as executable machine code in the RAM of a computer. There is a clear correspondence between the mental idea of the algorithm and the physical machine code, because the rules governing computer software and hardware guarantee that the information in the algorithm is preserved in its physical implementation. The mental algorithm is more abstract than machine code: the same algorithm may be compiled differently on many different machines, producing physically different machine code in each case. Yet there is a reliable process that takes us from the abstract mental idea to concrete physical results.

One of the features of information that makes its transmission so effective is what engineers and computer scientists call "control abstraction." Control abstraction separates the "what" from the "how," so that we gain the ability to use sophisticated technology without having to know how it works. For example, one may press a single button to engage cruise control on a car, knowing *what* it will do, but without knowing *how*, from an engineering perspective, it works. This phenomenon of control abstraction is not confined to the use of technology, but is ubiquitous in our experience, and suggests a way to understand mind-body interaction. Consider some examples.

5.1 Giving and Taking Orders

Suppose that Tim is working late, and his boss, Sarah, says, "Tim, go home!" Through the medium of her voice, Sarah's command is transmitted to Tim's mind. A mental command has been translated into physical sounds, and in the presence of learned conventions for interpreting those sounds (mastery of spoken English), those sounds are converted back into a command Tim understands. Now suppose that Tim decides to go home, and does so. It seems he has taken Sarah's command and turned it into a self-command (a volition), and just as Sarah's command was translated into physical sounds bearing certain information, so Tim's self-command is translated into a physical instruction to implement a motor-control program that moves his body.

Going home is a complex action: there are many more basic actions by means of which one goes home, including leaving the office, getting in the car, driving the car, etc. But take a simpler case. A police officer might order me "Hands up!" and I might immediately raise my hands without intending to do any more basic actions as a means of doing so. It seems there is a direct, automatic translation from the police officer's command to my self-command, to my bodily movements. Interestingly, this self-command is highly abstract: there is an indefinite number of physically different muscle contractions and nerve signals that could implement this same command, and I do not need to consciously intend any of these specific details. Rather, it seems that the volition is automatically converted into a motor control program that handles the details in the "background," details which vary enormously from case to case due to subtle differences in the initial state of the brain and body, but which are nonetheless shaped and targeted to compose the same type of action (hand-raising) on each occasion.

Once we conceive of volitions as commands, we become aware that the phenomenon of an abstract command producing a highly specific result is commonplace. There are numerous examples of hierarchical command structures that work in just this way. Consider a military analogy. The commander-in-chief says, "Occupy this city!" The generals take this command and devise a more detailed strategy, including air power and amphibious assault, which is then passed on to officers who determine how to organize rank and file soldiers, who in turn figure out how to make their contributions. This works because each lower level of the command hierarchy is capable of *translating* a more abstract command into a less abstract (more concrete) means of implementation until one gets to the lowest level of individual soldiers obeying orders whose actions collectively implement the higher-level commands all the way up to the top.

5.2 Computer Programming

This is analogous to control abstraction in computer programming. One can use a procedure or "method" by knowing an abstract description of what it does but without knowing the details of how it works. Today's computer programs contain abstract method calls, such as "Sort Table," "Search Table," etc. Here, a simple line of code (the method call) is analogous to a command, and when the program is compiled, this command is associated with a particular block of code which implements that command in detail. For example, a programmer can make use of a pre-written sort method using the recursive quicksort algorithm without knowing how the quicksort works, because the name of the method—Sort Table—is associated with the address of the particular lines of code which implement the quicksort.

This is very suggestive, as it seems that people can command themselves to raise their arms with no knowledge of the detailed, neurophysiological sequences required to produce an arm-raising, and one explanation would be that there is a *means of association* between the abstract command and a more detailed motor program which implements it. Just as the compiler translates a method call into an instruction to jump to specific lines of code (a subroutine), it seems plausible that if our volitions have anything to do with the movements we make, it is because the abstract prescriptive information in the volition is translated into a physical instruction to execute a motor program.

5.3 Writing and Reading

That such a translation is possible is plausible because, in our experience, information is a suitable intermediary between the abstract realm (e.g., ideas and commands), and the concrete realm (e.g., physically encoded data and instructions). Like the Roman god Janus, who had two faces and thus could look into the past and the future at the same time, information points us both to abstract forms of information (like concepts) and to concrete vehicles of that information (like speech and text). On account of its Janus-faced, abstract-concrete duality, it is plausible that information serves as an intermediary in paradigm cases of psycho-physical causation. For example, when we write down our ideas, information in the abstract sense (our mental ideas) is translated into concrete forms (physically encoded marks), and in the presence of an established public convention for interpreting them, these marks can be said to bear the information which previously existed only as abstract ideas in our minds. Again, when we read a book, our subjective awareness of those marks, and our largely unconscious integration of the public conventions for interpreting them, allow us to form abstract ideas corresponding to those physical marks. In these cases, it is the existence of the public convention for interpretation which makes translation between abstract mental information and physical information possible.

5.4 Memory

Now consider the example of memory. When we memorize something, most neuroscientists think that abstract conscious experiences and ideas are somehow translated into physical traces stored in the brain (engrams), and when we recall an item, those traces are somehow translated back into conscious memories. By analogy with the cases of writing and reading, I suggest that a plausible hypothesis is that there is a convention for translating between conscious states and physical brain traces. The key difference is that this convention was not developed by human ingenuity but is an innate feature of the mind-brain interface. The reasons to postulate it are: (1) memorization and recall are closely analogous to writing and reading; and (2) normally at least, the former processes are too reliable to occur without some innate, systematic means of correlation. It is simply not the case that people normally learn to memorize or recall anything by learning how to make or interpret physical traces in their brains—even if we can imagine a neuroscientist (with appropriate scanning technology) doing so!

At the same time, an information-theoretic account does not require the idea that memories are translated into physical traces in the brain. An alternative view is that the brain stores ways of accessing memories, while the memories themselves are stored in a non-material mind. Rather than conscious memories being translated into physical brain traces, perhaps these memories are translated into keys or other mechanisms that allow access to those memories. One reason for preferring this view is simply that memories have intrinsic intentionality (they are about something), but physical traces (such as engrams) do not have intrinsic intentionality, and so the translation of memories into something purely physical would not preserve what makes them memories. Just as a written text requires a subject that can interpret the text to recover its meaning, an engram would not become a memory until that physical trace is interpreted by a subject. So, while one could speak of storing memory *vehicles* in the brain, it would be a confusion to say that the memory itself (a meaningful, intentional state) was stored there, whereas a memory could be stored intact in a non-material mind. While most neuroscientists seem to favor the view that memory involves storing traces in the brain, it is important to note that our information-theoretic account allows this alternative to be a live option.[15]

5.5 Automatic Information Translation Hypothesis

These examples suggest an automatic information translation hypothesis, according to which human beings have the innate capacity to translate between abstract mental information and concrete physical information, and this hypothesis explains how an abstract volition results in a specific bodily movement (like an arm-raising) and how a specific pattern of nerve signals results in a specific mental sensation (like pain).

The automatic information translation hypothesis entails two models, the Command Model of Action (CMA), and the Signal Model of Sensation (SMS). The CMA maintains that volitions are abstract commands translated into physical instructions to execute bodily motor programs, and the SMS maintains that nerve signals are translated back into particular sensations that give humans the information they need to guide future actions.

How do the CMA and SMS address the four elements of the mind-body problem?

(A) The Correspondence Problem

The CMA is most plausible in the case of basic acts, acts one does directly, not by means of doing any simpler acts. To be sure, there is some relativity in the notion of a "basic act." Thus, for an expert, ballroom dancing may consist of basic acts that, for the novice, are a frustrating medley of simpler acts. However, it is plausible to think there are at least some *primitive* basic acts for an agent: these are acts that the agent has never done by means of trying to do anything simpler. Even if all of our movements are initially a bit shaky, it seems likely that there is some built-in connection between trying to move and moving. If a volition to do a primitive basic act is a command and this command automatically results in some specific sequence of movements, it seems that the most reasonable explanation is that there is an innate protocol or convention that maps volitional commands to motor programs. For if some such ability is not innate, it would be very difficult to explain how an agent could go on to acquire a complex repertoire of non-basic acts. If one does not begin with the ability to do some primitive basic acts, it is hard to see how one could *learn* to do anything: by definition, one has not learned to do primitive basic acts by doing anything else, and one cannot do more complex acts except by doing basic ones.

A second reason for thinking that primitive basic acts exist is simply that for most ordinary people, consciousness does not disclose the neurophysiological details of how the movements constituting an action are produced, and consequently, those details are not even candidates for voluntary control. As a result, one cannot intend to do something by trying to produce those details. It follows that actions as they are presented to consciousness as objects of volition will reach a point where they cannot be further decomposed because no subset of the more detailed neurophysiological events constituting the action is ever presented to consciousness as a candidate for voluntary control. Admittedly, this might be different for the expert in neurophysiology, since studying the detailed events would disclose them to the expert's consciousness, and the expert might then deliberately try to produce those events in order to make those events register on a brain scan. (Whether such events *can* be consciously controlled is another, and obviously empirical, question.) Yet it would remain a fact that for each individual, some actions have never in fact been produced by trying to produce some more basic action. That suffices to show that even if the details of the list vary from one individual to another, each individual has a list of actions that are primitive basic for him.[16]

So, suppose for the sake of argument that clutching is not only a reflex action but also, at some point, a primitive basic act under voluntary control. It seems that the volition to clutch must be automatically translated into a physical instruction to execute the appropriate motor program. No doubt one can optimize even basic acts by improved muscle coordination, and the motor program can have a feedback mechanism so that clutching can improve in various ways. But it seems some basic clutching, however shaky and incompetent, must result automatically on the very first occasion that the volition to clutch is formed. If this is right, then there is an innate "interpreter" system, analogous to a computer's compiler, which translates volitions into physical instructions to activate the associated motor program.

In fact, this interpreter must work in both directions. Thus, when a person stubs his toe, the SMS claims that physical nerve signals are automatically translated into an abstract mental quale, an experience of pain. So we have a two-way translation between mental states and physical instructions or signals. This is comparable to physical devices that can translate between continuous and discrete forms of information. For example, a symphony produces continuous sound waves that are recorded in a discrete (digital) form. When the resulting CD (or mp3 file) is played, discrete information is converted back into continuous sound waves.

If this is the case, then it suggests that human beings are so designed that our bodies are obedient to mental commands and our minds are receptive to the meaning of at least some physical signals sent to the brain from the body. Again, computer science provides an apt analogy. It is perfectly possible to design a computer system such that when the software issues the command "Sort Table," we delete the table's contents instead: all we have to do is associate the abstract command with the wrong subroutine. Likewise, even if the correct subroutine is invoked, the table will only be sorted if the commands are in fact executed by the hardware, which won't happen if the compiler translates them into instructions which the hardware cannot execute (either because they cannot be done, like division by zero, or because the hardware is defective). So in order for abstract commands to work, they must be associated with the right subroutine, and this must be translated into physical commands, which the hardware can execute. This generally occurs because computer scientists design the programming language, compiler, and hardware precisely *so that* these commands work. By analogy, it seems that we are so designed that, at least at the minimal level of primitive basic acts, our mental intentions are automatically translated into physical instructions to run motor programs, which our bodies can (normally) execute, and certain events in our bodies are (normally) translated into appropriate sensations of pleasure or pain.

To be sure, we can intend to do many things that our body won't or can't do (such as flying like Superman), and it is obvious that many of our more sophisticated movements (judo throws, dancing, etc.) require a lengthy learning history. And neurophysiological damage may mean we sometimes do not feel pain or feel an inappropriate pain (e.g., phantom limb cases), and we may also need training in some of our more sophisticated responses, like those of the expert wine-taster. So the mind-body connection is neither absolute nor magical. Yet it is still a remarkable fact that our brain can be trained to execute our mental commands in rather like the way that a sheepdog can be trained to obey the commands of a shepherd and the mind of an expert wine-taster can be trained to discern subtle differences unknown to the basic white or red bloke.

(B) The Abstraction Problem

The CMA and SMS also address the problem of causal specificity. To use a computer analogy, the reason that an abstract method call (e.g., Sort Table) can have specific physical effects (the data in the table is physically rearranged in sorted order) is that there are more specific instructions inside the method itself. So when the method call is translated into the physical instruction to jump to a particular memory location, the specific physical instructions at that location are executed. Likewise, if an abstract command to raise my arm is translated into a physical instruction to execute a particular motor program, inside that motor program are the detailed physical instructions that produce the particular sequence of movements. This is why one can move one's arm in highly specific ways without having to consciously intend each and every muscle contraction and nerve signal involved. This illustrates control abstraction. One can make the command to perform a complex operation very simple by designing a method call (or a button) so that when it is invoked (or pushed) a whole cascade of complex operations happens automatically. The system is so designed that an abstract command is translated into the myriad of complex physical events required to implement that command.

Going the other way, the SMS suggests an obvious reason why an experience of pain is more abstract than the particular bodily event that causes it (many different bodily events could produce a qualitatively indistinguishable experience of pain, including phantom limb cases where the qualia point to a non-existent limb as the source of the pain). This is because, in the transmission of a signal, information can be lost, either accidentally, as in a bad cell phone connection, or by design, because irrelevant information is filtered out. For example, consider the iconic AOL message "You've got mail." This is triggered by any new mail received by your inbox, and yet it does not discriminate between an indefinite number of physically different emails. On the assumption that we are so designed that we can make quick decisions based on salient information, rather than being swamped by a detailed specification of everything that happens to our body, it would make sense if we had the capacity to quickly discern that our body is damaged in some way without having to know the complete neurophysiological story.

That is, our body-brain interface would be so constructed that our default is to be informed of bodily events on a "need to know" basis. Just as the general does not need to know if each particular soldier changed his or her underclothes before engaging in combat, we do not need to know which nerve receptors were active when we stubbed our toe. Like "You've got mail," the painful experience of stubbing a toe on the doorstep tells us all we need to know to guide subsequent actions.

Notice that since we do not need to know how our actions are executed to command them, the experience of a pain in the toe provides information at the *right level of abstraction* to issue such commands as "Take off your shoe!" "Ice the toe!" and "Next time, raise your foot when crossing the doorstep!" If our pain sensation did tell us about the detailed neurological events, it might be quite unclear to us what self-command would be relevant to avoiding a similar pain in the future, especially since, in all their specificity, those events would be very unlikely to recur, so it would be useless knowledge if we had it. What matters is that we might stub our toe on the doorstep again (in any of an indefinite number of physically different ways), and that since this is unpleasant, we don't want to do that *type* of thing again.

(C) The Pairing Problem

As previously noted, substance dualists must explain why particular mental events are paired with particular bodily events. In particular, how is it that a particular mind is integrated with a particular body, so that only they (and no other minds and bodies) interact directly? What is clear is that the integration is contingent. For example, in the case of paralysis, the mind issues a command to raise an arm but nothing happens; and, after a stroke, a slight movement may occur but not a normal arm-raising (these are analogous to computer hardware failures). In phantom limb cases, signals from further up the nervous system are indistinguishable to the brain from those originating in the lost limb and may produce a qualitatively indistinguishable pain. And neural pathways can even be scrambled so that an event in one part of the body is interpreted as an event in another part. However, one mind is normally integrated with one body, and since this is a contingent fact, it requires some explanation.

A partial explanation for this integration open to the dualist is to claim that the mind does have spatial extension and is (roughly) spatially coincident with the body. That is why the mind is in direct interaction only with that body. Still, one may ask, why is the mind restricted in this way?

At its base, the CMA assumes that we are so designed that abstract mental commands are automatically translated into detailed physical instructions to move the body, and the SMS assumes that physical signals from the body are automatically translated into abstract sensations. For this system to work, the body must be responsive to just one set of commands. Imagine what it would be like if four or more people were independently in control of producing a single body's movements and had interests in soccer, reading, driving a car, and ballroom dancing. Whether or not hospitalization ensued, it seems unlikely that these agents would be successful in their individual projects. Similarly, if events in one body (say a toe-stubbing) produced the same conscious experience of pain in four different minds, then presumably three of them would have misleading information about their bodies, since these three minds would be embodied by bodies other than the one to which the toe-stubbing occurred. So it seems that it is in terms of the *functioning and efficiency of the mind-body whole* that one must seek an explanation of their integration, and this assumes that the mind-body whole is a *designed system* (a "union," as Descartes said). In other words, given the flow of information between mind and body, they must be designed to function as a unified, integrated system.

It is helpful here to consider the traditional Christian teaching about persons, known as psychosomatic holism: there is one unified person consisting of both mind and body. From within such a theistic framework, one solution to the pairing problem is that God designs human beings so that they normally function as integrated mind-body wholes. A reason for this is that God governs the world hierarchically through people and He assigns souls to particular spheres of influence. This is analogous to the way that the King might make a particular man, Jeffrey say, the Earl of Surrey. Nothing intrinsic to Jeffrey makes him an Earl, just as there is no intrinsic connection between a particular soul and a particular body: it is an entirely contingent assignment. Just as the King might have made some other man Earl, God might have assigned a soul to a different body; just as the King may promote or demote Jeffrey, God might increase or decrease a soul's sphere of influence; just as the King might remove Jeffrey from office, so God may separate the soul from the body on physical death. As Earl, Jeffrey has the sphere of influence he has (Surrey) simply because that is the sphere the King has assigned to him, much as God assigns different gifts and different vocations to different people. So, just as what Jeffrey decides directly affects only Surrey (and only indirectly affects Sussex or Hampshire), so a soul directly affects only one body (and can affect other bodies only through that body).

(D) The Causal "Oomph" Problem

But, says the physicalist, the real question is more fundamental: Given substance dualism, how can mind and matter make any causal difference to each other at all? However, our previous answers to the correspondence and abstraction problems surely help: the *transmission of information* must be a key part of the dualist response. Information has the remarkable Janus-faced quality required to help account for mind-body causal interaction because it can exist in both subjective (personal) and non-subjective (impersonal) forms. Thus, we speak of "getting information out of books" and "writing down some information we have learned," implicitly accepting that there is a correlation between physical encodings and subjective ideas made possible by the intermediary of an interpretive key or convention. To be sure, subjectivity and intentionality are not intrinsic to physical encodings, but they arguably do not need to be, because the interpretive key functions as a surrogate for subjectivity and intentionality. This, after all, is why a computer can compile a program (i.e., translate it into machine code) and execute it without the computer being conscious of the program and without the program meaning anything to the computer. So it is plausible that what happens when I will to raise my arm is that the content of this volition is automatically translated into a physical instruction which, in its neurophysiological context, is physically interpreted (analogous to the way compiled computer programs are physically interpreted) as the invocation of a motor control program, which produces movements that, as a matter of fact, implement the volition.

Going the other way, the SMS claims that when I stub my toe, this signal is physically interpreted in such a way that the resulting brain state gives rise to an appropriate subjective experience of pain. In this reverse process, we gain a mental state that does not need an interpretive key, because the interpretation is provided directly by the conscious subject: the content of a quale is just what it appears to a subject to be. This experience is not metaphysically necessary, but contingent on the way God has designed us as body-soul unions.[17]

So, if I am right, psychophysical causation typically involves translation between a subjectively and an objectively interpreted instruction, between an instruction that has *intentional meaning* and means something to the subject that interprets it, and an instruction that has what might be called *non-intentional meaning*: in its physical context, this is what the physical instruction "means" in the non-subjective and non-intentional sense that this is what the instruction *does*. Besides the analogy of the physical "meaning" of a compiled computer program, another analogy is the "meaning" of DNA instructions; neither the DNA, nor the cell, is conscious, but the cellular machinery provides a non-intentional meaning to those instructions: it physically executes those instructions by assembling proteins and protein machines.

For all of this to work, there must be an ability to associate mental ideas with corresponding physical encodings (spoken or written words, memory traces, etc.), and since this is done in such a regular and reliable fashion, and typically without conscious attention to the means of conversion between these mediums, an encoded, interpretive key is the most likely explanation. So, since some acts, and some conscious responses, appear to be primitive basic, the most reasonable hypothesis is a Tunable Channel Model (TCM).

According to TCM, there is real psychophysical causation through the intermediary of information, but the built-in correspondence between the mental and the physical is limited to primitive basic acts and untutored qualia. However, in addition to this correspondence, our faculties are gifted with the ability to refine basic acts and learn non-basic ones, and to refine our untutored qualia. What we have, I believe, is a psychophysical information channel which has some innate basic defaults, but which is also tunable by experience to become sensitive to more sophisticated signals. No-one is born able to dance the flamenco or pass a Parisian wine-tasting contest. But there is an open, dynamically tunable channel which can be made sensitive to these more refined signals. So we can understand TCM as a presupposition of the CMA and the SMS, which accounts for both an innate mind-body connection, so that psychophysical causation can get started, and the ability to refine that connection through experience (learning).

6. Limitations and Future Developments

As THE reader may surmise, there are several important limitations to the account developed here. First, the CMA and SMS are developed by way of *analogies*, especially an analogy between mind-body interaction and the transmission of information in computer systems. As always, analogies have their limitations. For example, in current computers, all information is stored in a discrete fashion, ultimately as bits (zeroes or ones), but neither the CMA nor the SMS is committed to an exclusively discrete format for information. While some models of cognition use computational models, it is entirely possible (and surely an empirical question, not one for armchair speculation) that some important information in the brain is stored in a continuous or analog (not discrete or digital) format. The models presented here are agnostic about whether the best models of information in either abstract mental or concrete physical forms are discrete or continuous.[18]

The important point is that there seems to be a clear distinction between information in the abstract form (commands, qualia) and information in the highly concrete and specific form (motor programs, nerve signals), and it seems that there is a generally reliable means of translating between the two forms. Further, it may be noted that I cast the net of analogies wider to show the ubiquity of this phenomenon, even outside of computational settings. It applies, for example, to the issuing of orders translated by a hierarchy of officers all the way down to rank-and-file soldiers. The salient feature of the computer models, also present in non-computer cases, is control abstraction, the fact that, via careful organization, whether of a computer system or of a hierarchy of soldiers, a very simple command can be translated into very complex results, and very complex signals can be translated back into a simple resulting message. So it is important not to put too much weight on the computational analogy, but rather to see it as illustrative of the more general phenomenon of control abstraction.

But there are two other, even more important limitations to the models outlined here, one philosophical, the other scientific. Philosophically, it cannot be claimed that the CMA and SMS "solve" the mind-body problem, because there is more to mind-body interaction than transformation of information. I do think the idea of automatic translation between abstract and concrete forms of information helps to make a substance dualist response to the mind-body problem more plausible, because it helps to address the otherwise strange discrepancy between abstract mental commands and qualia on the one hand and the details of motor control programs and nerve signals on the other hand. Yet it is a fair criticism that in illuminating this mystery, it leaves another, which may seem just as challenging as the original mind-body problem.

Even if is true that human beings are gifted with a tunable, automatic means of translating between abstract mental forms of information and concrete physical forms of it, still this does not tell us *what it is that effects the transition* from mental to physical or physical to mental information. This is the aspect of the mind-body problem—perhaps the hardest—which my models do not address. For example, when we "write down our ideas," a mental idea is somehow translated into physical marks on paper, and, according to many neuroscientists, when we consciously study some item in our mind, physical engrams are produced in our brain. But where is the boundary between mental and physical information, and what is it that accounts for crossing that boundary in either direction? My models do not answer these questions, but rest simply on the apparent facts that this boundary must exist and is routinely crossed. While the models do hypothesize an automatic translation system, they do not articulate *how* the boundary from mental to physical is traversed.

In partial response, I would reiterate that if the demand is for a physical mechanism that effects the translation, then it tacitly reintroduces a physicalist model of causation and is therefore question-begging against the dualist. At some point, the dualist is within his rights to appeal to Reidian basic powers of individuals that cannot be further analyzed.[19] But there are also other possibilities. Perhaps information should be thought of as an entity that is intrinsically neither mental nor physical, but a third kind of thing, one ontologically neutral between mental and physical entities but capable of manifestation in either form.[20] Note here there is an analogy with numbers, which apply effortlessly to both modes of existence (e.g., we have two hands, and we may also have two ideas), but themselves seem to be neither purely mental, since we seem to discover their properties, nor purely physical, since we cannot locate them in space and time. The nature of information itself might then help us to understand why it is apt to flow freely between minds and bodies, leaving us with the more tractable problem of explaining why it is more abstract in the former than in the latter mode. Again, as several contributors to this volume have suggested, maybe some form of idealism provides a framework which dissolves the mind-body problem by getting rid of mind-independent matter. This approach would still be dualist in the sense that it distinguishes spirits (minds) and ideas, but ideas would not even need to be translated into a material mode (there is none), but only translated between abstract and concrete forms. Regardless of which of these paths is taken, a promising line of inquiry is to make the metaphysics of information fundamental, and to use it to illuminate (and perhaps even dissolve) the mind-body problem. (For a leading example of this approach, see William Dembski's chapter in this volume.[21]) So, philosophically, what I am contributing here is (I hope) a piece of the puzzle that philosophers holding a variety of positions may find congenial to their approaches to the mind-body problem.

Scientifically, it may also be observed that the CMA and SMS are not sufficiently developed to deserve the title "theory," in the sense of a well-articulated account with clear testable implications. To some extent, I do not apologize for this. I am a philosopher, not a working neuroscientist, and it would be both presumptuous and amateurish on my part to develop the connections to practical neuroscience in any detail. That said, I do find some congenial and helpful pointers to possible neuroscientific applications in the pioneering work of Wilder Penfield. While somewhat dated, Penfield's approach is marked by an unusual openness to fundamental philosophical questions, and many of his basic insights have withstood the passage of time quite well (though experts in neuroscience are free to challenge this assertion). I will give just three examples of ways in which Penfield's ideas might help to build a bridge between my philosophical speculations and testable neuroscience.

First, in his work on "awake brain patients" (conscious patients undergoing brain surgery with a local anesthetic), Penfield pioneered a technique of using electrodes to induce a variety of responses. Yet, to his surprise, he discovered an important distinction between responses he could elicit automatically and ones he could not. This suggests some empirical approaches to discerning when the mind serves as an origin of new information, and this may help us better to understand the transition from mental to physical forms of information. Penfield found that, by using electrodes, he could induce movements, sensations, emotional interpretations, and memories. Yet the patient clearly knew that these responses did not originate within his own agency: "Invariably his response was: 'I didn't do that. You did.'.... 'I didn't make that sound. You pulled it out of me.'"[22] This distinction between passive responses to external agency and responses originating within a person's own mind was further supported, albeit negatively, by Penfield's failure to induce decisions and beliefs: "There is no place in the cerebral cortex where electrical stimulation will cause a patient to believe or to decide."[23] This observation provides a clearly falsifiable basis for exploring the differences between information that is actively originated by the mind, and information that is passively stored or executed by neurophysiological mechanisms.

Second, in work specifically devoted to a particular class of epileptic patients, Penfield discovered some intriguing differences between the powers of the conscious mind as distinguished from our automatic responses. In what are known as "epileptic automatisms," affected individuals may lose consciousness in the middle of an action that was initiated voluntarily. As many routine behaviors are automated through learning, the individual may be able to complete the originally intended behavior unconsciously. Observed examples include completing a piano piece, walking home, and even driving a car![24] For those physicalists that hold that the mind essentially is nothing but an organic computer, one might predict that there would be no observable difference in the behaviors themselves.[25] However, this is not the case. What Penfield found was that while affected individuals could complete stereotyped automatic sequences, without consciousness, they were not capable of innovation or adaption to unpredictable environmental cues. For example, though the driver could continue to drive, "he might discover later that he had driven through one or more red lights."[26]

More generally, in the process of learning anything new (whether facts or behaviors), it is well-known that initially a great deal of conscious attention is required until the fact is safely stored or the behavior largely automated: "every learned-reaction that becomes automatic was first carried out within the light of conscious attention and in accordance with the understanding of the mind."[27] So consciousness seems to be critical in generating new information to be stored in the brain, whether in acquiring a memory or behavioral routine or in adapting it in new ways (innovations, refinements, adaptation to environmental cues). As Jeffrey Schwartz has shown, conscious selective attention is also critical in breaking out of a harmful automatic routine, such as obsessive-compulsive disorder, by refocusing the mind on an alternative behavior, until there are detectable changes in neural pathways.[28]

Related to this active power of the mind to change the brain, Penfield's third observation is that conscious attention is vital to memory: "The imprint of memory's engram is somehow added during neuronal action. Conscious attention seems to give to that passage of neuronal impulses permanent facilitation for subsequent passage of potentials along the neuronal connections in the same pattern. Thus, a recall engram is established."[29]

There are many things which we react to or experience that we do not remember. Conscious attention plays a vital role in determining which stimuli actually produce memory traces in the brain, and this provides further evidence of its important role in explaining our subsequent behavior.

Taken together, these and other observations led Penfield to propose that the mind is to the brain much as a programmer is to a computer.[30] The computer is an amazingly versatile system that can be reprogrammed in indefinitely many different ways to solve problems automatically. But the creativity of the programmer does not seem to be automatic. Computers and computer programs did not assemble themselves via the automatic processes going on within nature: they are not, like repeating patterns of crystals, the result of natural laws. Rather, they are singularities, entities brought into existence for the first time by the creativity of their inventors and designers. Likewise, while there are many pre-programmed reflex behaviors, one of the extraordinary features of the human brain is how much of it is initially (at birth) uncommitted, yet capable of being programmed to develop novel memories, abilities, and behaviors. Like the vast reaches of RAM and auxiliary storage in a computer, it seems much of the brain is "designed" to be programmed, based on an individual's unique developmental history. Conscious attention seems to play a vital role in developing these new memories, abilities, and behaviors, which suggests it is a key source of this novel programming.

So the best way to develop the scientific implications of the CMA, SMS, and TCM would be to focus closely on how it is that conscious attention programs (or reprograms) the brain. Promising areas to research would include the neuropsychological development of children, and cognitive therapies for a range of phobias and cognitive disorders.[31] It might be that researchers in these (or other) areas will uncover a purely physical mechanism that accounts for the translation from conscious attention to new memories, abilities, and behaviors (although, given the hard problem of consciousness, it is difficult to conceive of how this could happen). That would tend to vindicate the physicalist view of human beings, and might suggest that consciousness itself is ultimately physical, philosophers' objections notwithstanding. But it might also be that Penfield's conjecture is vindicated: after repeated and exhaustive search for such a physical mechanism in all of the places it is reasonable to expect to find it, neuroscientists may conclude that most likely it does not exist.[32]

That would be analogous to the discovery in physics that electromagnetic radiation does not require the physical medium of ether to propagate. If no physical mechanism bridges consciousness and the brain, then given the fact that novel memories, abilities, and behaviors appear to require conscious attention, this provides evidence that information can exist in an immaterial form, yet also have real effects on the brain. We may then conclude that information, like the ether, does not require a physical medium to transition between mental and physical forms; rather, this transition reflects inherent powers of the mind, or of information, or of both. William Dembski's view that information is ontologically fundamental in structuring reality might also give us a better way to understand this result (see his chapter in this volume). It may even be that we are led to a view where the transmission of information itself is viewed as the basic structure from which both mind and body can be reconstructed.

7. Conclusion

Although substance dualists should not yield to the question-begging demand for a purely physical mechanism for psychophysical causation, it is not unfair for physicalists to press dualists for a more illuminating account of that relation. In my view, what is required to move forward in our understanding of the mind-body problem is the postulation of a basic means of translation between information in its subjective and intentional form and its non-subjective and non-intentional form. In this way, the ability of children to produce primitive basic acts and to have appropriate, yet untutored reactions of pleasure and pain can be explained, and if we further suppose that the interpreter or translator is a tunable channel, we can also account for the ability to learn more sophisticated actions and responses. The CMA, SMS, and TCM can help substance dualists develop an illuminating explanation of psychophysical interactions that addresses the worthiest of the physicalist challenges, and it encourages non-physicalist scientists to apply cutting-edge information theory to psychology and neuroscience.

Notes

1. Most discussion of Descartes's views of the mind-body problem are confined to what he says in the Sixth Meditation of his *Meditations on First Philosophy*. For a nice edition, see *Meditations on First Philosophy / Meditationes de prima philosophia: A Bilingual Edition*, edited and translated by George Heffernan (Notre Dame, IN: University of Notre Dame Press, 1990). In fact, Descartes had much more to say on the matter. See, for example, his "The Passions of the Soul," in *The Philosophical Works of Descartes*, vol. I, trans. E. S. Haldane and G. R. T. Ross (Cambridge: Cambridge University Press, 1967).
2. See also Stewart Goetz and Charles Taliaferro, *A Brief History of the Soul* (Malden, MA: Wiley-Blackwell, 2011), and Jonathan Loose, Angus Menuge, and J. P. Moreland, eds., *The Blackwell Companion to Substance Dualism* (Oxford: Wiley-Blackwell, 2018).
3. See Jaegwon Kim, *Philosophy of Mind*, 3rd ed. (Boulder, CO: Westview Press, 2011), 50–56. For an excellent discussion and response to the pairing problem, see Stewart Goetz and Charles Taliaferro's *Naturalism* (Grand Rapids, MI: Eerdmans, 2008), 57–64.
4. John Searle, *The Rediscovery of the Mind* (Cambridge, MA: MIT Press, 1992), 3.
5. At least, this is the normal case. There are possible exceptions in quantum physics, where particles appear to affect each other simultaneously. But whether these correlations are causal is controversial, and even if they are, it is not clear they are relevant to our understanding of more standard cases of physical causation.
6. John Searle, "Minds, Brains and Programs," *Behavioral and Brain Sciences* 3 (1980): 417–457.
7. See Goetz and Taliaferro, *Naturalism*, 64–70, and also their *A Brief History of the Soul* (Malden, MA: Wiley-Blackwell, 2011).
8. Richard Swinburne, *Mind, Brain, and Free Will* (Oxford: Oxford University Press, 2013), 105.
9. David Hume, "Sceptical Doubts about the Operations of the Understanding," section 4 of his *An Enquiry Concerning Human Understanding* (Cambridge: Cambridge University Press, 2007).
10. William Hasker, *The Emergent Self* (Ithaca, NY: Cornell University Press, 1999), 150.
11. E. J. Lowe, *Personal Agency: The Metaphysics of Mind and Action* (Oxford: Oxford University Press, 2008), 151.
12. Lowe, *Personal Agency*, 153–154.
13. See Lowe, *Personal Agency*, 102–107, and Angus Menuge, *Agents Under Fire* (Lanham, MD: Rowman and Littlefield, 2004), ch. 2.
14. Lowe, *Personal Agency*, 178.
15. I am grateful to a reviewer of an earlier draft of this chapter for alerting me to this possibility.

16. A reviewer suggests that rather than thinking in terms of basic acts, one might appeal to Reidian primitive powers. I am sympathetic to this idea, and such powers might explain the existence of primitive basic acts: we are simply designed in such a way that there are some things the mind has the immediate power to do.
17. This is why the physicalist, unlike the dualist, faces the problems of absent and inverted qualia. It is plausible that physicalists must maintain that supervening mental states are metaphysically necessitated by their physical causes. Yet it is surely possible that the same signals that in fact produce pain could cause no qualia or pleasant qualia, and in general, there are conceivable "zombie" worlds where our physical duplicates are unconscious, and "invert" worlds, where our physical duplicates feel pleasure when we feel pain and *vice versa*.
18. One concern about computational models of mind is that their elegance and mathematical tractability may lead one to embrace them as realistic descriptions of reality and not acknowledge their limitations. For example, if many aspects of human cognition are in fact grounded in continuous neurological phenomena, discrete systems may be, at best, approximate and simplified models of reality.
19. The idea of a "Reidian basic power" traces back to Thomas Reid. In his philosophy of common sense, he argued that we find ourselves in possession of certain basic powers, and that, since these powers are presupposed by any form of rational inquiry, it would be absurd and self-refuting to conduct an inquiry aimed at showing that these powers do not exist. Arguably, the entire project of science presupposes that scientists have the basic powers to translate their ideas of experiments into actual experiments and translate the results of those experiments into experiences and logical conclusions. Thus, even if we cannot explain how this translation occurs, it would be self-refuting to use science to undermine the possibility of that translation.
20. There is a well-known worry that this may "solve" the problem of mental-physical interaction only by creating problems of the same kind: How does the neutral interact with either the mental or the physical?
21. For more on this, see also William Dembski's *Being as Communion: A Metaphysics of Information* (Surrey, UK: Ashgate, 2014).
22. Wilder Penfield, *Mystery of the Mind: A Critical Study of Consciousness and the Human Brain* [1975] (Princeton, NJ: Princeton Legacy Library, 2015), 76.
23. Penfield, *Mystery of the Mind*, 77.
24. Penfield, *Mystery of the Mind*, 39.
25. For example, Daniel Wegner thinks that automatisms provide evidence that the conscious mind is epiphenomenal (that it merely previews, but does not cause, our physical behavior). See Daniel Wegner, *The Illusion of Conscious Will* (Cambridge, MA: MIT Press, 2002).
26. Penfield, *Mystery of the Mind*, 39.
27. Penfield, *Mystery of the Mind*, 59.
28. Jeffrey M. Schwartz and Sharon Begley, *The Mind and the Brain: Neuroplasticity and the Power of Mental Force* (San Francisco, CA: Harper, 2002).
29. Penfield, *Mystery of the Mind*, 66.
30. See especially Penfield, *Mystery of the Mind*, 57–61.
31. See for example, Mario Beauregard, "Mind Does Really Matter: Evidence from Neuroimaging Studies of Emotional Self-Regulation, Psychotherapy, and Placebo Effect," *Progress in Neurobiology* 81, no. 4 (March 2007): 218–236.
32. This need not be an objectionable "argument from ignorance." It is objectionable to argue that amoebae do not exist in a room one has searched with the naked eye, because the naked eye is not apt to detect amoebae even if they are present. But it is not objectionable to argue that elephants do not exist in a room after such a search, as they would be detected if present. Likewise, neuroscientists will be able to determine where a physical interface between mind and brain should most likely be found if it exists, and if after exhaustive search of all such locations, it is not found, this is a good (albeit defeasible) reason to conclude that there is no such mechanism.

20. Consciousness and Quantum Information

Bruce L. Gordon

This is an online-only chapter. You can access it at MindingtheBrain.org.

21. Human Creativity Based on Naturalism Does Not Compute

Eric Holloway and Robert J. Marks II

1. Introduction

Is the human brain capable of originating the enormous volume of creative prose we observe? Probability arguments are often given to support the impossibility of randomly generating a specific string of words. For the unguided random generation of text, a single specified target, like the first few hundred words of *Oliver Twist*, will have a minuscule probability. This target is much too specific to address the more general problem of creativity. A more interesting and relevant problem is to evaluate the probability that *any* meaningful phrase can be generated by chance. This provides a more reasonable measure of the cost of creativity. Using frequency of occurrence data and English dictionary data, we show that generating a meaningful phrase a few hundred letters in duration is not possible even when (a) the definition of "meaningful" is reduced to any sequence of words found in a dictionary,[1] and (b) generous estimates of the probabilistic resources of the universe are conceded.

Since the total space-time information capacity of the universe falls significantly short of the ability to generate meaningful text of only a few hundred letters, the origin of human creativity cannot be explained under naturalistic assumptions about the origin of the human mind. On the basis of conservation of information, we can conclude that the brain, understood as a merely material product of evolutionary chance, would be incapable of creativity, and that it must have been crafted by an external source.

2. Background

When a belletristic novel is written, what is the source of the underlying creativity? Materialists assume the prose arises from past naturalistic processes that have crafted the brain to be creative. The focus of this chapter is to demonstrate that prose creativity is not possible without external influence.

When considering the source of creativity, analysis of the naturalistic brain's capabilities alone is not sufficient for an explanation. The question of the origin of the brain's capability, by whatever means, must also be addressed. The larger problem of overall process thus focuses the creativity question squarely onto the final prose. What processes could possibly have developed the observed creativity?

The generation of meaningful phrases for purposes of communication is only a small component of creativity, but it is an essential one, and it is the probability of such generation that is the topic of this chapter.

The cost of meaningful phrases can be measured by information.[2] A steep mountain is being climbed and the generation of meaningful phrases is at the mountain's peak. There are many paths to the mountain's top. In the parlance of conservation of information, every path that reaches the top achieves the same *endogenous information*[3] as measured from the base to the summit. The harder the problem, the taller the mountain. The endogenous information measures the final elevation of the journey and is independent of the path traveled.

Externally applied information in the form of *active information*[4] can help ease the climb. Active information, measured in bits, is the degree to which a search is assisted by domain expertise.

Here is a simple example. Consider a lock with ten toggled up-down positions. An up position can be represented as a logical 1 and a down position as a logical 0. A ten-bit binary sequence like 1101001001 therefore corresponds to an up-down positioning of the toggles on the lock. If only one combination opens the lock, the endogenous information that measures the difficulty of the problem is ten bits. A locksmith comes along and tells you the first three digits of the combination are 110. If he is right, the locksmith has given you three bits of active information in your search for the correct lock combination.

Other examples are less obvious. Anyone who has played tic-tac-toe knows that the chances of winning increase when you are the first player and claim the middle square. You have acquired this expertise by examining the game and/or by repeated playing. When the first player is chosen randomly and all moves are made randomly by both players, there is an endogenous information associated with blindly playing tic-tac-toe. By applying domain expertise and choosing to go first in the center square, chances of winning have increased. Active information has been applied. Measurement of the active information here is less obvious than in the ten-bit lock example, but still the information can be measured.[5]

Imagine active information as up-escalators built on a mountain's surface that allow easier access to the peak. Building the escalators to make the ascent easier requires application of domain expertise about the problem being solved. If there are up-escalators, they are built by externally applied intelligence, not by chance. Analogously, for the generation of meaningful phrases, we will show that chance is not a viable source.

The principle of *conservation of information*,[6] built on the No Free Lunch theorem[7] and popularized by Wolpert and Macready,[8] is a surprising principle applicable to all of the paths to the mountain peak. Up-escalators on paths help but some mountain paths have escalators that go down, making the trip to the summit more difficult. On the average, one randomly chosen path to the summit works as well as any other.[9]

Active information is the amount of assistance provided to the creative process, and it can be measured in bits. Up-escalators are the active information in the mountain-climbing metaphor. As endogenous information measures the mountain's elevation, active information can be viewed as the vertical rise length of the escalators along the path.

The metaphoric mountain to be climbed can be any we choose. Thus, the question of how to reach the summit of a mountain from the bottom can be replaced by the more general question of how to traverse the path from a point A lower on the mountain to a point B closer to the summit. In relation to language, an ultimate goal might be the utterance of a meaningful phrase in *any* given language, but below this peak is the ability to generate meaningful phrases in English alone. Analysis of the informational cost of climbing this lower level is the topic of this paper.

2.1 Avoiding Preconceptions

As pointed out by Dembski and Marks,[10] active information can be introduced subtly. We must be ever vigilant in informational bookkeeping.

To illustrate the ease with which entrenched suppositions can be innocently introduced into our thinking, let us take an example from a different field of inquiry, the field of cosmology. Consider the common representation of the cosmological "Big Bang" as a "creation from nothing."

A common visualization of the Big Bang starts with a big empty space where a type of explosion happens. Yet "a big empty space" is *something*, and supposedly *nothing* existed before the Big Bang.

With some effort, the image of "a big empty space" can be removed from the visualization. Often the next description, with the idea of space removed, is that suddenly there was an explosion. Yet this doesn't work either. "Suddenly" assumes time, and time is *something*. And supposedly *nothing*, including time, existed before the creation. So we are left with trying to understand the meaning of a "Big Bang" creation void of space and time. Imagining the existence of nothing can be difficult. Doing so can tax the imagination.[11] Hidden biases, e.g., our inclination to assume the eternal existence of "time" and "space," easily sneak in.

Likewise, don't be fooled by artificial intelligence (AI). Consider *large language models* (LLMs) such as Open-AI's ChatGPT (Generative Pre-Trained Transformer) and Google's LaMDA (Language Model for Dialogue Applications). Given a prompt, these models can generate impressive prose. But the prose is written on top of a mountain of human intelligence. ChatGPT, for example, is trained on a half a trillion words in sentences all generated by humans.[12] Linguist Noam Chomsky has called ChatGPT a form of 'hi-tech plagiarism.'[13] Artists and programmers agree and are challenging similar AI in federal court for plagiarism.[14] Programmers claim their code has been stolen and artists claim the value of their art has been devalued and diluted amid a flood of similar-looking AI-generated images.

The impressive performance of AI chatbots is built on the back of almost all the literature in English ever written. And that's a lot of active information.

2.2 Averaging All Advice Is No Advice at All

A similar danger of hidden bias applies to the mountain-climbing problem. We often miss seemingly innocent hints that actually assist in the climb, because we are numbed by familiarity. But as in the case of the Big Bang example, proper illumination exposes such hints as assuming something is known about the solution of the problem. Every iota of guiding information must be entered in the accounting ledger. We cannot ignore, for example, the assistance of "memes," i.e., socially transmitted ideas.[15] Memes are externally applied guidance to traverse the mountain and must be counted as positive active information. If used, they are up-escalators. Likewise, human baby brains are pre-wired to nurse and recognize faces. This pre-wiring comes from somewhere and must be entered into the ledger tallying active information, because no such innate talent comes free of an information cost.

Another illustration of active information concerns hiding an Easter egg somewhere in the state of Wyoming. The higher the endogenous information, the more difficult the search problem and the more need there is for helpful active information. Searching for a single Easter egg in Wyoming is a hard problem and is therefore a search problem with a lot of endogenous information. Suppose, though, oracle Uncle Don knows where the Easter egg is hidden and during the search, gives you hints. "You are getting warmer!" says Uncle Don. "Warmer, colder, warmer. You are red hot." Uncle Don is adding a lot of active information to the search to make it easier to accomplish.

Alternatively, instead of Uncle Don, Uncle Joe might suffer from dementia and have no idea where the Easter egg is hidden. He offers random, made-up instructions about whether you are getting warmer or colder. The advice is helpful about fifty percent of the time. The remainder of Uncle Joe's advice takes you farther from the egg. Overall, Uncle Joe is not helpful in the Easter egg quest.

Or take a cantankerous oracle, Aunt Nancy, who knows where the Easter egg is hidden, but might purposefully give wrong directions, and say you are getting warmer when you are getting colder, and vice versa. Following Aunt Nancy's advice would never get you near the Easter egg.

Averaged over all possible uncles and aunts, one set of oracle advice is as good as another if you can't vet the validity of the information given. Uncle Don makes the search better. Aunt Nancy makes it worse. But the recipient of the advice has no idea which is which. It turns out that all advice averages out to no advice at all. This is the essence of the conservation of information. What is needed in practice is an entrusted oracle like Uncle Don with domain expertise who knows where the Easter egg is hidden and can be trusted to give proper advice.

Measurement of active information makes use of a baseline of Bernoulli's principle of insufficient reason (PrOIR)[16] where nothing is known at all. Bernoulli's PrOIR states that without prior information about the occurrence of one out of a set of possible mutually exclusive events, one should assume that all events are equally probable.[17] The principle is familiar. When a six-sided fair die is rolled, a probability of one sixth is assigned to each possible outcome. Bernoulli's PrOIR is equivalent to an assumption of maximum entropy.[18] In the mountain-climbing illustration, neither up- nor down-escalators are encountered in the trip to the summit when Bernoulli's PrOIR is applied. According to conservation of information, any other path to the summit is, on average, as good.

2.3 Knowing the Search Constraint

We have just shown that a consequence of conservation of information is that creation with no guidance by an algorithm (procedure) is as good on average as guidance by any search algorithm.[19] Trusted Uncle Don's suggested movements in the Easter egg hunt are countered by contrarian Aunt Nancy's always incorrect advice. They average out to no advice at all, or equivalently to demented Uncle Joe's random hints, which in the problem considered amounts to no guidance at all.

There is another angle. The egg hunt example thus far assumes the search for an Easter egg is over the large two-dimensional expanse of Wyoming. Suppose, instead, another search scenario where the Easter egg is the prize for traversing a large maze with many branches. Here, one can be close to the egg but the boundaries of the maze prohibit direct access. The distance information "as the bird flies" used by Uncle Don is no longer as useful.

Such a scenario would apply if the Wyoming egg search had to be performed along highways. Consider the case where the egg is on the other side of a building. Not being able to drive through the building, a trip around the building is required. In such a search, Uncle

Don's advice is not as useful. Indeed, going around the building requires moving away from the egg. Uncle Don would be yelling "You're getting cold!" and then "You're getting colder!" This is bad advice if you are on the right path to your goal that might temporarily take you further from the search target.

A less obvious example of the maze search is the solution of the Rubik's cube, where the goal is getting all six faces of the cube the same color. If during the search the colors are all the same except for a few, is the search close to a solution? No. To get to the final goal the cube must be twisted so that there are more and more mismatching colors before a point is reached where success can be achieved. Therefore, counting the number of color mismatches is not a good metric for determining how close a solution is. For Uncle Don to advise, "You are six colors off. You're getting warm..." followed by "You are only four colors off. You're getting warmer..." would not be useful. Uncle Don's tips only work well for the right problem. The counsel of demented Uncle Joe who offers random advice is as helpful here as in the case of the Easter egg hunt over flat Wyoming. Indeed, following Uncle Joe's advice in a maze can prove more useful than following Uncle Don's.

Flat land and maze searches are but two of an infinite number of search problem structures. Conservation of information dictates that, with no knowledge of the search structure, one set of advice (from, say, Uncle Don) is on average as good as any other set of advice. In fact, with no knowledge of the search structure, one can dispense with advice altogether, and simply repeatedly guess the final location of the egg without any intermediate steps. Conservation of information dictates that, on average, such a blind search works as well as any other search if there is no domain expertise to supply useful active information. And, we repeat, any useful assistance to the search, including knowledge about the search type, must be tallied as active information.

2.4 Climbing to the Summit

Under Bernoulli's PrOIR, chances of creation of any specified page from Charles Dickens's *Oliver Twist* by random letter selection is small. An easier problem illustrating low-level creativity is assessing the chance of creation of *any* meaningful page of text. This goal will be the mountain summit. The mountain's height is a function of the length of a phrase and the definition of "meaningful."

The searches we propose are not purely *ex nihilo*. There is an assumption of the existence of the English alphabet and of the ability to announce whether a word is meaningful or not. The climb to the mountain's summit is therefore not from the base but from an elevated base camp where these resources are available.

There are additional active information tools that get us even closer to the summit. In two of the phrase models considered, letters are not chosen randomly in accordance with Bernoulli's PrOIR but in accordance with their frequency of occurrence. This provides an up-escalator on the climb. Another up-escalator model makes use of the distribution of word sizes in the dictionary. Active information is being applied to make the problem easier. Surprisingly, the chance of generating anything creative, although better than using uniformly random letters according to Bernoulli's PrOIR, still rapidly exhausts any reasonable estimate of available computational resources in the universe or, indeed, some models of the multiverse.[20] Even when the phrase generation is assisted by modest active information, meaningful phrase generation is limited to fewer than a hundred letters. In concert with conservation of information, the inability of unguided searches to generate a modicum of literary creativity indicates that human creativity involves something beyond any unguided natural process.

2.5 Contents

This chapter undertakes to prove that the human mind requires substantive active information in order to generate meaningful phrases. It is true that the active information cost can be partly met with resources of space and time, but the amount of active information required exhausts even the most expansive estimates of available resources.

The chapter is in three sections:

1. *Analysis.* The chances of generating *any* meaningful text at random is evaluated using tools from Shannon information theory. Meaningful text is broadly defined as any sequence of words found in a specified dictionary. The expected bit cost (asymptotically) increases exponentially with respect to the text length in characters.
2. *Computational resource bounds.* The computational resources of the universe are next estimated. Seth Lloyd's assessment of the available bits of information available from our universe[21] is expanded to include the information resources using a common model of the multiverse. Even here, the chances of creating meaningful text of any significant length are minuscule.
3. *Implications about human creativity.* Lastly, we examine the implications of these results for the claim that the generation of meaningful text can be explained on naturalistic grounds, i.e., on the supposition that the creative process is nothing but the running of various search procedures by a human brain conceived of in wholly material terms as an organic computer.

3. The Chances of Generating Any Meaningful Phrase

In this section, the chances of generating a meaningful phrase in English is examined. The results require use of the mathematics of Shannon information theory, the specifics of which may be outside the interest range of some readers. For this reason, all mathematical details have been relegated to the appendix at the end of this chapter. Suffice it here to simply review the computed cost of creativity in the case of generating meaningful prose.

For the generation of meaningful phrases, there is a history of man's development that can be innocently infused to bias the search. The goal of this chapter, though, is to ignore that history and assume we know nothing about this development. What is the chance of generating meaningful phrases using any and all methods? The average performance of all of these methods is the same as repeatedly guessing the target. Once analyzed, the current human capability of forming meaningful phrases can be assessed and the difference in information must be chalked up to necessarily supplied active information.

Random sampling procedure allows evaluation of the cost in bits required to generate a meaningful phrase of a fixed length. Each query has a cost in bits. Evaluation of the expected number of queries prior to a success then allows computation of the final accumulated bit cost for creativity. In the three cases considered, the bit cost increases exponentially with respect to the length of the meaningful message as measured in letters or words.

3.1 The Cost of Being Meaningful

A *meaningful phrase* is herein defined simply as any sequence of characters and spaces that results in a sequence of words found in a dictionary. Coherence and grammar are not examined in this lax definition. Only 26 capital letters and the space are allowed. Doing so increases probabilities above the more detailed generation of phrases where both upper- and lower-case letters as well as punctuation and numbers are used. Including these other characters would severely worsen the chances of generating a meaningful phrase.

A short example of a meaningful phrase is

JESUS WEPT

It contains $L = 9$ characters, all taken from the English alphabet, plus a space. Both words are in the dictionary.

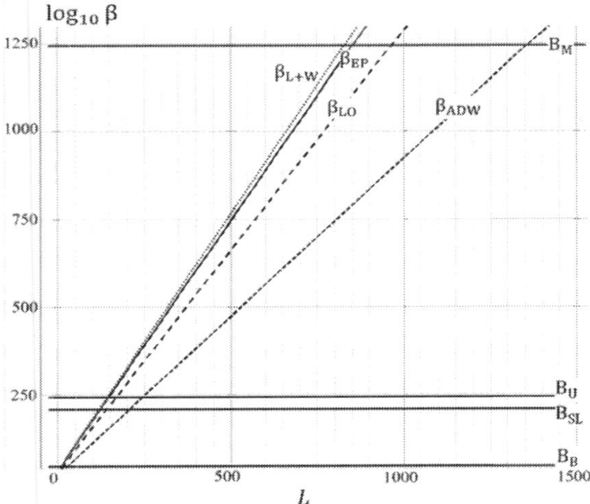

Figure 21.1 Plots of the bit cost count β (in bits) versus L, the number of letters in a meaningful phrase. The bit count is in $\log_{10}$ so that 500 corresponds to 10^{500} bits. Information Bounds (for the referents of the numerals in parentheses, see Table 21.1): The naturalistic brain bound of B_B (4), the Seth Lloyd bound of BSL (1), the universal bit bound B_U (2) and the multiverse bit bound BM (3) are indicated with horizontal lines. Models: The slopes of the lines in bit decades per character[22] are 1.4538 for β_{L+W} (36), 1.4150 for β_{EP} (29), 1.2840 for β_{LO} (33), and 0.8986 for β_{ADW} (48). The intercepts of the sloped lines with the horizontal information resource lines are tallied in Table 21.1.

The main results of this chapter are summarized in Figure 21.1. The horizontal axis contains the number of characters, L, in a meaningful phrase. The vertical axis in Figure 21.1 denotes information (β) in bits required to generate a phrase. Note that this axis is on a log scale so that 500 corresponds to 10^{500} bits. To appreciate such large numbers, there are about 10^{80} atoms in the universe. So 10^{500} is the number of atoms in 10^{420} of our universes.

Exponents can require scrutiny to appreciate. The number 10^{500} might intuitively look close to the number 10^{509} but, in fact, 10^{509} is a billion times larger than 10^{500}. In this light, the bit count on the vertical axis in Figure 21.1 is seen to increase extraordinarily fast.

There are four solid horizontal lines slicing Figure 21.1. They are bounds, in bits, of various information capacity models explained more completely in the appendix. From the bottom horizontal line to the top:

- B_B is the estimated bit capacity of the human brain.
- B_{SL} is Seth Lloyd's estimated bit capacity of the universe.[23]
- B_U is the number of Planck cubes in the universe times the Big Bang model estimate of the age of the universe in units of Planck time. A Planck cube is one Planck length on each side.
- The Planck length is *very* small. To give a relative size perspective, if a Planck length were scaled to an inch, then a proton would have a diameter of more than 5,000 light-years. Planck lengths are unbelievably short.[24]
- The unit of Planck time is likewise very short. It is the time it takes light to travel one Planck length.
- The purpose here is to generate a large number and speak of it as the number

of available bits. A Planck cube at some Planck time in history is obviously not a bit. But the derivation of the number is such that the corresponding number of bits inarguably exceeds the computational resources of the universe.

- B_M assumes there are 10^{1000} parallel universes in a multiverse.[25] Each universe is assumed to have the same Planck bit capacity as our universe.

4. Information Capacity Model Details

WHAT IS the information available to form meaningful phrases? Here are the models:

a. *Seth Lloyd.* Based on physics, the universe's computational capacity is 10^{120} operations on 10^{90} bits. We can combine these measures into one. Doing so no longer distinguishes a bit operated on from one being used in computation and gives a single bit cost bound for each model considered. The bit bound of all subsequent models should be thus interpreted.

The corresponding bit cost bound for the Lloyd model is

$$B_{SL} = 10^{120} \times 10^{90} = 10^{210} \text{ bits.} \quad (1)$$

If an event's cost is greater than this bit bound, it is not possible if Lloyd's model is correct. To set up an even more inarguable bound, we will use a capacity based on the universe that dwarfs that of Lloyd.

b. *Universal Planck cube-Planck time product:* The strings in string theory are on the order of a Planck length = 1.6 $\times 10^{-35}$ meters. If we make a cube one Planck length on the side, the volume is 4×10^{-105} cubic meters. We'll call this a *Planck cube*. Assuming our spherical universe has a radius of 47 billion light-years, it takes over 10^{184} Planck cubes to fill it. But maybe 10^{184} bits is too small a number, so let's think bigger. One Planck time unit (5.4×10^{-44} seconds) is the time it takes light in a vacuum to travel one Planck distance. A 14-billion-year estimate of the age of the universe translates to 8.2×10^{60} Planck time units. Assuming 10^{184} Planck cubes every Planck time interval in the history of the universe gives the space-time product of 10^{244}. We'll call this B_U. Thus,

$$B_U = 10^{244} \text{ bits} \quad (2)$$

is the universe's information bound. The Seth Lloyd bound in (1) above is dwarfed by this number.

c. *Multiverse Planck cubes-Planck times:* To remove all doubt, let's assume an even bigger number. Models of the multiverse vary. Even within a model, numbers vary.[26] A total of 10^{100}, 10^{500}, or even 10^{1000} parallel universes are hypothesized in one model.[27] Let's work with 10^{1000} parallel universes in the so-called multiverse and, for convenience, assume these universes are the same size and age as ours. Then we have the multiverse's information bound of

$$B_M = 10^{1000} \times B_U = 10^{1244} \text{ bits.} \quad (3)$$

That's a lot of bits![28] Say "trillion, trillion, trillion" over a hundred times and you'd still be shy.

4.1 Models

The number of bits required for generation of a meaningful phrase depends on the model used. Four models are considered and correspond to the four sloping lines shown in Figure 21.1. By providing active information, each case is an attempt to nudge meaningful results towards being more probable. These hints are more favorable than the clueless, unguided assumption of Bernoulli's PrOIR. They provide *a priori* structure to the task of forming meaningful phrases.

For two of models, letters are generated according to the *frequency of occurrence* (FOO) of letters in the English language. The letter E is the most commonly used letter. The letter Q is used the least. Many more E's are therefore generated than Q's. The FOO is used to structure the manner in which letters are chosen. In the first model, no spaces are used in phrase generation. In the second, spaces are used.

Another model uses the distribution of word lengths in the dictionary. Once a word length is chosen, say four letters, viability is assessed by evaluating the probability for randomly selected letters to form a word in the dictionary.

The dashed lines in Figure 21.1 correspond to the four models. They are labeled β_{EP}, β_{LO}, β_{L+W}, and β_{ADW}. Each plot denotes the cost in bits of generating a meaningful phrase of length L. Here is an explanation of the subscripts in each case.

a. EP used in the subscript β_{EP} uses letters chosen at random using Bernoulli's PrOIR. EP means "equal probability." No spaces are used in the meaningful phrase. A specific favorable parsing of the character string into a meaningful phrase is considered a success. This leads to ambiguities where, for example, MANSLAUGHTER can be interpreted either as either "manslaughter" or "man's laughter." Likewise, ATHEIST can be interpreted as "a theist" or its antonym "atheist." This does not matter here. Both are in the dictionary and therefore the meaningful phrase test is passed.

b. LO used in the subscript β_{LO} means "letters only" and no spaces are used. In this case, as a reference, a specific sequence of characters (not any meaningless phrase) is sought for comparison with any meaningful phrase. Unlike EP, instead of using randomly selected letters, letters are chosen in accordance with their frequency of occurrence (FOO). The β_{LO} plot in Figure 21.1 specifies the cost in bits versus phrase length L for generating a "letters only" meaningful phrase using the FOO of English letters. The slope of this line is more shallow and thus lower in cost than the EP curve. The assumption of a FOO allows a short escalator to be built on the path to the summit.

c. The subscript L + W in β_{L+W} stands for "Letters plus Words." In addition to FOO-generated characters, a word spacing generator places spaces in the phrase. The spaces must divide character bursts into dictionary words to generate a meaningful phrase. Doing so is more difficult than when no spaces are used. A long meaningful word, for example, can be broken into two meaningless words by a space and the meaningful phrase test is not passed. As is evident from the plots in Figure 21.1, the L + W results are less favorable than the EP (equal probability) case where no FOO is used. The up-escalator provided by the FOO in L + W is therefore shorter than the down-escalator imposed by requiring correct insertion of spaces. Solving the L + W case is interestingly more difficult than generating words without spaces using the uniform letter selection in EP.

d. ADW stands for *Any-Dictionary-Word*. This model provides the most active information of those considered. Rather than judiciously sprinkling spaces in a stream of letters, spaces can be determined from statistics of English word lengths. A dictionary contains words of different length. The ratio of the number of three-letter words divided by the total number of words in the dictionary serves as the probability that a randomly chosen word from the dictionary has three letters. This information is included in the model. For ADW a word length is first identified. Letters are chosen according to their FOO to fill out each word. The meaning of the generated word is then assessed. Is it in the dictionary? As seen in Figure 21.1, the ADW (Any-Dictionary-Word) assistance in finding a meaningful phrase is the most helpful of the three cases considered. For any bit bounds (horizontal lines) chosen, the ADW has the capability of generating longer phrases (larger *L*) than the other two models.

4.2 Putting It All Together

The intersections of the dashed line bit requirements with the horizontal line bit resources graphically shown in Figure 21.1 are summarized in Table 21.1. The numerical results are jaw-dropping. Using the most favorable model, ADW, the intercept with the Seth Llyod model horizontal line (B_{SL}) has the capacity to generate meaningful messages of only length $L = 230$ letters. The Planck bit count of the universe (B_U) increases this to only 268 letters. Remarkably, using the multiverse bound (B_M), an unspecified meaningful phrase of at most 1380 letters can be generated.

The number of letters able to be generated by such a large resource of information is unexpectedly small. Even with FOO hints and word length distribution information, the creation of meaningful phrases with no additional external guidance is beyond the computational resources of the multiverse.

Bit Bounds → ↓ Bit Cost ↓	$B_B = 10^{37}$ (4)	$B_{SL} = 10^{37}$ (1)	$B_U = 10^{244}$ (2)	$B_M = 10^{1244}$ (3)
β_{LO} (30)	27	165	192	989
β_{L+W} (34)	24	142	165	853
β_{ADW}(44)	38	230	268	1380

Table 21.1 Intercepts of the sloped lines with the horizontal lines in the plots in Figure 21.1. The bit bounds axis lists various scenarios of informational resources available. B_B is an upper bound of the resources in the human brain, B_{SL} is within the universe, B_U is a higher estimate within the universe, and B_M is an estimate within the multiverse. The bit cost axis lists models of bits per letter required to generate English text. β_{LO} is derived from only letter frequency. β_{L+W} adds information regarding word length. β_{ADW} incorporates the word distribution in an English dictionary. The numerals in the cells are the number of sequential English words that can be expected given the generation model and available informational resources.

Specific examples of familiar documents are listed in Table 21.2 as examples. Using the best model considered (ADW), generation of a phrase as long as the 31-word American *Pledge of Allegiance* requires 10^{164} bits. Messages as long as the seven-volume French novel *À la recherche du temps perdu* (*Remembrance of Things Past*) require a staggering $10^{58,000,000}$ bits.

Document	M	β_{LO} (33)	β_{L+W} (37)	β_{ADW} (48)
Pledge of Allegiance	31	227	263	164
Gettysburgh Address	272	1,978	2,293	1,419
Declaration of Independence	1,137	8,260	9,576	5,921
...with signatures	1,458	1.1×10^4	1.2×10^4	7,592
Genesis	3.8×10^4	2.8×10^5	3.2×10^5	2.0×10^5
King James Bible	7.8×10^5	5.7×10^6	6.6×10^6	4.1×10^6
À la recherche du temps perdu	1.1×10^7	8.1×10^7	9.4×10^7	5.8×10^7

Table 21.2 *M* is the number of words in the document to the left. The other columns are query bit costs for the equations referenced in the parentheses, each equation representing a word-generation model.

The columns list how many bits must be queried within the respective model in order to generate the number of words in the document. The numbers in the columns are $\log_{10}$ so that 1,978 corresponds to $10^{1,978}$ queries. For a comparison, see Table 21.1, which provides the number of bits available within different physical estimates.

5. Implications for Human Creativity

HUMAN BRAINS are wonderfully pre-wired,[29] allowing quick learning resonance in such areas as language,[30] facial recognition,[31] and even letter recognition.[32] Pre-wiring requires enormous active information.

The brain is a small portion of the entire universe, yet is responsible for an enormous prose output, including the nearly 1.8 million books listed on Amazon.com. Yet the universe's information capacity is far exceeded by the demands of a single book. Is there something special about the brain that can account for this remarkable result?

Sir Roger Penrose proposes quantum fluctuations as a source of nonalgorithmic creativity in the brain.[33] The brain contains about 10^{11} neurons, and each neuron contains about 10^{7} tubulins. Tubulins are structures that are hypothesized to produce coherent signals from quantum fluctuations. This increases the estimated number of computational operations per neuron from 10^{6} per second to 10^{16} per second, giving the brain an upper bound of 10^{27} operations per second. Multiplied by an average human life span of 76 years we get an upper computational bit count of

$$B_P = 10^{37} \text{ operations per person.} \quad (4)$$

According to the Population Reference Bureau, over 117 billion people have been born up to the present,[34] which gives us an upper bound of

$$B_B = 10^{49} \text{ brain operations in history.} \quad (5)$$

The multiverse helps. Multiplied by the 10^{1000}, (5) gives

$$B_{MB} = 10^{1049} \text{ brain operations in the multiverse} \quad (6)$$

which, being much less than the information bound B_M calculated in (3) above, is insufficient to produce meaningful English text of significant length. This is seen by the information demands of the familiar documents listed in Table 21.2.

It thus seems that human literary creativity cannot be explained by mere brain activity, where that activity is understood as purely naturalistic computationalism. Creativity requires the infusion of active information.[35]

6. Conclusions

WE HAVE analyzed the probabilities of generating documents through chance and determined whether these probabilities are possible given the resources of the human brain, the universe, and the multiverse. We saw that when the target is any meaningful phrase, any document whose length is as small as the *Gettysburg Address* becomes impossible to generate by undirected chance.

Recall the mountain-climbing illustration where the elevation of the peak is analogous to endogenous information. The endogenous information of the human brain can be thought of as at its current state. There are many paths to the endogenous information at this mountain's peak. With some modest active information sources such as FOO, we have illustrated that the mountain cannot be successfully climbed using the informational resources of the universe or even a common model of the multiverse. The information gap between needed and available informational resources is expansive. The resolution to this discrepancy can arise only from the identification of some guiding information source external to the brain in both its development and use. The guidance must come from a reliable oracle, from an information source like Uncle Don in the Easter egg hunt in Wyoming.

7. Appendix

HERE WE present detailed mathematics substantiating the results presented in Section 2.

7.1 Definitions

In order to obtain numerical results, English letter *frequency of occurrence* data and Webster dictionary word length data are used. Dictionaries are also modeled by a Poisson distribution. Shannon information theory is then used to establish probability values for generating any meaningful phrase.

To facilitate discussion of available probabilistic resources, we start with some definitions. Figure 21.2 illustrates defined sets using a Venn diagram.

- Shannon self-information, I, is related to the underlying probability p of an event by

$$I = -\log p.$$

When the base of the logarithm is 2, information I is measured in bits. The *entropy* in bits of an event is the average information of all possible outcomes. Thus

$$H = \sum_n p_n I_n = -\sum_n p_n \log_2 p_n.$$

Where p is the probability of the outcome n.

- *Alphabet Size.* An alphabet consists of N letters. Throughout the paper, we will assume $N = 26$ letters in the English alphabet. Doing so restricts words to be composed of only capital letters. No numbers, lower case letters or other symbols are allowed. If they were, the bit cost will increase markedly.

- *Letter FOO.* The entropy of an alphabet has frequency of occurrence (FOO)

$$\eta[n] = \Pr[\ n\text{th Letter }]; 1 \leq n \leq N.$$

If letters are equiprobable, then $\eta[n] = 1/N$.

- *Alphabet Entropy.* From the FOO distribution, we can compute the FOO entropy

$$H_N = -\sum_{n=1}^{N} \eta[n] \log_2 \eta[n]. \tag{7}$$

For equiprobable letters, $H_N = \log_2 N$.

- *Set of All Words.* Any finite sequence of letters from the alphabet will be called a word, w. A word may or may not be in the dictionary. Define the set of all words by

$$\mathcal{W} = \{w|\text{all collections of letters of finite length}\} \tag{8}$$

Clearly, the set's cardinality, $|\mathcal{W}|$, is countably infinite.

- *Word Length.* Let $|w|$ denote the length of the word w. The set of all words with length k is

$$\mathcal{W}_k = \{w \in \mathcal{W} \mid |w| = k\}. \tag{9}$$

- *Stationary Random Letter Source.* Let letters in the alphabet be generated using a source with entropy H_N per letter. We will assume the cardinality of the set $\mathcal{W}_k$ to be

$$|\mathcal{W}_k| = 2^{kH_N}. \tag{10}$$

- *Word Length Distribution.* The distribution of word lengths in randomly generated text is

$$\omega[k] = \Pr[|w| = k] \tag{11}$$

where $|w|$ is the length of a word in characters for all $w \in \mathcal{W}$. The mean word length is

$$\overline{|w|} = \sum_{k=1}^{\infty} k\ \omega[k] \text{ letters} \tag{12}$$

and the word length entropy is

$$H_{|w|} = -\sum_{k=1}^{\infty} \omega[k]\ \log_2 \omega[k].$$

- *Geometric word length generator.* Let α denote the probability of the occurrence of a space. The geometric word length distribution is

$$\omega[k] = \alpha(1-\alpha)^{k-1}; 1 \leq k \leq \infty. \tag{13}$$

Letters are added to a word with probability $(1-\alpha)$ until the occurrence of a space. For the geometric word length generator, the mean is

$$\overline{|w|} = \frac{1}{\alpha} \text{ letters} \tag{14}$$

and the word length entropy is

$$H_{|w|} = \frac{H(\alpha)}{\alpha} \text{ bits} \tag{15}$$

where the entropy of a Bernoulli trial with probability of success α is

$$H(\alpha) = -\alpha \log_2 \alpha - (1-\alpha)\log_2(1-\alpha) \tag{16}$$

- *Dictionary Word Length Distribution.* The set of words in a specified dictionary is

$$\mathcal{D} = \{w \mid w \in \text{ dictionary }\}. \tag{17}$$

The set of words of length k in a specified dictionary is $\mathcal{D}_k$ so that

$$\mathcal{D}_k = \mathcal{D} \cap \mathcal{W}_k. \tag{18}$$

The distribution of dictionary word lengths is $\pi[k]$ so that

$$\pi[k] = \frac{|\mathcal{D}_k|}{|\mathcal{D}|}. \tag{19}$$

Visualization of the sets of words and dictionary words is aided by the Venn diagram in Figure 21.2.

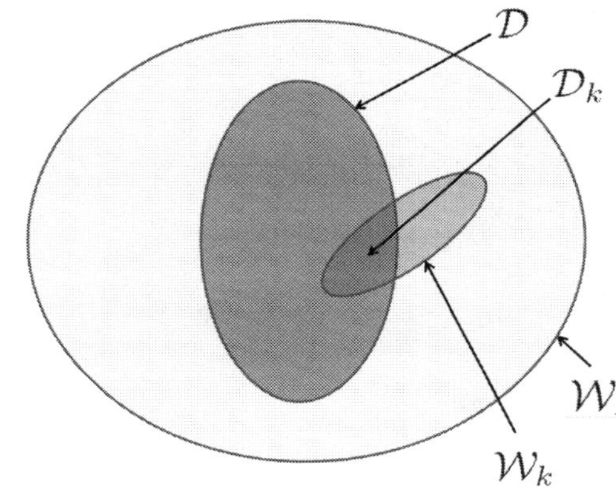

Figure 21.2 Venn diagram helpful in the visualization of sets of words discussed in Section 7.1.

7.2 Data

Here is some data that proves useful in assigning numerical values to our analysis.

7.2.1 English Letter FOO

The *frequency of occurrence* (FOO) probabilities, $\eta[n]$, for the $N = 26$ letters in the English alphabet are tabulated in Table 21.3. The space is not included. The corresponding FOO entropy, using (7), is

$$H_N = 4.1655 \text{ bits}. \tag{20}$$

From Table 21.4, we can also estimate the

probability of a space as number of spaces

$$\alpha = \frac{\text{number of spaces}}{\text{number of characters}} = \frac{743.8}{4307.4} = 0.1727. \quad (21)$$

The corresponding Bernoulli trial entropy using (16) is

$$H(\alpha) = 0.6638 \text{ bits}. \quad (22)$$

For the geometric word length generator distribution in (13), the mean, from (14), follows as

$$\overline{|w|} = 5.7911 \text{ letters} \quad (23)$$

and the entropy, from (15), is

$$H_{|w|} = 3.8441 \text{ bits}. \quad (24)$$

Table 21.5 lists the parameters used to calculate $H_{|w|}$.

7.2.2 Word Length Distribution

A histogram of dictionary word lengths, $|\mathcal{D}_k|$, is tabulated in Table 21.4 along with the corresponding distribution of dictionary word lengths, $\pi[k]$. The histogram is plotted in Figure 21.3. (The other plots in Figure 21.3 are discussed later under the topic of Poisson dictionary.) The number of entries in the dictionary is

$$|\mathcal{D}| = 234,371 \quad (25)$$

and the average word length is

$$\lambda + 1 = 9.5917 \text{ letters}. \quad (26)$$

Note that the average word length of a word in a phrase, $|w|$, is not the same as the average length of a word in the dictionary, $\lambda + 1$.

n	Letter	Count	$\eta[n]$
-	Space	743.8	-
1	E	445.2	0.1249
2	T	330.5	0.0927
3	A	286.5	0.0804
4	O	272.3	0.0764
5	I	269.7	0.0757
6	N	257.8	0.0723
7	S	232.1	0.0651
8	R	223.8	0.0628
9	H	180.1	0.0505
10	L	145.0	0.0407
11	D	136.0	0.0382
12	C	119.2	0.0334
13	U	97.3	0.0273
14	M	89.5	0.0251
15	F	85.6	0.0240
16	P	76.1	0.0214
17	G	66.6	0.0187
18	W	59.7	0.0168
19	Y	59.3	0.0166
20	B	52.9	0.0148
21	V	37.5	0.0105
22	K	19.3	0.0054
23	X	8.4	0.0024
24	J	5.7	0.0016
25	Q	4.3	0.0012
26	Z	3.2	0.0009

Table 21.3 Histogram values in this table are calculated from a sampling of 743 billion words [30]. The total number of characters, including spaces, is 4,307.4 billion. The letter count, excluding spaces, 3,563.6 billion. Entries in the Count column are in billions. Note that the letter FOO probabilities are in the column on the far right.

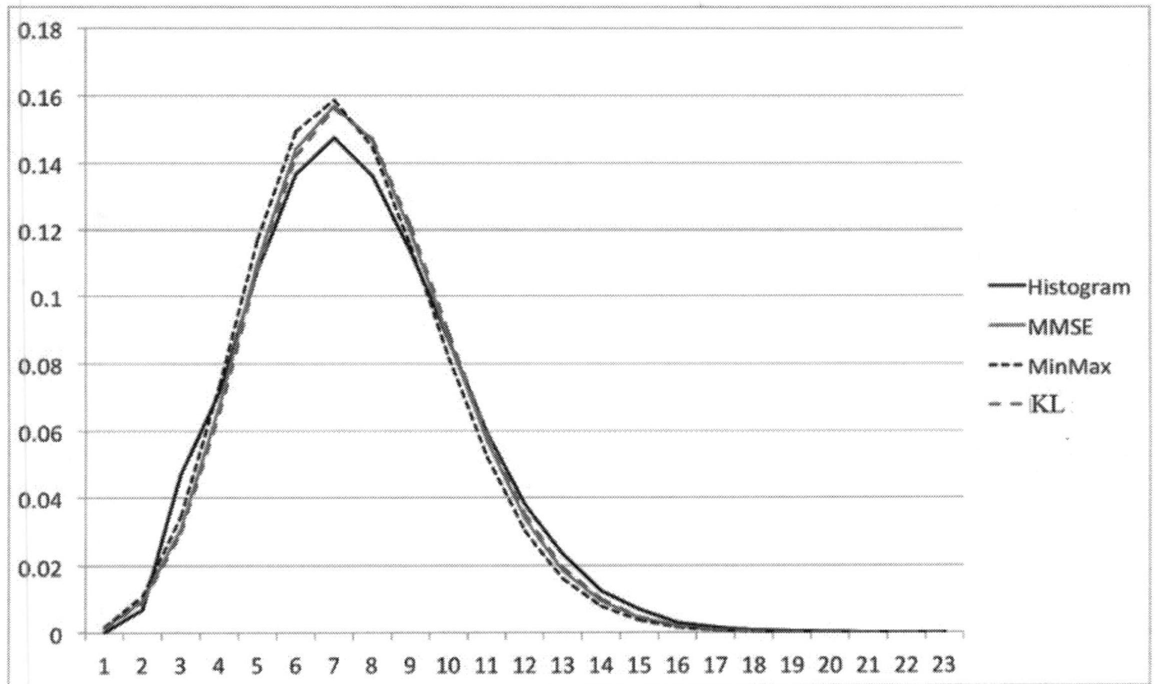

Figure 21.3 Probability mass functions of word lengths. The probability of a randomly chosen word of length 7, for example, is between 0.14 and 0.16. The dictionary word lengths from Webster's 2nd Edition tabulated in Table 21.4 is labeled Histogram. Three fits are shown to the Poisson random probability mass function. MMSE minimizes the mean squared error, MinMax minimizes the maximum error and KL minimizes the Kullback-Leibler distance. The three fits are almost graphically indistinguishable and, as shown by the listing in Table 21.7, the optimal λ's are all nearly identical.

k	$\|\mathcal{D}_k\|$	$\pi[k]$
1	26	0.000009
2	139	0.000593
3	1294	0.005522
4	4994	0.021310
5	9972	0.042552
6	17462	0.074513
7	23713	0.101188
8	29842	0.127341
9	32286	0.137770
10	30824	0.131531
11	25963	0.110789
12	20447	0.087251
13	14923	0.063679
14	9761	0.041652
15	5922	0.025270
16	3377	0.014410
17	1813	0.007736
18	842	0.003593
19	428	0.001826
20	198	0.000845
21	82	0.000350
22	41	0.000175
23	17	0.000073
24	5	0.000021
25	0	0.000000

Table 21.4 Dictionary word lengths, $|\mathcal{D}_k|$, from Webster's 2nd Edition. There are a total of 234,371 words in the dictionary. The column $\pi[k]$ contains the empirical probability derived from normalizing the $|\mathcal{D}_k|$ histogram. These values are plotted in Figure 21.3. The distribution of dictionary word lengths, $\pi[k]$ (see (19)), is in the right column. The average dictionary word length is listed in Table 21.5.

7.3 The Bit Cost Function

How many bits are expended on average in a blind search for a single specified event with success probability p? The answer is the bit cost, which can be evaluated from a *bit cost function.* Repeating a Bernoulli trial until a success occurs is a geometric random variable with a probability distribution of

$$p_n = p(1-p)^{n-1}; 1 \leq n < \infty.$$

The mean number of queries to achieve a success is the mean of the geometric random variable.

$$\bar{Q} = \frac{1}{p}.$$

Parameter	Numerical Value	Equation
H_N	4.1655	(20)
α	0.1727	(21)
$H(\alpha)$	0.6638	(22)
$\overline{\|w\|}$	5.7911	(23)
$H_{\|w\|}$	3.8441	(24)
λ	8.5917	(26)

Table 21.5 A list of numerical parameters derived from the data in Tables 3 and 4. From Table 21.3, H_N is the entropy per letter, α is the frequency of spaces, and its reciprocal is $\overline{|w|}$, the average word length. $H(\alpha)$ is the entropy of spaces used to derive $H_{|w|}$, the entropy per word in Equation (15).

If two dollars are spent on candy each time during twelve trips to the corner store, a total of $24 is spent overall. Likewise, if I bits of information are spent on each of the $\bar{Q}$ queries, the total expected number of expended bits is given by the *bit cost function*

$$\beta = \bar{Q}I = -\frac{\log_2 p}{p} = I\ 2^I \text{ bits.} \quad (27)$$

		Webster	Poisson
Cross Entropy	$H_{\omega \times \pi}$	7.1599	7.0738
Dictionary Size	$\lvert\mathcal{D}\rvert$ and $\lvert\mathcal{D}\rvert\text{max}$	234,371	181,516
Log Bitrate	$\frac{\log_2 \beta_{ADW}}{L}$	2.985	3.0339

Table 21.6 Derivation of log bitrates for Webster's dictionary and a Poisson dictionary for Equations (48) and (50).

7.4 Scenarios

Here are three models (7.4.1–7.4.4) using the bit cost function in the random generation of phrases. A summary of the asymptotic log bit count per character derived in this section is summarized in Table 21.8.

7.4.1 LO: Letters-Only

First, consider choosing a specified sequence of L letters from an alphabet of N characters.

7.4.1A. EQUIPROBABLE ALPHABET. If each character has the same probability of being chosen, then

$$p = N^{-L}$$

corresponding to

$$I = \log_2 N^L \text{ bits}$$

so that the bit cost, from (27), is[37]

$$\beta = N^L \log_2 N^L. \quad (28)$$

(This can be interpreted as N^L operations on $L \log_2 N$ bits.) This function gets large quickly. A more manageable equivalent is

$$\log_2 \beta = L \log_2 N + \log_2 L + \log_2 \log_2 N. \quad (29)$$

Asymptotically,

$$\log_2 \beta \underset{L\uparrow}{\rightarrow} L \log_2 N.$$

7.4.1B. STATIONARY ALPHABET.If the character generating source has entropy H_N then $I = H_N$ and the bit cost function is

$$\beta_{LO} = 2^{LH_N}(LH_N) \quad (30)$$

where, as previously noted, the subscript LO denotes *letters only* and the target is not anything meaningful but is a specified sequence. The log plot of β_{LO} is shown in Figure 21.1. Equivalently, (30) can be written as

$$\log_2 \beta_{LO} = LH_N + \log_2 L + \log_2 H_N. \quad (31)$$

Equation (29) is a special case at maximum entropy when $H_N = \log_2 N$. Asymptotically,

$$\log_2 \beta_{LO} \underset{L\uparrow}{\rightarrow} LH_N. \quad (32)$$

The asymptotic log bit count per character follows as

$$\frac{\log_2 \beta_{LO}}{L} \underset{L\uparrow}{\rightarrow} \frac{d}{dL} \log_2 \beta_{LO} \underset{L\uparrow}{\rightarrow} H_N.$$

For the FOO letter frequency in Table 21.3,

$$\frac{\log_2 \beta_{LO}}{L} \underset{L\uparrow}{\rightarrow} \frac{d}{dL} \log_2 \beta_{LO} \underset{L\uparrow}{\rightarrow} 4.2655. \quad (33)$$

7.4.2 L+W: Letters-Plus-Word-Spaces

Next the case of a phrase with randomly chosen words is considered. Let M words be generated using the distribution $\omega[k]$ in (11). The generated word lengths vary and can be viewed as cups of different sizes that must be filled. After the selected number of characters for the M words are selected, the character generator fills in characters like a soda dispenser fills different sizes of cups. The number of characters used is approximately

$$L = M\overline{|w|} \text{ characters}$$

where $\overline{|w|}$ is the word length distribution mean in (12). For a sequence of L letters, the nth letter in the alphabet will occur about $L\eta[n]$ times. The probability of this occurrence is $\eta[n]^{L\eta[n]}$. For all N letters, the probability is

$$\Pr[\text{ correct letters }] = \prod_{n=1}^{N} \eta[n]^{L\eta[n]}.$$

Likewise, the chance that all M word spacings are correct is

$$\Pr[\text{ correct word spacing }] = \prod_{k=1}^{\infty} \omega[k]^{M\omega[k]}.$$

The probability the phrase of M words is correct is the product of these probabilities.

$$\begin{aligned} p &= \Pr[\text{ correct phrase }] \\ &= \Pr[\text{ correct letters }] \\ &\qquad \times \Pr[\text{ correct word spacing }] \\ &= \prod_{n=1}^{N} \eta[n]^{L\eta[n]} \prod_{k=1}^{\infty} \omega[k]^{M\omega[k]}. \end{aligned}$$

The corresponding bit cost function is

$$\begin{aligned} \beta_{L+W} = M \left(\overline{|w|}H_N + H_{|w|}\right) \\ \times 2^{M\left(\overline{|w|}H_N + H_{|w|}\right)} \end{aligned} \quad (34)$$

where the subscript $L + W$ denotes *letters plus words*. The log plot of β_{L+W} is shown in Figure 21.1. Equivalently,

$$\begin{aligned} \log_2 \beta_{L+W} &= L\left(H_N + \frac{H_{|w|}}{\overline{|w|}}\right) \\ &\quad + \log_2 L + \log_2\left(H_N + \frac{\overline{H}}{\overline{|w|}}\right). \end{aligned} \quad (35)$$

Asymptotically,

$$\log_2 \beta_{L+W} \underset{L\uparrow}{\rightarrow} L\left(H_N + \frac{\overline{H}}{\overline{|w|}}\right)$$

and

$$\frac{\log_2 \beta_{L+W}}{L} \underset{L\uparrow}{\rightarrow} \frac{d}{dL}\log_2 \beta_{L+W} \underset{L\uparrow}{\rightarrow} 4.8293. \quad (36)$$

We have added the space constraint so that, unsurprisingly, this value is larger than that in (33), where spaces are not used.

7.4.3 Special Case: Gemetric Word Length Generator.

For the geometric word distribution in (13), we can use (14) and (15) to respectively write (35) and (36) as

$$\begin{aligned} \log_2 \beta_{L+W} &= L\left(H_N + H(\alpha)\right) \\ &\quad + \log_2 L + \log_2\left(H_N + H(\alpha)\right) \end{aligned}$$

and

$$\begin{aligned} \frac{\log_2 \beta_{L+W}}{L} &\underset{L\uparrow}{\rightarrow} \frac{d}{dL}\log_2 \beta_{L+W} \\ &\underset{L\uparrow}{\rightarrow} H_N + H(\alpha). \end{aligned}$$

For the FOO letter data in Table 21.3 and Bernoulli trial entropy in (22),

$$\frac{\log_2 \beta_{L+W}}{L} \underset{L\uparrow}{\rightarrow} \frac{d}{dL} \log_2 \beta_{L+W} \underset{L\uparrow}{\rightarrow} 4.8293. \quad (37)$$

Because the extra constraint of spacing has been added, this exceeds the letters-only number in (33).

7.4.4 ADW: Any Dictionary Words

By a "meaningful phrase," we mean a sequence of words each of which is resident in a specified dictionary. The sequence of words need not form a sentence or have coherent meaning. Using this criterion, the bar for "meaningful" has been set very low. The numerical values that result are therefore correspondingly generous.

Choose M words of random length k in accordance to the distribution $\omega[k]$. Given the word length, the chance a word is in the dictionary is

$$\Pr[w \in \mathcal{D} | w \in \mathcal{W}_k] = \frac{|\mathcal{D}_k|}{|\mathcal{W}_k|}.$$

Thus,

$$\begin{aligned}\Pr[w \in \mathcal{D}, w \in \mathcal{W}_k] &= \Pr[w \in \mathcal{D} | w \in \mathcal{W}_k] \\ &\quad \times \Pr[w \in \mathcal{W}_k] \\ &= \Pr[w \in \mathcal{D}_k, w \in \mathcal{W}_k] \\ &= \frac{|\mathcal{D}_k|}{|\mathcal{W}_k|}\,\omega[k]. \quad (38)\end{aligned}$$

Let the distribution of word lengths in the dictionary follow the distribution $\pi[k]$ as in (19). (Values of $\pi[k]$ from Webster's dictionary are listed in Table 21.4.) Then (38) becomes

$$\Pr[w \in \mathcal{D}_k] = \frac{\pi[k]\;\omega[k]\;|\mathcal{D}|}{|\mathcal{W}_k|}.$$

For M words, the number of words of length k is about $M\omega[k]$. The probability that all $M\omega[k]$ of these words are in the dictionary is

$$(\Pr[w \in \mathcal{D}_k])^{M\omega[k]} = \left(\frac{\pi[k]\;\omega[k]\;|\mathcal{D}|}{|\mathcal{W}_k|}\right)^{M\omega[k]}$$

and the probability of getting all M words from the dictionary is

$$p = \prod_{k=1} \left(\frac{\pi[k]\;\omega[k]\;|\mathcal{D}|}{|\mathcal{W}_k|}\right)^{M\omega[k]}.$$

Consequently,

$$\begin{aligned}I &= -\log_2 p \\ &= M \sum_{k=1} \omega[k] \log_2 \left(\frac{\pi[k]\;\omega[k]\;|\mathcal{D}|}{|\mathcal{W}_k|}\right) \\ &= M\Big(H_{|w|} - \log_2 |\mathcal{D}| \\ &\qquad + H_{\omega\times\pi} + \sum_k \log_2 \mathcal{W}|\omega[k]\Big) \\ &= M\Big(H_{|w|} - \log_2 |\mathcal{D}| \\ &\qquad + H_{\omega\times\pi} + \overline{\log_2 \mathcal{W}|}\Big) \quad (39)\end{aligned}$$

where

$$H_{\omega\times\pi} = -\sum_k \omega[k] \log_2 \pi[k]$$

is the *cross-entropy* between ω and π. Therefore, using the subscript ADW for *any dictionary word*, we have

$$\begin{aligned}\beta_{ADW} &= M\Big(H_{|w|} - \log_2 |\mathcal{D}| \\ &\qquad + H_{\omega\times\pi} + \overline{\log_2 |\mathcal{W}|}\Big) \\ &\quad \times 2^{M\left(H_{|w|} - \log_2 |\mathcal{D}| + H_{\omega\times\pi} + \overline{\log_2 |\mathcal{W}|}\right)} \quad (40)\end{aligned}$$

and

$$\log_2 \beta_{ADW} = M\Big(H_{|w|} - \log_2 |\mathcal{D}| + H_{\omega\times\pi} + \overline{\log_2 |\mathcal{W}|}\Big) + \log_2\Big(M\Big(H_{|w|} - \log_2 |\mathcal{D}| + H_{\omega\times\pi} + \overline{\log_2 |\mathcal{W}|}\Big)\Big). \quad (41)$$

Therefore,

$$\log_2 \beta_{ADW} \underset{L\uparrow}{\rightarrow} \frac{d}{dL} \log_2 \beta_{ADW} \underset{L\uparrow}{\rightarrow} \frac{L}{\overline{|w|}}\Big(H_{|w|} - \log_2 |\mathcal{D}| + H_{\omega\times\pi} + \overline{\log_2 |\mathcal{W}|}\Big). \quad (42)$$

Equivalently,

$$\frac{\log_2 \beta_{ADW}}{L} \underset{L\uparrow}{\rightarrow} \frac{d}{dL} \log_2 \beta_{ADW} \underset{L\uparrow}{\rightarrow} \frac{1}{\overline{|w|}}\Big(H_{|w|} - \log_2 |\mathcal{D}| + H_{\omega\times\pi} + \overline{\log_2 |\mathcal{W}|}\Big). \quad (43)$$

Special Case: Geometric Word Generator and Stationary Letter Source. Assume a stationary random letter source with entropy H_N. Then, from (10),

$$\overline{\log_2 |\mathcal{W}|} = H_N \sum_{k=1}^{\infty} k\, \omega[k] = H_N\, \overline{|w|}.$$

If we use the geometric word length generator distribution in (13), then (40) becomes

$$\beta_{ADW} = L\Big(H_N + H(\alpha) - \alpha \log_2 |\mathcal{D}| - \alpha H_{\omega\times\pi}\Big) \times 2^{L\left(H_N + H(\alpha) - \alpha\log_2 |\mathcal{D}| - \alpha H_{\omega\times\pi}\right)} \quad (44)$$

and (41) becomes

$$\log_2 \beta_{ADW} = L\Big(H_N + H(\alpha) - \alpha\left(\log_2 |\mathcal{D}| + H_{\omega\times\pi}\right)\Big) + 2^{L\left(H_N + H(\alpha) - \left(\alpha\log_2 |\mathcal{D}| + H_{\omega\times\pi}\right)\right)}. \quad (45)$$

Likewise (42) becomes

$$\log_2 \beta_{ADW} \underset{L\uparrow}{\rightarrow} L\Big(H_N + H(\alpha) - \alpha\left(\log_2 |\mathcal{D}| + H_{\omega\times\pi}\right)\Big) \quad (46)$$

and (43) becomes

$$\frac{\log_2 \beta_{ADW}}{L} \underset{L\uparrow}{\rightarrow} \frac{d}{dL} \log_2 \beta_{ADW} \underset{L\uparrow}{\rightarrow} H_N + H(\alpha) - \alpha\left(\log_2 |\mathcal{D}| + H_{\omega\times\pi}\right). \quad (47)$$

Special Case A: Webster's Dictionary. For the Webster's dictionary, we use $|\mathcal{D}|$ in (25) and the $\pi[k]$ data in Table 21.4 and calculate

$$\frac{\log_2 \beta_{ADW}}{L} \underset{L\uparrow}{\rightarrow} \frac{d}{dL} \log_2 \beta_{ADW} \underset{L\uparrow}{\rightarrow} 2.9851. \quad (48)$$

Special Case B: The Poisson Dictionary. A common model for the the distribution of word lengths in English is the shifted Poisson:[38]

$$\pi[k] = \frac{e^{-\lambda}\lambda^{k-1}}{(k-1)!} \;; 1 \le k \le \infty.$$

Fits of the Poisson for different error metrics using the data from *Webster's 2nd edition* in Table 21.4 are shown in Figure 21.3. Best fit values for λ are listed in Table 21.7 for each case. The values are about the same. We will use the default value of λ in (26).

	$\lambda + 1$
Histogram	9.59
MMSE	9.59
MinMax	9.66
KL	9.59

Table 21.7 Values of $\lambda + 1$ for the curves shown in Figure 21.3 resulting from minimizing different measures of error. λ is the average length of a dictionary word. MMSE is the minmax square error, and KL is the Kullback–Leibler divergence. This table shows that the different methods arrive at equivalent estimates of the average length of a dictionary word.

For a Poisson dictionary

$$|\mathcal{D}_k| = \frac{e^{-\lambda}\lambda^{k-1}}{(k-1)!}\,|\mathcal{D}|, 1 \le k < \infty. \qquad (49)$$

Then, for the geometric word length generator, the cross-entropy is

$$\begin{aligned} H_{\omega\times\pi} &= -\alpha\sum_{k=1}^{\infty}(1-\alpha)^{k-1}\log_2\frac{e^{-\lambda}\lambda^{k-1}}{(k-1)!} \\ &= -\alpha\sum_{k=0}^{\infty}(1-\alpha)^{k}\log_2\frac{e^{-\lambda}\lambda^{k}}{k!} \\ &= \left(\alpha\sum_{k=0}^{\infty}(1-\alpha)^{k}\log_2 k!\right) \\ &\quad - \log_2 e^{-\lambda} - \frac{\log_2\lambda}{\alpha} \\ &= \alpha\varphi(1-\alpha) - \log_2 e^{-\lambda} - \frac{\log_2\lambda}{\alpha} \end{aligned}$$

where the series

$$\varphi(z) := \sum_{k=0}^{\infty} z^k \log_2 k!$$

converges for $|z| < 1$.

A Bound on the Number of Words in the Poisson Dictionary. There are at most N^k words of length k. Thus,

$$|\mathcal{D}_k| \le N^k.$$

Using (49),

$$\begin{aligned} |\mathcal{D}| &\le |\mathcal{D}_k| \Big/ \left(\frac{e^{-\lambda}\lambda^{k-1}}{(k-1)!}\right) \\ &\le N \Big/ \left(\frac{e^{-\lambda}\lambda^{k-1}}{(k-1)!}\right) \end{aligned}$$

or, for all k

$$|\mathcal{D}| \le N \Big/ \left(\frac{e^{-\lambda}\left(\frac{\lambda}{N}\right)^{k-1}}{(k-1)!}\right).$$

The smallest this bound gets is when k maximizes the denominator. This occurs at the mode of the Poisson distribution which is

$$k_{\text{mode}} = \left\lceil \frac{\lambda}{N} \right\rceil .$$

Thus,

$$|\mathcal{D}| \le |\mathcal{D}|_{\max} = N \Bigg/ \left(\frac{e^{-\lambda} \left(\frac{\lambda}{N}\right)^{\lceil \frac{\lambda}{N} \rceil - 1}}{(\lceil \frac{\lambda}{N} \rceil - 1)!} \right) .$$

If $N \ge \lambda$, then $k_{\text{mode}} = 1$ and

$$|\mathcal{D}|_{\max} = N e^{\lambda}.$$

For $N = 26$ characters and, from (26), $\lambda = 8.5917$, we have

$$|\mathcal{D}|_{\max} = 181,516.$$

For this dictionary size, (47) becomes

$$\begin{aligned} \frac{\log_2 \beta_{ADW}}{L} &\underset{L\uparrow}{\rightarrow} \frac{d}{dL} \log_2 \beta_{ADW} \\ &\underset{L\uparrow}{\rightarrow} H_N + H(\alpha) \\ &\quad - \alpha \left(\log_2 |\mathcal{D}|_{\max} + H_{\omega \times \pi} \right) \\ &= 3.0339 \end{aligned} \tag{50}$$

which is almost identical to the value for Webster's dictionary in (48). A comparison can be seen in Table 21.6.

7.5 Plots

Plots of the three scenarios for the bit cost function derived in Section 7.3 are in Figure 21.1.

Scenario	$\lim\limits_{L\uparrow} \frac{d}{dL} \log_{10} \beta$	Numerical ($\log_{10}$)	Equation
EP (Equal Prob)	$\log_{10} N$	= 1.4150	
LO (Letters Only)		= 1.2840	(33)
L+W (Letters + Words)	$H_N + H(\alpha)$	= 1.4538	(37)
ADW (Any Dictionary Word)	$H_N + H(\alpha) - \alpha \left(\log_{10} \lvert\mathcal{D}\rvert - H_{\omega\times\pi} \right)$		
•Webster's		= 0.8986	(48)
•Poisson		= 0.9133	(50)

Table 21.8 Table of asymptotic log bit count per character derived in Section 7.4. Numerical values in this table are $\log_{10}$. These are the slopes of the lines in Figure 21.1. Details: The Webster $\log_2$ value of 2.9851 in (48) translates to 0.8986 in $\log_{10}$. The conversion factor is $\log_2 10$. Similarly, 3.0339 in (50) translates to 0.9133 for the Poisson case. $H_N = 4.1655$ in (20) translates to 0.6197 in base 10 and 4.8293 in (37) becomes 1.4538.

Notes

1. The sequence of dictionary words can be redundant. The criteria are so relaxed that a repetition like "a a a a a a a... a" would pass our definition of meaningful, since "a" is in the dictionary.
2. There are numerous ways to define information. For five different definitions, see Robert J. Marks II, "Information Theory & Biology: Introductory Comments," in *Biological Information: New Perspectives*, eds. Robert J. Marks II, Michael J. Behe, William A. Dembski, Bruce L. Gordon, and John C. Sanford (Singapore: World Scientific, 2013), 1–10. See also Robert Marks, "Robert Marks on Information and AI (Part I)," interview by Michael Egnor, *Mind Matters*, December 3, 2020, podcast, audio, 25:39, https://mindmatters.ai/podcast/ep111/. Transcript, https://mindmatters.ai/wp-content/uploads/sites/2/2020/12/Transcript-Mind-Matters-111-Robert-Marks.pdf. [AND THEN continue with "Note that in the present ...] chapter, *information* refers to Shannon information as either the number of bits required for computation or the number of bits of memory, as in the storage capacity of RAM expressed in gigabytes.
3. See William A. Dembski and Robert J. Marks II, "Conservation of Information in Search: Measuring the Cost of Success," *IEEE Transactions on Systems, Man, and Cybernetics - Part A: Systems and Humans* 39, no. 5 (Sept. 2009): 1051–1061; Robert J. Marks II, William A. Dembski, and Winston Ewert, *Introduction to Evolutionary Informatics* (Singapore: World Scientific, 2017); Daniel Andrés Díaz-Pachón, Juan Pablo Sáenz, J. Sunil Rao, and Jean-Eudes Dazard, "Mode Hunting through Active Information," *Applied Stochastic Models in Business and Industry* 35, no. 2 (2019): 376–393; Jonathan Bartlett, "Measuring Active Information in Biological Systems," *BIO-Complexity* 2020, no. 2 (2020): 1–11; Daniel Andrés Díaz-Pachón, Juan Pablo Sáenz, and J. Sunil Rao, "Hypothesis Testing with Active Information," *Statistics & Probability Letters* 161 (2020): 108742; Daniel Andrés Díaz-Pachón and Robert Marks II, "Generalized Active Information: Extensions to Unbounded Domains," *BIO-Complexity* 2020, no. 3 (2020): 1–6; Daniel Andrés Díaz Pachón and Robert Marks II, "Active Information Requirements for Fixation on the Wright-Fisher Model of Population Genetics," *BIO-Complexity* 2020, no. 4 (2020): 1–6.
4. See Dembski and Marks, "Conservation of Information in Search."
5. The endogenous information of winning a game of random moves with a randomly chosen first player is about 1.2 bits, corresponding to a probability of about 44%. (Ties count as losses.) If you go first and choose the center square as your first move, the active information for an otherwise random game is 0.42 bits (a 58% chance of winning). See RJMarksIII, "19—Intro to Computational Intelligence—Tic-Tac-Toe/Combs Control," *YouTube*, October 28, 2021, video, 1:20:53, https://www.youtube.com/watch?v=7iQZD8ZmQlQ.
6. Dembski and Marks, "Conservation of Information in Search"; see also Marks, Dembski, and Ewert, *Introduction to Evolutionary Informatics*.
7. See Tom M. Mitchell, "The Need for Biases in Learning Generalizations," originally published as Rutgers CS tech report CBM-TR-117, Laboratory for Computer Science Research, Department of Computer Science, Rutgers University (New Brunswick, NJ), May 1980, available online at https://www.cs.cmu.edu/~tom/pubs/NeedForBias_1980.pdf; see also Cullen Schaffer, "A Conservation Law for Generalization Performance," in *Machine Learning: Proceedings of the Eleventh International Conference*, eds. William W. Cohen and Haym Hirsh (New Brunswick, NJ: Morgan Kaufmann, 1994), 259–265.
8. David H. Wolpert and William G. Macready, "No Free Lunch Theorems for Optimization," *IEEE Transactions on Evolutionary Computation* 1, no. 1 (April 1997): 67–82; see also William A. Dembski, *No Free Lunch: Why Specified Complexity Cannot Be Purchased without Intelligence* (Lanham, MD: Rowman & Littlefield, 2002).
9. Some paths to climb a mountain encounter so many down-escalators that success is not possible.
10. Dembski and Marks, "Conservation of Information in Search."
11. There are models trying to explain the causality of the Big Bang, but in our view they remain highly speculative.
12. Rob Toews, "GPT-3 Is Amazing—And Overhyped," *Forbes*, July 9, 2020, https://www.forbes.com/sites/robtoews/2020/07/19/gpt-3-is-amazingand-overhyped/?sh=4ae392cc1b1c
13. John Naughton, "ChatGPT Isn't a Great Leap Forward, It's an Expensive Deal with the Devil," *The Guardian*, February 4, 2023, https://www.theguardian.com/commentisfree/2023/feb/04/chatgpt-isnt-a-great-leap-forward-its-an-expensive-deal-with-the-devil.

14. "GitHub and Copilot Intellectual Property Litigation," Joseph Saveri Law Firm, November 3, 2022, https://www.saverilawfirm.com/our-cases/github-copilot-intellectual-property-litigation. "AI Image Generator - Copyright Litigation," Joseph Saveri Law Firm, January 13, 2023, https://www.saverilawfirm.com/our-cases/ai-artgenerators-copyright-litigation. At this writing, these suits are still under consideration.
15. For a discussion of memes, see the remarks on Daniel Dennett by Michael Egnor, in his chapter "Neuroscience and Dualism" in this volume.
16. Ivo Schneider, "Jakob Bernoulli, *Ars Conjectandi* (1713)," in *Landmark Writings in Western Mathematics 1640–1940*, ed. Ivor Grattan-Guinness (Amsterdam: Elsevier, 2005), 88–104; Pierre-Simon de Laplace, *Théorie analytique des probabilités* (Paris, 1814). See also Eric W. Weisstein, "Principle of Insufficient Reason," Wolfram MathWorld (2002), https://mathworld.wolfram.com/PrincipleofInsufficientReason.html, and Marks, Dembski, and Ewert, *Introduction to Evolutionary Informatics.*
17. Variation of assumed number of outcomes is used to criticize Bernoulli's PrOIR. What is the chance getting two heads when flipping a fair coin twice? The classic answer of one chance out of four assumes outcomes of HH, HT, TH and TT. However, if the problem outcomes are posed as "HH" and "not getting HH" then there are only two outcomes, and blind application of Bernoulli's PrOIR says the chance of HH is one-half. See Hans-Werner Sinn, "A Rehabilitation of the Principle of Insufficient Reason," *The Quarterly Journal of Economics* 94, no. 3 (1980): 493–506. Such criticisms are not applicable to use of Bernoulli's PrOIR in this chapter. See, e.g., William A. Dembski and Robert J. Marks, "Bernoulli's Principle of Insufficient Reason and Conservation of Information in Computer Search," in *2009 IEEE International Conference on Systems, Man and Cybernetics* (Red Hook, NY: Curran Associates, 2009), 2647–2652, and Marks, Dembski, and Ewert, *Introduction to Evolutionary Informatics.*
18. Thomas M. Cover and Joy A. Thomas, *Elements of Information Theory* (Hoboken, NJ: John Wiley & Sons, 2012).
19. There is structure assumed in our analysis in that the building blocks of prose creativity are initially assumed, namely English characters.
20. Robert J. Marks II, "Diversity Inadequacies of Parallel Universes: When the Multiverse Becomes Insufficient to Account for Conflicting Contradistinctions," *Perspectives on Science and Christian Faith* 71, no. 3 (Sept. 2019): 146–152.
21. Seth Lloyd, "Computational Capacity of the Universe," *Nature* 406 (Aug. 2000): 1047–1054.
22. A bit decade is the increase in characters when the number of bits increases by a factor of ten.
23. Lloyd, "Computational Capacity of the Universe."
24. Katrin Becker, Melanie Becker, and John H. Schwarz, *String Theory and M-theory: A Modern Introduction* (Cambridge, UK: Cambridge University Press, 2007).
25. Marks, "Diversity Inadequacies of Parallel Universes."
26. A. Linde and V. Vanchurin, "How Many Universes Are in the Multiverse?," *Physical Review D* 81, no. 8 (Apr. 2010).
27. Becker, Becker, and Schwarz, *String Theory and M-theory*; B. G. Sidhart and R. Joseph, "Different Routes to Multiverses and an Infinite Universe," *Journal of Cosmology* 4 (2010): 641–654; Peter Woit, *Not Even Wrong: The Failure of String Theory and the Continuing Challenge to Unify the Laws of Physics* (New York: Penguin Random House, 2011).
28. 10^{1244} = 100, 000, 000, 000, 000, 000, 000, 000, 000,
000, 000, 000, 000, 000, 000, 000, 000, 000, 000,
000, 000, 000, 000, 000, 000, 000, 000, 000, 000,
000, 000, 000, 000, 000, 000, 000, 000, 000, 000,
000, 000, 000, 000, 000, 000, 000, 000, 000, 000,
000, 000, 000, 000, 000, 000, 000, 000, 000, 000,
000, 000, 000, 000, 000, 000, 000, 000, 000, 000,
000, 000, 000, 000, 000, 000, 000, 000, 000, 000,
000, 000, 000, 000, 000, 000, 000, 000, 000, 000,
000, 000, 000, 000, 000, 000, 000, 000, 000, 000,
000, 000, 000, 000, 000, 000, 000, 000, 000, 000,
000, 000, 000, 000, 000, 000, 000, 000, 000, 000,
000, 000, 000, 000, 000, 000, 000, 000, 000, 000,
000, 000, 000, 000, 000, 000, 000, 000, 000, 000,
000, 000, 000, 000, 000, 000, 000, 000, 000, 000,
000, 000, 000, 000, 000, 000, 000, 000, 000, 000,
000, 000, 000, 000, 000, 000, 000, 000, 000, 000,
000, 000, 000, 000, 000, 000, 000, 000, 000, 000,
000, 000, 000, 000, 000, 000, 000, 000, 000, 000,
000, 000, 000, 000, 000, 000, 000, 000, 000, 000,
000, 000, 000, 000, 000, 000, 000, 000, 000, 000,
000, 000, 000, 000, 000, 000, 000, 000, 000, 000,
000, 000, 000, 000, 000, 000, 000, 000, 000, 000,
000, 000, 000, 000, 000, 000, 000, 000, 000, 000,
000, 000, 000, 000, 000, 000, 000, 000, 000, 000,
000, 000, 000, 000, 000, 000, 000, 000, 000, 000,
000, 000, 000, 000, 000, 000, 000, 000, 000, 000,
000, 000, 000, 000, 000, 000, 000, 000, 000, 000,
000, 000, 000, 000, 000, 000, 000, 000, 000, 000,

000, 000.

29. Lorenzo Magnani, Abductive Cognition: *The Epistemological and Eco-Cognitive Dimensions of Hypothetical Reasoning* (Berlin: Springer Science & Business Media, 2009).
30. Andrew Kertesz, "Is Language Prewired in the Brain?," *Journal of Neurolinguistics* 3, no. 1 (1988): 29–37.
31. Kate Elgar and Ruth Campbell, "Annotation: The Cognitive Neuroscience of Face Recognition: Implications for Developmental Disorders," *Journal of Child Psychology and Psychiatry* 42, no. 6 (2001): 705–717.
32. Stanislas Dehaene and Ghislaine Dehaene-Lambertz, "Is the Brain Prewired for Letters?," *Nature Neuroscience* 19, no. 9 (2016): 1192–1193.
33. Stuart Hameroff, "Quantum Computation in Brain Microtubules? The Penrose-Hameroff 'Orch OR' Model of Consciousness," *Philosophical Transactions of the Royal Society of London Series A—Mathematical Physical and Engineering Sciences* 356 (1998): 1869–1896. Note that Penrose's model is not without its critics. See Scott Aaronson, *Quantum Computing since Democritus* (Cambridge, UK: Cambridge University Press, 2013).
34. Toshiko Kaneda and Carl Haub, "How Many People Have Ever Lived on Earth?," PRB (Population Reference Bureau), Nov. 15, 2022, https://www.prb.org/articles/how-many-people-have-ever-lived-on-earth/. This article is an updating of Haub's original article from *Population Today* 23, no. 2 (1995): 4–5. Note that we are unable to find similar figures for all mammals, all vertebrates, or even all eukaryotes. However, considering B_U, inclusion of these larger numbers would not affect our conclusions.
35. Mark H. Bickhard, "Troubles with Computationalism," in *The Philosophy of Psychology*, eds. W. O'Donohue and R. F. Kitchener (London: Sage, 1996), 173–183; Gualtiero Piccinini, "Functionalism, Computationalism, and Mental States," *Studies in History and Philosophy of Science Part A* 35, no. 4 (2004): 811–833.

22. The Human Mind's Sophisticated Algorithm and Its Implications

Winston Ewert

1. Introduction

Is THE human mind a computer? If not, what is it? Before we can answer, we must first clarify, what exactly is a computer? Historically, the term computer actually referred not to machines but to humans. Typically, these were teams of people working together to perform long and tedious calculations. They helped with such tasks as computing the positions of planets, producing mathematical tables, and simulating fluid dynamics. What made them computers was that they were following a procedure. They were not expected or allowed to engage in creative thinking or problem-solving; instead, every action they took was guided by the procedure given to them. All that our modern computers do is automate this procedure-following activity. Human computers and machine computers are similar in that they operate strictly by following a procedure.

What exactly constitutes a procedure? A procedure provides a step-by-step method for solving a particular class of problems. The procedure defines how to proceed at every step of the task, leaving no decision up to the judgment of the person or machine following the procedure. In the context of computers, these procedures are typically called algorithms.

If we ask whether a task can be accomplished by a computer, we are in essence asking whether it can be reduced to such a procedure or algorithm. Is it possible to devise a procedure which can accomplish the given task? It can readily be seen that some tasks, such as adding or multiplying numbers, can be reduced to following a procedure. After all, children are taught to follow a procedure in order to perform these tasks. Other tasks may seem harder, such as solving a high-school algebra equation or finding the best Scrabble word. Humans do not consciously follow a procedure when performing these tasks; nevertheless, these tasks can be and have been reduced to a procedure. A software engineer's job is essentially to reduce tasks to procedures that can then be encoded into computer software. As such, software engineers have reduced many tasks to procedures which humans perform without consciously following a procedure.

However, not every task can be reduced to a procedure. Researchers working in theoretical computer science have proven that a number of tasks cannot be reduced to a procedure. There is no procedure that can be written that will reliably perform these tasks. For example, there is no procedure that determines whether a logical statement, in first-order or higher logic, follows from a given set of premises. (First-order logic allows statements such as that there exists a man named Socrates or that all men are mortal.) No procedure is able to determine whether any Diophantine equation, a polynomial equation restricted to integer solutions, is solvable. No procedure can determine whether any other procedure will correctly finish or will instead get stuck in an infinite loop. See Chaitin's book *Meta Math!*[1] for more discussion of tasks which algorithms or procedures cannot solve.

Do the capabilities of the human mind fall into the first or second categories? Is everything that the human mind can do reducible to a procedure or program, even if we are not consciously aware of the procedure? Could we, in principle, duplicate the abilities of the human using a computer program? Or are there at least some tasks that the human mind can accomplish which cannot be reduced to a procedure? Are there things that the human mind can do which could not be duplicated by any procedure or program?

The human mind includes at least two aspects: phenomenal consciousness and problem-solving cognition. Other aspects of the mind certainly exist, but we will focus on these two aspects. Phenomenal consciousness refers to our ability to experience *qualia.* "Qualia" are properties of things (e.g., redness) experienced apart from the things having those properties (e.g., a red ball). Qualia are clearly not reducible to a procedure. A procedure can no more experience consciousness than a formula can. In contrast, problem-solving cognition is the ability to think, to decide, and to draw conclusions in answer to a particular question or problem. We will argue that the human mind's problem-solving cognitive abilities are consistent with that of a highly sophisticated procedure. We propose that the human mind consists of a computational component responsible for problem-solving cognition as well as a non-computational component responsible for phenomenal consciousness.

Regarding the first component, a computational process is only as good as the procedure or program that is running on that computational process. In order to match human abilities, that program must be highly sophisticated, far more sophisticated than the most advanced programs we have devised. Furthermore, in order for a computational process to devise such an advanced program, it would have to be running a still more sophisticated program. As will be shown here, this implies limitations to the abilities of artificial intelligence, eliminates the possibility of a singularity (a point where artificial intelligence surpasses and then far exceeds human intelligence[2]), and points to a transcendent origin of human intelligence.

2. Phenomenal Consciousness

Phenomenal consciousness is perhaps the most salient aspect of the human mind. As humans, we experience. We know sadness, happiness, anger, and joy. We experience the flash of insight and frustration of failure. We have seen redness, brightness, and roughness. All of these experiences are called qualia. Explaining how it is that the human mind experiences qualia is a very difficult problem, so much so that it has been called the hard problem of consciousness.[3]

This problem is too difficult; we will not attempt to solve it. However, we will observe that consciousness cannot be explained by a procedure. Procedures are abstract ideas. In particular, the sort of procedure which can be run on a computer is strictly a transformation of information, essentially a formula. Unless one is willing to claim that mathematical formulas are conscious, it must be concluded that procedures or programs are not conscious either. Whatever explains consciousness must be something non-computational. The mind must incorporate something beyond computation or procedure-following in order to account for consciousness.

A common approach to explaining consciousness is that it is somehow caused by the physical instantiation of computation. That is, it is not computation in the abstract that is conscious or generates consciousness, but the physical instantiation of that computation. Whatever the merits of this explanation, it remains true that consciousness is something in addition to computation. It is not the procedure itself, but something about the nature of the world which is responsible for consciousness.

However, some argue that we humans do not have consciousness, at least not in the way that we think we do.[4] Instead, some say, our consciousness is an illusion. Our mind is simply deluded into believing that it experiences consciousness. Or perhaps it does experience consciousness but is deluded into believing that this is different from computation. If we are willing to call into question the accuracy of human reasoning, it is possible, no matter how unassailable they may seem to us, that our conclusions about the reality of consciousness are simply wrong, an artifact of a faulty brain. However, by the same token every conclusion of the human mind may be wrong. Perhaps we are all insane and our every thought complete nonsense. Nevertheless, for practical purposes we must assume that our minds are accurate and thus that consciousness is not an illusion.

Consciousness may not be computational, but is it physical? This question cannot be answered without a definition of physicality. What actually makes something physical? What would it mean for something to fall outside of the domain of physics? Even if something is currently outside the domain of physics, will it remain so? For example, the action at a distance of gravity was considered non-physical when it was first discovered but is now considered easily within the domain of physics.

Certainly, no accepted theory in physics can now explain consciousness. It is, in some sense, outside of the domain of physics right now. However, the domain of physics has been expanded in the past to include concepts such as action at a distance, observer effects, and indeterminacy. It seems that nothing would prevent that domain from being expanded again to include entities required to explain consciousness. Given this possibility, it does not seem to be a useful question to ask whether consciousness is physical.

Theories of consciousness tend to fall into one of two categories. According to panpsychism or emergentism, consciousness is a byproduct of the brain.[5] It is not an integrated part of the mind, but rather some sort of side-effect that is caused by the brain due to some strange quirk in the laws of nature. This is the sort of explanation mentioned a few paragraphs above, in which the physical instantiation of computation somehow generates consciousness. Alternatively, consciousness might be a deliberate and integrated component of the mind: the mind might be a complex system of multiple components, one of which is the entity responsible for consciousness.

One reason to prefer the integrated component view is that that our consciousness appears to exhibit at least some control over our actions. If our consciousness is a byproduct of the brain, we would expect that there would be a purely one-way connection, i.e., the consciousness experiences what happens in the brain, but does not in turn influence the actions taken by the brain. On this view, our conscious minds may be deluded into thinking they control our actions, but they are simply along for the ride, experiencing the illusion of control. Yet we write and talk about consciousness. If our conscious minds do not in any way influence our brains, how is that our brains "know" about the conscious mind to talk about it? The brain must be aware in some sense of our conscious experiences. Conceivably, the brain might be deluded into thinking it has conscious experiences separately from our actual consciousness experience. But this seems a coincidence quite beyond belief. It seems much more probable that the brain is controlled to some degree by consciousness.

Another reason is the limitation of our consciousness. Our consciousness extends to a subset of our whole mind. It is as though our consciousness was designed to be aware of our rational thought and emotions while other aspects of our mental processes and body were deliberately excluded. This makes sense if our minds are a deliberately designed system where consciousness constitutes one part of the whole. If consciousness is a mere side-effect, this delineation makes much less sense. Why do we have this particular scope of consciousness? Why are we not conscious of every aspect of the human mind? Why does the consciousness not extend to our other internal organs? Why does each person have a separate consciousness instead of belonging to a hive mind? Alternatively, why is our mind not divided into many separate consciousnesses? The "side-effect" understanding of consciousness fails to account for the particular limitations of our conscious mind.

It therefore seems most plausible that consciousness is not computation and is not a simple byproduct of physical computation. Instead, our conscious mind is an integrated part of overall mind. It is a key component in the overall system of the human mind. Beyond that, the nature of the causes of consciousness are unknown. We think it is better not to speculate on their nature and to move on to problem-solving cognition.

3. Problem-Solving Cognition

3.1 The Übertask: The Halting Problem

There are many different cognitive tasks that humans undertake. We calculate mathematical operations, prove theorems, identify patterns, predict future observations, design solutions, etc. All of these can be understood as solving different kinds of problems or answering different kinds of questions. These would all seem to be quite different aspects of human cognitive abilities. However, they are not completely different cognitive tasks, but are instead closely related.

In fact, all of these different cognitive abilities can be subsumed under one ability, which I will call the *übertask*. If you had a technique or tool that allowed you to efficiently and reliably solve the übertask, you could use that tool to perform any of the other cognitive abilities. As such, all cognitive abilities are, in some sense, simply special cases of the übertask.

What is this übertask? It is the halting problem.[6] Given a particular program or procedure, does that program halt or loop forever?

As an example, consider the following program or procedure:

1. Twiddle your thumbs.
2. Go to Step 1.

If you think about this procedure, you should quickly see that it will never finish. A human or machine that seeks to follow this procedure will unavoidably move back and forth between Step 1 and Step 2 for all eternity. The procedure will never be finished. We say that such a program loops and does not halt.

You may be inclined to object that realistically no person or machine will follow the procedure forever. Eventually the person will give up or the machine breaks down. This is true. But that is a practical issue, and here we are interested in the theoretical one. If one followed the procedure unswervingly, would the procedure eventually end? Regardless of the practical issue, that question has a yes or no answer.

In contrast, consider another program:

1. Dance a jig.
2. Repeat Step 1 a million times.

If you think about this procedure, you should be able to determine that it will eventually finish. The poor follower of this procedure will have to dance a million jigs but will eventually reach the end of the procedure. Unlike the first program, this procedure will not go on forever. Instead, this program halts.

Every program must either halt or loop. There is no other option. A program cannot halt-loop or loop-halt. By the nature of a program or procedure, there cannot be any paradoxical program which neither halts nor loops. Furthermore, the resolution is one of objective truth; there is no subjective element to whether or not a program halts. There is an objective answer to the question: Does this program halt? The answer is always either yes or no.

It may seem a straightforward matter to determine whether or not a program halts. After all, it was done easily enough for the first two examples presented. However, not all programs are so straightforward. Consider an example:

1. Start with the number 4.
2. Check if the number is prime.
3. If it is prime, halt.
4. Otherwise, add 2 to the number and go to Step 2.

Determining whether this program halts requires a little more thought and requires more familiarity with math. If you can follow the logic of the program and know about prime numbers, you will be able to determine that it will never halt. The program checks for prime numbers, adding two each time. This means that the program will only look at even numbers. However, two is the only even prime, and thus this program is doomed to futility, searching for a non-existent even prime.

You might think that you could determine whether a program loops or halts simply by running it. Obviously, if you run the program and find that after some short period of time, it finishes, then you can conclude that it halts. However, if after running the program for a long period of time, it still has not halted, what can you conclude? You may run the program for hours, days, years, or even millennia. But you can never be certain that the program would not halt if run for just a little longer.

For the moment, let's assume we have a technique that enables us to reliably determine whether or not a given program halts. It does not matter whether this is by human intuition, inspecting goat entrails, or careful analysis. For now let's pretend that we could reliably and efficiently solve this sort of problem. How could we use that in order to solve other cognitive tasks?

3.2 Mathematical Reasoning

In mathematics there is a conjecture, or proposed rule, known as Goldbach's conjecture. It was originally proposed in the eighteenth century, and states that "every even number greater than 2 can be expressed as the sum of two primes." For example, eighty-eight is seventeen plus seventy-one; sixteen is eleven plus five; 120 is 67 + 53.

This rule has been checked for all numbers up to four quintillion. No exceptions have been found. For every even number up to four quintillion, prime numbers have been found such that their sum is equal to that number. However, no proof has been found for the conjecture. As such, mathematicians regard it as neither proven nor disproven. It is an unsolved problem.

Consider the following program:

1. Start with the number 4.
2. Determine whether it can be expressed as the sum of two primes.
3. If it cannot, halt.
4. If it can, add two to the number and go to Step 2.

This is a program that would systematically go through the even integers (from 4 up) to see if there is an exception to Goldbach's conjecture. In principle, it is the same sort of procedure that was used to verify the lack of exceptions up to four quintillion. This program will halt if it finds an exception. However, if there are no exceptions, it will loop forever. As such, to ask whether this program halts is equivalent to asking whether or not the Goldbach conjecture is true. If we had the ability to determine whether a program halts, we could use it on this program and determine the truth or falsity of the Goldbach conjecture.

In fact, it should be apparent that many similar postulates could be checked in this manner. There is nothing particularly special about the Goldbach conjecture. In any case where we can systematically count through possible counter-examples and verify each one, an analogous program can be constructed. If we could determine whether or not that program halted, we could determine whether or not the conjecture was true.

However, we can also consider another conjecture, the twin prime conjecture. This conjecture states that there is an infinite number of twin primes, primes which differ by two. Examples of twin primes include: 11 and 13, 17 and 19, and 137 and 139. This conjecture cannot be tested in the same way as Goldbach's conjecture, because no simple single counter-example can show the conjecture to be false.

However, mathematicians are not actually primarily concerned with whether these conjectures are true. They are concerned with whether or not they can be proven. Mathematicians search for proofs or disproofs of propositions of interest. Mathematical reasoning is not essentially about attempting to divine the truth or falsity of mathematical propositions but about exploring provable inferences.

A proof is a logical chain of argument where each proposition necessarily follows from the previous propositions. Mathematicians argue from propositions they are very certain are true to other propositions that they are uncertain about.

Consider the following program:

1. Start with the shortest possible proof, that is, the one with the fewest and simplest steps in it.
2. Determine whether the proof is a valid proof for the twin primes conjecture.
3. If it is, halt.
4. If it is not, move onto the next shortest proof and go to Step 2.

This is a program which will halt if there is a valid proof for the twin primes conjecture; otherwise it will loop forever. If we had our magic halting detection device, we could use it to determine whether or not there was a proof for the twin primes conjecture.

The astute reader will have realized that this is not restricted to the twin primes conjecture. Instead, any mathematical proposition might be substituted in this program. The same basic technique could be used to determine whether any mathematical proposition is provable. Likewise, the claim could be reversed to determine whether the same proposition is provably false.

If we had the ability to determine whether or not a program halted, all tasks of mathematical reasoning would be readily solved. For any given proposition, we could classify it as provably true, provably false, or unprovable either way. Since math is the task of proving and disproving propositions, math is, in some sense, a special case of solving the halting problem.

3.3 Pattern Detection

Humans identify patterns all around us. It is colder in the winter and warmer in the summer. Tree trunks contain concentric rings. Economies tend to have upswings and downswings. In fact, humans are arguably overly prone to identify patterns and see them where none exist.

Consider the following three sequences. Can you identify which have patterns and what those patterns are?

0101010101010101010101010101010101
1001111100011111001101010000110000
0110111001011101111000100110101011

The first has a pretty obvious and easily identifiable pattern. The second is a random string of ones and zeros generated by my computer. The third has a pattern, but identifying this pattern is left as an exercise to the inquisitive reader.

One way to characterize a pattern is to describe it using a program. The idea is that if there is a pattern, it will be possible to use that pattern to produce a shorter description. For example, the first binary string can be represented in the Python programming language as:

"01" * 17

Whereas the second string must be represented as something longer:

"1001111100011111001101010000110000"

Such differences in length give rise to the algorithmic information theory of Kolmogorov complexity, which characterizes the complexity of a binary string as the length of the shortest program which outputs that string.[7] The idea is that if a short program suffices to reproduce a string (which is presumed to encode the information of interest), the string is less complex and highly patterned, while if a longer program is required, the string is more complex and less patterned. Thus, short programs are associated with patterns and long programs with randomness.

The basic question is: Does this observation exhibit a pattern? An equivalent question is: Can one write a program which reproduces this observation in a shorter fashion?

Consider the following program *P*:

1. Run all programs of length less than some arbitrary threshold *l*. All programs should take turns running for a small amount of time so that they run in an interleaved or parallel fashion.
2. If any program (from Step 1) halts and outputs the desired output *x*, then halt *P*.
3. Otherwise, wait for further programs (from Step 1) to halt.

This program will halt if there is a program of less than a particular size that has the desired output. Otherwise, assuming there are one or more programs less than length *l* that loop forever, program *P* will loop forever. If we could reliably and efficiently determine whether or not a program halts, we could use that to determine whether a short program exists that produces the desired output *x*. This means that if we could determine whether programs halt, that same technique would be applicable to determining whether any data we have has structure (i.e., low Kolmogorov complexity) or is just random noise (i.e., high Kolmogorov complexity).

3.4 Predictions

Humans are often seeking to predict the future. We wish to know the outcome of an election or sporting event before it happens. We want to predict the weather tomorrow. We want to predict the outcome of mixing two particular chemicals. In practice, we wish to make these predictions on the basis of past observations.

Any such prediction is based on identifying a pattern and making a prediction on the basis of that pattern. For example, we predict the weather on the basis of past observations of patterns of weather. We assume that such patterns will continue and make a prediction on that basis.

However, this brings us to the problem of induction. On what basis do we assume that past patterns will continue? This is rendered particularly difficult because in many cases patterns do not continue. For example, if you were to observe that the days are getting shorter and shorter, you might reasonably conclude that the days would continue to get shorter and shorter until it is always dark. However, in fact, this is only the onset of winter, and the pattern will reverse itself after the winter solstice.

A famous example is that of asking whether or not the sun will rise tomorrow. We regard this as a virtually certain fact, on the basis of our observations of the sun rising every day for as long as we can remember. However, a turkey observes that he is fed by the farmer every day until that one day when he is killed and eaten. The observation of a consistent pattern is certainly no guarantee that the pattern will continue.

A common proposed resolution is Bayesian reasoning. Bayesian reasoning centers on the use of Bayes's theorem:

$$Pr[T|E] = \frac{Pr[E|T]}{Pr[E]}\ Pr[T]$$

According to Bayesian reasoning, three factors matter when determining how probable it is that a certain theory is true. Firstly, $Pr[T]$, how probable was the theory (T) before considering the evidence (E). Secondly, $Pr[E|T]$, how likely are we to observe the given evidence assuming that that theory was true. Thirdly, $Pr[E]$, how probable is the given evidence regardless of whether or not the theory was true. A theory is most probably true if the theory was probable before considering evidence, the evidence is highly probable assuming that the theory was true, and the evidence is not highly probable if the theory is false.

The fundamental difficulty of Bayesian reasoning is the determination of the probability of the theory before considering the evidence, known as the prior probability. (Note that since $Pr[T|E]$ is the probability of the theory after considering the evidence, it is known as the posterior probability.) The probability that a theory is true depends on how likely that theory was to be true before considering any evidence. Sometimes, the importance of this prior is dismissed by arguing that it does not matter. After considering sufficient evidence, the contribution of the prior ought to be insignificant. However, this depends on the prior.

Consider, for example, the case of determining whether or not the sun will rise tomorrow. Suppose that we have observed the sun to rise every day for a thousand days. Will it rise on day one thousand and one? Consider as the prior, the perfect uniform uninformed prior: all possible sequences of rising and not rising are equally probable. This seems fair as it makes no assumptions about the pattern.

There are two theories of interest. One theory says that the sun will rise every day for all one thousand and one days. The second theory says that the sun will rise every day for one thousand days but will fail to rise on the thousand-and-first day. All other possibilities have been contradicted by the evidence. But the observed sequence of a thousand days of rising suns is 100% compatible with either of the remaining theories. Both theories are thus equally probable. On our perfect uniform uninformed prior, Bayes's theorem will give each theory 50% probability. Bayesian inference did not help at all!

But this does not mean that Bayesian inference is hopeless. Rather, it means that Bayesian inference alone is insufficient. Bayesian inference only works if we additionally adopt some sort of assumption about the nature of the priors. In particular, anyone engaged in Bayesian inference will implicitly or explicitly consider the arbitrary pattern of the sun rising for one thousand days before ceasing as having less prior probability than the simple pattern of the sun rising every day.

One possible prior to adopt is that generated by selecting random computer programs and running them. The idea is that we program a computer by a sequence of coin flips and run the result to see what it does. Simple patterns which can be generated by short programs will be very common. Complex patterns which required long programs will be relatively rare. Programs that produce particular random sequences of data will be extremely rare. This is called Solomonoff induction.[8]

In the case of the sun rising (which we may indicate by the digit 1), there are many different programs that will output a sequence of 1's. On the other hand, it would be rare to find a program that outputs a sequence of exactly 1000 1's and then a 0 (indicating the non-rising of the sun). On that basis, we can conclude that even though both theories are consistent with observation, the probability of the sun consistently rising has a higher prior probability and thus a higher posterior probability. Numerous other programs will have been eliminated for not following our observations.

However, actually engaging in induction of this nature requires computing the probability of a particular outcome assuming it was generated by a randomly selected computer program. This is easier said than done. However, if we had the ability to solve the halting problem reliably and efficiently, we could use that solution in order to perform this calculation. For example:

1. Run all programs in parallel.
2. If any program gives the correct output, add the probability of selecting that program at random to the overall probability tally.
3. If the probability exceeds some threshold, halt.
4. Otherwise, continue running the programs in parallel.

The astute reader may wonder how we can run the infinite set of programs in parallel. There is a technique for doing this called interleaving, which gradually increases the set of programs being run so that at one point in time only a finite number of programs are run, but that eventually runs any given program given enough time.

The above program halts if and only if the probability of a particular output is greater than or equal to a particular threshold. If we could solve the halting problem for this sort of program, we could determine whether or not the probability is greater than any given value. This would allow us to determine how much probability we should assign to a particular outcome. As such, predicting the future is also a special case of solving the halting problem.

Furthermore, even if one is disinclined to adopt Solomonoff induction or Bayesian reasoning, the same conclusion follows. However induction might be performed, it must involve the identification of patterns. However, as shown in the previous section, the identification of patterns is a special case of solving the halting problem. Thus, any predictions based on that identification must also involve solving instances of the halting problem.

3.5 Search

Many human tasks can be characterized as selecting from a large set of possible options. For example, writing is the selection of words to convey particular meanings from the large set of all possible combinations of words. A circuit designer selects a particular configuration of wires and components from the large set of all possible circuits. An interior designer selects a particular combination of colors, furniture, and decorations from another large set.

Often, humans will have to investigate multiple possible choices before finding the right one. This is sometimes called "search," as the person is searching through possible choices in order to determine which one achieves the desired outcome. The "search space," the conceptual range of all possible choices, is usually much too large to look at every possibility. Instead, judicious use must be made of resources in order to effectively identify a good choice.

The No Free Lunch theorems[9] show that if we assume a uniform probability, all search strategies will work equally well. That is, if we assume nothing about the search space, it is impossible to make judicious use of resources. Therefore, we must make some sort of assumption about the nature of the search space in order to make judicious use of resources. We must assume some probability distribution other than the uniform.

But this is effectively the same problem as discussed in predicting the future. In this case, we wish to predict the quality of choices not yet investigated. That way it is possible to focus our resources on the choices most likely to be of high quality. We must identify the patterns in the data, and then use those patterns to predict unexplored areas of the search space. Just as in the case of predicting the future, this is a special case of solving the halting problem.

3.6 Summary Regarding the Halting Problem

We have seen that mathematical reasoning, identifying patterns, making predictions, and exploring search spaces are all special cases of the halting problem. That is, if we could reliably and efficiently solve the halting problem, we could use that to solve all of these other problems. This is why I have called the halting problem the übertask. In many cases, the reverse procedure is also possible: the ability to reliably and efficiently perform some, perhaps all, of these tasks could be exploited to solve the halting problem. However, the halting problem is well studied, and thus it is more helpful to view the other tasks as special cases of the halting problem. All of these different cognitive tasks are part of the same general cognitive ability. We can study the halting problem, and the conclusions will apply to these other kinds of cognitive tasks as well.

4. Limitations on Solving the Halting Problem Algorithmically

4.1 Can the Halting Problem Be Solved?

If we could reliably and efficiently solve the halting problem, that could be exploited to solve a wide array of cognitive tasks. However, we have no technique or tool which solves the halting problem. In fact, we can prove that no procedure or program could possibly exist which solves the halting problem for every possible program. Consider the following program which certainly cannot have access to a reliable halting detector:

1. Use the assumed available halting detector to determine whether this program halts.
2. If it halts, loop forever.
3. If it loops forever, halt.

This is a contrarian program. Whatever the halting detector claims will happen, this program does the opposite. Therefore, it must be the case that any procedure invoked to determine whether or not this program halts must be giving the wrong answer. If no procedure can determine if this program halts, then clearly no procedure can exist which solves the halting problem in general.

However, a slight variation changes the conclusion. If we restrict the programs allowed to those which require a limited amount of memory, then it is possible to determine whether or not those programs halt. Since a program is deterministic, its actions only depend on the state: which part of the program is running and what is in memory. Given a memory limit, there can only be a finite number of different states. A looping program must eventually repeat the same state. But since the same state will cause the program to take the same action, this will result in the program looping, repeating a cycle of states indefinitely. Thus, we can determine whether such a program halts by running it until it either halts or repeats a state.

It may seem that that this conclusion contradicts the proof that a halting detector cannot exist. The proof demonstrates that no halting detection procedures works for all possible programs. The proposed algorithm determines whether a particular subset of programs halts, namely those programs which use a restricted amount of memory. We see that the halting problem is not absolutely unsolvable. We can, in fact, solve subsets of the halting problem. More generally, we can speak of "partial halting detectors," procedures which indicate that a program either halts, loops, or is of unknown status. This avoids the proof of impossibility because a partial halting detector can simply indicate unknown for itself. However, such a detector may be able to classify a wide variety of other programs.

4.2 Partial Halting Detectors

Consider any partial halting detector, no matter how sophisticated. There are always programs that it is unable to classify. A partial halting detector might be successful in classifying programs which use a limited amount of memory, follow certain common patterns, or halt after being run for a short time. Nevertheless, it will be unable to classify other programs which use more memory, more unusual patterns, or take a long time to halt.

It is easy to augment the partial halting detector to identify additional programs. At the very least, it could check for a specific known program and indicate whether or not it halts. More generally, it could check for some property from which the status of the program can be proven. However, even after augmentation, such a detector is still a partial halting detector. There still remain programs that it cannot classify. It could thus be augmented indefinitely. The consequence of this is that we can infer the existence of an infinite sequence of increasingly powerful partial halting detectors.

The partial halting detectors we have constructed are only a little way along this infinite sequence. We know that much further down the line there exist, in principle, programs which classify many more possible programs. Even further down the line, there are still more sophisticated programs which classify vastly more possible programs. No matter how far down the infinite line we go, there are still more sophisticated programs which correctly classify larger and larger subsets of possible programs.

4.3 Identifying a Halting Detector

How do we know that a given partial halting detector is accurate? Any trivial program might assign labels of halting, looping, or unknown to given inputs but this is not useful unless those labels are accurate. We need some way to verify whether a particular halting detector is actually giving accurate answers. We can, in fact, use a halting detector in order to verify it. Consider the following program:

1. Run the following steps over all possible programs in parallel.
2. Run a partial halting detector to determine whether the testee partial halting detector will halt on the given program.
3. If it will not halt, halt the overall program.
4. If it will halt, run the testee detector.
5. If the testee detector claims the program halts, determine how long the testee detector claims it will take to halt, and verify that it does halt at that point.
6. If it does not halt, halt the overall program.
7. If the testee detector claims the program loops, run the program; if it ever halts, halt the overall program.

This admittedly complex program searches for counter-examples to the testee detector's claims. It will halt if the internal partial halting detector cannot determine whether the testee detector will halt, if the testee detector claims that a program halts when it does not, or if the testee detector claims that a program loops when it does not. If this program loops, it means that the detector is good and accurate. Thus, if we can verify that this program loops, we are able to verify the correctness of the testee detector.

However, we see that this procedure runs all of the programs which the testee detector claims to either halt or loop. As such, determining whether or not the overall program halts or loops will require determining whether these individual programs halt or loop. The testing partial halting detector must therefore be able to determine the status of at least all of the programs that the testee detector is able to classify. Furthermore, it must be able to also classify the halting status of the testee detector itself. No halting detector can accurately classify itself without running afoul of the proof of the impossibility of the halting detector.

Consequently, the testing halting detector must be further along the infinite line of halting detectors than the testee. The only way to verify a partial halting detector is with a more powerful partial halting detector. Any technique for deriving a partial halting detector will need to be able to verify the correctness of the partial halting detector. This would suggest that the only way for an algorithm to obtain a partial halting detector is using a more powerful partial halting detector. The only other alternative is that the partial halting detector derives from something other than an algorithmic process. However, it should be acknowledged that this is a sketch of an argument for this and not a formal proof.

5. The Halting Problem and the Nature of the Human Mind

5.1 Comparing Human Abilities to Algorithms

How DOES the ability of humans compare to these algorithms? Humans are not reliably and efficiently able to solve halting problems in general. Humans cannot look at a procedure and always determine whether or not that procedure will halt. We are unable to consistently find out whether a proof is possible for a given statement. We cannot reliably determine whether there is a pattern in some data. We do not always offer ideal predictions or search optimally. This is true even when humans devote considerable time to resolving these questions.

Nevertheless, humans do have strong abilities in these areas. We can determine whether a wide variety of programs halt. We have proven many facts, and proven some others unprovable. We have identified many patterns, made accurate predictions, and found many high-quality solutions.

In each of these cases, programs and procedures have been devised which can do some of these tasks. We have programs that identify whether or not some procedures halt. We have automated theorem provers that can prove many mathematical statements. There are programs that identify patterns in data. A commonly used example is a compression program, which exploits these patterns to reduce file sizes. Search algorithms and inference algorithms also show wide usage. However, these programs inevitably pale in comparison to the abilities of humans. Humans are able to solve a much wider variety of these problems than any of these programs.

A wide variety of cognitive tasks can be thought of as special cases of solving the halting problem. It is impossible for a procedure to solve the halting problem in totality, but it is possible to solve partial versions of the problem. There is an infinite range of possible partial halting detectors, each being able to classify more than the last. The only way to obtain or verify a halting detector is using a more powerful halting detector. Humans are unable to solve the complete halting problem, but are able to solve a large and varied subset. They are able to solve a much larger set than our best algorithmic approaches. They also created those best algorithmic approaches.

Human cognitive ability is consistent with being a program very far down the infinite line of halting detectors. We know, mathematically, that very sophisticated programs exist far beyond the abilities of our current programs. Humans also display abilities far beyond those of computer programs, but this is consistent with human cognition being based in, or at least compatible with, a highly sophisticated computer program. Further, we observe human cognitive ability as being limited. We are not able to solve every instance of a halting problem, but only a large and varied subset. Therefore, it is most plausible that human cognitive ability is equivalent to highly sophisticated partial halting detector.

5.2 Gödelian Arguments

The position taken here, that human cognitive ability is equivalent to a highly sophisticated algorithm, is controversial. Many think instead that human cognitive ability exceeds what would be possible for any algorithm no matter how sophisticated. Some authors[10] have argued against a computational understanding of human cognitive ability on the basis of Gödel's incompleteness theorems.[11] The pure fact that humans can prove these theorems is not evidence that humans are not computational because automated proof systems can also prove these same theorems.[12] However, a more sophisticated argument has been made.

If human cognitive abilities are computational, this implies that there is some formally definable system of logic they are following. We can thus construct a statement of the form: "this statement cannot be proven according to the underlying human system of logic." This is called a Gödel statement. If we assume that the human system of logic is consistent, then it cannot prove this statement as it would be proving a false statement. The only alternative is that the statement is true. But this creates a contradiction, because humans were able to determine that a statement was true which did not follow from the assumed formally definable system of logic that humans follow. Ergo, human mental abilities are not computational.

A crucial step in the above argument is the construction of the Gödel statement. But this statement contains a complete formalization of the human system of logic. However, the contention of this chapter has been that we cannot do this. We are only capable of generating a formal system of less sophistication than the one we implicitly operate on. If we were provided with a formalization of our own logical system, we would not be able to verify its correctness. Put another way, we aren't smart enough to understand how our minds work. The consequence is that we also are not smart enough to construct our Gödel statement. Since humans cannot construct or prove their Gödel statement as correct, there is no contradiction.

Penrose[13] realizes this limitation of the argument. As he puts it, the argument only follows if the algorithm underlying human mathematical reasoning is "knowable." If, instead, mathematicians follow an "unconscious unfathomable algorithm," his argument does not work. He attempts to argue that that this is implausible due to the difficulty in accounting for the origin of such an algorithm. He concludes that "the only plausible way that our own [algorithm or formal system] could have arisen would have been by divine intervention." Penrose, it should be emphasized, is not endorsing divine intervention but arguing that it is the only viable alternative to his thesis.

5.3 Gradually Increasing Intelligence

The view that human intelligence is not computational does not explain the human inability to solve many instances of the halting problem. If human intelligence transcended computation, we might expect human beings to be able to solve all instances of the halting problem. However, as a matter of empirical fact, they are not able to do this.

One proposed resolution is to invoke some idea of gradually increasing intelligence. Human abilities at any point time are consistent with a program, but over time humans and the human race increase in their abilities. Effectively, their programs are augmented. Instead of operating on a fixed program humans operate on something like a program that is increasing in sophistication over time. On that infinite line of better and better halting detectors, humans are gradually crawling along it towards better and better detection.

As Bartlett explains, "Another option… is that humans are able to incrementally arrive at solutions to halting problems. This would mean that humans have access to an oracle which is more powerful than finitary computational systems, but less powerful than a halting oracle."[14]

Bringsjord describes it as follows: "The basic idea would appear to be that as human minds develop through time over generations, they invent new concepts and techniques, which in turn allow previously resistant problems to be solved. There seems to be no upward bound whatsoever to this ascension."[15]

It certainly is true that human knowledge has increased over time. Techniques and tools have been developed over time that allow us to determine facts that we previously did not know. In terms of programs, we are able to classify more programs as halting or not-halting than past generations could. However, the crucial question is whether this progress will be unending. On what basis do these authors think the progress will be unending?

Bartlett offers an interesting argument for his position on the basis of computer programmers:

> In software development, humans have to develop software programs on universal computation systems, and those programs must halt. If they do not, their programs will be broken. Therefore, they must solve problems on at least some subset of the halting problem in order to accomplish their tasks. In addition, the problems that they are given to solve are not of their own design, so it is not a selection bias. It is simply not true that programmers are only choosing the programs to solve based on their intrinsic abilities to solve them because someone else (usually someone without the computational background needed to know the difference) is assigning the programs. In addition, it is incorrect to assert that programmers are working around their inabilities to solve certain types of halting problems, because, while the programmer might add some extrinsic complexity to a program, the complexity of the problem itself has an intrinsic minimum complexity regarding a given programming language. Likewise, simply writing it in another language does not help, because there exists a finite-sized transformer from any language to any other language, so being able to solve it in one language is de facto evidence of being able to solve it in another.[16]

To restate his argument: programmers must, at least, be able to solve the halting problem for the programs that they write. Any given task might be solved with a number of different programs. However, in order for the programmer to be successful, he or she must be able to determine if at least one of those programs will halt. It seems an amazing coincidence that the tasks a programmer is given so regularly have solutions for which the programmer can solve the halting problem.

However, this is only compelling against the claim that humans have a very limited halting detection ability. If instead humans are designed with a wide and varied ability to detect halting programs and that ability is especially suited to the sorts of problems that humans have to face, it is not surprising at all that the programmer's tasks happen to fall within his abilities. This is not a reflection of the human ability to solve arbitrary halting problems, even on an incremental basis, but rather the well-suitedness of the human partial halting detector to the reality we find ourselves in.

Bringsjord bases his argument on the "busy beaver" numbers. The busy beaver problem is to determine the longest running program of a particular size. Determining this answer requires classifying all programs of that size into halting or looping. Humans studying the busy beaver problem have done this for some small sets of programs. Bringsjord argues that if the busy beaver problem can be solved for size n, it will eventually be solved for size $n + 1$. Thus, given enough time, any given busy beaver number will be known and the halting status of any program will also be known.

But why does Bringsjord think that success in finding n implies eventually success on $n + 1$? He writes, "Though we aren't claiming [this premise] is unassailable, it seems to us remarkable that one would maintain its denial: that is, that one would maintain that it is *impossible*, however long the analysis takes, to move from success on n to success on n + 1."[17]

However, this position is not self-evident. Humans are limited in every other aspect, so why not in our ability to determine mathematical truths?

The only argument put forward in favor of this assertion is relegated to a footnote: "It should also be noted that if anything can count as evidence in favor of [this premise], it surely must be that, as a matter of empirical fact, our race continues to climb from n to $n + 1$."[18]

This line of evidence might be compelling if there were a long chain of increasing busy beaver numbers that were found. Instead, we currently only have four busy beaver numbers. This is not the sort of long chain of unbroken success that would motivate thinking that that progress will be unending.

6. Conclusions

We have argued that human consciousness cannot be reduced to a procedure; that would be a category error. A procedure can no more exhibit consciousness than a formula can. Rather, something else must explain consciousness. It is not possible to determine whether such consciousness is physical or not, since we lack a widely accepted definition of physicality. However, it does seem that consciousness is a designed aspect of our mind, not some side-effect of a physical computational processes.

Human cognition is consistent with being generated by a very sophisticated algorithm. Human abilities far surpass that of current computation. The arguments which have been put forward for non-computational human cognition implicitly depend on the assumption that the algorithm underlying human cognition is either trivial or at least within the comprehension of humans. If we instead understand that algorithm as unfathomably sophisticated, in particular beyond the comprehension of humans, the arguments do not work.

Based on these conclusions, we have proposed that the human mind comprises both non-computational consciousness and computational cognition. That cognition is rooted in a very sophisticated, but ultimately finite, program. The brain is thus, most plausibly, the physical computer responsible for running the sophisticated program. Whether the whole mind, including our consciousness, is physical is not a well-defined or, in our opinion, interesting question.

We have sketched an argument that generating cognitive abilities requires greater cognitive ability. This has a number of interesting consequences:

First, human cognitive ability will never be matched by artificial intelligence. We have argued that the only way to obtain an accurate partial halting detector is using a more powerful halting detector. When humans devise artificially intelligent systems, they use their internal powerful halting detection abilities to verify and/or construct the implicit halting detection present in the artificially intelligent system. However, they are only capable of devising a halting detector less powerful than the one they have. As such, we would expect that while humans will get better at building artificial intelligence systems, they will never be able to match themselves.

Second, the "singularity" will not happen. The idea of the singularity is that an artificially intelligent system will be able to build a slightly more intelligent artificial intelligence (AI) system. That system will, in turn, devise an even more intelligent system. This process, repeated over and over, will culminate in artificially intelligent systems which will leave humans far behind. However, the only way to obtain a partial halting detector is using a more powerful partial halting detector. An AI system cannot build a slightly more intelligent partial halting detector. Thus, the singularity will not occur.

Third, the human mind has a transcendent origin. Standard evolutionary theory claims that the human mind was produced by natural selection operating on random mutations. However, this would be a case of a very computationally simple process constructing an accurate, highly powerful halting detector. This cannot happen if the only way to obtain a partial halting detector is by using a more powerful halting detector. Instead, the human mind must have derived from something with more powerful halting detection abilities. Yet we cannot explain the human mind by an infinite regress of increasingly powerful partial halting detectors. Rather, the human mind must eventually be explained by a non-computational form of intelligence for whom the halting problem is no obstacle.

Notes

1. Gregory J. Chaitin, *Meta Math!: The Quest for Omega* (New York: Vintage, 2006).
2. Ray Kurzweil, *The Singularity is Near: When Humans Transcend Biology* (New York: Penguin, 2005).
3. David J. Chalmers, "Facing Up to the Problem of Consciousness," *Journal of Consciousness Studies* 2, no. 3 (1995).
4. Daniel C. Dennett, "Facing up to the Hard Question of Consciousness," *Philosophical Transactions of the Royal Society B: Biological Sciences* 373, no. 1755 (September 2018): 20170342, https://doi.org/10.1098/rstb.2017.0342.
5. David J. Chalmers, "Panpsychism and Panprotopsychism," in *Panpsychism: Contemporary Perspectives*, eds. Godehard Brüntrup and Ludwig Jaskolla (New York: Oxford University Press, 2016): 19–47.
6. A. M. Turing, "On Computable Numbers, with an Application to the Entscheidungsproblem," *Proceedings of the London Mathematical Society* s2-42, no. 1 (1936): 230–65, https://doi.org/10.1112/plms/s2-42.1.230.
7. Gregory J. Chaitin, *Algorithmic Information Theory* (Cambridge, UK: Cambridge University Press, 1987).
8. R. J. Solomonoff, "A Formal Theory of Inductive Inference. Part I," *Information and Control* 7, no. 1 (March 1964): 1–22, https://www.sciencedirect.com/science/article/pii/S0019995864902232?via%3Dihub.
9. David H. Wolpert and William G. Macready, "No Free Lunch Theorems for Optimization," *IEEE Transactions on Evolutionary Computation* 1, no. 1 (April 1997): 67–82, https://doi.org/10.1109/4235.585893.
10. See, e.g., John R. Lucas, "Minds, Machines and Gödel," *Philosophy* 36, no. 137 (1961): 112–127, http://www.jstor.org/stable/3749270.
11. Kurt Gödel, "Über formal unentscheidbare Sätze der Principia Mathematica und verwandter Systeme I," *Monatshefte für Mathematik* 149, no. 1 (September 1930): 1–29, https://doi.org/10.1007/s00605-006-0423-7.
12. Kurt Ammon, "An Automatic Proof of Gödel's Incompleteness Theorem," *Artificial Intelligence* 61, no. 2 (1993), https://doi.org/10.1016/0004-3702(93)90070-R.
13. Roger Penrose, *Shadows of the Mind: A Search for the Missing Science of Consciousness* (New York: Oxford University Press, 1994).
14. Jonathan Bartlett, "Using Turing Oracles in Cognitive Models of Problem-Solving," in *Engineering and the Ultimate: An Interdisciplinary Investigation of Order and Design in Nature and Craft*, eds. Jonathan Bartlett, Dominic Halsmer, and Mark R. Hall (Broken Arrow, OK: Blyth Institute Press, 2014), 113, https://www.academia.edu/61894364/Using_Turing_Oracles_in_Cognitive_Models_of_Problem_Solving.
15. Selmer Bringsjord et al., "A New Gödelian Argument for Hypercomputing Minds Based on the Busy Beaver Problem," *Applied Mathematics and Computation* 176, no. 2 (2006): 516–30.
16. Bartlett, "Using Turing Oracles," 111.
17. Bringsjord et al., "A New Gödelian Argument," 525.
18. Bringsjord et al., "A New Gödelian Argument," 525n15.

23. Mathematical Objects Are Non-Physical, So We Are Too

Selmer Bringsjord and Naveen Sundar Govindarajulu

1. Introduction

Chimpanzees, the chair in which we presently sit, and the chunk of aged cheddar cheese and wine on the table before the first author; these things are physical, clearly. Are there any *non*-physical things? Even those who would answer this question with an adamantine negative, if reflective, will agree that perhaps the best candidates for this category are not mental states had while enjoying such cheddar with fine Carménère (states which dualists since Descartes have long insisted are non-physical, since they are bearers of so-called—to use modern terminology—"qualia"[1]), but instead logico-mathematical objects with which plenty of human persons, from their very earliest elementary-school years, are acquainted. These immaterial objects range from the familiar to the exotic, and are the targets of study in the formal sciences.[2] We focus herein on two familiar and elementary classes of such logico-mathematical objects: (1) algorithms (such as the famous but simple Quicksort, discovered by Tony Hoare); and (2) inference schemata that form the foundation of rigorous reasoning in the formal sciences (such as *modus tollens*, e.g., from two declarative propositions "if ϕ then ψ" and "not-ψ" one can validly deduce "not-ϕ"). Inference schemata form this foundation because the formal sciences are theorem-driven, theorems are obtained by proofs, and proofs are sequences of propositions linked by inferences that must be sanctioned by such schemata (though often the schemata employed are left implicit and not called out by name). Of course, as the reader might imagine, many propositions *themselves* are logico-mathematical objects central to the formal sciences. For instance, the proposition that there are infinitely many primes, first proved by Euclid, would be such a specimen. (Below, we shall have occasion to discuss this proposition, and the general class into which it falls.)

Given the context created by the logico-mathematical objects referred to in the previous paragraph, we can now inform the reader that the overarching structure of our case for the proposition that human persons[3] are immaterial will have two steps. In Step 1 we adapt and focus prior reasoning from James Ross in order to show that such objects as algorithms and inference schemata are non-physical (= immaterial). Then, in Step 2, we show that humans interact with these objects in a certain crucial way: we *understand* them; specifically, we understand that we frequently validly implement them. We then argue, in part by appeal to Bringsjord's prior refinement of John Searle's famous Chinese Room Argument (CRA), for the proposition that such understanding can't be achieved by standard computing machines, that such understanding entails that we must ourselves be non-physical.[4] Of course, inevitably some will want to resist our ultimate conclusion. Accordingly, we consider and rebut some objections, including one based on the eponymous Benacerraf-Field Problem, which in a word says that we can't fathom how our justified belief in propositions regarding logico-mathematical objects could ever be explained. When we wrap up the paper, we briefly point to some more exotic formal objects than those routinely in play in K–12 math education; for example, infinite cardinals. Such exotica, it seems to us, are almost on face immaterial. (To keep things brief and simple, we refer only to the smallest infinite cardinals: $\aleph_0$ and $\aleph_1$.)

2. Logico-mathematical Objects, in General

SOME READERS may find the phrase "logico-mathematical object" to be a bit of a mouthful, and perhaps even pedantic. Actually, the idea is quite straightforward, and the objects in question are encountered and reasoned over by even very young schoolchildren, who usually continue in this regard for many years, and are along the way introduced to more and more such objects of increasing complexity. One of the first logico-mathematical objects young children come across in the classroom is $\mathbb{N}$, the set of all natural numbers

$$\{0, 1, 2, \ldots\}.$$

This object is referred to via a "number line" shown graphically, and of course before this object is introduced, the young mind will have been told about the numbers 1, 2, 3, and so forth with help from such physical things as fingers, and will often have been introduced to the arithmetic functions of addition and subtraction applied to natural numbers. In public education in the U.S. State in which the first author resides, New York, Grade 4 mathematics instruction introduces students to a new logico-mathematical object that sometimes causes a bit of intellectual turbulence: $\mathbb{Q}^+$, the positive rational numbers.[5] This introduction happens, of course, once these students are taught fractions, and how to add, multiply, and divide them. A bit later, students are introduced to the *real numbers*, $\mathbb{R}$. If they are lucky enough to reach study of the differential and integral calculus in Grade 11 or 12, they are taught how motion and change can be understood with help from functions over $\mathbb{R}$. In addition, in today's world, it's likely that our young student will be introduced as well to logico-mathematical objects that are part and parcel of computer programming and computer science—objects such as *arrays*, *algorithms*,[6] and so on. In the course of learning mathematics, or computation, inevitably the student will also be introduced to the basic Boolean operators: *and* ($\wedge$), *or* ($\vee$), *not* ($\neg$), *if* ... *then* ... ($\rightarrow$), and ... *if and only if* ... ($\leftrightarrow$).[7] And, finally, our student will be taught how to check and create some proofs, since this is standard fare in Algebra 2 and Geometry, both required in 47 of the 50 U.S. States (in public education). The role of proof and proof creation in secondary mathematics education in the U.S. can be clearly seen by turning to the Common Core textbooks for Algebra 2.[8] We mention all of this just to ensure that the reader understands that logico-mathematical objects, and human interaction with them, are routine, extensive, and persistent.

Although we shall need to be more specific below, the classes of logico-mathematical objects cited to this point will pretty much suffice for the remainder of the present essay.

3. Narrowing the Focus with Simple Exemplars

To MAKE matters more concrete, let us focus on just a few particular logico-mathematical objects, and then anchor subsequent discussion to them. What specimens should we select as exemplars? Well, mildly put, there are quite a few logico-mathematical objects. How many? Any serious attempt to answer this question would overwhelm all the space available in the present chapter. Let's rest content with the simple observation that the universe of such objects is infinite, and with the helpful follow-on observation that this universe can be to a degree rationalized by approaching it in accordance with the sub-parts of the universe that are associated with particular disciplines within the formal sciences, and sub-parts of these disciplines. We now pass to our exemplars.

3.1 Exemplar 1, an Algorithm: Quicksort

Quicksort, which we denote by $\mathcal{Q}$, first presented to the world by Hoare,[9] is deservedly well-known in the computational formal sciences, and this shall be our first exemplar. The algorithm itself is at bottom a simple recursive one. (There are now numerous variants, but we ignore this for efficiency.) The algorithm is to receive an array of ordered objects, for example

$$\left\langle \boxed{5}\ \boxed{9}\ \boxed{10}\ \boxed{7}\ \boxed{4}\ \boxed{3}\ \boxed{11}\ \boxed{8}\ \boxed{6} \right\rangle,$$

and to then produce as output the sorted version of this input, which in this case is:

$$\left\langle \boxed{3}\ \boxed{4}\ \boxed{5}\ \boxed{6}\ \boxed{7}\ \boxed{8}\ \boxed{9}\ \boxed{10}\ \boxed{11} \right\rangle.$$

So, what's the algorithm? In order to answer this question, we can't avoid resorting to what we can call *embodiments* or *tokens* of the general and abstract *type* $\mathcal{Q}$.[10] This terminology, and the associated concrete practice, is easy to grasp. For an example, we give one high-level embodiment /token $\hat{\mathcal{Q}}_1$ of $\mathcal{Q}$ that views the algorithm as a three-stage one.[11] Before supplying the example in question, we draw the reader's attention to what we just did with a bit of suggestive notation: we used "hat" $\hat{O}$ to indicate that what is being referred to is an embodiment of the thing O (in this case, of course, an algorithm). Hence, the hat in '$\hat{\mathcal{Q}}_1$' says that we have here an embodiment of the algorithm $\mathcal{Q}$ itself. Very well, and now to the embodiment in question itself:

I Pick the rightmost element in the array as the *pivot*.

II Partition the array so that all elements in the array less than the pivot are before it, and all elements greater than the pivot are placed after it.

III Recursively apply both I and II to the sub-array now before the pivot, as well as to the sub-array now after the pivot.

This is said to be "high-level" for obvious reasons. $\hat{\mathcal{Q}}_1$ doesn't tell us how to carry out partitioning, and it relies on an understanding of what recursion means—or at least what it means in this context. But no worries: Stage II can be further specified by saying that we simply move to the left one entry at a time, and decide whether to move an entry to the right of our pivot, or else leave it where it is. And how to decide? Simple: If what we find is greater than our pivot, append it to whatever sub-sequence is to the right of the pivot; otherwise just leave what we find alone. Using a double-box to indicate our pivot, the result of executing Stage I and then Stage II in $\hat{Q}_1$ on the initial input array will result in this configuration:

$$\left\langle \boxed{5}\ \boxed{4}\ \boxed{3}\ \boxed{\boxed{6}}\ \boxed{8}\ \boxed{11}\ \boxed{7}\ \boxed{10}\ \boxed{9} \right\rangle.$$

Now the algorithm calls for Stage III in $\hat{Q}_1$, which means that the sub-array to the left of $\boxed{\boxed{6}}$ with $\boxed{3}$ as the pivot of this sub-array is processed; ditto for the sub-array to the right of $\boxed{\boxed{6}}$ with $\boxed{9}$ as the pivot of this sub-array. In the case of the right sub-array, here's the result of running Stage I, which is to be passed to Stage II to be processed (we once again indicate the pivot by a double-box):

$$\left\langle \boxed{8}\ \boxed{11}\ \boxed{7}\ \boxed{10}\ \boxed{\boxed{9}} \right\rangle.$$

Stage II applied to the input to it immediately above then results in this:

$$\left\langle \boxed{8}\ \boxed{7}\ \boxed{\boxed{9}}\ \boxed{10}\ \boxed{11} \right\rangle.$$

We continue in this way until we reach sub-arrays composed of but one element, which are by definition sorted, and hence processing is guaranteed to terminate.

It should be obvious to the reader that an infinite number of embodiments or tokens of Quicksort are available.[12] Many of these embodiments call upon programming languages used today. We shall assume, going forward, that $\hat{Q}_2$ refers to an embodiment of Quicksort = Q that is expressed in the modern functional programming language known as Clojure.[13]

3.2 *Exemplar 2, an Inference Schema: Modus Tollens*

Next, we use a variant of the famous "Wason Selection Task" (WST) [14] to anchor our presentation of *modus tollens* = MT, the gist of which, intuitively, can be thought of as the kernel of a kind of *disconfirmation*, in which if it is claimed that ϕ implies ψ, and one observes that ψ isn't the case, one can safely infer that ϕ doesn't hold either. We can be a bit clearer about what *modus tollens* is by way of the following oft-used token of it:

$$\frac{\phi \rightarrow \psi,\ \neg\psi}{\neg\phi}$$

The token written immediately above, which—following our "hat" technique explained and introduced above—we shall denote by '$\widehat{MT_1}$,' tells us that if we have two formulae of the form indicated by the two expressions above the horizontal line (the first a conditional and the second the negation of the consequent of that conditional), then the inference schema in question allows us to infer what's below the horizontal line, namely that the antecedent in the conditional can be negated.

Now here's our selection-task challenge: Imagine that, operating as a teacher of mathematics trying to transition one of our students to proof (from mere calculation), we have a deck of cards, each member of which has a digit from 1 to 9 inclusive on one side, and a majuscule Roman letter A, B, . . . K on the other. From this deck, we deal onto a table in front of one of our students the following four cards:

E	T	4	7
c1	c2	c3	c4

Now suppose that we inform the student that the following rule R is absolutely guaranteed with respect to the entire deck, and hence specifically also for the four cards c1–c4 now lying in front of the student: "Every card with a vowel on one side has an even positive integer on the other side." Next, we issue the student the following challenge:

> "Does card 4 have a vowel on its other side? Supply a proof to justify your answer."

What should the student do in order to succeed? It should be clear that the student should answer in the negative, and provide a proof that makes use of *modus tollens*, such as in the following sequence, which we trust will be readily understood by all our readers, after a bit of inspection:[15]

Line	Proposition	Justification
1.	$\forall c\,[Vowel(c) \rightarrow Even(c)]$	Rule R
2.	$\neg Even(c4)$	from observation
3.	$Vowel(c4) \rightarrow Even(c4)]$	from 1. by instantiation
4.	$\neg Vowel(c4)$	from 2. & 3. by $\widehat{MT_1}$

In this proof, the final step to yield line 4., as indicated, makes use of *modus tollens*. Note the match of lines 2., 3., and 4. with the token of *modus tollens* that is $\widehat{MT_1}$, given above.

3.3 Additional Exemplars: Proof-by-Cases Schema, and a Theorem Via It

While below we devote a dedicated section (§6) to anticipating and disarming objections to the main theses we advance in the present essay, we now take a few moments to quickly dispose of a weak objection that will inevitably come to the minds of some readers. The objection is simply that *modus tollens* is in fact not explicitly used in the proofs given by practicing formal scientists. While we fail to see why a failure to call out *modus tollens* explicitly by name raises any question for our case for the immaterial nature of logico-mathematical objects, let us simply admit that it's true that while *modus tollens* is routinely taught in introductory formal logic,[16] rarely do professional formal scientists cite it explicitly. But this is irrelevant, for two reasons. The first reason is that the relevant professionals *do* in fact use *modus tollens*—they just don't cite it by name. This practice is the same as that followed when theorems having the form of a biconditional $\phi \leftrightarrow \psi$ are proved, because often the proofs in question are divided into two phases, one the so-called "left to right" direction in which ϕ is assumed, and then the so-called "right to left" or "converse" direction in which ψ assumed and, eventually, ϕ deduced. The second reason why *modus tollens* is perfectly fine as an exemplar is that we could just as well use inference schemata that are more robust, and which *are* explicitly called out and used in more sophisticated formal-science deduction. One nice exemplar in this regard is "proof by cases." Unlike *modus tollens*, proof by cases is a schema employed in some fairly famous, indeed in some cases even "immortal," deductive reasoning. The schema says that if we have some disjunction, and in addition perceive that each disjunct leads by some reasoning to our goal γ, then we can conclude from the disjunction itself that γ holds. Let us call this expression of the inference schema "$\widehat{PBC_1}$."[17]

Let us end this section by referring to another type of logico-mathematical object, one that will be necessary to have in play when we arrive at §6, i.e., axioms and theorems. All readers will be well-acquainted with the fact that the formal sciences make crucial use of both of these things. In early math education, when students first learn arithmetic, they are essentially learning how to calculate on the basis of axioms and theorems in nothing less than the branch of mathematics called *number theory*, but such calculation is rarely described in terms of axioms and theorems. However, all students who pay attention and progress through basic secondary mathematics education are explicitly exposed to Euclid's axioms for plane geometry, and are asked to deduce some simple theorems from these axioms. What we need for present purposes is just a simple, single exemplar from this category, and without loss of generality we choose what is commonly called "Euclid's Theorem," a rather famous result that says that there are infinitely many prime numbers. Conveniently, part of what's clever in the Euclidean proof of the theorem is a use of proof by cases, and as a matter of fact *modus tollens* too. For use below, let us denote the theorem by "ET," and leave the clever-but-not-difficult proof itself in an endnote.[18]

4. The Two-Step Argument for the Immateriality of Us

As THE reader will recall, our overarching argument for the proposition that humans (more precisely, to remind the reader, human *persons*; see note 3) are immaterial is a two-step one. In Step 1 we show that algorithms and inference schemata (and, as an obvious consequence, also axioms and theorems) are non-physical, by building atop some seminal prior work by James Ross. In Step 2, we then show that the humans who (sometimes) correctly use such algorithms and schemata are themselves non-physical (= immaterial). Let's move directly to Step 1 now.

4.1 Step 1: Algorithms and Inference Schemata are Non-Physical

Step 1 builds upon an insightful argument given by James Ross for a related proposition.[19] This related proposition is that "formal thinking" isn't a physical process. Here is an encapsulation of Ross's argument for this proposition in his own words:

> Some thinking (judgment) is determinate in a way no physical process can be. Consequently, such thinking cannot be (wholly) a physical process. If all thinking, all judgment, is determinate in that way, no physical process can be (the whole of) any judgment at all. Furthermore, "functions" among physical states cannot be determinate enough to be such judgments, either. Hence some judgments can be neither wholly physical processes nor wholly functions among physical processes.[20]

As the reader can see, Ross's objective is to establish that the process of thinking about, and thinking guided by, formal structures isn't physical. (As seen in the quoted passage, he terms this process "judging.") In contrast, our subobjective (= the objective of Step 1) is to establish that the *things* bound up with such formal thinking are non-physical (and of course the overarching goal, again, is to show—in Step 2—that *we* are not physical things). However, it's easy enough to adapt the argumentation that Ross gives, and to sharpen it. We do this now.

To begin, recall the exemplars and related things that we have at our disposal at this point: the Quicksort algorithm $\mathcal{Q}$ and any number of embodiments thereof (we have $\hat{\mathcal{Q}}_1$ and $\hat{\mathcal{Q}}_2$ on hand; the latter embodiment is the Clojure code we provide in the relevant endnote); the inference schema *modus tollens* = MT and any number of embodiments thereof ($\widehat{MT}_1$, given by us above, is on hand for use); and in addition we have on hand other inference schemata of greater complexity as needed (in particular *proof by cases* and the embodiments $\widehat{PBC}_1$ and $\widehat{PBC}_2$, and the theorem **ET** (with the token of this theorem shown in note 17). Next, here is the proposition that we establish in Step 1 (where of course Smith is an arbitrary stand-in for any human agent):

($\star$) If Smith validly implements $\mathcal{Q}$ via for instance the aforementioned Clojure program $\hat{\mathcal{Q}}_2$, or validly instantiates inference schema MT via for instance a proof having an inference conforming to $\widehat{MT}_1$, then in neither case is this validity due to satisfaction of some relation R holding between Smith and some physical embodiment of the abstract types in question.

We can introduce a more perspicuous variant of this proposition by representing the ternary validity relation in it by *Val*, and the referenced relation posited to constitute satisfaction of the *Val* relation by R_1 in the case of our algorithm, and R_2 in the case of our inference schema. Then we have two more precise propositions derived from ($\star$), respectively. Note that we give an English reading in both of these sub-cases, which immediately follow, and which make use of elementary logic, with the existential quantifier $\exists$—as our English renditions show—for "there is at least one . . . ." The structure of the sub-cases is to assert that the validity relation holds exactly when some other relation between our agent and a material embodiment holds.

($\star_1'$) $\neg\exists R_1[Val(s, \mathcal{Q}, \hat{\mathcal{Q}}_1)]$ if and only if $R_1(s, \hat{\mathcal{Q}}_1)$
English: It's not the case that there is a relation R_1, holding *only* between Smith and the material embodiment $\hat{\mathcal{Q}}_1$ (of $\mathcal{Q}$), in virtue of which agent Smith validly implements Quicksort = $\mathcal{Q}$.

($\star_2'$) $\neg\exists R_2[Val(s, MT, \widehat{MT}_1)$ if and only if $R_2(s, \widehat{MT}_1)]$
English: It's not the case that there is a relation R_2, holding *only* between Smith and the instantiation $\widehat{MT}_2$ (of MT), in virtue of which agent Smith validly instantiates *modus tollens* = MT.

Next, we need to articulate a Rossian argument that can do the job of establishing both ($\star_1'$) and ($\star_2'$). In this argument, we essentially rely upon what Ross relies upon (see note 20), but we give our own rationale to make the present essay self-contained. Here's the argument:

The Step-1 Argument

Suppose without loss of generality that our agent Smith, using paper and pen, writes down the program $\hat{\mathcal{Q}}_2$ as a physical token/embodiment of $\mathcal{Q}$, and also writes down a proof that instantiates $\widehat{MT}_1$, the token of MT. For fixity, but without loss of generality, the embodiments in both cases can be pieces of paper upon which Smith writes. How can we be sure that the validity of what Smith does here cannot be due to the satisfaction of some relation holding exclusively between Smith and these physical embodiments; that is, how do we apprehend that the negation of ($\star$) doesn't hold? Well, let's temporarily suppose otherwise; that is, we assume $\neg(\star)$. Now, we ask a simple question: How many embodiments of $\mathcal{Q}$ and MT are there? Our parable has featured only one, and our prose above has in the case of both Quicksort and *modus tollens* presented a total of two—but obviously there are many, many more. How many, then? Clearly, there are an *infinite* number of such embodiments. Minimally, there are as many embodiments as there are natural numbers, so we can index embodiments to these numbers. But in each case, the embodiment E_j, where $j \in \mathbb{N}$, is different, indeed often radically so. So the relation *Val* must be one holding between our Smith, the particular embodiment $\hat{\mathcal{Q}}_i$, and $\mathcal{Q}$, in virtue of some *ternary* relation, one that includes ranging over $\mathcal{Q}$ itself/MT itself! For if this is not the case, what serves to unite the infinite embodiments E_j? So, now, are we to take $\mathcal{Q}$ and MT to be physical, or non-physical? It cannot be the former. For then we are right back to where we started from, since we then have merely a relation between our agent and a particular embodiment, say E_k, by definition (since every physical thing is embodied). We thus arrive at having to affirm (⋆), the opposite of what embroiled us in this trouble.

4.2 Step 2: Why We *Are Immaterial*

Now we come to Step 2. This is the harder, more intricate step. In particular, it's one that involves an understanding of the limitations of machine intelligence, or what is called "AI," for "artificial intelligence." In particular, we refer here to limitations on so-called "Strong" AI, which aims to produce intelligent machines able not only to *behave* like human persons, but to quite literally *be* human persons (with cognition and consciousness at, and possibly above, the human level). The distinction between Strong and Weak AI is discussed at length elsewhere by one of the present authors,[21] but we take the distinction at this point to be sufficient for present purposes, which we now continue to advance.[22]

To begin Step 2, we assert that whether or not the vast majority of scientists whose professional business is to study human persons and their brains believe that these persons are physical things, the vast majority of such thinkers *say* that they believe this, when in conventional scientific venues.[23] For the sake of argument in the present paper we shall take these scientists at their word. But what word, exactly? We doubt we can find a better case in point than John Anderson, and the mentor he venerates, Allen Newell, one of the illustrious founders of AI, and a substantive contributor at the famous Dartmouth conference that marks the birth of modern AI (as explained in another of our works).[24] Specifically, we can consider Anderson's *How Can the Human Mind Occur in a Physical Universe*?[25] The title of this book should send a clear message to our readers. The core, driving ideas in the book are: first, whatever the mind (i.e., the human person; see again note 3) is and however it might work, it's certainly physical; and second, we need an account of how the mind works that at least fits what physics tells us, and preferably an account that itself is, if you will, "physics-ish."

Step 1 in our overarching argument relied significantly upon analysis and argument given by Ross. Step 2 relies upon insights from John Searle, first given in his landmark paper "Minds, Brains, and Programs."[26] There, Searle introduces his famous "Chinese Room Argument" (CRA). The Chinese Room is a room (some refer to it as a "box") that he enters, accompanied only by a "rulebook" that allows him (in concept: he has to work fast!) to output symbol strings in response to such strings coming into the room—but all the while Searle has no idea what the symbols in question mean, because they are in Chinese, a language he doesn't at all understand. Subsequently, Bringsjord refined these insights and provided the relevant argumentation,[27] which he later further refined along with Noel.[28] There is insufficient space to rehearse the Searlean argument in any detail. It suffices to report here simply that the argument's conclusion is that human understanding of what symbolic inscriptions mean cannot possibly be achieved by a standard computing machine. Such a machine, for instance a modern high-speed digital computer, can process all sorts of symbolic inscriptions, but it cannot have human-level understanding of these inscriptions. Why? The essence of the Searlean answer, given in its most mature form in the aforementioned Searle article, is: "Because when we consider whether understanding is conveyed by our own mere symbol manipulation à la computing machine (of any symbols, with processing governed by any program), we see that we have no understanding thereby whatsoever. But we *would* have such understanding if a computing machine could achieve understanding by its mere computing."[29]

Now, given this brief background, what is the Step-2 argument for the conditional that if logico-mathematical objects are immaterial, we are as well? Here is a brief articulation:

The Step-2 Argument

As humans, we can *understand* Quicksort = Q and *modus tollens* = MT, clearly. In fact, you, the reader, are in this fortunate group of understanders. This understanding, as we appreciate in the light of Searlean argumentation (see above), happens only when we do more than move around particular symbols and diagrams (that can be used to token the relevant types). These things are all mere embodiments, by definition.[30]

Now, if the immateriality of the logico-mathematical objects in question don't imply that we are ourselves immaterial, what are we? We would be exactly what John Anderson says we are: biological, and hence physical, computers.[31] But then, even as such we can understand Quicksort and *modus tollens*, this despite the fact that as such computing machines we are restricted to interaction consisting of our manipulation of *particular* embodied symbols and diagrams. Indeed, such manipulation is the essence and full reach of what a computing machine is. In other words, on this line, we all as humans become nothing more than Searles-in-the-room! But this is inconsistent with what is possible for things-in-the-room: these things, as mere symbol manipulators, can't have understanding therefrom. Hence, the assumption of our materiality has led us to contradiction, and we therefore in fact can't be material.

5. Regarding Related Work

Angus Menuge has written an excellent paper that relates in interesting ways to the two-step argument given above.[32] His paper is an attempt to "show that materialism is incapable of explaining a large and important area of human knowledge,"[33] i.e., knowledge of abstracta, including specifically some of the formal objects upon which we focus here. Menuge appears not to be aware of the reasoning of Ross to which we crucially and centrally appeal in our Step-1 argument;[34] this is noteworthy because some of what Menuge writes about "understanding" a rule of inference is directly in line with Ross's thinking.[35] While we avowedly rely upon Ross, Ross in turn relies upon what he says are certain long-established "jewels of analytic philosophy."[36] To really flesh out our Step-1 argument, we would analyze and tap directly into its roots in the work of those Ross builds upon, most prominently Nelson Goodman.[37] The power of Ross's paper inheres in no small part in the fact that though his position on the nature of thinking is that of an immaterialist and a theist, his support is found in many of those who are neither.

Many additional points could be written about the relationship between Menuge's paper and ours; we shall rest content with the following additional two:

1. In general, we do not at all wish to go in the "knowledge connection" direction. Notice that we are not talking about knowledge in the case of the agents in our parables. Take a look at the key propositions we set off typographically in the Step-1 argument given above. There is no reference in those to knowledge or belief in said agents. In particular, there is no mention of knowledge at all in the agents that validly follow an algorithm or an inference schema. We see talk of knowledge as a bit of a potential quagmire, to be avoided. Note that most physicalists who are computer scientists will simply

maintain, *contra* Menuge, that while it's true for instance that they know that every positive integer n is greater than 0, they *don't* know that $1 > 0$, that $2 > 0$, that $3 > 0$, *ad infinitum*.[38] And most of these folks will not agree that, when Jones validly codes or follows Quicksort = $\mathcal{Q}$, Jones knows that something holds of an infinite number of cases, or knows an infinite number of propositions. An argument like that is, as we see things, much better to make directly of professional mathematicians and logicians—but even then you run into epistemic finitists. We want to avoid this rabbit hole (which is not to say that we agree with such skeptics). At a minimum, because we are inclined to require formal argumentation/proof in sorting out such matters, we would need to demonstrate, formally, that an agent's knowing some proposition ϕ entails this agent's knowing at least as many propositions $\psi_1, \psi_2, \ldots$ as there are natural numbers. We confessedly find such a need daunting—despite the fact that we have worked on infinitary knowledge and belief from an at-once formal and computational perspective.[39]

2. While Menuge articulates an abductive argument for the existence of a transcendent being, our two-step argument eventuates in a conclusion about the nature of human persons, and stops there. One could certainly promisingly explore linking from what we conclude, to Menuge's abductive reasoning about God—but that exploration requires a separate, future day.

6. Refuting Two Objections

WE NOW anticipate and refute two objections.

6.1 *"We Simply Legislate Logico-mathematical Objects!"*

Here's how the present objection can be expressed: "Your case for the immateriality of human persons is vitiated by a simple fact: these persons *legislate* the logico-mathematical objections upon which your case is built. That this is so is shown by Lakoff and Nuñez."[40]

There are at least two problems with this objection, both of which are fatal.

The first problem is that the objection is flatly self-refuting: if it's sound, it's unsound. This defect relates to the notion that logic is invincible, which can be encapsulated by way of the following short parable:

> Jones holds that belief fixation by rational agents should be based upon the construction and assessment of arguments. In particular, Jones holds that rationally believing some proposition ϕ can happen only if there is some argument for ϕ of which the rational agent in question is aware, and if, in addition, that agent understands the formal validity of the argument in question. For instance, ϕ might be some expression of this in-English proposition: "The cardinal number $\aleph_0$ is a non-physical object." Smith challenges Jones, by expressing his dripping disdain for logic, which he (Smith) regards to be worthless, or at least just plain wrong. But if Smith's reasoning succeeds, then it does so on the strength of making use of logic. It then follows that if Smith's reasoning succeeds it fails, since that reasoning is aimed at establishing that logic is worthless and so on.

The problem here can of course be expressed without a story, and brought to bear on the present objection, as follows. The critic here appeals to the book *Where Mathematics Comes From* by Lakoff and Nuñez, in which the authors argue that human persons are *physi-*

cal creatures that create mathematical objects—including therefore the particular objects upon which we place weight in the present essay: i.e., algorithms and inference schemata. For ease of exposition, let's say that Lakoff and Nuñez offer only one argument for this claim, and let's label that argument 'α'. The present objection has force only if α is formally valid; that is, only if the inferences in α conform to inference schemata that regiment normative correctness in reasoning (including, most certainly, *modus tollens*). Now suppose that these authors are correct. Then it follows that the inference schemata that undergird α have themselves been legislated by humans. But then α is not really valid at all, for what's to stop someone from advancing an argument for the falsity of the very inference schemata that Lakoff and Nuñez have employed in their α? Nothing (assuming that Lakoff and Nuñez are correct). Hence, if Lakoff and Nuñez are correct, they end up refuting themselves.

The reader should rest assured that a close analysis of *Where Mathematics Comes From* more than fully supports our claim that the case given in this book is self-refuting. The reason, in short, is that Lakoff and Nuñez appear to be completely unaware of the fact that their own argumentation hinges on the non-arbitrariness of inference schemata. They seek to show, for instance, that what they call the "laws of arithmetic" arise from particular "cognitive mechanisms" that run in "embodied minds," but they are blithely unaware of the fact that the inference schemata—conformity to which is a *sine qua non* for the soundness of their argumentation, if merely arising directly from the physical mechanisms in question—are entirely arbitrary, and hence unsound.

Oddly enough, Lakoff and Nuñez do ask this question, and we quote: "And why, in formal logic, does every proposition follow from a contradiction?"[41] Leaving aside the fact that there is no such thing as "formal logic" as a single monolithic system,[42] and leaving aside the fact that the inference schema to which Lakoff and Nuñez here allude, i.e., explosion, is not included in all formal logics, the question we press is: What about the "law" labeled *modus tollens*? Lakoff and Nuñez rely upon it, and upon many other such "laws." But if the laws just emerge from the particular physical things that Lakoff and Nuñez are, and the laws are themselves physical things arising adventitiously from the physical mechanism of cognition on the part of Lakoff and Nuñez, then what's to stop other agents from showing up and saying that they reject *modus tollens* in favor of some preferred schema of their own that marks a rejection of *modus tollens*?[43]

6.2 *"The Benacerraf-Field Problem Refutes You!"*

The objection here can be stated in compressed form as follows: "Your position, and the reasoning you give to support it, presuppose a solution to an unsolvable problem: the so-called Benacerraf-Field Problem. Hence your position shouldn't be affirmed."

We suspect that a number of our readers may be unaware of the Benacerraf-Field Problem (which we will label "B-FP"), or of the earlier, less refined version of the problem (which we will label "BP"), set forth in 1973 by Paul Benacerraf, writing alone. Here is an encapsulation of Part 1 of the problem as Benacerraf summed it up:

> [O]n a realist (i.e., standard) account of mathematical truth our explanation of how we know the basic postulates must be suitably connected with how we interpret the referential apparatus of the theory. . .[But]

> what is missing is precisely... an account of the link between our cognitive faculties and the objects known... We accept as knowledge only those beliefs which we can appropriately relate to our cognitive faculties.[44]

Given the foregoing context we have laid out, a "postulate" can be understood to be an axiom or theorem in the formal sciences, and in particular our exemplar **ET**, Euclid's Theorem, can be used without loss of generality, and without begging any questions against Benacerraf. Now, Part 2 of BP is the claim that there can in fact not be a "suitable connection" between the "basic postulates" of mathematics and the "cognitive faculties" of human persons. But why is Benacerraf pessimistic in this regard? What supports the claim? The answer is that he insists that (i) any suitable connection must be a *causal* one, and that (ii) there can't be a causal connection between such an agent and a postulate. The idea, specifically tied to our context, would be that between you, the reader, and **ET**, which you can be assumed to now fully understand (by virtue e.g. of having assimilated note 18 below), there must be some sort of causal connection. We can confirm this interpretation, from the words of Benacerraf himself:

> I favor a causal account of knowledge on which for X [= you] to know that S [= ET] is true requires some causal relation to obtain between X and the referents of the names, predicates, and quantifiers of S ... [But]... combining *this* view of knowledge with the "standard" view of mathematical truth makes it difficult to see how mathematical knowledge is possible ... [T]he connection between the truth conditions for the statements of number theory [such as **ET**] and any relevant events connected with the people who are supposed to have mathematical knowledge cannot be made out.[45]

It's at this point easy to see that BP poses not the slightest problem for the view we have advanced about logico-mathematical objects, the nature of human mentation regarding them, and the nature of the agents who enjoy such mentation. How? Well, the BP is baldly based on the particular theory of the interaction between these agents and things like Euclid's Theorem and Quicksort and *modus tollens* that Benacerraf himself happens to like. For confirmation, simply look again at the passage quoted above. He informs us that he happens to "favor a causal account" of knowledge; and it's only on this account that the relation between agents and logico-mathematical objects becomes problematic. The idea seems to be that there must be some sort of *physical causal* connection between a human and, say, **ET**. Benacerraf is simply begging the question against anyone like Ross, or the two of us, who believe and seek to establish that there is no such causal connection to be had. And why not? Because we hold that the objects are immaterial, and that the agents are too; and obviously then the immediate implication is that there's no (physical) causal relation between agents and the objects to be had![46]

Our case for the immateriality of human persons is not yet clear of the general idea expressed by Benacerraf. For so far we have considered only the BP, the original version of the problem as specifically defined by Benacerraf; we have not yet considered and disposed of the variant of the BP introduced by Field in 1989, i.e., the variant B-FP. As Clarke-Doane reports, philosophers of mathematics today invariably take what's at issue to be the B-FP, not the original BP.[47] Well then, what *is* the B-FP? We do not have the luxury of giving a full presentation of the problem, and then proceeding to a detailed refutation. But no matter, for the B-FP, in broad strokes, is quite easy to convey: it's the problem that "it appears in principle impossible to explain" in *any* way how it is that our beliefs align so perfectly with logico-mathematical objects.[48] Field doesn't demand a causal explanation; he just demands an explanation, insists that there simply isn't one to ever be had, and then says that because of the absence of such an explanation our belief in mathematical entities is—to use his word—"undermined."[49]

In response, we offer what is, as far as we are aware, a new counter-objection to the BP/B-FP, one rooted in work we have carried out in the intersection of self-belief and AI.[50] In this work we have among other things presented axioms for characterizing, precisely, what we call *cognitive consciousness*, including in particular cognitive *self*-consciousness. From this work we only need here a sliver of one of the axioms in question: namely, that at least in the case of human self-consciousness we have beliefs about our own occurrent mental states, and those beliefs are true. For instance, all of us every now and then believe that we are angry (beyond just simply *being* angry), and sometimes we believe that we are angry and really shouldn't be (perhaps because of some ethic we subscribe to, but have nonetheless violated after succumbing to temptation), and so on. To focus things, consider not anger, but pain, acute pain. Suppose that Tommy believes at some time t that he is in acute pain at t. Is Tommy's belief correct? Well, how could he possibly be mistaken about such a thing? When you believe that you are in excruciating pain at some particular time, you *are* in excruciating pain.[51] This fact is what makes such beliefs—to use a term sometimes used by philosophers of mind—"incorrigible."

But notice then how the Benacerraf-Field Problem is obliterated. This happens because we absolutely, positively cannot give an explanation for why our beliefs that we are in excruciating pain are veridical. It is not as if we have some argument in support of such beliefs, or empirical evidence from some experiment we have run, or the testimony of some other agent who assures us that we are indeed in pain; no, nothing of the sort, at all. When we believe that we are in pain, we are right; and the absence of some explanation for this alignment does not in any way impugn the brute fact that our beliefs are correct. It follows from this that Field's objection melts away, for we have a case where the absence of an explanation puts not the slightest dent in the reliability and correctness of our beliefs. This shows that the general premise employed in the B-FP against the immateriality positions we have presented and defended herein—the premise that lack of explanation for correct belief about abstract, seemingly non-physical things (e.g., pain) provides reason to doubt the accuracy of those beliefs—is destroyed.[52]

7. Conclusion, The Cardinals, and Beyond

So, IN sum, we are immaterial—if we're right. We have little doubt that some physicalists will remain obdurate, despite the two-step argument we have given, and defended. We also have little doubt that additional objections will be brought against our case for the proposition that human persons are immaterial. Given this reality, it seems prudent for us to point out that the two-step argument given above employs only some exceedingly simple formal objects. Put in terms of modern mathematical education, everything we've done above uses no more than pre-college mathematics. We point this out because our case will grow in power as the robustness of the logico-mathematical objects to which we appeal grows.[53] What, specifically, have we in mind? The most fertile area to mine in order to articulate even more powerful versions of the argument given above is likely to be the world of the very, very, very large. This means that turning to set theory should prove productive—and this move can be taken by building seamlessly upon the elementary elements introduced above. Specifically, subsequent refinements and extensions of our case can start with a more serious look at the progression of sets—

$$
\begin{aligned}
\mathfrak{0} &:= \{0\} \\
\mathfrak{1} &:= \{0,1\} \\
\mathfrak{2} &:= \{0,1,2\} \\
\mathfrak{3} &:= \{0,1,2,3\} \\
&\vdots \\
\mathfrak{n} &:= \{0,1,2,3,\ldots,n\} \\
&\vdots
\end{aligned}
$$

—and then two of the sets identified above that enter into childhood math education: $\mathbb{N}$, the natural numbers; and $\mathbb{Q}^+$, the positive rational numbers. Specifically, the more serious look is undertaken in order to deeply understand the *size* of some $\mathfrak{n}$ in this progression, versus the size of $\mathbb{N}$. A first step in achieving this deeper understanding is to prove and thereby understand that $\mathbb{N}$ is of an infinite size, whereas each $\mathfrak{n}$ is merely finite. A second step is to understand that the size of $\mathbb{N}$ corresponds to the first infinite size-indicating number: the cardinal $\aleph_0$. And a third step is to understand that even though the set of positive rationals $\mathbb{Q}^+$ has all the natural numbers as a proper subset, $\mathbb{Q}^+$ is nonetheless also of size $\aleph_0$. These first few steps in the line of inquiry we sketch here will require inference schemata rather more nuanced than *modus tollens*, and likewise algorithms a bit trickier than Quicksort![54]

These are assuredly immaterial things, and, by application of the reasoning pattern given above, we as beings who understand these immaterial objects must ourselves be immaterial.

Of course, we can move on to sets that are larger than $\mathbb{N}$ and $\mathbb{Q}^+$, to the size of $\mathbb{R}$, which corresponds to the next infinite cardinal, $\aleph_1$; and from here we can continue. As we do this, it will, we believe, begin to seem simply preposterous that the logico-mathematical objects in mental play are not immaterial, and we predict it will be harder and harder to see how deep human understanding of these objects can be obtained by processing that is no more than standard mechanizable manipulation of embodied symbols (i.e., no more than Turing-machine-level computation). Notice, finally, that there is an empirical prediction that clearly emerges from our suggested extension and refinement of the line of reasoning given above. Our prediction is that AI carried out on the basis of its textbook definition now in force for nearly three quarters of a century[55] will perpetually fail to match great human achievements in the formal sciences.

Notes

1. First introduced in the early twentieth century by C.I. Lewis: "There are recognizable qualitative characters of the given, which may be repeated in different experiences, and are thus a sort of universal; I call these "qualia." But although such qualia are universals, in the sense of being recognized from one to another experience, they must be distinguished from the properties of objects . . . The quale is directly intuited, given, and is not the subject of any possible error because it is purely subjective." See Lewis's *Mind and the World-Order: Outline of a Theory of Knowledge* (New York: Charles Scribner's Sons, 1929), 121.
2. Pure mathematics, mathematical/theoretical physics,

formal logic, decision theory, game theory, theoretical computer science, etc.

3. Notice that we specifically speak of human persons. Human beings have physical bodies, of course; and physical bodies are (obviously) not non-physical. Hence it is dangerous to speak of human beings as simply "immaterial." This danger is often dodged by speaking of "the mind," or "the human mind"; but because this way of speaking seems to multiply entities beyond what seems reasonable (see R. Chisholm, "Is There a Mind-Body Problem?," *Philosophic Exchange* 2 (1978): 25–32), we prefer to speak of "human persons." And for ease of exposition in the present essay, from this point on, we shall feel free to say "humans" instead of "human persons."
4. Some readers might wisely ask whether circumspection dictates that we say instead, "we must ourselves, at least in part, be non-physical." We are of the opinion that further analysis would inevitably reveal that this more conservative language is superfluous. The reason is that since we as humans are individuated persons enjoying a unity of consciousness and thinking, we admit of parts, once our argument herein is appreciated, only insofar as we make use of things like hands or feet or eyes or brain parts. See again Chisholm, "Is There a Mind-Body Problem?"
5. Public K–12 mathematics education in New York State follows so-called "Common Core" standards (at least presently). These standards, which revolve around a progression of increasingly tricky logico-mathematical objects, can be scrutinized here: https://www.engageny.org/common-core-curriculum.
6. In mathematics, students in Grade 1, in New York State public education, are taught algorithms for addition, subtraction, and multiplication, and for simple "algebraic reasoning" in which unknowns in equations are determined.
7. Sometimes the symbols used to denote the Boolean operators/connectives will be different. E.g., one sometimes sees '$\supset$' instead of '$\rightarrow$' for material implication, and sometimes '$\equiv$' for '$\leftrightarrow$', etc.
8. See, for example, A. Bellman, S. Bragg, and W. Handlin, *Algebra 2: Common Core* (Upper Saddle River, NJ: Pearson, 2012).
9. See C. A. R. Hoare, "Algorithm 64: Quicksort," *Communications of the ACM* 4, no. 7 (July 1961): 321. The discovery of the algorithm was in 1959.
10. In prior work that must be left aside here, we have made considerable use of the type-vs-token distinction for rendering talk of algorithms and computer programs precise. See, e.g., K. Arkoudas and S. Bringsjord, "Computers, Justification, and Mathematical Knowledge," *Minds and Machines* 17, no. 2 (2007): 185–202; S. Bringsjord, "A Vindication of Program Verification," *History and Philosophy of Logic* 36, no. 3 (2015): 262–277, preprint available at: http://kryten.mm.rpi.edu//SB_progver_selfref_driver_final2_060215.pdf; S. Bringsjord and N. S. Govindarajulu, "Are Autonomous-and-Creative Machines Intrinsically Untrustworthy?," in *Foundations of Trusted Autonomy*, eds. H. Abbass, J. Scholz, and D. Reid (Berlin: Springer, 2017), 317–335. In the present paper, we will use both "embodiment" and "token" freely, with just the general understanding that these are physical things that vary among themselves, but which all stem from the general type they refer to.
11. Many would classify what we give here as an "algorithm-sketch."
12. To ward off any potential confusion arising here in the mind of the reader, we state (but relegate to this endnote so as not to break the main flow), that the previous sentence doesn't refer to the (surely; after all, there are at least as many such inputs as there are natural numbers) countably infinite inputs that $\mathcal{Q}$ can receive, but rather to the fact that the set $E_{\mathcal{Q}}$ of all embodiments of $\mathcal{Q}$ is countably infinite. (Note that we are not making any claim about the size of a set of *algorithms*; for present purposes there is only *one* algorithm in play here: Quicksort = $\mathcal{Q}$, as discovered and then communicated by Hoare. Yes, there are variants of the Hoare-given algorithm, and those are different algorithms (within a family), each of which can be embodied in countably infinite different ways. But this can and should be left aside in the present essay.) Why, it might be asked, is it the case that $E_{\mathcal{Q}}$ is countably infinite? A full answer to this question is beyond the scope of our essay, but we can encapsulate a response, in the form of a brief non-technical argument, one we trust is sufficiently cogent despite being compressed:

    Argument: Each embodiment $e \in E_{\mathcal{Q}}$ need only have a genuine notational difference relative to its predecessor, however small that difference might be, in order to count as an embodiment distinct from that predecessor. Specifically, if each e can be expressed in a different grammar for a different alphabet, or even in a different font for a given alphabet, distinctness is obtained. Clearly such expression is possible. Font-wise, evidence for such expression can be found in D. Hofstadter and G. McGraw, "Letter Spirit: Esthetic Perception and Creative Play in the Rich Microcosm of the Roman Alphabet," in D. Hofstadter et al., *Fluid Concepts and Creative Analogies: Computer Models of the Fundamental Mechanisms of Thought* (New York: Basic Books, 1995), 407–488, in which even just the Roman alphabet seems endlessly embodiable. The size of the collection of the distinct fonts and languages/alphabets to which we allude is clearly infinite; hence this collection is minimally the size of the natural numbers.

To conclude the present endnote, as far as we can see, nothing substantive hinges on the particular issue at hand, but we nonetheless, for the record, assert that a more robust treatment making use of either fonts or languages/alphabets could include not merely an argument, but an outright proof.

13. The token $\hat{\mathcal{Q}}_2$ immediately follows. This is a function defined in the Clojure programming language of today. All readers, regardless of background, will understand upon a bit of inspection that this is a definition (hence the string **defn**) of a function called **quick-sort**, and will note that this function is recursive, since it calls itself. This is in keeping with the abstract algorithm $\mathcal{Q}$ discovered by Hoare.

```
(defn quick-sort [coll]
 (if (not-empty coll)
  (let [pivot (rand-nth coll)]
   (concat (quick-sort
     (filter #(< % pivot) coll))
     [pivot]
     (quick-sort
       (filter #(> % pivot) coll))))))
```

14. See P. Wason, "Reasoning," in *New Horizons in Psychology I*, ed. Brian M. Foss (Harmondsworth, UK: Penguin, 1966), 135–151.
15. It is crucial for our overarching argument that the proof now given is a *formal* one, because as such we have before us a specimen that, in ironclad and directly observable fashion, makes use of the token of the immaterial *modus tollens* we have introduced. Nonetheless, we point out (and this may be a bit helpful for those less familiar with the conventional symbols of elementary formal logic), the proof can be given informally:

    Proof: We are given the rule R, viz., that for any card c whatever ('c' operating here as a variable), if c has a vowel on one side, then c has an even number on the other. We observe that Card 4 does not have an even number on it. Instantiating the variable in Line 1, we have that: If Card 4 has a vowel on one side, it has an even number on its other side. But now by *modus tollens*, we deduce that Card 4 does not have a vowel on the other side. **QED**

16. Where it's sometimes given a different name, e.g. "conditional elimination"; see, e.g., J. Barwise and J. Etchemendy, *Language, Proof, and Logic* (New York: Seven Bridges, 1999).
17. A more precise specification is this one:

$$\frac{\begin{array}{c}\phi_1 \vee \phi_2 \vee \ldots \vee \phi_k \\ \{\phi_1\} \vdash \gamma, \ldots, \{\phi_k\} \vdash \gamma\end{array}}{\gamma}\ \widehat{PBC}_2$$

18. Euclid's indirect proof of his theorem (= **ET**) can be couched in terms of proof-by-cases and *modus tollens*, as follows:

    Proof: Suppose that $\Pi = p_1 = 2, p_2 = 3, p_3 = 5, \ldots, p_k$ is a finite, exhaustive consecutive sequence of prime numbers. Next, define $\mathbf{M}_\Pi$ as $p_1 \times p_2 \times \cdots \times p_k$, and set $\mathbf{M}'_\Pi = \mathbf{M}_\Pi + 1$. Then either $\mathbf{M}'_\Pi$ is prime, or not; we thus have two (exhaustive) cases to consider. Both cases lead by *modus tollens* to the negation of our supposition:

    C1 Suppose $\mathbf{M}'_\Pi$ is prime. In this case we immediately have a prime number beyond any in Π, and our supposition is negated.

    C2 Suppose on the other hand that $\mathbf{M}'_\Pi$ is *not* prime. Then some prime p divides $\mathbf{M}'_\Pi$. (Why?) Now, p itself is either in Π, or not; we hence have two sub-cases. Supposing that p is in Π entails that p divides $\mathbf{M}_\Pi$. But we are operating under the supposition that p divides $\mathbf{M}'_\Pi$ as well. This implies that p divides 1, which is absurd (a contradiction). Hence the prime p is outside Π, and once again the starting supposition is negated.

Hence for *any* such list Π, there is a prime outside the list. That is, there are infinitely many primes. QED

19. James Ross, "Immaterial Aspects of Thought," *The Journal of Philosophy* 89, no. 3 (Mar. 1992): 136–150.
20. Ross, "Immaterial Aspects of Thought," 137. In seeking to show that some of our thinking, when we validly follow inference schemata and algorithmic functions, is immaterial, Ross appeals in his own case to prior work upon which he intends to build. He writes: "Now we need reasons why no physical process or function among physical processes can determine "the outcome" for every relevant case of a "pure" function. Those considerations mark some of the most successful in analytic philosophy, from W. V. Quine, to Nelson Goodman, to Saul Kripke" (Ross, "Immaterial Aspects," 140).
    Ross's reader is supposed to be familiar with what he is specifically appealing to here, and it's beyond the scope of the present paper for us explain the role that work by the trio Ross cites plays in Ross's reasoning. In the case of Goodman, the motivated reader can consult Nelson Goodman, *Fact, Fiction, and Forecast* [1955], 4th ed. (Cambridge, MA: Harvard University Press, 1983).
21. See S. Bringsjord, *What Robots Can and Can't Be* (Dordrecht: Kluwer, 1992).
22. Of late, some have taken to using "AGI" (for "artificial general intelligence") to refer to Strong AI. We herein stick with the original, older terminology, because in our experience some today mean by their use of "AGI" to refer to a category of artificial agents that have general-purpose cognitive powers cutting across many (perhaps all) human-relevant domains, but *not* necessarily to artificial agents that are subjectively aware/conscious. The phrase "Strong AI" unmistakably refers to creatures that have full-blown—as it's called—phenomenal consciousness.
23. In our personal experience, outside such venues and in the flow of real discussion in real life, at the lunch or dinner table and not in the "official" environments of academic papers and presentations (which are undeniably more than colored by careerist ambition), the belief in question is often admitted, sometimes unwittingly, to be, minimally, weak, and maximally, simply absent.
24. S. Bringsjord and N. S. Govindarajulu, "Artificial Intelligence," The Stanford Encyclopedia of Philosophy, https://plato.stanford.edu/entries/artificial-intelligence, published Jul 12, 2018, last modified Jul 7, 2022.
25. J. Anderson, *How Can the Human Mind Occur in the Physical Universe?* (Oxford, UK: Oxford University Press, 2009).
26. John Searle, "Minds, Brains and Programs," *Behavioral and Brain Sciences* 3 (1980): 417–424.
27. Bringsjord, *What Robots Can and Can't Be.*
28. S. Bringsjord and R. Noel, "Real Robots and the Missing Thought Experiment in the Chinese Room Dialectic," in *Views into the Chinese Room: New Essays on Searle and Artificial Intelligence*, eds. J. Preston and M. Bishop (Oxford, UK: Oxford University Press, 2002), 144–166.
29. For a more recent defense of this line, in connection with whether modern robotics can endow a robot with human-level understanding, see S. Bringsjord, "The Symbol Grounding Problem… Remains Unsolved," *Journal of Experimental & Theoretical Artificial Intelligence* 27, no. 1 (2015): 63–72, doi:10.1080/0952813X.2014.940139.
30. In our experience as educators, a hallmark of a human's understanding an algorithm such as $\mathcal{Q}$ is that he/she can grasp that various particular embodiments $\hat{\mathcal{Q}}_1$, $\hat{\mathcal{Q}}_2$, and so on of $\mathcal{Q}$ all token the same type $\mathcal{Q}$.
31. For another book-length defense of the view that this is what we are, see S. Pinker, *How the Mind Works* (New York: Norton, 1997).
32. Angus Menuge, "Knowledge of Abstracta: A Challenge to Materialism," *Philosophia Christi* 18, no. 1 (2016): 7–27. We are indebted to reviewers of an earlier version of our paper for bringing Menuge's paper to our attention.
33. Menuge, "Knowledge of Abstracta," 7.
34. Our Step-2 argument of course marks a debt to Searle's CRA and subsequent improvement achieved by Bringsjord, and the CRA line of reasoning does not make use of Menuge's reasoning.
35. Menuge, "Knowledge of Abstracta," 23.
36. Ross, "Immaterial Aspects of Thought," 137. Among these jewels, for technical reasons too far afield for treatment herein, the "jewel" that is most relevant to our two-step case is the "Grue Paradox" (GP) seminally introduced by Nelson Goodman in *Fact, Fiction, and Forecast* in 1955; this is work that Ross specifically cites. In the view of the first author, the only escape from GP is to require of the gemologist featured in Goodman's famous parable that he specify any and all inference schemata used to

support conclusions about scientific laws regarding emeralds — but once these schemata are specified, it's the following of them as algorithms that becomes central, and allows GP to be avoided. This is philosophically in line with purely mathematical treatments of inductive logic, even when such treatment involves no argumentation, but—following Carnap, the founder of formal inductive reasoning—uses the machinery of probability calculation to adjudicate competing scientific hypotheses; see, for example J. Paris and E. Vencovská, *Pure Inductive Logic* (Cambridge, UK: Cambridge University Press, 2015).

37. Nelson Goodman, *Fact, Fiction, and Forecast* [1955], 4th ed. (Cambridge, MA: Harvard University Press, 1983).
38. For exploration of exactly these issues from the perspective of computational formal logic and AI, see K. Arkoudas and S. Bringsjord, "Metareasoning for Multi-agent Epistemic Logics," in *Proceedings of the Fifth International Conference on Computational Logic In Multi-Agent Systems (CLIMA 2004)*, eds. Joao Leite and Paolo Torroni (Springer-Verlag: Berlin and Heidelberg, 2005), 111–125, doi:10.1007/11533092_7. A pdf, with different pagination, is available at: http://kryten.mm.rpi.edu/arkoudas.bringsjord.clima.crc.pdf.
39. See, for instance, Arkoudas and Bringsjord, "Metareasoning for Multi-agent Epistemic Logics."
40. G. Lakoff and R. Nuñez, *Where Mathematics Comes From: How the Embodied Mind Brings Mathematics into Being* (New York: Basic Books, 2000).
41. Lakoff and Nuñez, *Where Mathematics Comes From*, xiii.
42. The fact left aside here is brought to the attention of the cognitive-science community in S. Bringsjord and N. S. Govindarajulu, "Rectifying the Mischaracterization of Logic by Mental Model Theorists," *Cognitive Science* 44, no. 12 (2020), e12898.
43. The fatal problem of arbitrariness infecting the case given by Lakoff and Nuñez for the embodied and legislative nature of mathematics is related in interesting ways to the argument given by C. S. Lewis against naturalism. See Lewis, *Miracles* (New York: Macmillan, 1947). A great place for interested readers to start is V. Reppert, "The Argument From Reason," in *The Blackwell Companion to Natural Theology*, eds. W. L. Craig and J. P. Moreland (West Sussex, UK: Wiley-Blackwell, 2009), 344–390. However, we inform, and indeed caution, our readers that Reppert affirms, for the assessment of arguments, a "Bayesian model with a subjectivist theory of prior probabilities" (p. 346), and we wholly reject Bayesian frameworks, since they are insufficiently expressive [they have base formal languages that are purely extensional (e.g., zero-order and first-order logic)], and are committed to standard Kolmogorovian probability calculi interpreted subjectively—a position we take to be decisively overthrown by, e.g., J. Pollock, *Thinking about Acting: Logical Foundations for Rational Decision Making* (Oxford, UK: Oxford University Press, 2006). We consider any premeditated framing of competing arguments in natural theology within Bayesianism and/or its underlying formalisms to almost instantly be fatal to natural theology. Along this line, see the above-cited S. Bringsjord and N. S. Govindarajulu, "Rectifying the Mischaracterization of Logic by Mental Model Theorists."
This is as good a place as we can find to inform the reader that while we see (as just indicated) a connection between the view that mathematics is legislated and the argument originated by Lewis (and improved by Reppert), the connection between our two-step argument for the immateriality of human persons and what Reppert calls "The Argument from Reason" is one we find painfully obscure. We say this knowing full well that Reppert classifies the argument of James Ross (upon which we of course heavily rely herein) as belonging in the same family as his own (see Reppert, pp. 365–366). Of note for motivated readers and scholars is the fact that Reppert's explicit distillation of Ross's argument (Reppert calls the distillation a "formalization") says that this argument's ultimate conclusion is that some human mental states are non-physical, e.g., "the mental states involved in mathematical operations are not and cannot be identical to physical states" (p. 366). As we have made clear, the key proposition yielded by Step 1 in our two-step argument is that certain *objects* are non-physical, and as the title of the present paper makes plain, the idea is that if these objects are non-physical, then the objects that we are, are non-physical too.
44. P. Benacerraf, "Mathematical Truth," *Journal of Philosophy* 70 (1973), 674.
45. Benacerraf, "Mathematical Truth," 671–673.
46. We add, by the way, that further investigation of the relevant literature on a causal account of knowledge

is not helpful to Benacerraf. Alvin Goldman, arguably the leading proponent of such causal accounts in the twentieth century, and someone in fact credited with originating such accounts, is disinclined to apply them in the case of knowledge of logico-mathematical propositions. See A. Goldman, "A Causal Theory of Knowing," *The Journal of Philosophy* 64, no. 12 (1967): 357–372.

47. J. Clarke-Doane, "What is the Benacerraf Problem?," in *New Perspectives on the Philosophy of Paul Benacerraf: Truth, Objects, Infinity*, ed. F. Pataut (Cham, Switzerland: Springer Verlag, 2017), p. 20.
48. H. Field, *Realism, Mathematics, and Modality* (Oxford, UK: Basil Blackwell, 1989), p. 233.
49. Field, *Realism, Mathematics, and Modality*, p. 233.
50. S. Bringsjord and N. Govindarajulu, "The Theory of Cognitive Consciousness, and Λ (Lambda)," *Journal of Artificial Intelligence and Consciousness* 7, no. 1 (2020): 155–181.
51. Someone might somehow doubt this, though we can't fathom how. But if so, we can simply retreat to something that will get the job of overthrowing the B-FP done just fine: we can focus on such self-beliefs as that one *seems* to be in acute pain. No one can possibly be mistaken in a belief that one seems to be in acute pain—and this despite the fact that no one can rigorously explain such perfect knowledge. Note that the first of the present authors purports to have proved long ago (in 1992) that human persons are neither Turing-level computing machines nor material, in light of the infallibility just described, which human persons possess; see Bringsjord, *What Robots Can and Can't Be.*
52. Much more could be explored, and an anonymous reviewer has brought to our attention a fascinating portal to the "wild" terrain that could be explored. This reviewer understands us to be maintaining that there is no causal connection between human persons and the immaterial formal objects with which they are "directly acquainted." This, the reviewer says, fits "the assumption of many philosophers that abstract entities are causally powerless." But the reviewer alertly notes that there is another option, one suggested by Alvin Plantinga: perhaps there is a non-physical, non-standard kind of causation that holds between immaterial entities. While we are far from familiar with this line (but busy now learning!), it seems to at least the first of the present authors that the relationship described by Plantinga (1980) between God's intellect and immaterial formal objects—see A. Plantinga, *Does God Have a Nature?* (Milwaukee, WI: Marquette University Press, 1980)—is quite relevant, and perhaps by analogical reasoning illuminating. Yet, for present purposes, we refrain from further analysis, let alone taking a stand. We are quite happy to rely upon the "facts of the case" in the incorrigibilist phenomena brought to the reader's attention above.
53. Part of the reason in turn why this is so is that any degree of plausibility associated with the view that logico-mathematical objects can be in some way tied directly and exclusively to physical objects quickly erodes to zero, or at least—and the irony of picking the purely formal concept employed by both Leibniz and Newton for their invention of the calculus is intended—to an infinitesimal.
54. We shall need to employ, e.g., the inference schema *mathematical induction*, and from its use another inference schema often referred to as *the pigeonhole principle*. Nice coverage is provided in D. Goldrei, *Classic Set Theory* (Boca Raton, FL: Chapman & Hall, 1996).
55. According to which AI is the field devoted to building artificial agents that compute functions from what they perceive to the actions they take, where these functions are Turing-computable ones that match in their nature what Searle-in-the-room has as a resource. For such textbooks, see, e.g., S. Russell and P. Norvig, *Artificial Intelligence: A Modern Approach*, 4th ed. (New York: Pearson, 2020), and G. Luger, *Artificial Intelligence: Structures and Strategies for Complex Problem Solving*, 6th ed. (London, UK: Pearson, 2008).

24. Can Consciousness Be Explained by Integrated Information Theory or the Theory of Cognitive Consciousness?

Naveen Sundar Govindarajulu and Selmer Bringsjord

This is an online-only chapter. You can access it at MindingtheBrain.org.

25. How Informational Realism Dissolves the Mind-Body Problem

William A. Dembski

1. Making Information Real and Fundamental

To see how informational realism dissolves the mind-body problem, we need first to be clear on what informational realism is and why it is credible. Informational realism is not simply the view that information is real. We live in an information age, so who doesn't think that information is real? Rather, informational realism asserts that the ability to exchange information is the defining feature of reality, of what it means, at the most fundamental level, for any entity to be real.

In consequence, informational realism constitutes a relational ontology. In a relational ontology, things exist not in themselves but insofar as they relate to other things. Or, as Wesley Wildman puts it, "The basic contention of a relational ontology is simply that the relations between entities are ontologically more fundamental than the entities themselves. This contrasts with a substantivist ontology in which entities are ontologically primary and relations ontologically derivative."[1]

Informational realism does not deny the existence of things (i.e., entities or substances). But within informational realism, what defines things is their capacity for communicating or exchanging information with other things. Things are inferred from the information they communicate. Information, as the relational glue that holds reality together, thus assumes primacy in informational realism.

In informational realism, things make their reality felt by communicating or exchanging information. Thus, things that are not in immediate or mediate informational contact with other things might just as well not exist and so, in fact, don't exist within informational realism. Informationally isolated or disconnected entities thereby become nonentities.

Existence conferred by informational connectedness resonates with idealism's *esse est percipi* (to be is to be perceived). Yet with informational realism, the fundamental proposition would read instead *esse est informari* (to be is to be informed), or better yet *esse est informare et informari* (to be is to inform and be informed).

In substituting information for perception, informational realism is able to preserve a commonsense realism that idealism has always struggled to preserve. If a tree falls in a forest and no one is there to perceive it, does it make a sound? For idealism to preserve our commonsense intuitions in such situations, it needs an omnisentient God (or some comparable device) that is everywhere as perceiver and so is there to record the sound when the tree falls (God hears it).

Informational realism, on the other hand, does not need an omnisentient God to preserve our commonsense realism. Specifically, the tree, in its fall, communicates information to its immediate surroundings, which then ramifies through the whole of reality, reality being an informationally connected whole. So yes, within informational realism, the tree's fall makes a sound even if no sentient being is in immediate informational contact with it.

Informational realism is incompatible with materialism, the view that matter in its various motions and modifications is fundamentally all there is. Within materialism, information is a byproduct of matter, arising from the way matter is structured. Biologist Christian de Duve made this point explicit in laying out his "ages" of the history of life. First, according to him, is "The Age of Chemistry," which is "ruled entirely by the universal principles that govern the behavior of atoms and molecules." Only then, built immediately on top of chemistry, comes "The Age of Information," which arises "thanks to the development of special information-bearing molecules."[2]

By contrast, within informational realism, matter is a dispensable abstraction. That's not to say that physical things, and especially bodies, don't exist within informational realism. And that's not even to say that ordinary language ought obsessively to expel all references to matter. Matter is a convenient shorthand, as we'll see, for referring to certain types of physical things, especially those composed of parts that interact mechanistically. Informational realism is therefore fine with materiality insofar as it keeps its metaphysical pretensions in check. And it is even more fine with physicality. Some things operate in space and time, exhibit an observable structure and dynamics, and make themselves evident to our senses. They are physical and, where suitably confined, have bodies. All good.

But we infer physical things from their particular type of informational behavior. By contrast, matter, as we'll see, conceived as some sort of substrate for physicality, is superfluous and even misleading. Laplace famously remarked to Napoleon that he didn't need God to account for the structure and dynamics of the solar system. Similarly, informational realism has no need for matter in its account of reality.

Informational realism's rejection of matter differs from that of Berkeleyan idealism. Berkeley's immaterialism flowed from his combination of idealism and empiricism. The material things that minds perceived were for Berkeley unnecessary inferences from phenomena that impressed themselves as ideas on minds. And since within Berkeleyan idealism all that exists are minds and ideas residing in minds, to posit or infer material things as somehow mind-independent realities made no sense. Within Berkeley's philosophical system, Occam's razor therefore neatly excised matter.

Informational realism, on the other hand, doesn't dispense with matter on such a priori grounds. The problem with matter for informational realism, as we'll see, is that matter always resolves, on closer inspection, into information sources whose self-revelation is, without remainder, informational. Informational realism in this way dispenses with any need for a material substrate to anchor information. Instead of viewing matter as a medium for information, informational realism says that any such medium (i.e., one acting as a carrier for information) is wholly informational. This point will become clearer as we proceed.

The push to put information at the foundation of reality has come more from scientists than philosophers. Perhaps the most prominent scientist to take this view has been the late physicist John Archibald Wheeler (1911–2008). In his autobiography, Wheeler describes being successively in the grip of three metaphysical ideas: Everything Is Particles, Everything Is Fields, and then, at the end of his career, Everything Is Information. Elaborating on the last of these ideas, Wheeler wrote: "The more I have pondered the mystery of the quantum and our strange ability to comprehend this world in which we live, the more I see possible fundamental roles for logic and information as the bedrock of physical theory."[3]

To say that everything is information calls to mind those joke ontologies in which reality rests on the back of a turtle, which in turn rests on the back of another turtle, with turtles all the way down to infinity, only now it's information rather than turtles all the way down. In fact, as we'll see momentarily, Wheeler's full ontology requires quite a bit more than just information (notably, observers to monitor the information).

In general, informational realists like Wheeler and myself assign a primacy to information that nonetheless falls short of making every single thing an item of information. My own approach to informational realism would say that anything is what it is by the way it makes its reality felt through information. But that's not to say that everything is, without remainder, information.

How did Wheeler arrive at his informational realism? Wheeler, in promoting information as the primal entity of the universe, invented the catchphrase "it from bit":

> Every "it"—every particle, every field of force, even the space-time continuum itself—derives its function, its meaning, its very existence entirely—even if in some contexts indirectly—from the apparatus-elicited answers to yes-or-no questions, binary choices, bits. "It from bit" symbolizes the idea that every item of the physical world has at bottom—a very deep bottom, in most instances—an immaterial source and explanation; that which we call reality arises in the last analysis from the posing of yes-no questions and the registering of equipment-evoked responses; in short, that all things physical are information-theoretic in origin...[4]

Wheeler's "it from bit" approach to informational realism has always struck me as problematic. It arises from his Participatory Anthropic Principle, in which observers create reality by measuring it. To underscore that this was indeed Wheeler's view, I purposely left off just now the end of the passage above. Here is the full conclusion: "in short, that all things physical are information-theoretic in origin *and that this is a participatory universe.*"[5] It's one thing to say that measurement requires information. It's another thing to say that the thing being measured is created by the observer doing the measuring. That seems a bit much, and the ontological status of these observers raises thornier questions than it resolves.

These observers promise to resolve the quantum mechanics measurement problem (the problem that, at the quantum scale, observers invariably interact with and thus perturb the systems they measure, thereby precluding any pure observation of nature). But any success in this regard comes at the cost of explaining how you could get conscious observers that collapse wave functions and do quantum measurements in the first place. That, in my view, is an even thornier problem than the quantum measurement problem.

Setting aside Wheeler's participatory understanding of the universe, which seems like a huge pill to swallow, his "it from bit" approach to informational realism suggests a discredited operationalist view of science that sees the meaning of scientific concepts in their methods of measurement.[6] Since any measurements can always be captured with a sufficient number of bits (bits can represent any numbers, and numbers can represent measurements to any precision), there's a sense in which all scientific measurements reduce to bits. But does that mean that reality itself reduces to bits? Or does it mean instead that science, insofar as it is quantified, reduces to bits? It seems Wheeler only established the latter.

Paul Davies, in questioning Wheeler's full-blown informational realism, agrees with him that "science is... an interrogation of nature" in which "bits" are used to characterize "its." But Davies then notes that those bits will constitute "data that we try to link into a coherent scheme."[7] It's such a coherent scheme—an empirically adequate and theoretically satisfying understanding of nature—that remains the goal of science. Thus, in Davies's view, the use of information to study nature falls short of establishing that nature, the object of scientific study, is at bottom informational.

It's this more ambitious view of nature that I seek to justify in this chapter. Given that the main task of this chapter is to explain and make plausible informational realism, we will therefore need to define information and show how it works, clarifying, in particular, what it means to exchange or communicate information.

2. The Contingency-Constraint View of Information

What, then, is information? Information arises from the interplay between *contingency* and *constraint*. Contingency refers to the different ways things might be. Note the plural: way*s* things might be. If there's only one way a thing can be, there's no contingency and thus, as we shall see, no information. Constraint refers to a narrowing of these contingencies, including some and excluding others. Suppose I tell you that it's raining outside. The contingencies here are that it's raining outside or that it's not raining outside (two contingencies). The constraint here is that by telling you it's raining outside (focusing on one contingency), I'm excluding it's not raining outside (ruling out the other contingency).

The constraint, if it is to be informative, must neither include all contingencies nor exclude all contingencies. If I tell you that it's raining or not raining outside, I haven't given you any information because you already knew that it had to be raining or not raining outside (I would thus have simply stated a tautology). Likewise, if I tell you that it's raining and not raining outside, I haven't given you any information because you already knew it couldn't be both raining and not raining outside (I would thus have simply stated a contradiction).

Writing about human communication, philosopher Robert Stalnaker defined information using different terms but to the same end: "To learn something, to acquire information, is to rule out possibilities [contingencies]. To understand the information conveyed in a communication is to know what possibilities [contingencies] would be excluded [constrained] by its truth."[8] Going beyond human communication, philosopher Fred Dretske made the same point: "No information is associated with, or generated by, the occurrence... of events [constraint] for which there are no possible alternatives [contingency]."[9]

Information always presupposes a range of contingencies or possibilities, and then a focusing on or constraint of some portion of that range to the exclusion of the rest of that range. Information thus always presupposes both a *yes* and a *no*, a *yes* to what the constraint admits or includes, and a *no* to what the constraint dismisses or excludes. In fact, there's a sense in which *no* is more fundamental to information than *yes*, because simply saying yes to everything guarantees that no information can be generated. The constraint that generates information always involves saying no to contingencies, ruling things out, excluding possibilities. Information requires negation.

What enables this contingency-constraint approach to information to become part of science? Two things:

1. quantification by means of probabilities;
2. computability by representing information in bits.

Regarding point (1), the range of contingencies under consideration typically invites a probability distribution, and the constraint, by identifying a subset of those contingencies, yields a probability. So, for instance, there are 2,598,960 distinct poker hands. These constitute the relevant range of contingencies. One constraint would be hands with exactly one pair. These number 1,098,240. So, assuming all hands are equiprobable, the probability of getting exactly one pair is roughly 0.42. But consider instead a royal flush, of which there are only four. Assuming again equiprobability, the probability of getting a royal flush is roughly 0.00000154, a much smaller probability.

Regarding point (2), in many situations, and all situations involving scientific measurement, information can be represented as character strings, which is to say, as linear sequences of characters drawn from a finite alphabet (or finite set of symbols). Because such characters can in turn always be represented by bits (think of the way UTF or ASCII encodes the characters on a keyboard into bits), bit strings have become the universal convention within science to represent information. Moreover, once this is done, it is straightforward to do computation on such bit strings, which can serve as stored memory as well as programs.

To say that information emerges through the interplay between contingency and constraint requires some elucidation. In this regard, I find it helpful to think of information as a *verb* rather than as a *noun*. If you will, information happens rather than simply is. Because most of the information we deal with takes the form of alphanumeric characters arranged in a linear order (this can be anything from bit strings *in silico* to text resulting from ink on paper), we tend to think of information as *items of information*. But items of information are really the end product of an informational happening. Ask yourself this: How do we get an item of information in the first place?

Suppose you are reading Shakespeare's *Hamlet*. Shakespeare, in writing the play, applied ink to paper to form an item of information, namely, the first finished manuscript of the play. Simply looking at that manuscript would not suggest information's defining interplay between contingency and constraint. Nevertheless, the activity of Shakespeare in writing the play does suggest this interplay. While composing *Hamlet*, Shakespeare might have written any number of other plays. In fact, he might have written any sequence of letters and spaces onto sheets of paper (most of which would be gibberish). All these different texts that he might have written constitute the relevant range of contingencies. And then, by writing *Hamlet*, Shakespeare drastically constrained that range of contingencies.

That's how the information that constitutes *Hamlet* was born, as a constraint on contingencies. Moreover, the text of *Hamlet* that you are reading comes through successive exchanges of information starting from the first finished manuscript and moving through typesetting and printing presses, and in our day through digital texts and eBooks. Each of these information exchanges occurs via constraints on contingencies. As the information is copied, it could be copied in any number of ways, only a small proportion of which are close enough to the original to count as *Hamlet*. The same principles apply to the MP3 music files you're listening to or the MP4 video files you're watching. In all such cases, there's a causal chain of information exchanges leading back to the author/artist/composer of the information from the point of origin.

Information is therefore not just dynamic rather than static but also inherently relational, connecting back to the originating source of the information. Accordingly, we can think of information not just as a verb but also as a *transitive verb*. As a transitive verb, information is thus about connecting a subject to an object. The subject is the giver or sender of the information, and the object is the receiver or target of the information. This implies a directionality to the information, going from the sender to the receiver, and explains why a synonym for *information theory* is *communication theory*.

In an exchange of information, information is communicated from a sender to a receiver. Yet we need to be careful here about identifying sender and receiver too rigidly and seeing the directionality of information too much as a one-way street. As a practical matter, we are often concerned with one direction in the information exchange (such as sending an email from writer to reader). But we need to be sensitive to information flowing also in the other direction. Often there can be information blowback, where the receiver of information sends information back to the original sender (whether deliberately or inadvertently), thus flipping their roles.

That said, information exchange is not like Newton's third law, in which for every action there is an equal and opposite reaction. In many inquiries, such as tracking the flow of information on the internet, we may be concerned with information flow in only one direction. But we always need to be aware that the exchange can be bidirectional. The case of an observer with a quantum mechanical system is a case in point: in the act of measurement, the system is sending the observer information, but likewise the observer is perturbing, and thus inputting information, into the system to elicit the measurement. Quantum measurements are inherently bidirectional. Closer to home, Shakespeare, as sender, gives *Hamlet* to the public, as receiver; yet the public's enthusiastic reception of *Hamlet*, now acting as sender, is not lost on Shakespeare, as receiver (the public's enthusiasm, communicated to Shakespeare as information, likely influenced subsequent plays that he wrote, such as *King Lear* and *Macbeth*).

3. Philosophical Resonances with Information

The contingency-constraint view of information resonates with certain ideas that philosophers in the past have developed. Nonetheless, I don't see this view as fully developed at any point in the history of philosophy. Contingency and constraint are certainly not new ideas. But their role has, in philosophical inquiry, tended to be peripheral or secondary. It might therefore be helpful to review some of these philosophical antecedents that resonate with the contingency-constraint view of information and see where this view promises to chart new paths.

3.1 Aristotelian Forms

Aristotle was not a materialist, so for him information was not simply a byproduct of matter.

The forms that for him generated information were real. And yet, they inhered in matter. The contingency-constraint view of information thus arises for Aristotle in his theory of *hylomorphism*. Aristotle thought of substances (things) as combining matter (Greek *hulē*) and form (Greek *morphē*).[10] Matter thus gets stamped with form. Contingency here consists of the different ways matter might get stamped with form, and constraint of how one form rather than another gets stamped on a given piece of matter. Aristotle distinguished two ways that matter acquires form or information:

1. information arising by nature (Greek *phusis*) through potentialities already inherent in things (like an acorn expressing its inherent information by growing into an oak tree);
2. information arising by design (Greek *technē*) imposed externally on things (like pieces of wood crafted by artisans into a ship, the pieces of wood left to themselves having no such power of self-organization).

What's wrong with hylomorphism? True, in much of our experience, information resides in a physical medium, which we are apt to think of as material. But it's simply untrue that such materiality is an essential feature of information, as Aristotelian hylomorphism would have it. All that the contingency-constraint view of information requires is a constraining of contingencies, whatever the nature of those contingencies. Within Christian theism, for instance, God creates through a spoken word. That word, existing logically prior to creation (the creation depending for its existence on this word), resides in the mind of God. Moreover, God, within Christian theism, is free and so could have spoken other words (hence contingency as well as constraint). And finally, whatever we might mean by the word "material," the Christian God is a non-material being, so those thoughts and words as they exist in the mind of God are likewise non-material.

But one doesn't have to look to theology to see that information doesn't presuppose materiality. Consider computation. I'm still partial to Windows XP, which is now about twenty years behind the times. But I don't need to run Windows XP on a dedicated material machine. I can run it in simulation mode on a Windows 7 machine. And I can run Windows 7 in simulation mode on a more current Windows 10 machine. And, because all computation can be done on Turing machines, which are abstract machines arising from a purely mathematical model of computation, I could in principle run the Windows 10 machine in simulation mode on a Turing machine. And just to be clear, mathematical objects like Turing machines don't require materiality. In fact, because Turing machines require "infinite tapes," they cannot be materially realized.

Aristotelian hylomorphism is thus mistaken in requiring information to have a material, or even physical, medium. Materiality, and even physicality, is therefore a nonessential feature of information. Informational realism therefore turns the tables on hylomorphism and matter generally: whatever we mean by matter must itself be thoroughly informational. For the informational realist, there is no matter as such but only sources of information that in their contingencies and constraints may reveal material aspects. Yet, insofar as anything discloses itself as material, informational realism would say that it never reaches a final resting place of pure materiality. Informational realism thus regards materiality as evanescent whenever a thing's supposed material underpinnings are probed with sufficient depth.

Unlike ancient materialists, modern materialists are in a better position to appreciate the evanescence of materiality. It's one thing for the ancient materialists, such as Democritus, to see everything as consisting of atoms that cannot be cut or divided (the word *atom* comes from the Greek, with the *a* being the alpha privative indicating negation and the *tom* referring to cutting). Democritus's atoms, as indivisible units, could plausibly define the nuts and bolts of matter and thus help render materialism plausible.

But such atoms don't exist. In modern physics, matter consists of particles and fields, and in our experiments and theorizing these never quite seem to terminate in some rock-bottom materiality. Like an infinitely layered onion, higher and higher energies are required to dig deeper and deeper into matter, and yet with no end in sight. More to the point, all attempts to reach down to and understand deeper layers of materiality yield not in the first instance matter but instead information. Or, to recast a famous statement by the economist Milton Friedman, "*matter is always and everywhere an informational phenomenon*."[11]

This fact about matter being at base informational becomes especially evident when one considers the Higgs boson. Dubbed "the God particle," the Higgs boson peels back matter as far as we can go given the current state of science and technology. The Higgs boson was conjectured to exist back in 1964, but it was not discovered to be real until 2012. But how was it discovered to be real? Fundamental particles like the Higgs boson are not observable in any straightforward sense. Physicists in fact talk of "creating" such particles by concentrating high energies in precise ways and then recording scatter plots of interactions involving these and other known particles, with the detected particle exhibiting a characteristic pattern or "signature" predicted by, in this case, the standard model of particle physics.

This all sounds, dare I say, *informationy*. And at the end of the day, that's all the physicists trying to discover the Higgs boson had, namely, information. To determine whether the Higgs boson was real, they had to consider a range of different possible scatter plots that a particular adjustment of the Large Hadron Collider (LHC) might deliver. There's the contingency. And then they had to determine whether the actual scatter plot that was found matched the characteristic pattern predicted by the standard model of particle physics. That's the constraint. It was this pattern match that confirmed for the scientists that the Higgs boson was real. Note that what I'm describing here is not an idealized rational reconstruction of what the physicists at CERN did in running the LHC to discover the Higgs boson. *This is exactly what they did.*[12]

In taking information seriously, Aristotelian hylomorphism is therefore a step in the right direction compared to materialism. Nonetheless, hylomorphism gives matter too independent an existence, and in the end matter cannot be maintained as a necessary condition for information.

3.2 Platonic Forms

Plato's theory of forms, like Christian theism, promises to give information a place outside the purely physical realm. That said, the contingency-constraint view of information plays at best an incidental role in Plato's theory of forms. For Plato, what is most real are the forms, which exist in a non-physical world, sometimes called a Platonic heaven (think of the perfect circle vs. an approximation of a circle drawn on a whiteboard). The physical world of daily experience is for Plato a world of shadows, a world of imperfect representations of the perfect forms.

In Plato's theory of forms, there is, strictly speaking, no contingency at the level of the forms. The forms are what they are, eternally. Even the Neoplatonism of Plotinus denies any contingency at the level of forms. Plotinus posited the One, from which all things, forms included, emanate. But these emanations are not free actions by an agent that could have done otherwise. The One of Neoplatonism is not an agent at all and admits no more contingency than the forms of the original Platonism of the Academy. Neoplatonic emanations proceed by necessity from the perfection of the One down through stages of decreasing perfection.

Platonism throughout its history has placed a premium on the forms and distrusted their physical manifestation. Thus, insofar as contingency arises for Plato, it is in the importation of the perfect forms into the imperfect world of experience, which is a physical world and which is imperfect precisely because its physicality obscures the forms. The form of chair is thus represented in, but also to varying degrees obscured by, the different physical chairs. Contingency resides in the different possible chairs that might embody the form of chair, and constraint arises in the particular chairs that arise to the exclusion of others. Unlike the Christian doctrine of creation, where the physical world is created in order to reveal the creator (whether the existence, attributes, or power of the creator), within Platonism there's always the sense that the physical world is substandard and that to peer into heaven one must bypass it. The world is not a window or springboard into heaven, as in orthodox Christian theology.

Karl Popper makes a similar point about Plato's philosophy, though in reference not to Christianity but to Aristotle's philosophy, and specifically how their differing theories of forms led to differing views on progress. Thus, in Plato's theory, writes Popper, "all change, at least in certain cosmic periods, must be for the worse; all change is degeneration. Aristotle's theory admits of changes which are improvements; thus change may be progress. Plato had taught that all development starts from the original, the perfect Form or Idea, so that the developing thing must lose its perfection in the degree in which it changes and in which its similarity to the original decreases."[13]

All the important informational flow for Plato is from the forms downward to the imperfect world of physicality. It's a one-directional flow, and for souls entrapped in this nether world of physicality, their only remedy consists not in transmitting information back to the world of forms from below but in simply escaping the physical world. Aristotle had a much more "live in this world and exchange information in it" mentality. Raphael's famous fresco *The School of Athens* captures the difference between the two, with Plato depicted as pointing up (to the non-physical world of forms) and Aristotle spreading out his hand horizontally (to the physical world we inhabit).

3.3 Idealism

Idealism, at least in some of its varieties, is compatible with informational realism. Bruce Gordon, in his chapter on idealism for this volume, reviews the varieties of idealism and advocates a neo-Berkeleyan ontological idealism. I have sympathies for this form of idealism and, if forced to choose, would choose it over other forms of idealism. That said, while capable of being shoehorned into idealism, informational realism strikes me as having a very different flavor.

The basic idea of idealism is that reality is a construction of mind and that the things that minds construct are ideas, thereby putting ideas at the foundation of reality. The information sources that within informational realism become real by exchanging information could thus be certain types of ideas. With powerful enough minds and rich enough ideas, any configuration of informational relationships, and thus any informational reality, could in principle be actualized.

At the center of Berkeleyan idealism is the Christian God, whose mind can entertain any and all ideas, and thereby underwrite a full-throated informational realism. In agreement with informational realism, Berkeleyan idealism regards matter as a dispensable abstraction, with matter doing no important metaphysical work in either ontology (Berkeley himself regarded matter as a fiction). Moreover, with the Christian God at the center of Berkeleyan idealism, common objections to idealism vanish, such as that a tree falling in a forest won't make a sound if no one is there to hear it. As we saw earlier, it will make a sound because God is omnisentient, and thus there to perceive the tree falling and making a sound, even if no other sentient being is there to witness the event.

Nonetheless, even with this broad compatibility between idealism and informational realism, the two strike me as very different. Consider the thoughts of Michelangelo's David, a unicorn, and a circuit board schematic. What about Michelangelo's David makes it real? What about the unicorn makes it unreal? And what accounts for the transition "from thought to thing" in taking the circuit board schematic and building the actual circuit board? Informational realism, as we'll see, is able to answer such questions in terms of externalizing and reifying information.

But how does idealism answer such questions? In the end, one is left with saying that these are different ideas with different properties (after all, within idealism everything is minds thinking ideas). For instance, different logical predicates will need to attach to Michelangelo's David (e.g., "Michelangelo's David exists in the actual world") versus the unicorn (e.g., "unicorns exist in other possible worlds but not in the actual world"). And the thought of the circuit board as a schematic will need to be distinguished from the thought of the circuit board as an operational thing.

I have no doubt that one can get logical consistency between idealism and informational realism. And yet, logical consistency, in simply requiring freedom from contradiction, sets a low bar for any philosophical position. Other desiderata such as beauty, insight, and fruitfulness will also come into play. In any case, any attempted rapprochement between the two ontologies has always left me unsatisfied. Granted, that's not an argument. But as William James pointed out, most philosophical allegiances are decided less by argument than by the temperament of the philosophers holding their various positions.

Reality, in idealism, is not merely a construction of mind but the thinking of thoughts or ideas by mind. But the thinking of ideas doesn't seem very thing-like. Where is the autonomy or independent sphere of activity that we customarily associate with things? Minds that think thoughts are free to think other thoughts and can shift their attitude toward existing thoughts. We are free to formulate thoughts and give them the shape we like. A mind thinking thoughts strikes me as similar to a child playing with army figures. The figures don't move on their own, so the child, in playing with them, needs at every point to position and move them. The army figures are thus in essence puppets. Likewise, ideas, to the degree that they lack independence from mind (as seems unavoidable in idealism), strike me as hard-pressed to avoid becoming puppets of the mind.

Or, in the same vein, it's like the lock of Belle's hair in the Disney animation *Beauty and the Beast*. That lock moves as Belle moves, but it's not because the hair has any independent existence. Rather, the animator at each point intentionally has to reposition the image of the lock of hair against the animation background. The animation represents the mind of the animator, and the lock of hair is real against the storyboard of the animation. But that animated lock of hair, so compatible with idealism, falls short of the lock of hair of a real-life Belle as she experiences the movement of her hair in our real (unanimated) world. Again, I'm not saying that idealism and informational realism cannot be made compatible as a matter of logic (perhaps we are living in a *Matrix*-like simulation, so it's all one big animation). But our gut-feeling about the world we inhabit, the world we experience as real, seems different.

The puppet and animation metaphors for me capture a serious problem with idealism even if disclaimers can be attached to idealism to minimize the lack of autonomy that seems inherent in it. Interestingly, Bruce Gordon, as he develops idealism in his chapters for this book, combines libertarian free will with a limited occasionalism. In a full-throated occasionalism, any activity in the world is in fact an occasion for God acting, thus making God the only true actor in reality. Occasionalism takes the puppet and animation metaphors to their logical extreme: the child is the only actor playing with army figures (*qua* puppets), and the Disney animators collectively constitute the only actor in *Beauty and the Beast*.

Gordon, guided by his understanding of quantum mechanics at the microscale as well as his commitment to libertarian free will at the macroscale, ends up advocating a limited form of occasionalism in his chapter on quantum information for this book: "God is the sole efficient cause of everything happening in that portion of nature not subject to the influence of creatures with libertarian freedom." Gordon thus exemplifies idealism's natural affinity for occasionalism. True, Gordon modulates the full force of occasionalism by making a place in his idealism for libertarian free will. Yet to my taste, idealism in all its varieties gives insufficient autonomy to the things that make up reality, a reality that both Gordon and I regard as the creation of the Judeo-Christian God.

Perhaps the best statement of my uneasiness with idealism comes from G. K. Chesterton in his reflections on the Christian doctrine of creation. According to Chesterton, creation demands separation. Created entities are not prosthetic extensions of the creator (such as toy soldier puppets or locks of hair belonging to animated characters). Rather, created entities require a separating, or divorcing, or setting free, from the creator. As Chesterton put it:

> It was the prime philosophic principle of Christianity that this divorce in the divine act of making (such as severs the poet from the poem or the mother from the new-born child) was the true description of the act whereby the absolute energy made the world. According to most philosophers, God in making the world enslaved it. According to Christianity, in making it, he set it free. God had written, not so much a poem, but rather a play; a play he had planned as perfect, but which had necessarily been left to human actors and stage-managers, who had since made a great mess of it.[14]

In idealism, the creator is always a mind and the created things are in the end always ideas residing in the creator's mind, and so there is no natural way to recover the independence, or separation, or autonomy, inherent in the act of creation as described by Chesterton. It may be recoverable by unnatural means, but that's not to the credit of idealism. Informational realism, on the other hand, lends itself quite naturally to this separation between creator and created things.

In informational realism, information sources communicate information. Those information sources may have minds and the information communicated may arise from thoughts in minds. But the information sources, as communicators of information, cannot merely be minds, and the information communicated, even if arising from thoughts, cannot merely be thoughts.

The distinction between idealism and informational realism is the distinction between a mind thinking thoughts and a speaker (who has a mind) uttering words (that articulate thoughts). In our experience, thoughts residing solely in minds are inert, but thoughts given utterance by speakers able to articulate them can be active and powerful. Informational realism captures this speaker-speech act dynamic in a way that idealism doesn't.

Once a word is given utterance, it is out of the bag and can't be taken back. It has an independent existence, and it can create new realities. J. L. Austin's linguistic theory of performative utterances, or performatives, details this fact about human language.[15] Thus, when a preacher announces, "I now pronounce you married," or an employee announces, "I resign," or an employer announces, "You're fired," new realities are created simply in the announcement or utterance. Here again we see the independence and autonomy we naturally associate with things but not with mere thoughts.

The utterance of words externalizes information, and with performative utterances the externalized information in fact creates new realities. But often information needs to be not just externalized but also reified. Performative utterances create in the very act of saying, but a lot of saying is just words without creative potency. Thus it can be one thing to say something (externalization) but quite another to build or implement it (reification). Fortunately, the contingency-constraint view of information is general enough to cover not just the externalization of information in language but also the reification of information in the making of things. This point is elaborated in what follows.

4. Shannon Information

We've seen that informational realism does not presuppose matter. But does informational realism presuppose minds? In fact, informational realism presupposes only information sources capable of interacting by the exchange of information with other information sources. Mental properties can be inferred for some of these information sources based on the types of information they communicate. A randomly shaped, stationary rock might convey no information characteristic of mind. On the other hand, through his plays, a writer like Shakespeare would. It thus seems reasonable for both non-mindlike and mindlike things to be regarded as information sources.

Such considerations, however, raise the question whether the very concept of information, and not just informational realism, presupposes minds? Perhaps non-mindlike information sources are capable of conveying information, but the information so conveyed, to be information, requires a mind to interpret it as information. Science writer John Horgan takes this view. As he sees it, information is inherently semantic and presupposes minds. In an article for *Scientific American* whose title—"Why Information Can't Be the Basis of Reality"—takes direct aim at informational realism, Horgan claims:

> The concept of information makes no sense in the absence of something

> to be informed—that is, a conscious observer capable of choice, or free will... If all the humans in the world vanished tomorrow, all the information would vanish, too. Lacking minds to surprise and change, books and televisions and computers would be as dumb as stumps and stones. This fact may seem crushingly obvious, but it seems to be overlooked by many information enthusiasts.[16]

Yet despite his appeal to crushing obviousness, Horgan is wrong about information being inherently semantic and presupposing minds. In fact, his point is not obvious at all. Certainly for theists—holding to a God who has a mind, knows all things, and is eternal—the elimination of humans would do nothing to eliminate God and would thus retain the type of agent that Horgan deems as necessary for information. But leaving theism aside, Horgan's claim still fails. If all humans suddenly vanished, as Horgan asks us to imagine, biological cells with their genetic machinery would still exist and communicate information (Horgan was not asking us to imagine the disappearance of all life). And even if all life vanished, contingency and constraint would still operate in nature to produce information. True, it would not be semantic information. But it would still be information.

The contingency-constraint view of information is a generalization of Shannon information, the theory of information that Claude Shannon developed in the late 1940s and that to this day defines the field. In this section, I want to review Shannon's contribution to information theory because it elucidates the contingency-constraint view of information and also because it demonstrates, contra Horgan, that information is not inherently semantic.

Shannon's masterwork on information was his 1949 book *The Mathematical Theory of Communication*.[17] Warren Weaver, in his introduction to that book, succinctly described Shannon's concept of information:

> The word *information*, in this theory [i.e., Shannon's theory of information or communication], is used in a special sense that must not be confused with its ordinary usage. In particular, *information* must not be confused with meaning [i.e., semantics]. In fact, two messages, one of which is heavily loaded with meaning and the other of which is pure nonsense, can be exactly equivalent, from the present viewpoint, as regards information. It is this, undoubtedly, that Shannon means when he says that "the semantic aspects of communication are irrelevant to the engineering aspects." But this does not mean that the engineering aspects are necessarily irrelevant to the semantic aspects.[18]

It's therefore clear that Shannon's theory is, out of the gate, not a semantic theory of information. Indeed, Shannon's theory is at heart a syntactic theory. As such, Shannon's theory allows for semantic aspects because syntax can be a carrier for semantics (as noted in Weaver's introduction). To say that Shannon's theory is a syntactic theory is simply to say that it is concerned with the linear arrangement of symbols (bits, alphanumeric characters, strings of these, etc.). Shannon was, after all, a communications engineer, and in formulating information theory, he was attempting to solve an engineering problem. The details of his theory do not concern us, but the gist does, and it is captured in what Shannon called the "schematic diagram of a general communication system":[19]

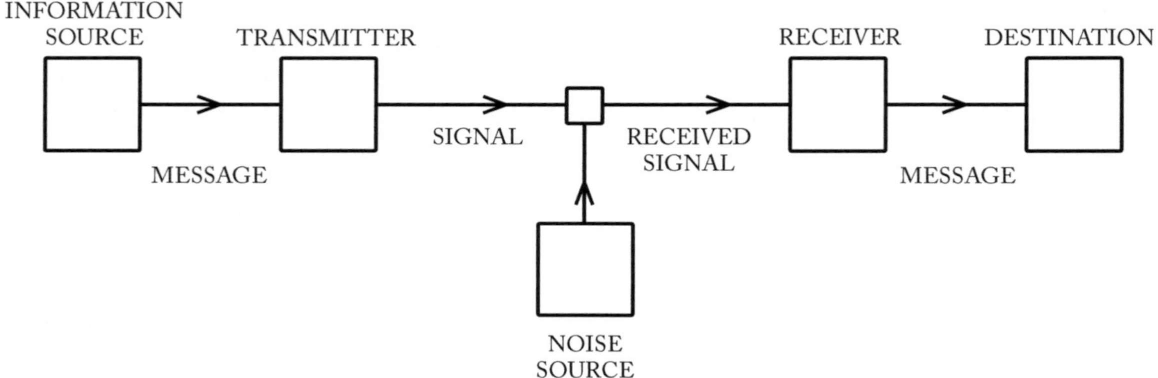

Figure 25.1

The engineering problem that Shannon faced was this: how to send a sequence of symbols drawn from a finite set of possible symbols across a communication channel so that the sequence received matched the sequence sent. Shannon worked for Bell Labs at the time, so he was thinking in particular about transmitting such symbol or character strings across phone lines (similar to sending them these days across the internet). But the phone lines, like our internet lines (whether these be wires, fiber optics, or wireless), were all subject to noise that could corrupt the communications, altering the message received from the message sent. Shannon's reference to "message" here can be confusing, as underscored by Weaver in his introduction, because Shannon did not intend, in the first instance, the meaning of a message but rather its arrangement as a character string. We tend to think of messages as inherently meaningful, but Shannon, in using the term, didn't.

As far as Shannon was concerned, the important thing was to send character strings, whatever they might be, reliably across the communication channel so that sender and receiver would be looking at the same string ("message"). It didn't matter whether the strings were random or meaningful, gibberish or coherent, boorish or sophisticated. Shannon's theory decisively resolved the problem of getting character strings down communication channels without alteration despite the disrupting effects of noise. Given a strong enough signal and given effective mathematical techniques that he developed for overcoming noise (such as encoding schemes and error correction), Shannon showed that the character strings could with arbitrarily high probability be transmitted from sender to receiver so that both would be looking at the same string.

Focused as Shannon's theory is on character strings, it's clear that his theory is, in the first instance, a syntactic theory and that semantics is incidental to it. It's also clear that his theory exemplifies the contingency-constraint view of information, contingency signifying the different character strings that might be sent across the communication channel, constraint signifying the one string actually sent as well as those that with high probability would be received.

Even though information within Shannon's theory always starts in a syntactic space, it readily extends to a semantic space by focusing on the meaning of syntactical arrangements. Consider the symbol string "It's raining outside." Treated as a symbol string, it resides in a syntactic space with characters drawn from a typewriter keyboard and arranged linearly. But as a meaningful statement, it describes a fact about the weather, namely, that it's raining outside. As such, it resides in a semantic space, narrowing, and thus constraining, contingent weather patterns down to those that exhibit rain.

But consider next the symbol string "It's raining outside or it's not raining outside." As a symbol string, it is longer than "It's raining outside" and thus has greater complexity and therefore, when considered syntactically, conveys more information than the shorter string (its representation requires more bits). But as a meaningful statement, it constitutes a tautology. As residing in a semantic space, it provides no constraint on contingent weather patterns, and therefore conveys no information. Thus, even though "It's raining outside or it's not raining outside" is more information-rich than "It's raining outside" when it is considered syntactically, the opposite is true when it is considered semantically.

Informational realism is readily compatible with Shannon's theory of information, both in its syntactically pure form and in its expanded semantic form (as a rider atop syntax). That's because Shannon's theory in either form exemplifies the contingency-constraint view of information that lies at the heart of informational realism. Insofar as Shannon's theory applies directly to information sources in the actual world, it is largely confined to linguistic systems that convey meaning, engineering systems that send and receive alphanumeric characters, and living systems whose chemical constituents behave like alphanumeric characters (such as nucleotides and amino acids).

Shannon's theory therefore applies directly to only limited portions of reality as we know it. Yet it applies indirectly much more widely. The contingency and constraint that are at the heart of information, and that occur outside of linguistic, engineering, and biological contexts, can often be coded syntactically in ways that allow an application of Shannon's theory (as when a jpeg image of an arbitrary physical object can be analyzed in terms of pixels).

Informational realism as a metaphysical position is not wedded to Shannon's theory and could be formulated without it. But Shannon's theory provides a particularly clear approach to information and, where applicable, turns the study of information into an exact science. Informational realism thus readily accommodates it.

5. The World According to Informational Realism

If, as informational realism asserts, reality is fundamentally informational, what does reality look like? Because we live in an intellectual climate pervaded by materialism, it may be useful to approach this question by first seeing how informational realism turns the basic intuition of materialism on its head. Materialism is always a bottom-up affair. The materialist attempts to understand matter at its most basic (atoms, electrons and protons, quarks, Higgs bosons, etc.) and then attempts to reconstruct reality from such elemental constituents. In contrast, information is always a top-down affair. It always starts with contingencies, and then constrains those contingencies. Materialism is about starting with less (a few basic types of particles), and then aggregating them to produce more (the panoply of objects built out of them). Information is about starting out with more (the contingencies) and then realizing only some to the exclusion of others (the constraint).

Given that information is a top-down affair, we may ask if there's a very top. The very top would be the collection of all possible worlds. This collection would constitute the ultimate range of contingencies. Against the collection of all possible worlds, the ultimate act of information would then be to realize the actual world, the world we inhabit, in an act of constraint that excludes all other worlds. The late philosopher David Lewis described the actual world as "the way things are, at its most inclusive."[20] The actual world omits nothing that is the case. The actual world thus needs to be understood in relation (and contradistinction) to other possible worlds because, as Lewis adds, "things might have been different, in ever so many ways."[21] The actual world is all that is the case. Other possible worlds describe counterfactual states of affairs that might have been but are not.

For Lewis, because possible worlds constitute the way things are *at their most inclusive*, they are self-contained, and any deity present to a possible world would therefore need to reside entirely within that world rather than spanning it and other worlds. Hence, for Lewis, there is no being that can take a God's-eye view of the collection of all possible worlds and then, in an informational act at the level of all possible worlds, constrain all such contingencies down to the actual world. This absence of any being beyond or outside the collection of all possible worlds is, however, a defect of Lewis's metaphysics. Because Lewis allows any non-self-contradictory state of affairs to be actualized somewhere (i.e., in some possible world), it follows that an infinitely powerful and all-knowing God capable of bringing into existence any particular world is possible. Such a being entails no self-contradiction, and some would even argue that such a being is necessary, and therefore would exist in every possible world.[22]

In any case, for Lewis, such a being will reside in at least one possible world. But such a world would then contain all possible worlds, if only as conceptualizations in the mind of the almighty God who inhabits that world. But in an ultimate act of self-reference, such a being, who is capable of thinking all possible worlds and choosing among them, could then actualize one of these possible worlds within the collection of all possible worlds, this world becoming not only the actual world but also the world in which this almighty God resides. In consequence, this God would of necessity actualize the world that this God inhabits.

The prospect of a world containing a deity capable of producing every world may seem mind-bending, but informational realism combined with Judeo-Christian theism takes us readily to such a conclusion. For the Judeo-Christian theist, the actual world is realized through an act of creation by an infinite personal creator-God. The creation of the actual world is thus the ultimate act of creation, and God in creating it both transcends the actual world and is immanent in it. Combined with informational realism, God, in the act of creation, becomes the ultimate source of information. Moreover, the world God creates reflects this key fact about God, namely, that God is an information source. The actual world reflects this fact by consisting in its entirety of information sources that exchange information with each other and ultimately with God. All information sources other than God thus become derived information sources, deriving their reality from the ultimate information source, God.

It may now be asked whether informational realism presupposes a being like the Judeo-Christian God. In fact, informational realism may be developed without assuming such a being, limiting itself to information sources that are in immediate or mediate communication with each other. Nevertheless, I find the Judeo-Christian God to be an invaluable inspiration for informational realism, especially through its doctrine of creation. The Judeo-Christian God creates by speaking. Created things thus become effected spoken words. Eastern Orthodox theologian Christos Yannaras elaborates this point:

> The world in its entirety and in its every detail is an effected word (*logos*), a personal creative activity of God. According to the account of Genesis, God created everything only by his word: "He spoke, and it came to be" (Ps. 33:9 NIV). The word of God does not come to an end, but is hypostasized in an effected event, "immediately becoming nature." As the human reason of a poet constitutes a new reality, which is the poem, outside of himself but at the same time a consequence and manifestation of his own reason, so also the word of God is given effect dynamically "in the ground and formation of creation."[23]

Speech acts thus become the prime metaphor for creation and creativity within informational realism. In this way, informational realism constitutes a significant advance, in my view, over idealism. In idealism, ideas or thoughts residing in minds are what's real. But within informational realism, ideas or thoughts residing in minds need to be externalized and articulated in order to count as real and make any difference in the world. Externalization means that the thought needs to get outside the mind and articulation means that the thought must be given a definite form ("this not that," again illustrating contingency and constraint).

The need for thoughts to be articulated as information in order to count as real gets at the heart of creativity, which is about going from thought to thing. Recall in an earlier section the examples of Michelangelo's David and a circuit diagram, the one drawn from art and aesthetics, the other from engineering and function. Consider the circuit diagram first. The mere thought of a circuit diagram capable of fulfilling a given purpose is not enough. The circuit diagram must be fully articulated if it is going to do any good. But even once it is articulated in an act of information, it needs to be further realized as an actual circuit, which will require a still further act of information.

The circuit board example illustrates how creation within informational realism is always *double creation*, the articulation of a plan (perhaps a blueprint, perhaps less detailed) as information, and then its implementation in a further act of information. This conception of creation as double creation is evident in Shannon's communication diagram, in which the sender first articulates a message (first creation) and then the receiver receives it (second creation). Without both the first and second creation working in tandem, with the first making a clean handoff to the second, the creative process ends stillborn. Christian theology supplies the preeminent example of double creation: God creates the world (second creation) according to a divine plan (first creation), with that plan fully articulated and eternally present in the mind of God.

But what about Michelangelo's David? How does it illustrate creation as double creation? Unlike the circuit board example, there's no blueprint for Michelangelo's David, and that's true even if Michelangelo was attempting to make his statue of David look like some human model (any such model would still have left plenty of free play to Michelangelo and thus fall shy of a point-for-point blueprint). As a sculptor, Michelangelo was chipping away and buffing marble to form his David. All those actions with his tools constituted information output by Michelangelo (as sender), with the effects on the marble block that became his David constituting information input (as receiver). Double creation is still evident here in such improvisational design in that we infer a plan (perhaps a changing and adaptive plan) on Michelangelo's part (first creation, perhaps only in his mind, but nonetheless articulated in his mind) that gets implemented in bringing into existence the statue of David (the second creation).[24]

Because informational realism is a relational ontology and because graph theory provides a common visual way to model relationships, it's natural to think of reality within informational realism as a graph consisting of nodes connected by edges, such as the following:

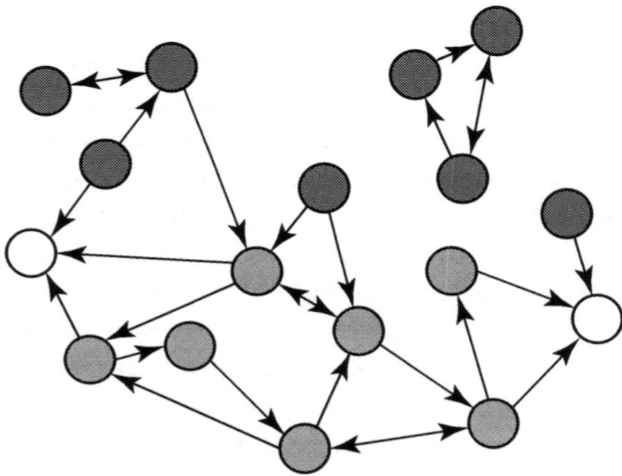

Figure 25.2 In such a graph, the nodes would be information sources and the edges would be communication channels facilitating the exchange of information.

Such a graph-theoretic representation to informational realism is, however, problematic in three ways:

1. It fails to represent the nesting of information sources as when information sources are hierarchically organized.
2. It suggests that informational exchanges only happen proximately between nodes that share an edge when in fact information readily allows for "action at a distance."
3. It says nothing about how the flow of information (its exchange or communication) can be rule-governed, with the rules themselves being informational (i.e., resulting from a constraint on contingency).

Regarding the first concern, take for instance a human body. It consists of cells. These cells communicate information among themselves. The cells themselves are groupings of organelles and other informational sources, which can be resolved still further to the level of molecular biology, biochemistry, and ultimately physics. At the same time, the cells can be grouped together to form tissues. These in turn can be grouped to form organ systems. And these, along with other structures, can be grouped to form the entire body. The body, as it interacts with its environment, communicates information. But the body consists of a hierarchy of nested information sources that communicate information within given hierarchical levels as well as across hierarchical levels. Graphs consisting of nodes and edges fail to capture this hierarchical nesting of information sources.

Regarding the second concern, we have a natural tendency to think of the communication channels that connect information sources in spatial or proximate terms. This is a mistake. The edges that connect nodes in a graph would have us think that adjacent nodes are in direct informational contact and that the greater the number of nodes that separate two information sources, the less their informational contact. But information is not a spatial relationship. In fact, any two information sources might exchange or communicate information. Communications channels can vary dynamically, so that any two information sources might be connected by an "edge." We identify an information exchange by the correlational pattern between the two information sources, which is the degree to which information at sender and receiver match up. We infer the communication channel from the correlation, and this can happen even in the absence of any clearly defined or physically specifiable communication channel. It's the correlation that's crucial. Action at a distance is thus baked into informational realism.[25]

Regarding the third concern, the flow of information is often rule-governed or lawlike, and such constraints on the flow of information (there's that word again, i.e., "constraint") will, within informational realism, itself be the outcome of an informational act against a backdrop of different possible rules for governing the flow of information (implicit in those possible rules is that other word again, i.e., "contingency"). For example, in the physical world, much of the information that moves about is constrained in its transmission by the speed of light, though quantum entanglement allows for instantaneous communications among quantum states. The various coding protocols that govern the movement of information across the internet illustrate this point even more clearly in that the coding protocols are themselves clearly the result of human engineers putting constraints on different possible protocols, deciding to allow some forms of internet communication, ruling out others.[26]

Where does the physical world figure into informational realism? Even though informational realism as such doesn't require a physical world, combined with a biblical worldview, it does. Even matter seems to have a legitimate place in a biblical worldview. True, the Greek word for matter, *hulē*, hardly appears in the Greek New Testament or in the Greek translation of the Old Testament, known as the Septuagint. In the Septuagint, *hulē* appears only in Job 19:29 (where it is translated as "judgment"), Job 38:40 (where it is translated as "covert" or "thicket"), and Isaiah 10:17 (where it is translated as "brier"). In these three instances, it means nothing like what materialists mean as the basic stuff out of which physical objects are constructed. Moreover, the only appearance in the New Testament of *hulē* is in James 3:5: "Even so the tongue is a little member, and boasteth great things. Behold, how great a matter [i.e., *hulē*] a little fire kindleth!" I give here the King James translation because all the other English Bibles translate *hulē* as forest (forests being composed of wood and wood, classically, providing the prime metaphor for matter). So even in the New Testament, *hulē* is not being used to denote matter in the sense of materiality.

Does the concept of matter therefore play a negligible role in both the Greek version of the Old Testament and the Greek New Testament? No. Cognates of the word *phusis* (from which we get our word "physics"), and which translate to "nature" as well as "physical," appear frequently throughout the New Testament, as do references to bodies (Greek *sōma*), especially the human body (which the New Testament even identifies as the temple of the Holy Spirit). Moreover, bodies are typically described in the New Testament as composed of flesh (Greek *sarx*). And the Old Testament refers to humans as composed of dust (Greek *chous* in the Septuagint, which certainly seems matter-like; compare Ecclesiastes 3:20 and 12:7). In creating humans in Genesis 2:7, God shapes or molds (Greek *plasso*) this dust. Replacing *chous* or *sarx* with *hulē* and replacing *plasso* with *morphe*, this looks like Aristotelian hylomorphism!

Nonetheless, I stand by my critique of hylomorphism in section 3 of this chapter. Informational realism argues only against matter as an information-independent substrate for information, a view that Christian theology can readily accommodate. That said, Christian theology cannot dispense with or minimize physicality and embodiment. Indeed, of all Christian doctrines, none is more central than the Incarnation, which describes God as taking bodily form in Jesus Christ (every other key Christian doctrine, from the Cross to the Resurrection, depends on the Incarnation). Moreover, Jesus's body did not just appear to be a body. Rather, it was an actual physical body no different from any other human physical body. Perhaps the earliest Christian heresy was docetism, which said that Jesus only appeared to have a body (i.e., that his human form was an illusion). Christian orthodoxy has always taught that Jesus had a real physical body.

The language of matter and form, the language of nature and the physical world, and the idea that material or physical objects, when suitably localized, constitute bodies all have a legitimate place. Informational realism has no interest in policing the ordinary use of these terms. True, informational realism calls into question materiality (as an information-free substrate for information) in a way that it doesn't call into question physicality (that there is a physical world operating in space and time, exhibiting a certain constitution, structure, and dynamics, and open to observation by the senses and to measurement by technology). But at the end of the day, not much is achieved by distinguishing too carefully among the terms materiality, physicality, and embodiment. Dissolving the mind-body problem depends less on the analysis of these terms than on understanding how informational realism unseats their metaphysical misappropriation.

One final point before we turn to how informational realism dissolves the mind-body problem, and that's the classification of distinct types of information sources within informational realism. In this section, we've described God as the ultimate information source and all other information sources as derived information sources. Yet at the same time, I've stated that God is not strictly speaking necessary to informational realism, which is also true. Informational realism, as we'll see in the next section, is a minimalist ontology that allows other metaphysical positions to be grafted onto it.

What this means is that informational realism, taken in and of itself, allows no hard and fast classification of distinct types of information sources. Such classifications will instead depend on additional metaphysical assumptions. As a Christian, for instance, I would see angels as real and so, combined with informational realism, I would also see them as non-physical information sources. The classification or taxonomy of information sources is an interesting question, but informational realism as such is not in the business of classifying information sources but simply of giving adequate room for information sources, whatever they might be.

6. Dissolving the Mind-Body Problem

Informational realism dissolves the mind-body problem by doing one thing and doing it well: It gives information free rein. Informational realism insists on letting information be what it will be. It does not prejudge what types of information may be allowed to exist or what types of information sources may be allowed to operate. That's for reality to decide. Informational realism is a minimalist ontology, requiring information to be taken seriously and yet without preconceptions about what information may and may not do (hence the minimalism). The ontologies responsible for there even being a mind-body problem (notably materialism) are those that artificially restrict the creation and flow of information. These ontologies will deny, on a priori grounds, that certain types of information sources can generate certain types of information. Conversely, they will demand, again on a priori grounds, that certain types of information sources must be able to generate certain types of information.

Thus, for instance, materialism will deny that someone can have an NDE (near-death experience) in which the person's mind gains information of events or conversations even though the person's body could not have been exposed to that information via material causes. Indeed, the NDE literature is filled with accounts of people on the operating table privy to conversations in other hospital rooms even though there's no chain of material causation from the one place to the other. When they wake up after being on the operating table, they may know what family members during the operation were talking about in a nearby room even though they were totally anaesthetized and the room was completely out of earshot. Materialism says this is impossible, and that any information about those conversations known to the person who was operated on must be through some subsequent communication of the information via "ordinary channels." By contrast, informational realism neither requires nor disallows such NDEs, insisting that informational evidence rather than metaphysical presuppositions decide whether and to what degree such NDEs exist.[27]

On the flip side, materialism will demand that certain types of information sources must be able to generate certain types of information. Materialism, for instance, will demand that mind be reduced to mechanism because it sees all things as composed of parts that operate by mechanical principles. And since computing machines are the universal mechanism (as Alan Turing demonstrated), that means our minds are just computers, and it's only a matter of time, given Moore's Law, that the computers we build will get powerful enough to equal and then surpass human minds. In informational terms, that means computers will output information qualitatively indistinguishable from that generated by humans, and eventually will output information that's so beyond what we are capable of that it will put us to shame. For materialism, it's certain and bankable that computers will pass the Turing test.[28] To think otherwise is to think erroneously. By contrast, informational realism allows that humans could be more (indeed much more) than mechanisms, insisting that informational evidence rather than metaphysical presuppositions decide whether the human mind can plausibly be regarded as a mechanism.

Given its minimalist ontology, informational realism is not a metaphysical sledgehammer that resolves the mind-body problem simply because its ontology requires the problem to be resolved in one way or another. There's nothing in informational realism that requires the human mind to be reducible to or, for that matter, irreducible to a computing machine. Informational realism, because it gives information free rein, cannot preemptively close the door to attempts to reduce mind to mechanism (information, after all, can operate mechanistically, as when a digital computer outputs information). Yet at the same time, and again because it gives information free rein, informational realism deprivileges other metaphysical sledgehammers that specify with certainty how to resolve the mind-body problem, the worst offender here being materialism.

Informational realism makes neither mind nor matter nor their combination a key element of its ontology. It follows that for informational realism, the mind-body problem is not to show how a non-material mind and a material body can interact (which must be done to justify a substance dualism, for instance). On the flip side, because informational realism is so minimalist, it can accommodate metaphysical additions, which in turn may make for more traditional understandings of the mind-body problem.

I personally find compelling the account of the human person given in Genesis in which spirit and body combine to form a living soul (soul thus becomes a union of spirit and body). And so, grafting this Genesis account of the human person onto informational realism, I will see in humans the joining of dual information sources (a spiritual information source along with a bodily information source to form a coherent whole or soul). Does this make me a dualist? Perhaps, but I can't say for sure. We're still just talking information sources, and it's not clear to me whether spiritual and bodily information sources are ontologically distinct or just convenient ways of referring to different sides of the same unified reality.

In any case, for the purposes of this chapter, I want to focus on informational realism in its minimal sense and understand the mind-body problem from that vantage (rather than from some more top-heavy metaphysical vantage). In informational realism, information sources make themselves known by exchanging information. Any direct access to reality is therefore only through the information exchanged, though there is indirect access to the information sources responsible for the information exchanged. The information sources themselves as well as their characteristics are inferred from the information they give out.

Perhaps these information sources are generating information de novo. Perhaps they are merely, like a conduit, transferring information from other sources. Perhaps they are tuning into and retransmitting information like a radio. Informational realism has no a priori way of limiting the information that an information source is capable of producing or receiving. That's for inquiry and evidence to decide. So too, informational realism has no a priori way of demanding that an information source must be capable of producing or receiving certain types of information. That's likewise for inquiry and evidence to decide.

Informational realism's ontological minimalism may find few takers among those with deeply held prior metaphysical commitments. Certainly, materialists, enamored of matter and its vast potentialities, as they see it, will want information to be constrained by materialism's ontology. They will therefore be disinclined to embrace informational realism, at least in its pure, minimalist form, because a mechanistic view of life and mind are not built into informational realism's ontology. Vitalists, who, in contrast to materialists, regard living things as fundamentally different from machines, may be likewise disappointed with informational realism, though this time because an organismic (non-mechanistic) view of life is not built into informational realism's ontology.[29] Both materialism and vitalism would thus regard informational realism as not going nearly far enough in its metaphysical ambitions. By contrast, I would say informational realism goes just as far as it needs to in giving reality room to breathe, allowing the world to be itself as opposed to forcing it into preconceived molds.

What, then, is the actual mind-body problem within informational realism? It is the problem of strong AI (i.e., strong artificial intelligence), that is, whether machines can display the full intellectual capacities of humans. In informational terms, it's whether machines can output information qualitatively indistinguishable from that of humans, which is to say, pass the Turing test. This form of the mind-body problem goes back at least as far as the thirteenth-century logician Raymond Lully, who in turn inspired the seventeenth-century philosopher Gottfried Wilhelm Leibniz to formulate the problem in this way. It's this formulation of the problem that appears in the writings of atheist Enlightenment *philosophes*, such as Julien La Mettrie, who in *Man the Machine* affirmed that the human mind, by which he meant the human nervous system, is indeed a machine. It's this formulation of the problem that inspires Ray Kurzweil, strong AI's biggest contemporary cheerleader, to give his books titles such as *The Age of Intelligent Machines* and *The Age of Spiritual Machines*.

Informational realism does not rule out strong AI from the start. Informational realism in its pure, minimalist form carries no theological or metaphysical presuppositions about humans being irreducible to (or for that matter reducible to) their material aspects. Just as digital computers are machines that do in fact reduce to their material aspects (i.e., to the lawlike interaction of their material parts), so too informational realism leaves open that humans might reduce to their material aspects. Some information sources, like computers, are fully captured by their material aspects. Informational realism, at its most basic and minimalist, leaves open that possibility for humans as well.

But unlike materialism, informational realism does not demand that humans reduce to their material or physical aspects.[30] That's for philosophical and scientific inquiry to decide. Erik Larson has recently argued that the problem of strong AI remains for now completely unresolved because of the failure of computer scientists to understand, much less model computationally, the human capacity to perform abductive inferences (also known as inferences to the best explanation).[31] These inferences are completely basic to human intellectual function, and yet they remain utterly beyond the reach of current algorithms, certainly for now and possibly forever. It's not that we're making little progress on this problem. It's that we're making no progress at all.

Larson's is a strong argument. And it is entirely in line with informational realism, which by taking matter as a linguistic convenience feels no metaphysical compulsion to explain all things in terms of their matter-like properties. Where matter works, use it. Where it doesn't, don't feel an obsessive need to make it work.

Interestingly, Descartes, whose seduction by matter is largely responsible for this chapter as well as for this whole book, understood the challenge of strong AI and the grounds for rejecting it. As he put it in his *Discourse on Method*,

> Although machines can do many things as well as or even better than us, they fail in other ways, thereby revealing that they do not act from knowledge but solely from the arrangement of their parts. Intelligence is a universal instrument that can meet all contingencies. Machines, on the other hand, need a specific arrangement for every specific action. In consequence, it's impossible for machines to exhibit the diversity needed to act effectively in all the contingencies of life as our intelligence enables us to act.[32]

For Descartes, anything viewed strictly from the vantage of matter was a machine. Computers would thus for him be machines. Animals, because for him lacking in reason or intelligence, were for him also machines ("automatons"). And humans, when viewed simply as bodies, were for him machines. And so the challenge for Descartes was to connect (at the pineal gland, no less!) the human body to a mind capable of making the human person more than a machine. That challenge doesn't face informational realism, which doesn't default. It doesn't default to materialism and then try to salvage what it can from a non-materialism. Things are what they are as informational sources, and if the matter-like aspects of those informational sources don't adequately account for their information-communicating powers, then so much the worse for those matter-like aspects.

Think of it this way: When we try to understand a digital computer, we can look at it from two vantages. We can look under the hood, as it were, and see how its digital parts are interacting in lawlike ways (algorithmically). At the same time, we can look at the informational output, or performance, of the computer, such as the solutions it is giving to well-posed problems. Moreover, we can precisely track the correlation between what's going on under the hood and what information the computer is outputting, and from there we can legitimately conclude that we really are dealing with a machine.

But we have nothing like this correlation for humans. As a general rule, we can't get under the hood (such as by opening a person's skull and mapping their brain function). And even if we could (such as by surgery or with an fMRI), we don't know enough about the brain and its operation to show that brain states (what's going on under the hood) correlate algorithmically with the person's cognitive behavior (the informational performance output). To say that there's a tight algorithmic correlation between the two in the same way that there is for a digital computer thus becomes an article of faith. For materialism, it becomes an inviolable article of faith.

For informational realism, certain physical systems, like the human body, happen to generate informational behaviors that are not, for all we can know or determine, algorithmic consequences of its material aspects. But how can the human body (notably the brain and nervous system) generate such informational outputs or behaviors apart from the body acting as a machine to generate them? For materialism, that would be impossible. By contrast, informational realism, as a minimalist ontology, has no burden to answer this question and no burden to assuage materialism. Perhaps the human person combines a physical body and a spiritual mind as dual information sources that somehow unite, without contradiction, thus making some form of substance dualism true.

Regardless, informational realism's dissolution of the mind-body problem is achieved not by appealing to substance dualism, but by unburdening the mind-body problem of any compulsion to explain the body's informational behavior in terms of any presumed mechanistic processing of information by the body. Such compulsions to explain things mechanistically arise invariably from nonnegotiable metaphysical assumptions. Informational realism, in its very ontological minimalism, discredits such metaphysical assumptions and thereby dissolves the mind-body problem. Rather than offer a detailed solution to the mind-body problem, informational realism clears the metaphysical landscape of rubbishy so-called solutions that masquerade as conclusions of sound arguments when in fact they are just restatements of deeply held prejudices.

In essence, informational realism says: "Show me with full scientific and philosophical rigor why I should think that a mature science of computation and neurophysiology ought to underwrite a mechanistic reduction of human intellectual capacities (in particular, as such capacities express themselves informationally). And don't just tell me that it's got to be so because a materialistic (or naturalistic or physicalistic) worldview would have it so." Informational realism, as a minimalist ontology, is a fair arbiter, giving competing metaphysical positions no unfair advantages and no unfair disadvantages. Neither Aristotelian vitalism nor today's scientific materialism, for instance, receives from it preferential treatment. Instead, these positions must prove themselves on their own merits, and where they make scientific claims, they must justify their claims scientifically.

Throughout this chapter, my main foil has been materialism. In closing, I therefore want to draw a final distinction between materialism and informational realism. Materialism's great fault is that it artificially restricts the information sources that may be active in the world, ensuring that information sources unacceptable to materialism, even if they exist, will get ignored or misunderstood. In contrast, informational realism's great virtue isn't that it requires information sources unacceptable to materialism to exist but that it allows them to be discerned and properly understood if they do exist. Whereas materialism requires the world to fit its preconceptions, informational realism takes the world as it is, allowing the world to be itself. In so doing, informational realism defangs materialism and dissolves the mind-body problem.

Notes

1. Wesley J. Wildman, "An Introduction to Relational Ontology," typescript, May 15, 2006, available online at http://people.bu.edu/wwildman/media/docs/Wildman_2009_Relational_Ontology.pdf (last accessed May 11, 2021). See also Christos Yannaras, *Relational Ontology*, trans. N. Russell (Brookline, MA: Holy Cross Orthodox Press, 2011). Christian theology is often seen as underwriting a substantivist ontology (the focus being on the unity of the godhead), but insofar as God is conceived fundamentally as Trinity, it can also be seen as underwriting a relational ontology, the relations among Father, Son, and Holy Spirit taking priority over the existence of individual members of the Trinity or even their joint existence.
2. Christian de Duve, *Vital Dust: Life as a Cosmic Imperative* (New York: Basic Books, 1995), 10.
3. John A. Wheeler and Kenneth W. Ford, *Geons, Black Holes, and Quantum Foam: A Life in Physics* (New York: Norton, 1999), 64.
4. John A. Wheeler, "Information, Physics, Quantum: The Search for Links," in *Complexity, Entropy, and the Physics of Information*, ed. W. Zurek (Redwood City, CA: Addison-Wesley, 1990), 5.
5. Wheeler, "Information, Physics, Quantum," 5. Emphasis added.
6. See Hasok Chang, "Operationalism," *Stanford Encyclopedia of Philosophy*, Fall 2021, https://plato.stanford.edu/entries/operationalism. The philosophy of science has largely rejected operationalism because the operations that give us scientific measurements provide at best a springboard into the construction of theories that are supposed to give us a genuine understanding of nature. The measurements themselves don't provide that understanding, as operationalism insists by reducing the meaning of scientific concepts to their methods of measurement.
7. Paul Davies, "Bit Before It?," *New Scientist* 161 (January 30, 1999): 3.
8. Robert Stalnaker, *Inquiry* (Cambridge, MA: MIT Press, 1984), 85.
9. Fred Dretske, *Knowledge and the Flow of Information* (Cambridge, MA: MIT Press, 1981), 12.
10. Aristotle develops his theory of hylomorphism in both his *Physics* and his *Metaphysics*. See Thomas Ainsworth, "Form vs. Matter," *Stanford Encyclopedia of Philosophy*, Summer 2020, https://plato.stanford.edu/entries/form-matter/. Note that *hulē*, currently the more common transliteration of the word for "matter," was formerly rendered *hyle*, which explains the "y" in the English word "hylomorphic." The word is pronounced as two syllables, with accent on the "u," which is long, and the final ē pronounced like the "ey" in "they."
11. The original statement reads "*inflation is always and everywhere a monetary phenomenon*" (emphasis in the original). See Milton Friedman, "The Counter-Revolution in Monetary Theory," IEA Occasional Paper, no. 33 (London: The Institute of Economic Affairs, 1970), 11, available online at https://miltonfriedman.hoover.org/internal/media/dispatcher/214480/full (last accessed May 12, 2021).
12. For further details, see my book *Being as Communion: A Metaphysics of Information* (Surrey, UK: Ashgate, 2014), ch. 10, titled "Getting Matter from Information."

13. Karl R. Popper, *The Open Society and Its Enemies*, single volume edition (Princeton, NJ: Princeton University Press, 2013), 222.
14. G. K. Chesterton, *Orthodoxy*, in *Collected Works of G. K. Chesterton*, vol. 1 (San Francisco: Ignatius, 1986), 281–282.
15. J. L. Austin, *How to Do Things with Words* (Oxford: Clarendon Press, 1962).
16. John Horgan, "Why Information Can't Be the Basis of Reality," *Scientific American*, blog entry (March 7, 2011), https://blogs.scientificamerican.com/cross-check/why-information-cant-be-the-basis-of-reality/.
17. Claude Shannon and Warren Weaver, *The Mathematical Theory of Communication* (Urbana, IL: University of Illinois Press, 1949).
18. Shannon and Weaver, *The Mathematical Theory of Communication*, 8.
19. Shannon and Weaver, *The Mathematical Theory of Communication*, 34. Note that for copyright reasons, I had this diagram redone. It matches point for point Shannon's original diagram.
20. David Lewis, *On the Plurality of Worlds* (Oxford: Blackwell, 1986), 1.
21. Lewis, *On the Plurality of Worlds*, 1.
22. Alvin Plantinga, *The Nature of Necessity* (Oxford: Clarendon, 1974), ch. 10, where Plantinga revives the ontological argument.
23. Christos Yannaras, *Elements of Faith: An Introduction to Orthodox Theology* (Edinburgh: T&T Clark, 1991), 40.
24. For a fuller account of creation as double creation, see Dembski, *Being as Communion*, ch. 20.
25. Fred Dretske makes the same point this way: "It may seem as though the transmission of information… is a process that depends on the causal inter-relatedness of source and receiver. The way one gets a message from *s* [source] to *r* [receiver] is by initiating a sequence of events at *s* that culminates in a corresponding sequence at *r*. In abstract terms, the message is borne from *s* to *r* by a causal process which determines what happens at *r* in terms of what happens at *s*. The flow of information may, and in most familiar instances obviously does, depend on underlying causal processes. Nevertheless, the information relationships between *s* and *r* must be distinguished from the system of causal relationships existing between these points." Dretske, *Knowledge and the Flow of Information*, 26.
26. Cosmology considers the various constraints on the laws and constants of nature that allow for a life-permitting universe. Most of the ways of formulating these laws and specifying these constants lead to a lifeless universe. Such are the contingencies relevant to a life-permitting universe. So, even in the formation of the universe as we know it, the contingency-constraint understanding of information is central. See Stephen C. Meyer, *Return of the God Hypothesis: Three Scientific Discoveries That Reveal the Mind Behind the Universe* (New York: HarperOne, 2021), ch. 8.
27. For examples of NDEs where information is transferred in the absence of material causation, see Mario Beauregard and Denyse O'Leary, *The Spiritual Brain: A Neuroscientist's Case for the Existence of the Soul* (New York: HarperCollins, 2007), ch. 6.
28. To pass the Turing test, computers would need to output information so similar to the type of information that humans output that no one could tell which is which. Alan Turing first proposed this test in "Computing Machinery and Intelligence," *Mind* 59 (1950): 433–460. Inspired by the Turing test, the Loebner Prize is given annually to the computer program whose information output is deemed most human-like. See https://en.wikipedia.org/wiki/Loebner_Prize (last accessed July 13, 2021). No Loebner-Prize-winning computer program has to date even come close to passing the Turing test.
29. But note, unlike materialism, vitalism would suggest that the human mind is not a mechanism and that strong AI cannot in the end succeed. Michael Denton argues this point persuasively in responding to Ray Kurzweil's claim that the human mind is a machine: "[I]f we are incapable of instantiating in mechanical systems any of the 'lesser' vital characteristics of organisms such as self-replication, 'morphing,' self-regeneration, self-assembly and the holistic order of biological design, why should we believe that the most extraordinary of all the 'vital characteristics'—the human capacity for conscious self-reflection—will ever be instantiated in a human artifact?" Quoted from Michael Denton, "Organism and Machine: The Flawed Analogy," in *Are We Spiritual Machines?*, ed. Jay W. Richards (Seattle: Discovery Institute Press, 2002), 97.
30. By materialism, I mean reductive forms of it, where the whole point and appeal is that if you understand with full precision and detail what the material parts of a thing are doing, you understand with full precision and detail what the thing as a whole is doing. There are nonreductive varieties of materialism, but these, as far as I'm concerned, lose all the down-home charm of ordinary reductive materialism by, in effect, allowing any phenomena and capacities to ride atop matter as

unexplained, and indeed unexplainable, mysteries. Except for a ritualistic obeisance to matter, nonreductive materialism does no more to explain such human capacities as intellect, language, and consciousness than the most incoherent substance dualism (and I think there can be coherent substance dualisms).

31. Erik J. Larson, *The Myth of Artificial Intelligence: Why Computers Can't Think the Way We Do* (Cambridge, MA: Harvard University Press, 2021).

32. René Descartes, *Discourse on Method*, 1637, pt. 5, my translation. Note well Descartes's joint reference to contingencies and to the role of intelligence or reason in sorting through, and thereby constraining, those contingencies. Descartes is here giving expression to the contingency-constraint view of information!

Index

N

O

P

W

Y

Z

Contributors

Editors

Angus J. L. Menuge is Chair of the Philosophy Department at Concordia University Wisconsin. He was raised in England and became an American citizen in 2005. He holds a BA (Honors, First Class) in philosophy from Warwick University, and an MA and a PhD in philosophy from the University of Wisconsin-Madison. He has written many peer-reviewed and popular articles on the philosophy of mind, philosophy of law, and the foundation of ethics. He is author of *Agents Under Fire* (Rowman and Littlefield, 2004) and editor of *Legitimizing Human Rights* (Ashgate, 2013; Routledge, 2016), and *Religious Liberty and the Law* (Routledge, 2017). He is co-editor with Jonathan J. Loose and J. P. Moreland of *The Blackwell Companion to Substance Dualism* (Blackwell, 2018) and, with Barry W. Bussey, of *The Inherence of Human Dignity*, volume I and II (Anthem Press, 2021). Menuge is past president of the Evangelical Philosophical Society (2012–2018).

Brian Krouse is a software engineer with research interests in the philosophy of mind, computer science, and neuroscience. Krouse has a bachelor's degree in physics from Whitman College, a master's in computer science with a focus on artificial intelligence from Arizona State University, and a master's in applied mathematics with a focus on computational neuroscience from the University of Washington. He was an early employee at GoDaddy (from 1997 through 2013), the domain name registration and hosting company, and spent most of his employ there in software development and management positions, culminating in Vice President of Hosting Development.

Robert J. Marks II is Distinguished Professor of Electrical & Computer Engineering at Baylor University, and Director and Senior Fellow of the Walter Bradley Center for Natural and Artificial Intelligence. He has worked in the field of artificial intelligence for more than three decades, with research supporters that include NASA, JPL, NIH, NSF, Raytheon, the Army Research Lab, and the Office of Naval Research. He has authored several hundred peer-reviewed journal and conference papers, and his books include *Non-Computable You: What You Do That Artificial Intelligence Never Will* (Discovery Institute Press, 2022), *Neural Smithing* (MIT Press, 1999) and *Handbook of Fourier Analysis & Its Applications* (Oxford University Press, 2009).

AUTHORS

Douglas Axe holds the Rosa Endowed Chair of Molecular and Computational Biology at Biola University. After a BS in chemical engineering from UC Berkeley and MS and PhD degrees from Caltech, Axe held research scientist positions at the University of Cambridge and the Medical Research Council Centre in Cambridge. His work and ideas have been featured in many scientific journals, including the *Journal of Molecular Biology*, the *Proceedings of the National Academy of Sciences*, and *Nature*; in such books as *Signature in the Cell* and *Return of the God Hypothesis* by Stephen Meyer and *Life's Solution* by Simon Conway Morris; and in popular media ranging from *The New York Times* and *Newsweek* to *The Gospel Coalition* website. He is the author of the award-winning book *Undeniable—How Biology Confirms Our Intuition That Life Is Designed* (HarperCollins, 2016) and co-author (with Jay Richards and William Briggs) of *The Price of Panic—How the Tyranny of Experts Turned a Pandemic into a Catastrophe* (Regnery, 2020).

Selmer Bringsjord specializes in the logico-mathematical and philosophical foundations of artificial intelligence (AI) and cognitive science (CogSci), in collaboratively building AI systems/robots (including ethically correct ones) on the basis of computational logic, and in the logic-based modeling and simulation of rational, human-level-and-above cognition. His work in these areas has been expressed in over 250 publications; he has communicated/debated in person in myriad countries; and he has pursued his investigations via sponsored-research awards of over $26 million. Though Bringsjord spends considerable engineering time in pursuit of ever-smarter computing machines for his much-appreciated sponsors, he insists that "armchair" reasoning has enabled him to deduce that the human mind will forever be superior to such machines, chatbots notwithstanding.

Cristi L. S. Cooper is Dean of Academics at Summit Classical Christian School in Bellevue, Washington. She holds a BS in biology from Colorado State University, an MS in biology (with an emphasis on the neuroscience of taste and smell transduction) from the University of Denver, and a PhD in neurobiology and behavior from the University of Washington. She has co-authored several primary research papers and reviews on topics related to development and regeneration in peer-reviewed journals including *Cell*, *Genes & Development*, *Development*, and *Nature Genetics*. She studied the neural mechanisms of decision-making at the University of Washington and, although this wasn't her primary area of research, has kept abreast of the latest research in this field ever since. While science was her first love, she now focuses her energy on raising her three children with her husband and training K-12th grade students (at the school where she now works) to think critically, ask good questions, and approach the world as scientists from an early age.

Natalia Dashan holds a bachelor's in psychology from Yale University and was a Research Assistant at Harvard Business School and the Yale Emotional Intelligence Center. Her essay "The Real Problem at Yale is Not Free Speech" was one of the first works to shed light on the underlying processes driving political division during the years 2015–2019. She is the creator of the interpersonal dynamics program *Mutually Assured Seduction*, and is a Visiting Fellow at the Foundation for Research on Equal Opportunity.

William A. Dembski is an entrepreneur, researcher, and writer. As an entrepreneur, he has focused on building educational websites and technologies, notably AcademicInfluence.com, an alternative to the *US News* college rankings. A mathematician and philosopher, Dembski received a PhD in mathematics from the University of Chicago and a second PhD, in philosophy, from the University of Illinois at Chicago. He also holds an MDiv from Princeton Theological Seminary. He has published in the peer-reviewed engineering, mathematics, biology, philosophy, and theology literature. He is the author or editor of more than twenty-five books, many of them on intelligent design but also covering other topics, such as education and baseball. His best-known book, *The Design Inference*, was published with Cambridge University Press in 1998. A revised and expanded second edition is to appear in fall 2023. A man of wide interests and unending curiosity, Dembski enjoys attacking difficult problems and resolving them.

Michael Egnor is Professor of Neurological Surgery and Pediatrics at the Renaissance School of Medicine at Stony Brook University and is Program Director of the neurosurgery residency training program. He holds an MD from the Columbia University College of Physicians and Surgeons and completed a residency in neurosurgery at the University of Miami. Since 1991 he has been on the faculty of neurosurgery at Stony Brook. In addition to a full-time clinical practice focused on pediatric neurosurgery, he directs a scientific research program in intracranial dynamics and hydrocephalus. He has a strong interest in Thomistic philosophy, philosophy of mind, and neuroscience, and he has published and lectured extensively on these topics.

Winston Ewert is a software engineer and researcher with a passion for applying his skills as a computer scientist to uncovering the mysteries of life. He obtained a bachelor's degree in computer science from Trinity Western University, a master's in computer science from Baylor University, and a PhD in electrical and computer engineering from Baylor University. He works primarily in the field of intelligent design, exploring the implications of computer simulations of evolution, developing the theory of specified complexity, and understanding genomes as examples of sophisticated software.

Joshua R. Farris is a Humboldt Experienced Researcher Fellow and Visiting Researcher at the Ruhr Universität Bochum. He is also Visiting Professor at Missional University and London School of Theology. Previously, he was the Chester and Margaret Paluch Professor at Mundelein Seminary, University of Saint Mary of the Lake, Fellow at The Creation Project, and Fellow at Heythrop College. He has taught at several universities in philosophy, theology, and Great Books. He has published more than fifty peer-reviewed articles and chapters in a variety of journals in philosophy, philosophy of religion, analytic theology, systematic theology, historical theology, and interdisciplinary studies. He is also published in *The Imaginative Conservative*, *The Christian Post*, *The American Mind*, *Mere Orthodoxy*, and *Essentia Foundation,* among others. He has recently completed a new monograph entitled *The Creation of Self.*

David Gelernter (BA, MA Yale 1977, PhD SUNY/Stony Brook, 1983) is Professor of Computer Science, Yale University, and co-inventor of the first Cloud in 1980 (called a Tuple Space) with M. Arango and N. Carriero. He was first to build an online social network, in the form of a Twitter-style stream, called a *Lifestream* (in 1990 with E. Freeman). As a graduate student he invented Linda, a distributed programming language that has been taught and used from Scandinavia to South Africa. He was also among the first to explore the idea of online social networks, in his book *Mirror Worlds* (Oxford University Press, 1991), called "one of the most influential books in computer science" (*Technology Review*, 2007). In his 2016 book *The Tides of Mind* (W.W. Norton) he described a new model of the human mind, which the *Chicago Tribune*'s Nick Romeo called "a new paradigm for the study of human consciousness."

Stewart Goetz is Professor of Philosophy at Ursinus College and visiting scholar at St. Peter's College, Oxford. He has co-authored two books with Charles Taliaferro, *Naturalism* (Eerdmans, 2008) and *A Brief History of the Soul* (Wiley-Blackwell, 2011), and co-edited with Taliaferro the four-volume *Encyclopedia of Philosophy of Religion* (Wiley Blackwell, 2022). He is senior co-editor of the Bloomsbury series *Bloomsbury Studies in Philosophy of Religion*.

Bruce L. Gordon, ARCT, BSc (Honors), MA, MAR, PhD, is Professor of the History and Philosophy of Science at Houston Christian University and a Senior Fellow of Discovery Institute's Center for Science and Culture. His doctorate is in the history and philosophy of physics from Northwestern University. He has held positions at the University of Notre Dame, Baylor University, Discovery Institute, The King's College, and Houston Christian University. Widely published in academic journals and edited collections on subjects ranging from the foundations of physics to the philosophy of mind and philosophical theology, he co-edited *The Nature of Nature* (ISI Books, 2011) with William Dembski, and *Biological Information: New Perspectives* (World Scientific, 2013), with Robert Marks, Michael Behe, William Dembski, and John Sanford. He co-authored *Three Views on Christianity and Science* (Zondervan, 2021) with Michael Ruse and Alister McGrath. He lives in Houston, Texas, with his wife and daughter.

Naveen Sundar Govindarajulu is an AI and machine learning researcher with experience in both industry and academia. He has published in prestigious venues such as NeurIPS, IJCAI, and AAAI. Naveen has worked in various research labs, including Yahoo Research, Abacus.AI, TIFR, and HP Labs. A Fulbright scholarship awardee, he earned his PhD in computer science from Rensselaer Polytechnic Institute and holds degrees in physics and electrical and electronics engineering from BITS Pilani. He has been involved in a wide range of AI projects, from reasoning systems for AI ethics to developing state-of-the-art meta-learning methods for time series forecasting. Throughout his career, he has actively pursued research in the philosophical dimensions of AI.

Joseph Green (a pseudonym) is a postdoctoral researcher in system neuroscience. He has bachelor's and master's degrees in physics, as well as a PhD in neuroscience. He has published numerous peer-reviewed studies in the fields of system and computational neuroscience, as well as machine learning. He has worked as a reviewer for several peer-reviewed journals, including *Nature Neuroscience*, *eLife,* and *PlosCB*. As a postdoctoral fellow, his study centers on natural and machine intelligence, with an emphasis on neural data. He has chosen to remain anonymous due to his delicate academic stage and untenured employment.

Mihretu P. Guta teaches analytic philosophy at Biola University and Azusa Pacific University, and is an associate fellow at the Center for Bioethics and Human Dignity Academy of Fellows at Trinity International University. He earned his PhD in philosophy from Durham University (UK) and then worked as a postdoctoral research fellow at Durham University within the Durham Emergence Project set up with cooperation between physicists and philosophers and funded by the John Templeton Foundation. Guta's postdoctoral research focused on the nature of the emergence of the phenomenal consciousness taken from the standpoint of metaphysics, philosophy of mind, cognitive neuroscience, and quantum physics. Guta is the editor of *Selfhood, Autism, and Thought Insertion* (with Sophie Gibb, 2021); a special issue of *TheoLogica*, "E. J. Lowe's Metaphysics and Analytic Theology" (with Eric LaRock, 2021); *Consciousness and the Ontology of Properties* (Routledge, 2019); and *Taking Persons Seriously* (with Scott B. Rae, forthcoming). Guta is presently working on a single-authored manuscript entitled "The Metaphysics of Substance and Personhood: A Non-Theory Laden Approach."

Gary Habermas (PhD, Michigan State University) has written or edited fifty books, over half centering on various aspects of Jesus' resurrection. These include *Did Jesus Rise from the Dead?* with Antony Flew (Harper, 2003) and *The Risen Jesus and Future Hope* (Rowman & Littlefield, 2003). Other topics concern near-death experiences, religious doubt, and personal suffering. He has also contributed more than eight-five chapters and essays to additional books, plus close to 200 articles and reviews to journals and other publications. These publishers include Oxford, Cambridge, Blackwell, Harper, Dell, Fortress, Harcourt-Brace, Baker, InterVarsity, and Zondervan. He has been a Visiting or Adjunct Professor at numerous graduate schools and seminaries in the United States and abroad, having taught several dozen graduate courses for these schools. He is Distinguished Research Professor at Liberty University School of Divinity, where he has taught since 1981, teaching full-time in the PhD program.

Eric Holloway is a Senior Fellow with the Walter Bradley Center for Natural and Artificial Intelligence, and holds a PhD in electrical and computer engineering from Baylor University. A Captain in the United States Air Force, he served in the United States and Afghanistan. He is the co-editor of *Naturalism and Its Alternatives in Scientific Methodologies* (Blyth Institute Press, 2017).

C. Eric Jones is Professor of Psychology at Regent University. He holds a BA in psychology, an MA in experimental psychology, and a PhD in experimental social psychology from Florida Atlantic University. His work concentrates on the connections among humans' relational nature and human social thought and behavior within social and positive psychology. He is former Associate Director of the Society for Christian Psychology and was the 2018 Regent University Teacher of the Year.

Robert A. Larmer is Chair of the philosophy department at the University of New Brunswick, and received his PhD in philosophy from the University of Ottawa. He has published widely on the relationship between science and religious belief, notably on how divine agency in the world should be conceived. He is recognized as an expert on philosophical issues concerning the rationality of belief in miracles, having published several books in that area, the most recent being *The Legitimacy of Miracle* and

Dialogues on Miracle. Larmer has served on the executive boards of the Canadian Philosophical Association and the Evangelical Philosophical Society, and has twice served as President of the Canadian Society of Christian Philosophers.

Jonathan J. Loose was Senior Lecturer in Philosophy and Psychology at Heythrop College, University of London until 2018. Subsequently he continued his research through the Universities of London and Cambridge. Loose holds a BSc (Honors, First Class) and PhD, both in cognitive sciences, from Exeter University, UK. His recent interests have been in philosophy of mind and philosophy of religion. He has published a number of peer-reviewed articles and symposia in these areas and is co-editor with Angus J. L. Menuge and J. P. Moreland of *The Blackwell Companion to Substance Dualism* (Blackwell, 2018). Loose also has contributed to various volumes, including *The Naturalness of Belief: New Essays on Theism's Reasonability* (Lexington, 2018); *Christian Physicalism: A Philosophical and Theological Critique* (Lexington, 2017); and the *Dictionary of Christianity and Science* (Zondervan, 2017).

Jim Madden is a professor of philosophy at Benedictine College, and earned his BA from St. Norbert College, his BA from Kent State University, and his PhD from Purdue University. He is interested mainly in the philosophy of mind, phenomenology, and philosophy of religion. He is the author of *Mind, Matter, and Nature: A Thomistic Proposal for the Philosophy of Mind* (CUA Press, 2013) and *Thinking About Thinking: Mind and Meaning in the Era of Technological Nihilism* (Cascade, forthcoming).

J. P. Moreland is Distinguished Professor of Philosophy at Talbot School of Theology, Biola University. He has authored, edited, or contributed papers to ninety-five books, including *Universals* (McGill-Queen's, 2001), *Consciousness and the Existence of God* (Routledge, 2008), *The Blackwell Companion to Natural Theology* (Wiley-Blackwell, 2009), *The Blackwell Companion to Substance Dualism* (Wiley-Blackwell, 2018), *Debating Christian Theism* (Oxford University Press, 2013) and, most recently, (with Brandon Rickabaugh) *The Substance of Consciousness: A Comprehensive Defense of Contemporary Substance Dualism* (Wiley-Blackwell, 2023) He has published more than a hundred articles in journals such as *Philosophy and Phenomenological Research, American Philosophical Quarterly, Australasian Journal of Philosophy, MetaPhilosophy, Philosophia Christi, Religious Studies*, and *Faith and Philosophy*. Moreland was selected in August 2016 and 2022 by *The Best Schools* as one of the fifty most influential living philosophers in the world.

Charles Taliaferro (PhD Brown, MTS Harvard) is Professor Emeritus of Philosophy and Emeritus Overby Distinguished Professor, St. Olaf College. He is the author, co-author, or editor of forty books. He has co-authored two books with Stewart Goetz, *Naturalism* (Eerdmans, 2008) and *A Brief History of the Soul* (Wiley-Blackwell, 2011), and co-edited with Goetz the four-volume *Encyclopedia of Philosophy of Religion* (Wiley-Blackwell, 2022). He is the Editor-in-Chief of *Open Theology* and serves on the editorial boards for *Sophia* and *Philosophy Compass*.

Made in the USA
Coppell, TX
21 October 2023